Sustainable Development and Biodiversity

Volume 23

Series Editor

Kishan Gopal Ramawat
Botany Department, Mohanlal Sukhadia University, Udaipur, India

This book series provides complete, comprehensive and broad subject based reviews about existing biodiversity of different habitats and conservation strategies in the framework of different technologies, ecosystem diversity, and genetic diversity. The ways by which these resources are used with sustainable management and replenishment are also dealt with. The topics of interest include but are not restricted only to sustainable development of various ecosystems and conservation of hotspots, traditional methods and role of local people, threatened and endangered species, global climate change and effect on biodiversity, invasive species, impact of various activities on biodiversity, biodiversity conservation in sustaining livelihoods and reducing poverty, and technologies available and required. The books in this series will be useful to botanists, environmentalists, marine biologists, policy makers, conservationists, and NGOs working for environment protection.

More information about this series at http://www.springer.com/series/11920

Dinesh Kumar Maheshwari ·
Shrivardhan Dheeman
Editors

Field Crops: Sustainable Management by PGPR

Springer

Editors
Dinesh Kumar Maheshwari
Department of Botany and Microbiology
Gurukul Kangri University
Haridwar, Uttarakhand, India

Shrivardhan Dheeman
Department of Botany and Microbiology
Gurukul Kangri University
Haridwar, Uttarakhand, India

ISSN 2352-474X ISSN 2352-4758 (electronic)
Sustainable Development and Biodiversity
ISBN 978-3-030-30925-1 ISBN 978-3-030-30926-8 (eBook)
https://doi.org/10.1007/978-3-030-30926-8

This Springer imprint is published by the registered company Springer Nature Switzerland AG
The registered company address is: Gewerbestrasse 11, 6330 Cham, Switzerland

Preface

The sustainable management of field crops is a concern to cultivate them economically and eco-safe ways so as to reduce the abusive use of harmful agrochemical not only creating a 'soil-security' but also posing several health-related issues in the human race. Although, much progress has been made in understanding a technology-oriented sustainable agro-ecosystem, it is essential to involve the frontiers of knowledge of PGPR's roles to fill the gap between ever increasing population and crop productivity.

In the recent scenario, field crops are geared up to feed the phenomenally increasing population has largely become chemical based and input intensive. The plateauing crop yields and falling inputs response have severely reduced the sustainability of the crop ecosystem. In addition, there is a growing awareness to shift from chemical to organic agriculture. This will partially reduce the pressure on the chemicals demand to increase the credibility and sustainability to raise field crops healthy.

The present book is an endeavor to provide voluminous information on the role of PGPR in sustainable management of field crop production and enhancement of productivity. The main focus lies on different portrayals meant to establish significance of PGPR in minerals, nutrition and their assimilation, alleviation of abiotic stress of drought and salinity, influence of info-chemicals, in sustainable management of field crops. This not only ends with these crops but also expands horizons to the other vegetable species such as tomato, pepper, melon, radish, lettuce, etc. The plant–microbes relationships in soil-ecological system and their accurate benefits in concern to raise the question on the safety of food production become imperative to adapt biological fertilization strategy that may minimize the use of chemical inputs.

Plant growth-promoting microbes (PGPM) becomes an effective solution to the problem of mercury toxicity from the contaminated agricultural lands. These beneficial PGPR impose drought tolerance by the production of exopolysaccharides, phytohormone, 1-aminocyclopropane-1-carboxylate (ACC) deaminase, volatile compounds and by up- or downregulation of stress-responsive genes and by altering root morphology during water scarcity. PGPR have abilities to induce defense against insect herbivore. Some beneficial microbes also induce systemic resistance (ISR) against microbial pathogenic viruses, parasites, and herbivorous insects and some display direct insect pathogenicity.

This book will be valuable not only for the scientific community but also to the teachers, researchers, and students studying graduation and postgraduation in various streams of Agriculture, Horticulture, Biotechnology, Microbiology, Phytopathology, Agronomy, and Environmental Sciences.

We desire to acknowledge all the subject specialists, contributors, who were quite cooperative to spare their cooperation and patience to make this book a successful endeavor. Thanks are due to our research team members, who generously assisted in the compilation and completion of this task. We extend our sincere thanks to Dr. (Mrs.) R. Valeria and her colleagues for their valuable support to facilitate completion of this project. Support from MHRD-UGC BSR fellowship (DKM) is duly acknowledged.

Haridwar, Uttarakhand, India

Dinesh Kumar Maheshwari
Shrivardhan Dheeman

Contents

Contributors

Bukola Victoria Ailenokhuoria Institute of Agricultural Research and Training, Obafemi Awolowo University, Ibadan, Nigeria

Seemab Anjum COMSATS University Islamabad, Chak Shahazad, Islamabad, Pakistan

Antonino Baez Ecology and Survival of Microorganisms Group, Laboratorio de Ecología Molecular Microbiana (LEMM), Centro de Investigaciones en Ciencias Microbiológicas (CICM), Instituto de Ciencias (IC), Benemérita Universidad Autónoma de Puebla (BUAP), Ciudad Universitaria, Puebla, Mexico

Nitin Baliyan Department of Botany and Microbiology, Gurukula Kangri Vishwavidyalaya, Haridwar, Uttarakahnd, India

Asghari Bano Department of Biosciences, University of Wah, Wah Cantt., Pakistan

Eulogio J. Bedmar Department of Soil Microbiology and Symbiotic Systems, Estación Experimental del Zaidín, CSIC, Granada, Spain

J. K. Bisht ICAR–Vivekananda Institute of Hill Agriculture, Almora, Uttarakhand, India

Antonio Castellano-Hinojosa Department of Soil Microbiology and Symbiotic Systems, Estación Experimental del Zaidín, CSIC, Granada, Spain;
Pharmacy Faculty, Department of Microbiology, University of Granada, Granada, Spain

Jairam Choudhary ICAR–Indian Institute of Farming Systems Research Modipuram, Meerut, Uttar Pradesh, India

Mahipal Choudhary ICAR–Vivekananda Institute of Hill Agriculture, Almora, Uttarakhand, India

Mukesh Choudhary ICAR–ICAR-Indian Grassland and Fodder Research Institute, Jhansi, Uttar Pradesh, India

Reena Deshmukh Biotechnology Centre, Jawaharlal Nehru Agriculture University, Jabalpur, Madhya Pradesh, India

Shrivardhan Dheeman Department of Botany and Microbiology, Gurukula Kangri Vishwavidyalaya, Haridwar, Uttarakhand, India;
Department of Microbiology, Sardar Bhagwan Singh University, Dehradun, Uttarakhand, India

Sandhya Dhiman Department of Botany and Microbiology, Gurukula Kangri Vishwavidyalaya, Haridwar, Uttarakahnd, India

Joseph Disi Department of Entomology, University of Georgia, Athens, GA, USA

Ramesh Chand Dubey Department of Botany and Microbiology, Gurukula Kangri Vishwavidyalaya, Haridwar, Uttarakahnd, India

Atilla Dursun Department of Horticulture, Faculty of Agriculture, Atatürk University, Erzurum, Turkey

Noha K. El-Dougdoug Faculty of Science, Botany and Microbiology Department, Benha University, Benha, Egypt

Hassan Etesami Department of Soil Science, Faculty of Agricultural Engineering and Technology, University College of Agriculture and Natural Resources, University of Tehran, Tehran, Iran

Prakash Chand Ghasal ICAR–Indian Institute of Farming Systems Research Modipuram, Meerut, Uttar Pradesh, India

Iti Gontia-Mishra Biotechnology Centre, Jawaharlal Nehru Agriculture University, Jabalpur, Madhya Pradesh, India

Darren Heenan-Daly School of Biological Earth and Environmental Science, University College Cork, Cork, Ireland

Ahmed A. Hmed Faculty of Science, Botany and Microbiology Department, Al-Azhar University, Nasr City, Cairo, Egypt

Haluk Caglar Kaymak Department of Horticulture, Faculty of Agriculture, Atatürk University, Erzurum, Turkey

Rumana Keyani COMSATS University Islamabad, Chak Shahazad, Islamabad, Pakistan

Naeem Khan Department of Plant Sciences, Quaid-i-Azam University, Islamabad, Pakistan

Sandeep Kumar Department of Botany and Microbiology, Gurukula Kangri Vishwavidyalaya, Haridwar, Uttarakahnd, India

Ertan Sait Kurtar Department of Horticulture, Faculty of Agriculture, Selçuk University, Konya, Turkey

Dinesh Kumar Maheshwari Department of Botany and Microbiology, Gurukula Kangri Vishwavidyalaya, Haridwar, Uttarakhand, India

Vijay Singh Meena ICAR–Vivekananda Institute of Hill Agriculture, Almora, Uttarakhand, India

P. K. Mishra ICAR–Vivekananda Institute of Hill Agriculture, Almora, Uttarakhand, India

Dalia Molina-Romero Ecology and Survival of Microorganisms Group, Laboratorio de Ecología Molecular Microbiana (LEMM), Centro de Investigaciones en Ciencias Microbiológicas (CICM), Instituto de Ciencias (IC), Benemérita Universidad Autónoma de Puebla (BUAP), Ciudad Universitaria, Puebla, Mexico;
Laboratorio de Biología Molecular y Microbiología, Facultad de Ciencias Biológicas, BUAP, Ciudad Universitaria, Puebla, Mexico

Yolanda Elizabeth Morales-García Ecology and Survival of Microorganisms Group, Laboratorio de Ecología Molecular Microbiana (LEMM), Centro de Investigaciones en Ciencias Microbiológicas (CICM), Instituto de Ciencias (IC), Benemérita Universidad Autónoma de Puebla (BUAP), Ciudad Universitaria, Puebla, Mexico;
Laboratorio de Biología Molecular y Microbiología, Facultad de Ciencias Biológicas, BUAP, Ciudad Universitaria, Puebla, Mexico;
Facultad de Estomatología-BUAP, Puebla, Mexico

Jesús Muñoz-Rojas Ecology and Survival of Microorganisms Group, Laboratorio de Ecología Molecular Microbiana (LEMM), Centro de Investigaciones en Ciencias Microbiológicas (CICM), Instituto de Ciencias (IC), Benemérita Universidad Autónoma de Puebla (BUAP), Ciudad Universitaria, Puebla, Mexico

Rabia Naz COMSATS University Islamabad, Chak Shahazad, Islamabad, Pakistan

Mahaveer Nogia ICAR-NBSS&LUP, Reginal Centre, Udaipur, India

Asia Nosheen COMSATS University Islamabad, Chak Shahazad, Islamabad, Pakistan

Oladapo Titus Okareh Faculty of Public Health, Department of Environmental Health Sciences, College of Medicine, University of Ibadan, Ibadan, Nigeria

Adedayo Omowumi Oyedele Institute of Agricultural Research and Training, Obafemi Awolowo University, Ibadan, Nigeria

S. C. Panday ICAR–Vivekananda Institute of Hill Agriculture, Almora, Uttarakhand, India

Piyush Pandey Department of Microbiology, Assam University, Silchar, Assam, India

Manoj Parihar ICAR–Vivekananda Institute of Hill Agriculture, Almora, Uttarakhand, India

Arunav Pattanayak ICAR–Vivekananda Institute of Hill Agriculture, Almora, Uttarakhand, India

Laura Abisaí Pazos-Rojas Ecology and Survival of Microorganisms Group, Laboratorio de Ecología Molecular Microbiana (LEMM), Centro de Investigaciones en Ciencias Microbiológicas (CICM), Instituto de Ciencias (IC), Benemérita Universidad Autónoma de Puebla (BUAP), Ciudad Universitaria, Puebla, Mexico

Barbara Doyle Prestwich School of Biological Earth and Environmental Science, University College Cork, Cork, Ireland

Verónica Quintero-Hernández CONACYT, LEMM, CICM, IC-BUAP, Ciudad Universitaria, Puebla, Mexico

Winston Franz Ríos-Ruiz Universidad Nacional de San Martín–Tarapoto, Tarapoto, Perú

América Paulina Rivera-Urbalejo Ecology and Survival of Microorganisms Group, Laboratorio de Ecología Molecular Microbiana (LEMM), Centro de Investigaciones en Ciencias Microbiológicas (CICM), Instituto de Ciencias (IC), Benemérita Universidad Autónoma de Puebla (BUAP), Ciudad Universitaria, Puebla, Mexico;
Laboratorio de Biología Molecular y Microbiología, Facultad de Ciencias Biológicas, BUAP, Ciudad Universitaria, Puebla, Mexico;
Facultad de Estomatología-BUAP, Puebla, Mexico

S. K. Samal ICAR-Research Complex for Eastern Region, ICAR Parisar, Patna, Bihar, India

Swapnil Sapre Biotechnology Centre, Jawaharlal Nehru Agriculture University, Jabalpur, Madhya Pradesh, India

Meenu Saraf Department of Microbiology, School of Sciences, Gujarat University, Ahmadabad, Gujrat, India

Musa Seymen Department of Horticulture, Faculty of Agriculture, Selçuk University, Konya, Turkey

Jocelyn Simmons Department of Agriculture Food and Resources Sciences, University of Maryland Eastern Shore, Princess Anne, MD, USA

Ahmed R. Sofy Faculty of Science, Botany and Microbiology Department, Al-Azhar University, Nasr City, Cairo, Egypt

Mahmoud R. Sofy Faculty of Science, Botany and Microbiology Department, Al-Azhar University, Nasr City, Cairo, Egypt

Lateef Bamidele Taiwo Institute of Agricultural Research and Training, Obafemi Awolowo University, Ibadan, Nigeria

Sharad Tiwari Department of Plant Breeding and Genetics, Jawaharlal Nehru Agriculture University, Jabalpur, Madhya Pradesh, India

Önder Türkmen Department of Horticulture, Faculty of Agriculture, Selçuk University, Konya, Turkey

Renzo Alfredo Valdez-Nuñez Universidad Nacional de San Martín–Tarapoto, Tarapoto, Perú

Siva L. S. Velivelli Donald Danforth Plant Science Center, St Louis, MO, USA

Ram Prakash Yadav ICAR–Vivekananda Institute of Hill Agriculture, Almora, Uttarakhand, India

M. R. Yadav Rajasthan Agricultural Research Institute, Durgapura, Rajasthan, India

Humaira Yasmin COMSATS University Islamabad, Chak Shahazad, Islamabad, Pakistan

Simon Zebelo Department of Agriculture Food and Resources Sciences, University of Maryland Eastern Shore, Princess Anne, MD, USA

Chapter 1
Plant Growth-Promoting Rhizobacteria (PGPR) as Protagonists of Ever-Sustained Agriculture: An Introduction

Dinesh Kumar Maheshwari, Meenu Saraf and Shrivardhan Dheeman

Abstract The rhizosphere is a zone of soil surrounding to the plant roots, where the biology and chemistry of the soil is influenced by root metabolites pumped into the soil, called root exudates. The rhizosphere of ample quorum of microorganisms, thus, regarded as an ecology. The beneficial bacteria in the rhizosphere are recognized as plant growth-promoting rhizobacteria (PGPR) that influence plant growth and health promotion by several means of mechanisms. These can influence plant traits under fluctuating environmental conditions and improve yield productivity in a sustainable way. The use of PGPR in field crops has attended by enormous researches to enhance crop production and productivity in a sustainable manner. The inoculation strategies such as co-inoculation of two or more beneficial bacteria as bioinoculant apparently provided greater phytostimulation perhaps because of the synergistic and multifarious effects due to co-occurrence and co-interaction with field crops. PGPR are emerging tools of sustainable agriculture; also providing strategic avenues to combat biotic and abiotic stresses of crops. Further, this chapter introduces PGPR as a central candidate with multifaceted mechanisms and influences on sustainable management of field crops.

Keywords Rhizosphere · PGPR · Biofilm · Bacterial diversity · Field crops

---

D. K. Maheshwari (✉) · S. Dheeman
Department of Botany and Microbiology, Gurukula Kangri Vishwavidyalaya,
Haridwar 249404, Uttarakhand, India

M. Saraf
Department of Microbiology, School of Sciences, Gujarat University,
Ahmadabad, Gujrat, India

S. Dheeman
Department of Microbiology, Sardar Bhagwan Singh University,
Dehradun 248161, Uttarakhand, India

D. K. Maheshwari and S. Dheeman (eds.), *Field Crops: Sustainable Management by PGPR*, Sustainable Development and Biodiversity 23,
https://doi.org/10.1007/978-3-030-30926-8_1

1.1 Introduction

Environmental sustainability, specifically for sustainable agriculture must be free of abusive use of chemical fertilizers spoiling biological diversity and soil fertility simultaneously and had raised several concerns to find the alternatives, which are natural in origin. Plants comprise an excellent ecosystem with microorganisms that interact with different degrees under attraction towards secretions of diverse metabolites called root exudates. Since Hiltner in 1904 has understood about 'Rhizosphere', i.e. the layer of soil influenced by roots is much richer in bacteria than the surrounding bulk soil. The role of rhizosphere microbiology has been established as an arena of sustained agriculture and economy as well. In this quest, the search for desired beneficial microorganisms including bacteria, fungi, actinobacteria, protozoa and algae are by far the most common. Each gram of soil contains up to 10 billion bacterial counts, possibly due to their ability to grow rapidly. Such bacteria are bound with the surface of soil particles and soil aggregates; quite a good number of these interact with plant roots, hence rhizospheric in nature. The interaction between bacteria and plant may be harmful or neutral and sometimes proved beneficial to that of plants. Such benefits may occur either by symbiotic, associative and free-living relationship with the roots of plants (Weller and Thomashow 1994; Glick 1995; Maheshwari and Annapurna 2017) to provide them a healthy status.

In a healthy plant, all the biochemical reactions are in a state of equilibrium. If any deviation due to biotic or abiotic stresses comes in any way, it tends to disrupt the metabolic processes making the plant unhealthy. Such plants overcome this state of the affair by using various inputs in the form of biological ways either by adding organic compost or by crop-rotation. There is another alternative in the form of plant growth-promoting rhizobacteria, i.e. PGPR which outcompetes the stresses affecting plant growth and development. These free-living or endophyte bacteria are not only functional and diverse in habitat but also capable of supporting plant growth and development.

In the recent scenario, cultivation of field crops are geared up to feed the phenomenally increasing population which has largely become chemical based and input intensive. This has accelerated the process of resource depletion and environmental pollution. The plateau crop yields and falling input response have severely reduced the sustainability of the crop ecosystem.

Microbial inoculants, which are naturally occurring beneficial microorganisms, hold the promise because of self-replication, non-toxic inputs and increase the productivity and protection for the benefits of the crop production. They make important contributions to fertility and productivity both for plants and soil. Efforts made during the past decade have generated awareness about the cheap input amongst the growers. In addition, there is a growing concern to shift farming from chemical to organic which, partially reduce the pressure on the chemicals demand, increase the credibility and sustainability to raise healthy field crops.

1.2 Rhizosphere: A Home of PGPR

'Rhizosphere' denotes the area of intense microbiological activity that extends to several millimetres from the root system of growing plant. In other words, the rhizosphere is a hotspot of microbial interactions, as exudates released by plant roots are the main food source for microorganisms and contribute as a driving force to enhance their population density and activities (Raaijmakers et al. 2009). Rhizosphere encompasses the millimeters of soil surrounding a plant root, where complex biological and ecological processes occur (Bais et al. 2006). Soil which is not a part of the rhizosphere and does not penetrate by plant roots is known as bulk soil. Natural organic compounds and microbial populations are much lower in bulk soil than in the rhizosphere (Stotsky et al. 2000). The above facts illustrate the significance of plant growth-promoting rhizobacteria (PGPR) as a potential candidate to be used as beneficial bio-fertilizers having merit over conventional fertilizers (Maheshwari et al. 2015). But, before providing their beneficial impact on the plants living in rhizosphere, they must be colonized and should stay there for long term, which is understood as root colonization. In some cases, associative interactions between plant and bacteria are accomplished with the help of rhizodeposits, which also pose growth-stimulating effects for microbial survival. As symbiotic systems, associative interactions are of great interest because many crops show an increase in yield after inoculation (Höflich et al. 1994). On the other hand, rhizosphere is the infection court where soil-borne pathogens establish a parasitic relationship with the plant. To infect root tissue, pathogens have to compete with the member of the rhizosphere microbiome for available nutrients and micro-niche (Chapelle et al. 2016). Soil can be considered as a microbial seed bank (Lennon and Jones 2011). The carry-over effect on the assemblage of the 'rhizomicrobes', i.e. the proliferation of microbes in, on or around the roots has important implications for the co-evolution of plant–microorganism interactions in natural ecosystems. It includes dispersal of microorganisms from a source of inoculum to the actively growing root, multiplication or growth in the rhizosphere.

In fact, rhizosphere is responsible for the success of free-living bacteria which requires a proper process for root colonization. Nowadays, there is a possibility to manipulate these zones of root-colonization via 'rhizosphere engineering'. This process is quite important from different points of view in order to produce bacterial-mediated growth-promoting substance to support crop yield and productivity. A successful root/rhizosphere colonizer exclusively means the existence of features, which survive in fierce competition of the indigenous micro-flora in the growth enhancement processes and in improving homeostatic mechanisms upon stress challenges. The environmental homeostasis in the rhizosphere of crop plant protect it against a broader range of pathogens. Also, as a consequence, the plant switches on downstream signalling pathways and produces antimicrobial compounds to kill the pathogen and maintain homeostasis (Thormar and Hilmarsson 2007). This is a very precisely controlled and complex process involves a number of genes and signalling pathways (Zipfel 2009). It is this complexity of plant–

pathogen interactions, which makes it very difficult to discern, due to which anatomical features, metabolites and signalling pathways become activated. The traditional, biochemical and genetic experimental methods are inadequate tools for the task.

Rhizospheric competence is a pre-requisite for plant growth and development, PGPR play a major role in two different ways, indirectly or directly, both share support mechanisms to bring the sustainability for major crops. For this, the bacterial genera are used as inoculants for seed dressing and allow them to raise the plants resulting in growth promotion and increased yield. Rhizospheric competence affects positively on root and plant biology in relation to provisional nutrition, growth promotion, development and health (Aragno 2005). Hence, to advance the future of agriculture, in terms to find more precise bacteria, which can invade in the rhizosphere of other plants too can be found by studying free-living association in depth (Cook 2012).

1.3 PGPR and Rhizospheric Biofilm

The success of PGPR in agriculture has been sorted out if PGPR attributed effective root colonization with the traits and subsequently able to form biofilms in the rhizosphere, which assures their tenancy in a manner to establish a successful relationship with the plant (Saleh-Lakha and Glick 2006). Rhizospheric ecology comprised reciprocal influence of inter-alia plant–microbe interactions influenced by its biotic and abiotic factors. Among the community of beneficial as well as deleterious bacteria, PGPR plant growth-promoting bacteria were entirely well-thought-out as beneficial bacteria and play a multifarious role in the environment and agriculture and being an important germ-plasm. Plant-associated bacteria generally interact with host tissue surfaces during pathogenesis and symbiosis, due to and in commensal relationships. The bacteria associated with plants are aggressive to form increasingly revealing biofilm-type structures that vary from small clusters of cells to extensive biofilms in nature. This implies that PGPR competence strongly depends either on their abilities to take advantage of a specific environment or their abilities to adapt according to changing conditions. PGPR may be uniquely equipped to sense chemoattractants, e.g. rice exudates induce stronger chemotactic responses of endophytic bacteria as compared to non-PGPR present in the rice rhizosphere.

1.4 Rhizospheric Dynamics and Diversity

The rhizosphere harbours one of the most complex, diverse and active plant-associated microbial communities. Although, selection for the rhizosphere community is evident, the specific bacterial traits that make them able to colonize this environment are still poorly understood (Lopes et al. 2016). But, there is a quest that, is PGPR diversity quantifiable? Actually, to study microbial diversity, taxonomic and functional diversity are major concerns where taxonomic diversity integrates different aspects of diversity and provides a more complete picture with a deeper understanding of microbial interactions in soil ecosystems (Torsvik et al. 2002). The use of molecular methods for the study of genetic diversity primarily, the sensitive and accurate PCR-based genotyping methods enable differentiation among closely related bacterial strains and the detection of higher bacterial diversity (Doignon-Bourcier et al. 1999; Tan et al. 2001). Jha et al. (2010) reported that a good diversity index should encompass both Dominance indices (Example, Simpson Index) and Information indices (Example, Shannon-Wiener Index). To comprehend such diversity, it is advantageous to investigate the combined uses of species richness and diversity as well as to estimate the combinatorial effect of species richness and diversity in order to understand their role and distribution in their habitat. In another study focused on bacterial diversity that explored the Himalayan region as well-known biodiversity hotspots and rich in ecological diversity. However, it suffers from low agricultural productivity due to several climatic and agricultural limits. The research reveals the diversity of *Bacillus* population associated with *Eleusine coracana* (ragi). The study was carried out using two universal primers derived from the highly conserved region of 16S rRNA gene fD1 and rD1 for species identification. Overall, the site-specific diversity of the bacterial strains using the Simpson and Shannon–Wiener Index had positive correlation with the altitude gradient. Ragi, *Eleusine coracana* L., is one of the main food grain crops now being commercially exploited (Dheeman et al. 2017).

Microbial diversity is puzzled with co-existence of species that remains functional in their ecology and their degree of functionality has been studied under the term of functional diversity (Torsvik and Øvreås 2002). The functional diversity of species influences ecosystem dynamics, stability, productivity, nutrient balance and other aspects of ecosystem functioning (Tilman 2001). Embedded microbial community belongings to rhizosphere possesses functional diversity, which ultimately enhance agricultural production (Maheshwari 2014). Functional diversity in the rhizosphere considered the mechanistic behaviour of microbial guilds, which ultimately promotes the plant growth. It has multidimensional cloud of species trait and each trait represents an individual or a species (Fig. 1.1), e.g. Phosphate-Solubilizing Bacteria (PSB), whereas phosphate solubilization is a function of rhizospheric microorganisms to alleviate the nutritional requirement in soil, and further aid in plant growth promotion. The functionality of PGPR in agriculture is increased with its diversity (Tilak et al. 2005). Such groups of selective individuals are also considered as functionally diverse and are utilized to improve agricultural production as cited by Maheshwari (2014).

The transfer of nutrients from plant roots in the rhizosphere controls plant–microbe interactions and populations of bacteria have a functional role within communities that permit their survival. The relationship between microorganisms and plants confers the availability of nitrogen along with other nutritional requirements, viz. phosphorus, zinc, potassium, etc. (Maheshwari 2011; Dhiman

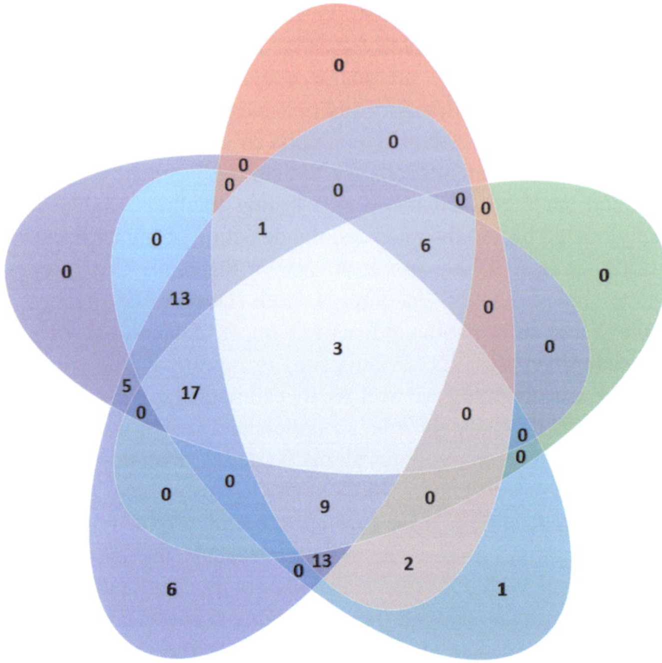

Fig. 1.1 A five-way Venn Set Diagram (VSD) showing multidimensional overlaps among 5 traits of PGPR and elucidating commonness in *Bacilli* on trait level (Author's Lab)

et al. 2019). Plant roots offer a niche to the proliferation of soil bacteria and sustain their life under the influence of root exudates and lysates (Lugtenberg et al. 2002). In the agro-ecosystem, several diverse genera of microorganisms are responsible for multifarious metabolic functions and interfere in soil fertility, plant health, growth, nutrient cycling, organic matter formation, decomposition and soil structure maintenance (Kennedy 1999). Plant growth is a multi-complex phenomenon and rhizobacteria acts as bio-stimulant, bio-fertilizer and plant-protective agent or sometimes in bioremediation of pollutants.

A large and ever-increasing number of genera in the community of PGPR have been discovered so far. These are different morphologically, however share nutritional and ecological preferences including ecology of rhizospheric niches. These were isolated from a varied range of habitats (from harsh to adaptive environments) such as rhizospheric soil, cow and buffalo dung, root nodules of various dicotyledonous plants and even from termitarium (Chauhan et al. 2016) and termite gut (Fröhlich et al. 2007)

Rhizospheric bacteria are communities heavily influenced by soil texture and rhizospheric ecosystem as illustrated in Fig. 1.2.

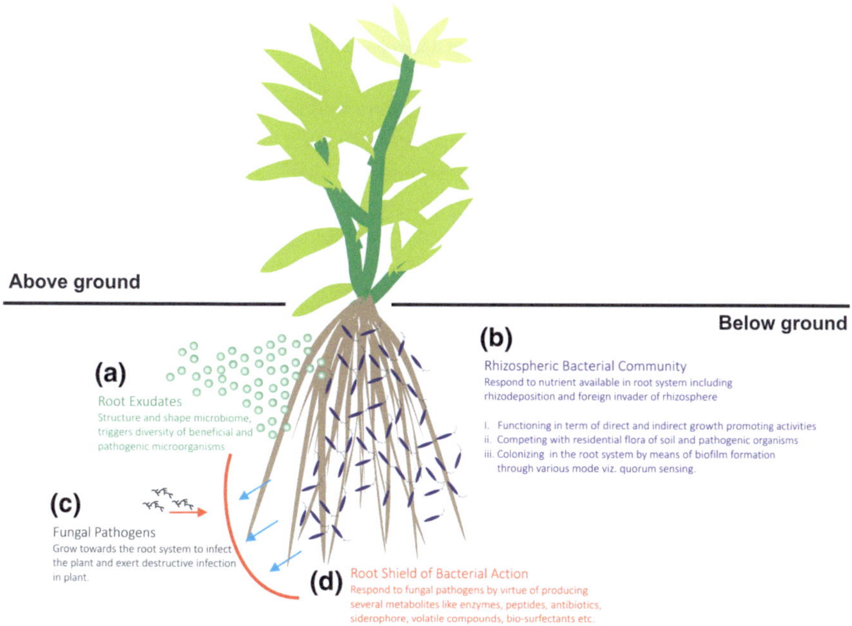

Fig. 1.2 Bacterial community in rhizosphere showing their sole importance from A to D

1.5 PGPR and Field Crops

Various countries spend tens of billions of dollars in importing food as outdated technologies, poor yields, shrinking farms and unreliable weather pattern inflict. However, the use of PGPR not only saves the biodiversity and is too cheap for farmers. Their wide contribution to growth and health promotion is now well established. PGPR enhance plant growth by a wide variety of mechanisms like biological nitrogen fixation, phytohormone production, phosphate solubilization, siderophore production, 1-aminocyclopropane-1-carboxylate deaminase production (ACC), exhibiting antifungal activity, production of volatile organic compounds (VOCs) promoting beneficial plant–microbe symbiosis, interference with pathogen toxin production, etc. (Ryu et al. 2003). Current advancements in the exploration of rhizosphere diversity along with PGPR and their functional ability and mechanism to facilitate plant growth promotion corroborated their application as a reliable phenomenon in the management of sustainable agricultural system. Use of PGPR strains have the potential to induce antioxidant enzymes, amino acids like proline under salinity stress and could serve as a useful tool for alleviating salinity stress in salt-sensitive plants (Patel and Saraf 2013).

Monitoring soil microbial life and their influence on the tripartite interaction of plant soil and microbes manage agricultural production. In understanding a technology-oriented sustainable agro-ecosystem, significant progress have been made, but it is essential to broaden the frontiers of knowledge of PGPR's roles in field crop and others to fulfill the gap between ever-increasing population and productivity. More recently, drone technology is proved to be beneficial for the spray of the plant protection agents, biostimulants, etc. In fact, precision monitoring of interaction of beneficial microbe with soil and plant proved to be instrumental for agriculture production (Baliyan et al. 2018). In addition, the chapters in the book expand the frontiers of PGPR in the arena of field crops cultivation and productivity enhancement of sustainable agriculture.

1.6 Conclusions

The plant growth-promoting rhizobacteria (PGPR) are emerging tool of sustainable agriculture and a central candidate with multifaceted mechanisms and influences on sustainable management of field crops. Further, a quantum and ample research is needed under the consideration of plant–microbe interaction to augment sustainable management of field crops.

Conflict of Interest The author(s) have no conflict of interest.

References

Aragno M (2005) The rhizosphere a hot spot of bacterial diversity. Microbial diversity current perspectives and potential applications. IK International Pvt. Ltd., New Delhi, pp 261–284

Bais HP, Weir TL, Perry LG, Gilroy S, Vivanco JM (2006) The role of root exudates in rhizosphere interactions with plants and other organisms. Annu Rev Plant Biol 57:233–266

Baliyan N, Dheeman S, Maheshwari DK, Dubey RC, Vishnoi VK (2018) Rhizobacteria isolated under field first strategy improved chickpea growth and productivity. Environ Sustain 1 (4):461–469

Chapelle E, Mendes R, Bakker PAH, Raaijmakers JM (2016) Fungal invasion of the rhizosphere microbiome. ISME J 10(1):265

Chauhan AK, Maheshwari DK, Kim K, Bajpai VK (2016) Termitarium-inhabiting *Bacillus endophyticus* TSH42 and *Bacillus cereus* TSH77 colonizing *Curcuma longa* L.: isolation, characterization, and evaluation of their biocontrol and plant-growth-promoting activities. Can J Microbiol 62(10):880–892

Cook RJ (2012) Management of the associated microbiota. In: Plant disease. An advanced treatise, pp 145–166

Dheeman S, Maheshwari DK, Agarwal M, Dubey RC, Aeron A, Kim K, Bajpai VK (2017) Polyphasic and functional diversity of high altitude culturable *Bacillus* from rhizosphere of *Eleusine coracana* (L.) Gaertn. Appl Soil Ecol 110:127–136

Dhiman S, Dubey RC, Maheshwari DK, Kumar S (2019) Sulphur oxidising buffalo dung bacteria enhance growth and yield of *Foeniculum vulgare* Mill. https://doi.org/10.1139/cjm-2018-0476

Doignon-Bourcier F, Sy A, Willems A, Torck U, Dreyfus B, Gillis M, de Lajudie P (1999) Diversity of bradyrhizobia from 27 tropical Leguminosae species native of Senegal. Syst Appl Microbiol 22(4):647–661

Fröhlich J, Koustiane C, Kämpfer P, Rosselló-Mora R, Valens M, Berchtold M, Kuhnigk T, Hertel H, Maheshwari DK, König H (2007) Occurrence of rhizobia in the gut of the higher termite *Nasutitermes nigriceps*. Syst Appl Microbiol 30(1):68–74

Glick BR (1995) The enhancement of plant growth by free-living bacteria. Can J Microbiol 41 (2):109–117

Höflich G, Wiehe W, Kühn G (1994) Plant growth stimulation by inoculation with symbiotic and associative rhizosphere microorganisms. Experientia 50(10):897–905

Jha CK, Patel D, Rajendran N, Saraf M (2010) Combinatorial assessment on dominance and informative diversity of PGPR from rhizosphere of *Jatropha curcas* L. J Basic Microbiol 50:211–217

Kennedy AC (1999) Bacterial diversity in agroecosystems. In: Invertebrate biodiversity as bioindicators of sustainable landscapes, pp 65–76

Lennon JT, Jones SE (2011) Microbial seed banks: the ecological and evolutionary implications of dormancy. Nature Rev Microbiol 9(2):119

Lopes LD, Pereira e Silva MDC, Andreote FD (2016) Bacterial abilities and adaptation toward the rhizosphere colonization. Front Microbiol 7:1341

Lugtenberg BJ, Chin-A-Woeng TF, Bloemberg GV (2002) Microbe–plant interactions: principles and mechanisms. Antonie Van Leeuwenhoek 81(1–4):373–383

Maheshwari DK (2011) Bacteria in agrobiology: plant nutrient management. Springer Science & Business Media, Germany

Maheshwari DK (2014) Bacterial diversity in sustainable agriculture. Springer Science & Business Media, Switzerland

Maheshwari DK, Annapurna K (2017) Endophytes: crop productivity and protection, vol II. Springer International Publishing, Switzerland

Maheshwari DK, Dubey RC, Agarwal M, Dheeman S, Aeron A, Bajpai VK (2015) Carrier based formulations of biocoenotic consortia of disease suppressive *Pseudomonas aeruginosa* KRP1 and *Bacillus licheniformis* KRB1. Ecol Eng 81:272–277

Patel D, Saraf M (2013) Influence of soil ameliorants and microflora on induction of antioxidant enzymes and growth promotion of *Jatropha curcas* L. under saline condition. Euro J Soil Biol 55:47–54

Raaijmakers JM, Paulitz TC, Steinberg C, Alabouvette C, Moënne-Loccoz Y (2009) The rhizosphere: a playground and battlefield for soilborne pathogens and beneficial microorganisms. Plant Soil 321(1–2):341–361

Ryu CM, Farag MA, Hu CH, Reddy MS, Wei HX, Paré PW, Kloepper JW (2003) Bacterial volatiles promote growth in Arabidopsis. Proc Natl Acad Sci 100(8):4927–4932

Saleh-Lakha S, Glick BR (2006) Plant growth-promoting bacteria. Modern Soil Microbiol, 503–520

Stotsky A, Egardt B, Eriksson S (2000) Variable structure control of engine idle speed with estimation of unmeasurable disturbances. J Dyn Syst Measur Cont 122(4):599–603

Tan ZY, Kan FL, Peng GX, Wang ET, Reinhold-Hurek B, Chen WX (2001) *Rhizobium yanglingense* sp. nov., isolated from arid and semi-arid regions in China. Int J Syst Evol Microbiol 51(3):909–914

Thormar H, Hilmarsson H (2007) The role of microbicidal lipids in host defense against pathogens and their potential as therapeutic agents. Chem Phys Lipid 150(1):1–11

Tilak KVBR, Ranganayaki N, Pal KK, De R, Saxena AK, Nautiyal CS et al (2005) Diversity of plant growth and soil health supporting bacteria. Current Sci, 136–150

Tilman D (2001) Functional diversity. Encyclo Biodiver 3(1):109–120

Torsvik V, Øvreås L (2002) Microbial diversity and function in soil: from genes to ecosystems. Curr Opin Microbiol 5(3):240–245

Torsvik V, Øvreås L, Thingstad TF (2002) Prokaryotic diversity–magnitude, dynamics, and controlling factors. Science 296(5570):1064–1066

Weller DM, Thomashow LS (1994) Current challenges in introducing beneficial microorganisms into the rhizosphere. Molecular ecology of rhizosphere microorganisms: biotechnology and the release of GMOs, pp 1–18

Zipfel C (2009) Early molecular events in PAMP-triggered immunity. Curr Opin Plant Biol 12 (4):414–420

Chapter 2
Bacterial Mixtures, the Future Generation of Inoculants for Sustainable Crop Production

Yolanda Elizabeth Morales-García, Antonino Baez, Verónica Quintero-Hernández, Dalia Molina-Romero, América Paulina Rivera-Urbalejo, Laura Abisaí Pazos-Rojas and Jesús Muñoz-Rojas

Abstract Plant growth-promoting bacteria (PGPB) have been extensively studied, because of different mechanisms to perform phytostimulation, as well as the abilities to colonize plants. The number of crop types and hectares of agricultural land, where PGPR are applied is low compared with the total crops or farming area.

Y. E. Morales-García (✉) · A. Baez · D. Molina-Romero · A. P. Rivera-Urbalejo ·
L. A. Pazos-Rojas · J. Muñoz-Rojas (✉)
Ecology and Survival of Microorganisms Group, Laboratorio de Ecología Molecular Microbiana (LEMM), Centro de Investigaciones en Ciencias Microbiológicas (CICM), Instituto de Ciencias (IC), Benemérita Universidad Autónoma de Puebla (BUAP), Ciudad Universitaria, Edificio IC-11, Colonia Jardines de San Manuel, Puebla, Mexico
e-mail: lissiamor@yahoo.com.mx

J. Muñoz-Rojas
e-mail: joymerre@hotmail.com; joymerre@yahoo.com.mx

A. Baez
e-mail: antonino.baez@correo.buap.mx

D. Molina-Romero
e-mail: molinardalia@gmail.com

A. P. Rivera-Urbalejo
e-mail: america99670@gmail.com

L. A. Pazos-Rojas
e-mail: yellowtweety194@hotmail.com

Y. E. Morales-García · D. Molina-Romero · A. P. Rivera-Urbalejo
Laboratorio de Biología Molecular y Microbiología, Facultad de Ciencias Biológicas, BUAP, Ciudad Universitaria, Edificio 112-A, Colonia Jardines de San Manuel, Puebla, Mexico

Y. E. Morales-García · A. P. Rivera-Urbalejo
Facultad de Estomatología-BUAP, Puebla, Mexico

V. Quintero-Hernández
CONACYT, LEMM, CICM, IC-BUAP, Ciudad Universitaria, Edificio IC11, Colonia Jardines de San Manuel, Puebla, Mexico
e-mail: vquinterohe@conacyt.mx

D. K. Maheshwari and S. Dheeman (eds.), *Field Crops: Sustainable Management by PGPR*, Sustainable Development and Biodiversity 23,
https://doi.org/10.1007/978-3-030-30926-8_2

However, the PGPB application in crop fields is increasingly becoming more accepted due to the advantages to crops and environment such as the increment in productivity, the diminution in the use of chemical fertilizers and toxic compounds such as pesticides and herbicides. These traits make beneficial bacteria formulations the 'right choice' in healthy agriculture since they are highly compatible with sustainable crop production. The co-inoculation of plants with two or more beneficial bacteria apparently provides greater phytostimulation than mono-inoculation, perhaps because of the synergistic and multifarious effects occurring when two or more microorganisms are co-interacting. There is a consensus that bacterial mixtures provide greater benefits to the plants, the number of formulations containing more than three species of microorganisms in consortium is still limited. Therefore, we believe that more research and investment is needed to design and formulate multi-species inoculants containing compatible bacteria and other beneficial microorganisms in order to be capable of coexisting both in the designed formulation and associated with plants for sustainable benefits.

Keywords Bacterial inoculants · Bacterial desiccation · PGPB · Beneficial bacteria

2.1 Introduction

Bacteria were the first colonizers of the planet and it is believed that they sustain life on earth (Lin et al. 2014; Pace 1997; Strom 2008; Zavarzin 2008). Only around 1% of bacterial diversity have been isolated (Amann et al. 1995; Curtis et al. 2002; Prashar et al. 2014), and majority of them could be beneficial for agriculture (Lugtenberg and Kamilova 2009; Mitter et al. 2017; Philippot et al. 2013), biotechnology (Broadbent et al. 2003; Chemier et al. 2009; Zhang 2018), biomedicine (Morales-García et al. 2007), bioremediation (Böltner et al. 2008; Fernández et al. 2012; Liu et al. 2017b), and other benefits.

Bacteria have the capability to interact with plants and increase their health and growth. The plant growth-promoting bacteria (PGPB) have been isolated from diverse plant sources and geographical places as given in Table 2.1. The source of isolation could be from soils, rhizosphere, rhizoplane, root nodules, endophytic zones and the epiphytic region of the plants (Ahmad et al. 2018; Cazorla et al. 2007; Kumar et al. 2014; Molina-Romero et al. 2015; Morales-García et al. 2011; Vandamme et al. 2002; War Nongkhla and Joshi 2014). Bacteria associated with maize plants also have been isolated and some genera reported include *Azospirillum*, *Burkholderia*, *Bacillus*, *Rhizobium*, *Enterobacter*, *Klebsiella* and *Arthrobacter* (Table 2.2).

A heterogeneous group of bacteria may be associated with plants, but not all of them have the capability to improve their growth. Therefore, rigorous studies should be performed to test their beneficial traits and the mechanisms involved to

Table 2.1 Bacterial strains isolated from diverse environments and mechanisms involved in the plant growth promotion

Isolated strain	Mechanisms for plant growth promotion implicated	Place of isolation	Geographical place for isolation	References
Bacillus subtilis Q3 *Paenibacillus* sp. Q6	Phosphate solubilization	Cotton rhizosphere	Bahawalpur and Haroonabad Punjab—Pakistan	Ahmad et al. (2018)
Enterobacter sp. EP2a JN653461	Phosphate solubilization, BNF	Epiphytic from *Houttuynia cordata*	India	War Nongkhla and Joshi (2014)
Gluconacetobacter diazotrophicus UAP-Cf05	BNF, IAA	*Coffea Arabica* rhizosphere	Puebla—México	Jimenez-Salgado et al. (1997)
Azotobacter sp. NAT 13	BNF	Cotton rhizosphere	Tolima—Colombia	Guzmán et al. (2012)
Pseudomonas fluorescens ABE66285	ACC deaminase, Phosphate solubilization	*Larrea tridentata* rhizosphere	Chile	Jorquera et al. (2012)
Burrkholderia tuberum STM678^{T}	ACC deaminase	Legume nodules	France	Vandamme et al. (2002), Onofre-Lemus et al. (2009)
Bacillus subtilis PCL1605	Antifungal activity	Rhizoplane from avocado	Malaga—Spain	Cazorla et al. (2007)
Klebsiella variicola T29A	IAA	Sugarcane endophyte	Puebla-México	Rosenblueth et al. (2004)

Abbreviation meaning: BNF (Biological Nitrogen Fixation), IAA (Indol Acetic Acid), VOCs (Volatile compounds), ACC (1-aminocyclopropane-1-carboxylate) deaminase

unravel the hidden characters that how bacteria promote the plant growth (Lugtenberg and Kamilova 2009; Molina-Romero et al. 2015).

Adhesion and colonization to the root zone are the first critical steps that bacteria must overcome to the right integration on the plant–host so as to establish and develop their beneficial functions (Albareda et al. 2006; Muñoz-Rojas and Caballero-Mellado 2003). A number of reports have been published on the concept of cellular adherence to the surface of maize seeds and colonizing the rhizosphere (Table 2.3). Adhesion occurred in the range of 10^{5}–10^{6} CFU/seed while rhizosphere colonization was in the order of 10^{3}–10^{8} CFU/gV [Colony Forming Units/ Vermiculite (g)], depending on different factors including bacterial species and experimental conditions. To demonstrate the PGPB capabilities, bacteria have to be bioprimed on seed surface or seed dressing and allow to grow them. The vegetative and reproductives such as seed germination index, length and diameter of stems,

Table 2.2 Bacterial strains isolated from maize and mechanisms involved in the plant growth promotion

Isolated strain	Mechanisms involved in the promotion of plant growth	Localization in the plant	Geographical place for isolation	References
A. brasilense UAP-154	IAA, siderophore production	Maize rhizosphere	Tlaxcala-México	Dobbelaere et al. (2001), Tapia-Hernández et al. (1990)
Arthrobacter sp. V54 *Bacillus* sp. V 39	Phosphate solubilization, BNF, Siderophore production	Maize rhizosphere	Ngaoundal, Adamawa, Cameroon	Tchakounté Tchuisseu et al. (2018)
Bacillus methylotrophicus M4-96	VOCs, Auxins	Maize rhizoplane	Morelia-Michoacán	Pérez-Flores et al. (2017)
Burkholderia tropica MTo 293	VOCs, BNF, Phosphate solubilization and siderophore production	Maize stem (Endophytic)	Oaxaca-México	Tenorio-Salgado et al. (2013)
Burkholderia tropica MCu-831	FBN	Maize rhizoplane	Morelos-México	Reis et al. (2004)
Burkholderia unamae MTl-641^T	BNF, ACC deaminase	Maize rhizosphere	Morelos-México	Caballero-Mellado et al. (2004)
Enterobacter sp. UAPSO3001	Unknown	Maize rhizosphere	Tlaxcala-México	Morales-García et al. (2011)
Klebsiella spp. (Zea)	BNF	Maize stem (Endophytic)	USA	Palus et al. (1996)
Rhizobium etli	BNF	Maize stem (Endophytic)	Morelos-México	Gutiérrez-Zamora and Martínez-Romero (2001)

Abbreviation meaning: BNF (Biological Nitrogen Fixation), IAA (Indol Acetic Acid), VOCs (Volatile compounds), ACC (1-aminocyclopropane-1-carboxylate) deaminase

number of leaves, height of stems, root and shoot length, fresh and dry weight of plants along with important physico-chemical characters such as chlorophyll content, total nitrogen, percentage of nitrogen and others (Ahmad et al. 2018; Molina-Romero et al. 2017d; Morales-García et al. 2011; Muñoz-Rojas and Caballero-Mellado 2003). In one of the study, Bashan et al. (2017) stated that fresh weight is less precise than the dry weight. The fresh weight is affected by numerous environmental and other parameters including relative humidity, temperature, air, soil environment, etc,. of the excised plant parts to interfere with excess moisture present in washed roots and shoots, besides, the total time involved in collecting and washing samples until weighing (Bashan et al. 2017). Therefore, differences in appearance observed between inoculated plants and non-inoculated one represent a

Table 2.3 Bacteria quantity associated with seeds or rhizosphere of maize

Bacterial strain	Cells number adhered to seeds (Log CFU/seed)	Cells number colonizing the rhizosphere of plants Log CFU/gV	Variety of maize	References
Azospirillum brasilense Sp7	5.71 ± 0.41	6.10 ± 0.16	Blue maize CAP15-1 TLAX	Molina-Romero et al. (2017b)
Azospirillum brasilense Sp7	6.95 ± 0.11	5.12 ± 1.40	Rojo-Criollo	Morales-García et al. (2013)
Pseudomonas putida KT2440	7.50 ± 0.14	6.40 ± 0.2	Blue maize CAP15-1 TLAX	Molina-Romero et al. (2017b)
Pseudomonas putida KT2440	7.61 ± 0.03	8.10 ± 0.18	Girona	Ramos Martin and Muñoz-Rojas (2006)
Pseudomonas putida KT2440	7.02 ± 0.15	3.90 ± 0.81	Rojo-Criollo	Unpublished results
Sphingomonas sp. OF178	6.55 ± 0.33	6.18 ± 0.71	Blue maize CAP15-1 TLAX	Molina-Romero et al. (2017b)
Sphingomonas sp. OF178	6.87 ± 0.16	4.74 ± 0.40	Rojo-Criollo	Morales-García et al. (2013)
Acynetobacter sp. EMM02	7.87 ± 0.20	6.70 ± 0.14	Blue maize CAP15-1 TLAX	Molina-Romero et al. (2017b)
Enterobacter sp. UAPS03001	7.02 ± 0.30	9.26 ± 0.55	Rojo-Criollo	(Morales-García et al. 2011)
Enterobacter sp. UAPS03001	7.23 ± 0.18	7.74 ± 0.25	Blue maize CAP15-1 TLAX	Pazos-Rojas et al. (2018)
Klebsiella variicola T29A	6.69 ± 0.03	7.4 ± 0.35	Blue maize CAP15-1 TLAX	Pazos-Rojas et al. (2018)
Klebsiella variicola T29A	7.85 ± 0.25	8.75 ± 0.31	White autochthonous from Papalotla-Tlaxcala, México	Unpublished results

good indicator for the capability of bacteria to promote the growth of plants (Fig. 2.1) but decisive conclusion comes from data of dry weight increments (Bashan et al. 2017; Molina-Romero et al. 2017b; Morales-García et al. 2011; Muñoz-Rojas and Caballero-Mellado 2003).

Various workers have suggested the lab's data do not coincide with field conditions (Çakmakçi et al. 2006; Mehnaz et al. 2010; Shahzad et al. 2013). Even a few have suggested to carry out experiments first, under well-controlled conditions at the plant chamber, then under greenhouse conditions and later under field conditions. Several research steps are involved to carry out in the three phases of experiments prior to developing a commercial formulation delivered to the market (Bashan et al. 2014). Numerous molecular mechanisms involved to participate and to demonstrate the capability or ability of one strain to promote the plant growth and health promotion (Lugtenberg and Kamilova 2009). The mechanisms could be classified in a general form as direct and indirect mechanisms (Goswami et al. 2016;

Fig. 2.1 Bacterial mixture (EMMIM-5) improving the growth of maize plants (Blue maize CAP15-1 TLAX) in comparison with non-inoculated plants (control), 45 days after the inoculation of germinated seeds

Jha and Saraf 2015; Molina-Romero et al. 2015; Paterson et al. 2017). Direct mechanisms involve the bacterial contribution to the assimilation of diverse nutrients or metabolites available for plants that have positive effects on their growth. The indirect mechanisms contribute to the health support to the plants that concomitantly enhance their growth and development. Among indirect mechanisms, induced systemic resistance by beneficial bacteria, antimicrobial production, etc., have been worked out.

2.2 Plant Growth Promotion by Direct Mechanisms

2.2.1 Biological Nitrogen Fixation

BNF refers to the ability of some bacteria to capture atmospheric nitrogen and transform it to combined nitrogen, usually in the form of ammonium (NH_4^+) (Annan et al. 2012; Dixon and Kahn 2004; Santi et al. 2013). Fixed nitrogen could be donated to the plants that in turn provide carbon source to bacteria. The interaction of different species of *Rhizobium* with legumes is one of the models more widely studied. The interaction of *Rhizobium*–legumes is highly specific and both populations are favoured both for the plant and bacteria during relationships such as mutualistic symbiosis (Simms and Taylor 2002). Although BNF is highly effective in the interaction of rhizobia–legumes, this phenomenon has not been as significant in non-legume plants except very few such as sugarcane inoculated with the free-living bacterium *Gluconacetobacter diazotrophicus* (Muñoz-Rojas and Caballero-Mellado 2003; Sevilla et al. 2001). However, some grasses showed high levels of nitrogen obtained through the BNF process (Herridge et al. 2008). Further investigations are yet to be carried on the bacteria with highly effective BNF capability associated with non-legumes or performing genetic engineering of non-legume plants to introduce genes related to nitrogen fixation (Saikia and Jain 2007). The free-living bacteria with potential to perform the BNF process include *Pseudomonas fluorescens*, *Beijerinckia* sp., *Azoarcus* sp., *Azotobacter* sp., *Burkholderia unamae*, *Gluconacetobacter diazotrophicus*, *Herbaspirillum* sp., *Azospirillum brasilense*, *Nostoc* sp. and *Rhizobium* sp. (Caballero-Mellado et al. 2004; Defez et al. 2016; Fibach-Paldi et al. 2012; Gutiérrez-Zamora and Martínez-Romero 2001; Guzmán et al. 2012; Kaschuk and Hungria 2017; Sevilla et al. 2001).

2.2.2 Phytohormones Production

This is a widely distributed mechanism among plant-associated bacteria (Costacurta and Vanderleyden 1995). Various groups of PGPR secrete gibberellins, auxins and cytokinins, the known phytohormones that promote plant growth (Bottini et al. 2004; Costacurta and Vanderleyden 1995; Kang et al. 2014). Indole acetic acid (IAA), an auxin molecule, is the most studied phytohormone wherein biosynthetic pathway is well known in the genus *Azospirillum* (Cassán et al. 2014; Patten and Glick 1996). IAA induces the elongation and division of root cells, included root growth and has greater root surface enabling plants with higher nutrient absorption and growth (Patten and Glick 2002). *Gluconacetobacter diazotrophicus*, *Azospirillum brasilense*, *Pseudomonas* sp., *Enterobacter cloacae*, *Klebsiella variicola* and *Bacillus amyloliquefaciens* have been reported as producers both induced and constitutive IAA in vitro (Defez et al. 2016; Fuentes-Ramirez et al. 1993; Idris et al. 2007; Malik and Sindhu 2011; Patten and Glick 1996).

2.2.3 Phosphate Solubilization

A fundamental element of plant metabolism is phosphorus since it is an important structural element in DNA, RNA, phospholipids and communication signals. Although this component is abundant in the soil, but remains in unavailable form hence cannot be taken up by the plants in most soil conditions. Only monobasic or dibasic form of P is soluble and hence bioavailable to the plants (Gyaneshwar et al. 2002). Certain bacterial and fungal genera have the capability to solubilize phosphate from soils and allow the phosphorus uptake by plants (Castagno et al. 2011; Gyaneshwar et al. 2002; Kang et al. 2002). The ability of phosphate solubilization in rhizobia has been previously reported (Pandey et al. 2005). However, various free-living aerobic bacteria were reported later for phosphate solubilization (Khan et al. 2007). Some PGPB solubilize phosphates from inorganic or organic compounds; involving the use of nonspecific phosphatases, C-P lyases, phosphatases and phytases (Lugtenberg and Kamilova 2009; Molina-Romero et al. 2015). On the other hand, releasing organic acids by different bacteria may chelate phosphorus, making it bioavailable to plants (Aeron et al. 2011; Vyas and Gulati 2009). Bacteria with the ability to solubilize phosphates include *Pseudomonas putida*, *Bradyrhizobium japonicum*, *Enterobacter agglomerans* and *Rhizobium leguminosarum* (Molina-Romero et al. 2015, 2017b; Rodríguez et al. 2006). Besides enzyme and organic acids of bacterial origin, other mechanisms of P solubilization include inorganic acids produced by chemoautotrophs and the H+ pump observed in *Penicillium rugulosum* (Khan et al. 2014).

2.2.4 ACC Deaminase Production

Under environmental conditions plants are exposed to adversities caused by abiotic and biotic factors influencing their growth (Morgan and Drew 2006); the climate, the amount of water available, solar radiation, attack by pathogens, pests and animals, and others. Changes in these biotic or abiotic factors represent intense stress on the plants which in turn trigger ethylene-mediated systemic defense response causing excessive energy expenditure to the plants. Some PGPB produce the enzyme 1-aminocyclopropane-1-carboxylate (ACC) deaminase and improve the growth of plants by decreasing ethylene levels (Bal et al. 2013; de Oliveira et al. 2006; Glick 2014; Glick et al. 2007). Following the discovery of the role of ACC deaminase in plant growth promotion (Glick et al. 1998), rhizobacteria able to synthesize ACC deaminase were reported by the pioneers in India (Aeron et al. 2017; Kumar et al. 2012; Maheshwari et al. 2015a). Kumar et al. (2012) reported the role of *Bacillus* sp. in ACC deaminase synthesis and other rhizobacteria were reported by Aeron et al. (2017). In the Indian research scenario, a beta class of Proteobacteria first reported as rhizobia, was identified as *Burkholderia* sp., having ACC deaminase activity (Pandey et al. 2005). The ACC deaminase enzyme breaks the ethylene precursor avoiding its synthesis and also releasing ammonium that can later be used by plants as a nitrogen source (Singh et al. 2015). Some PGPB with the capability to produce ACC deaminase are *Burkholderia unamae* MTl-641, *Azospirillum lipoferum*, *Pseudomonas fluorescens* TDK1, *Enterobacter cloacae*, *Bacillus* sp. AR-ACC1, and *Agromyces* ANR-ACC2 (Esquivel-Cote et al. 2013; Nadeem et al. 2007; Onofre-Lemus et al. 2009; Zahir et al. 2008). The ACC deaminase system can also be considered an indirect mechanism of plant growth promotion since it blocks out an intensive defense response avoiding unnecessary energy expenditure by plants.

2.3 Indirect Growth Promotion Mechanisms

Today, this kind of mechanisms are increasing in knowledge, especially the elimination of microbial pathogens by ISR, which shows an interesting mechanism to protect plants from being attacked by phytopathogens.

2.3.1 Antagonistic Mechanisms

Phytopathogenic microorganisms are major causal organisms of plant diseases and crop losses. Various genera of PGPB are capable of eliminating those pathogens, hence, they can be used as biocontrol agents (Beneduzi et al. 2012; Compant et al. 2005; Liu et al. 2017a). Inhibitory substances produce by PGPB with biocontrol potential include siderophores, bacteriocins, broad-spectrum antibiotics, lytic enzymes, lipopeptides and antifungal metabolites (Mohamed et al. 2017; Molina-Romero et al. 2015; Morales-García et al. 2007; Sivasakthi et al. 2014). There are several PGPB able to prevent the deleterious effects of phytopathogenic organisms by producing inhibitory substances, e.g. *Gluconacetobacter diazotrophicus*, *Burkholderia tropica*, *Bacillus amyloliquefaciens*, *Pseudomonas fluorescens*, *Lysinibacillus sphaericus*, *Bacillus subtilis*, *Bacillus altitudinis*, *Azospirillum brasilense, Rhizobium etli, Rhizobium leguminosarum*, *Kosakonia radicincitans* (Bardin et al. 2004; Cawoy et al. 2015; Krishnan et al. 2007; Lambrese et al. 2018; Naureen et al. 2017; Príncipe et al. 2018; Russo et al. 2008; Saravanan et al. 2008; Sivasakthi et al. 2014; Tenorio-Salgado et al. 2013; Tortora et al. 2011).

2.3.2 Induced Systemic Response (ISR)

The interaction of plants with pathogens or beneficial microorganisms triggers the systemic response of plants (van Loon 2007). Systemic acquired resistance (SAR) is triggered by pathogens while induced systemic resistance (ISR) is triggered by non-pathogenic microorganisms (Pieterse et al. 2014; van Loon 2007). SAR is a phenomenon where plants acquire an enhanced defensive response against subsequent pathogen attack as a result of a primary, limited infection (van Loon et al. 2006), this response is mediated by salicylic acid and it is quite aggressive for plants causing necrosis in some cases (Pieterse et al. 2014; Ramamoorthy et al.

2001). ISR is triggered by beneficial bacteria, it is mediated by ethylene and jasmonate, and this response prevents the establishment of pathogens in the plant (Pieterse et al. 2014; Su et al. 2017; van Loon 2007). Induced systemic resistance (ISR) is the phenomenon to immunize the plants by PGPR against phytopathogens and confer substantially enhanced levels of peroxidase (PO), polyphenol oxidase (PPO), phenylalanine ammonia lyase (PAL), β−1, 3 glucanase and chitinase as a part of the systemic response (Sharma et al. 2018). Field application of PGPB with the ability to induce a systemic response in plants might prevent the attack of pathogens resulting in an increase of the yield in different crop plants (Gkizi et al. 2016; Su et al. 2017). Not only PGPB but some other bacterial components are also able to induce the ISR response, for example, flagella, lipopolysaccharides, siderophores and cyclic lipopeptides (Ramamoorthy et al. 2001). Some examples of PGPB capable to trigger ISR are *Pseudomonas fluorescens* FB11, *Bacillus altitudinis, B. cereus, B. subtilis, B. amyloliquefaciens, B. pasteuri, B. pumila, B. mycoide, B. sphaericus, Burkholderia phytofirmans, Rhizobium leguminosarum bv. viceae* FBG05, *Pseudomonas putida* 89B-27, *Serratia marcescens* and *Paenibacillus alvei* K165, *Rhodopseudomonas palustris* (Bhattacharyya and Jha 2012; Elbadry et al. 2006; Gkizi et al. 2016; Kloepper et al. 2004; Su et al. 2017).

2.3.3 Volatile Organic Compounds Production

Volatile organic compounds (VOCs) are small and gaseous molecules produced by bacteria, identified as proficient signal molecules functioning as chemical attractants or repellents (Hernández-Calderón et al. 2018; Ortíz-Castro et al. 2009). VOCs interact with plants in the soil and may promote plant growth by inducing the ISR, suppressing phytopathogens, stimulating of photosynthesis and modulating phytohormone signalling (Santoro et al. 2015; Sharifi and Ryu 2018). Some VOCs identified include aldehydes, alcohols, ketones, indoles, terpenes, fatty acids and jasmonate (Pieterse et al. 2014; van Loon 2007).

An alternative function of VOCs produced by the PGPB is to increase the resistance of plants to abiotic stresses such as salinity, drought and heavy metals (Farag et al. 2013). VOCs can act directly as (i) phytohormones, (ii) helping the acquisition of iron, (iii) regulating the growth and morphogenesis of the plant and (iv) exerting biocontrol of phytopathogens either by antibiosis or by triggering ISR (Farag et al. 2013; Park et al. 2013; Zhang et al. 2009). Examples of bacteria that produce VOCs are *Bacillus subtillis* SYST2, *Bacillus subtilis* GB03, *Bacillus amyloliquefaciens* IN937a, *Erwinia carotovora*, *P. polimyxa* E681, *Pseudomonas fluorescens*, *Pseudomonas stutzeri* E25 and *Stenotrophomonas maltophilia* CR71 (Cheng et al. 2017; Ortíz-Castro et al. 2009; Park et al. 2013; Rojas-Solís et al. 2018; Santoro et al. 2015; Tahir et al. 2017; Zhang et al. 2007, 2009). VOCs can be exploited as an eco-friendly, cost-effective and sustainable strategy for agricultural practices (Kanchiswamy et al. 2015).

2.3.4 Elimination of Toxic Compounds for Plants

In modern agriculture, toxic compounds such as fungicides, pesticides and herbicides have been excessively used on crop fields. The indiscriminate and excessive applications of these compounds could have harmful effects for the development of plants, animals and human being (Benbrook 2016; Donham 2016; Igbedioh 1991; Park et al. 2015). These toxic compounds even modify the microbial diversity and microbial activity (Chen et al. 2001; Johnsen et al. 2001; Smith et al. 2000); which is fundamental for the proper development of plants (Berendsen et al. 2012). For this reason, it is necessary to eliminate these toxic compounds from the soil to allow the plants to grow in a sustainable manner an optimal growth of crop plants. There are several bacteria having the ability to eliminate toxic compounds from the soil while interacting with host plants. Among the most widely studied bacterial species are *Pseudomonas putida*, *Sphingomonas* sp. OF178, *Burkholderia unamae*, *Burkholderia tropica*, *Bacillus subtillis*, *Pseudomonas rhizophila* S211 etc. (Ahemad and Khan 2012; Böltner et al. 2008; Caballero-Mellado et al. 2004, 2007; Caballero et al. 2005; de Oliveira et al. 2006; Hassen et al. 2018). However, the ability of these bacteria to eliminate pesticides present in agricultural soils are yet to be elucidated.

Several published data related to PGPB stated that the actual mechanism comprising phytostimulation of PGPB neglected, and in some only the growth promotion characteristics have been reported. In fact, the transcription level of key genes involved in plant growth promotion or physiological characteristics such as ability to fix nitrogen, solubilize phosphates or produce phytohormones have been insignificantly tested in vitro (Ahmad et al. 2018; Caballero-Mellado et al. 2004; Fibach-Paldi et al. 2012; Fuentes-Ramirez et al. 1993; Molina-Romero et al. 2017b; Onofre-Lemus et al. 2009; Rosenblueth et al. 2004; Tenorio-Salgado et al. 2013). There is limited information about the actual genes involved in the plant growth promotion, making defective mutants and corroborating the lack of plant growth promotion characteristics in PGPB (Rajput et al. 2015; Schneider et al. 1991; Sevilla et al. 2001; Zhang et al. 2009). For example, the deletion of *nifH* gene in *Gluconacetobacter diazotrophicus* makes this bacterium no longer able to fix nitrogen both in vitro and in association with host plants, demonstrating that aforesaid mechanism has been involved in the promotion of growth (Sevilla et al. 2001).

2.4 The State of the Art of PGPB Publications and Patent Numbers

Literature on PGPB-type bacteria is broad, ranging from isolation, phenotypic, biochemical and molecular characterization, ability to colonize the rhizosphere of plants, plant growth promotion and studies of genes involved in plant–bacteria

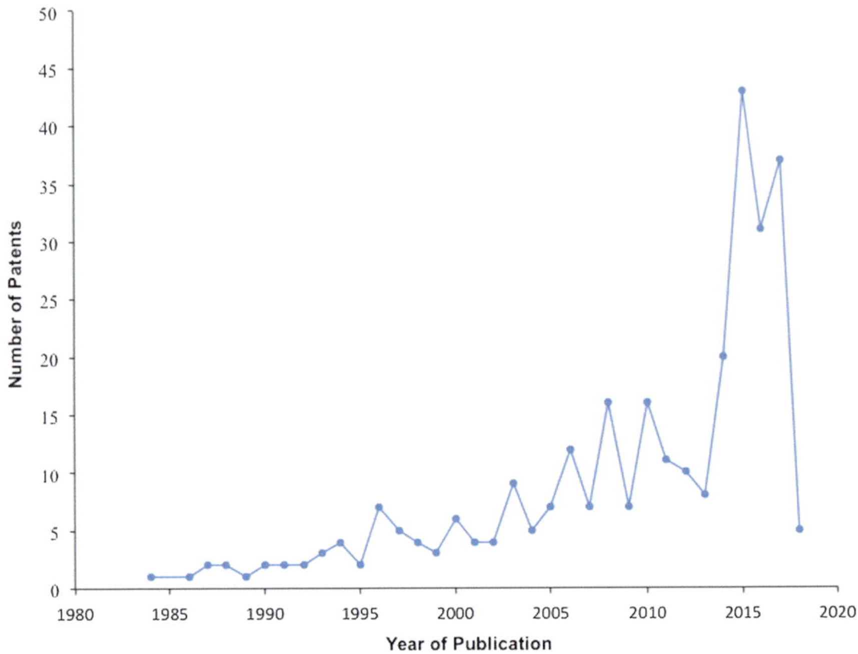

Fig. 2.2 Number of patents applied by year in the global world. All patents are related to the use of PGPB as crop plants inoculants. Source data were acquired from two prestigious platforms of patents search: Spacenet and Patent Inspiration

interaction (Huang et al. 2014; Lugtenberg and Kamilova 2009). Public databases such as NCBI or Google Scholar show an increasing trend in the number of articles published on issues related to the capacity of bacteria to promote plant growth. In fact, there are around 2742 published works in PubMed database related to this field up today (Table 2.3), but this data fluctuates according to the new publications that are stored in the system. Interestingly, the number of patents is approximately fivefold lower than that of published articles suggesting that very limited works have ended up in a potential application. Therefore the number of patents is lower than the number of publications and on the other hand, patent number continues to increase (Fig. 2.2). This could be due to the urgent need for the implementation of environmental-friendly technology provided to reduce the contamination of toxic compounds used in intensive agriculture (Baez-Rogelio et al. 2017). Countries with the major patent numbers related to the application of PGPB are USA, Germany, Canada, United Kingdom and Spain (Fig. 2.3).

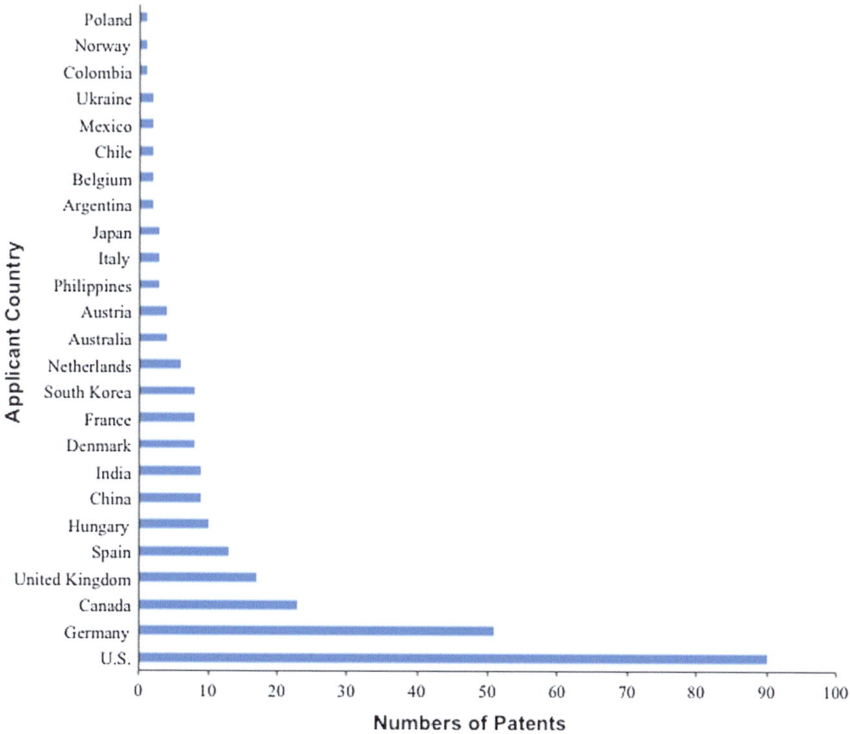

Fig. 2.3 Analysis of the number of patents carried out by applicant country. All patents are related to the use of PGPB as crop-plant inoculants. Source data were acquired from two prestigious platforms of patents search: Spacenet and Patent Inspiration

2.5 Bacterial Inoculants for Agriculture

Non-pathogenic bacteria with growth-promoting capacity whose mechanisms of phytostimulation are supported by molecular studies can be considered candidates to formulate bacterial inoculants. The number of crop types and hectares of agricultural land where PGPB are applied is quite low corresponded to that of the total crops or farming area. This is perhaps because of lack of a culture of adopting sustainable technologies, inconsistencies of yield production in yield and problems of technology associated to massive inoculation (Bashan et al. 2014; Carolan 2009; Cummings 2009). *Azospirillum brasilense* is one of the most studied bacteria applied in agricultural crops, it has been used in various crops around the world with successful results in more than 70% of cases in comparison to that obtained in less reported in other PGPB (Dobbelaere et al. 2001; Fuentes-Ramirez and Caballero-Mellado 2006; Okon and Labandera-Gonzalez 1994). Among the inoculants intended for sustainable crop production that are currently being marketed, we can find mono-inoculants formulated with a single bacterium such as

Azospirillum brasilense, *Azotobacter vinelandii*, *Rhizobium etli*, or *Bacillus* sp. (Baset Mia and Shamsuddin 2010; O'Callaghan 2016; Okon and Labandera-Gonzalez 1994). Countries that are adopting the use of these bacteria on crop fields are México, Argentina, USA, China, Belgium, Uruguay and to a certain extent in India (Bashan et al. 2014; Dobbelaere et al. 2001; Molina-Romero et al. 2015; Shen 1997).

In India, the first study on legume-*Rhizobium* symbiosis was initiated by N. V. Joshi in 1920, and commercial production began in 1956 (Barman et al. 2017; Ghosh 2004; Majumdar 2015; Mazid and Khan 2014). The development of technology for inoculants production in India evolved through the nineteenth century including some remarkable events: (A) the earliest documented production of *Rhizobium* inoculant by M. R. Madhok in 1934. (B) The discovery of nitrogen fixation by Blue-Green Algae (BGA) in a rice field and the report on the performance of *Azotobacter* in rice soil by B. N. Uppal, both events in 1939. (C) First commercial production of biofertilizer in 1956 by N. V. Joshi. (D) The study of microbial phosphate solubilization by Sen and Pal in 1957. (E) The first work for the quality standardization for legume inoculant in 1958. (F) Isolation of the first non-symbiotic N-fixing organism Derxiagummosa by P. K. Dey and R. Bhattacharyya in 1960. (G) Increment in the requirement of biofertilizers for soybean mainly in Madhya Pradesh in 1964. (H) Projects set up by the Indian Council of Agricultural Sciences (ICAR) in 1968, where *Rhizobium* study got priority. (I) Use of Indian peat as carrier reported by V. Iswaram in 1969. (J) Use of coal as an alternative carrier reported by J. N. Dube in 1975. (K) Indian standard specification for Rhizobium in 1976. (L) Use of ISI mark for *Rhizobium* in 1977. (M) In 1979, a coordinated project for BNF in all India took place and the ISI standardization was made for *Azotobacter* inoculant. (N) Setting up of National Project on Development and use of Biofertilizers by Ministry of Agriculture, Government of India, in 1983. (O) Five biofertilizers (containing: *Rhizobium*, *Azotobacter*, *Azospirillum*, Phosphate Solubilizing Bacteria and mycorrhiza) were incorporated in India's Fertilizer Control Order (FCO) in 1985, and the First National Productivity award on Biofertilizer was celebrated. (P) Setting up of National Facility Centre for BGA at IARI in 1988. (Q) The NIKU (National Input Complete and Utilization) Bio-Research Laboratory was established in 1997 at Pune. (R) The Tenth Plan document (2002–2007) that emphasize the use of biofertilizers, biocontrol agents, organic manures, etc. with infrastructure support (Barman et al. 2017; Majumdar 2015).

Through the history of biofertilizers in India, some adversities have limited the development of this important product (Ghosh 2004). Adoption of the technology has not consistently grown over time and has slowed down in the late 1990s. Although there have been more and more new entries in the market, the average capacity of bigger producers has come down, characterizing the industry by many small units. Lack or low diffusion of technology with the farmers and low participation of the private sector in the commercialization of inoculants (below 50%) such as problems in transportation and distribution represented the major problems in rural areas (Mazid and Khan 2014). Despite adversities, India is one of the most important countries in biofertilizer production (Mahajan and Gupta 2009). The first biofertilizer modern unit was started by the Gujarat State Fertilizer Company (GSFC), Vadodara, in 1985 and many more followed thereafter. Today the number of microbial inoculants manufactured in India has increased and products include Rhizonick (*Rhizobium*), Azonik (*Azotobacter chroococcum*), Spironik (*Azospirillum brasilense*), Phosphonive (Phosphate solubilizing bacteria), Sulphonik (Sulphur-oxidizing bacteria and fungi), Niku-2000 (Decomposing culture), Trichonik (*Trichoderma viridi*), Vermiculture (mixed with N-fixing inoculant and P solubilizers), Blue-Green Algae (BGA) (containing photosynthetic nitrogen fixers), Vesicular arbuscular mycorrhiza (VAM), K-solubilizer (*Frauteria aurantia*), etc. (Barman et al. 2017; Majumdar 2015; Singh et al. 2014b). Agro Industries Corporation has the maximum capacity to produce biofertilizers in India, followed per State Agriculture Universities and the private sector (Singh et al. 2014b). Since 2016, Indian Government has been promoting biofertilizers development through various schemes of National Mission of Sustainable Development (NMSA)/ Paramparagat Krishi Vikas Yojana, Rashtriya Krishi Vikas Yojana (RKVY) and National Mission on Oilseeds and Oil Palm (NMOOP) and Indian Council of Agricultural Sciences (ICAR) (http://pib.nic.in/newsite/PrintRelease.aspx?relid=137762).

On the other hand, among the principal companies dedicated to produce and commercialize biofertilizers in México are Biofabrica Siglo XXI, Biosustenta, Fertilizantes Mexicanos, Biokrone. Biofabrica Siglo XXI developed products such as Azofer, an inoculant formulated with *Azospirillum brasilense*; Rhizofer, formulated with *Rhizobium etli,* etc. (http://www.biofabrica.com.mx/about.html). Another Mexican company is Biosustenta, which is developing and producing biological inputs of Ferbiliq inoculant, based on *Azospirillum brasilense* and *Glomus intraradices*; Endomaz biofertilizers made from *A. brasilense* (http://cosustenta.com/catalogo.html). Fertilizantes Mexicanos markets biofertilizers formulated with nitrogen-fixing bacteria (Azoton AA Plus); 'Biomatrix + powder' formulated with nitrogen-fixing bacteria and phosphate solubilizers, other of its products are Bioespiril L and Raizinn Biol (http://biofertilizantes.mx/index.html). Similarly, Bio Organica Mexicana provides biofertilizers such as Ultralite AZO (http://www.bio-organica.com.mx). All above-mentioned companies offer to the farmers an ecological alternative to increase the production of their crops, for impacting positively their economy, due to the investment of lower costs of these products compared to chemical fertilizers. In Argentina, agrobiotechnology has been developed for the past 50 years, using PGPR isolated from their soils for the formulation of bacterial inoculants (Molina-Romero et al. 2015). These formulations were developed to improve growth and increase the productivity of leguminous plants and cereals of agricultural importance. Among the companies with more experience in the production of microbial inoculants are NITRASOIL ARGENTINA S.A. Co., which has developed an inoculant that contains bacteria of the genus *Azospirillum* sp. strain AZ39, recommended by the INTA (National Institute of Agricultural Technology) for being considered the best microbial inoculant (http://www.nitrasoil.com.ar). The company Rizobacter Argentina S.A.

Co. (http://www.rizobacter.com/argentina/) also has a variety of products whose formulation is used in soybean, corn and wheat. The bacteria used for the formulation of these inoculants are *P. fluorescens*, *Bradyrhizobium* sp., *Mesorhizobium ciceri*, *Sinorhizobium meliloti*, *Rhizobium leguminosarum* biovar *trifoli* and *Bradyrhizobium japonicum*. The firm Nitrasoil S. A., FPC Argentina S.A. (http://www.fpcinoculantes.com.ar), Granaries and Elevadores Argentinos de Colon Soc. Coop. Ltda (GEA) (http://geadecolon.com.ar) and bionet (http://www.bionetsrl.com/inoculantes-bionet-soja-premium.php) are dedicated to the formulation and commercialization of fertilizers and biological inoculants for the agricultural sector (Molina-Romero et al. 2015).

Although the use of these formulations is still moderate today but, are increasingly accepted by farmers, due to the improved growth of cultivars inoculated with PGPB, their application have enhanced the crop productivity (Lugtenberg and Kamilova 2009) decrease in the use of chemical fertilizers (Dobbelaere et al. 2001) and potential to decrease in the use of toxic compounds such as pesticides and herbicides (Myresiotis et al. 2012). Such characteristics make these microorganisms the first choice for organic agriculture being highly compatible for sustainable practices.

2.6 Co-inoculation Versus Mono-inoculation

In fact, microbial co-inoculation is considered to be an innovative approach and had been applied worldwide (Wang et al. 2018). Compared with mono-inoculation, co-inoculation of microorganisms has apparently been more effective in stimulating plant growth because of the synergistic effect that occurs when more than one microorganisms are co-interacting with same or their diverse genera (Atieno et al. 2012; Barea et al. 2002; Zoppellari et al. 2014). For example, the co-inoculation of lettuce with *Bacillus* sp. and *Glomus intraradices* make the use of water more efficient under stress conditions (Vivas et al. 2003). Similarly, co-inoculation of pea with *Rhizobium* spp. and *Bacillus megaterium* increased the biomass of roots and aerial region, the percentage of nitrogen and productivity (Elkoca et al. 2007). The consortium of *P. aeruginosa* KRP1 and *B. licheniformis* KRB1 were reported suppressive for the fungal phytopathogens *F. oxysporum* and *S. sclerotiorum* causing disease in *Brassica campestris* (Maheshwari et al. 2015b). Although, there are promising results of co-inoculations,, formulations on the market containing three or more species of microorganisms are limited. Research papers have shown that the inoculation of sugar cane with a mixture of more than three bacteria enhanced the production of sugarcane in soil characteristically having low to medium-level and nitrogen fertilization (Table 2.4) (Molina-Romero et al. 2017b; Morales-García et al. 2013; Oliveira et al. 2009).

The design, formulation and optimization of an effective mixture of bacteria to be used as inoculants, is not so easy because, it requires microbe–microbe interaction, bacterial ability of adhesion to seeds and root colonization in plants

Table 2.4 Search of keywords related to plant growth-promoting bacteria and some beneficial bacterial species. The search was made on 13 August 2018

Keywords searched	Pub Med-NCBI	Scholar google	Patents searched in Spacenet
Plant growth-promoting bacteria	2742	1,070,000	560
Plant growth-promoting rhizobacteria	845	46,900	83
Bacterial plant inoculants	331	36,800	32
Azospirillum brasilense	777	29,800	53
Gluconacetobacter diazotrophicus	107	4270	7
Rhizobium etli	472	11,700	4
Herbaspirillum seropedicae	136	4450	5
Burkholderia unamae	16	454	1
Burkholderia tropica	24	2400	1
Pseudomonas fluorescens	5733	160,000	877

(Baez-Rogelio et al. 2017; Singh et al. 2014a; Sundaramoorthy et al. 2012). In addition, studies on the antagonistic relationships among the strains involved in a mixture of bacteria should be conducted before design and application of a multi-species inoculant. Some antagonistic effects also occur among bacteria when associated with plants (Baez-Rogelio et al. 2017; Molina-Romero et al. 2017b; Oliveira et al. 2009).

2.7 Steps in the Design of Mixed Inoculants and Some Experiences of Inoculation

The design and formulation of multi-species inoculants represent furthered challenges since it requires additional assays to guarantee the coexisting of bacterial strains when they are in the formulation and associated with the plants, to raise their growth promotion (Baez-Rogelio et al. 2017; Molina-Romero et al. 2017b; Morales-García et al. 2013; Wang et al. 2018). Due to the characteristics and advantages of multi-inoculant preparations in comparison with conventional mono-inoculants, the term second-generation inoculants has been assigned and some multi-species formulations have been patented (Alatorre-Cruz et al. 2015; Morales-García et al. 2013; Olovaldo et al. 2017).

There are different strategies to design a multi-species inoculant, but all of them should overcome some general challenges in order to obtain the desired results. Besides all experiments and challenges faced during the development of mono-inoculants, five additional challenges should be sorted out during the design of polymicrobial inoculants.

Table 2.5 Hypothetical antagonism assays using the double agar layer method

	Bacterial strains explored as sensitive of inhibitory substances								
BSEPIS	A	B	C	D	E	F	G	H	I
A		+	+	+		+	+	+	+
B					+				
C		+		+				+	
D		+			+		+		
E		+		+					
F		+	+						+
G		+							
H		+							+
I		+					+		

BSEPIS means bacterial strains explored as a producer of inhibitory substances. In this hypothetical example, strain A eliminates to almost all tested strains and strain B is sensitive to all strain tested. Therefore, these strains cannot be used to design a bacterial mixture, because the first could eliminate the others in a natural environment and the second could be eliminated by the other strains. Examples of compatible strains to formulate a mixture could be E, G, H, F or I, C, E, G

(1) Carry out antagonism assays. Several articles about bacterial antagonism have been published with the purpose to identify producer bacteria of inhibitory substances (Beneduzi et al. 2012; Muñoz-Rojas et al. 2005; Tenorio-Salgado et al. 2013). However, for the formulation of polymicrobial inoculants, antagonism assays should be carried out with the purpose of identifying compatible strains (Molina-Romero et al. 2017b). In this sense, the development of a matrix of bacterial interaction is highly important (Table 2.5). The methods used to perform the antagonism test for building the interaction matrix are the double-layer agar and simultaneous inhibition methods (Molina-Romero et al. 2017a). Both the methods provide complementary information. In the agar double-layer assay, the first bacterium is grown alone on the first agar-media layer, after this, the colony is removed and the remaining bacteria are killed with chloroform vapours leaving all metabolites produced by the bacteria on the first agar layer. The second bacterium is then incorporated into the second layer of agar media before gelling point of agar, if the growth of second bacterium is inhibited by metabolites produced by the first one, which means that antagonistic metabolites were produced without the influence of the second bacterium. In counterpart, during simultaneous inhibition method, both bacteria are grown simultaneously on the same agar media competing for the same space and nutrients. An observed inhibition could be the result of the bacterial interaction. The major challenge is to find compatible strains, able to coexist not only in the antagonism assays but also in other environmental conditions like growing in liquid media, during the formulation, in adherence to seeds and plant rhizosphere (Molina-Romero et al. 2017b). But, we recommend to test the interaction at other environmental conditions until the definitive members of a

multispecies inoculant have been selected, because it will be necessary to develop selective culture media for each member of the multi-species inoculant before to be able to identify and quantify each bacteria interacting in the consortium. All inoculants of second generation contain strains capable of coexisting without antagonizing each other and inclusive with the ability to inhibit the plant pathogens causing diseases (Alatorre-Cruz et al. 2015; Baez-Rogelio et al. 2017; Molina-Romero et al. 2017b; Morales-García et al. 2013; Oliveira et al. 2009; Olovaldo et al. 2017).

(2) Bacterial tolerance to desiccation. Desiccation of organisms is defined as the loss of intracellular water until the equilibrium with water is present in the environment, extreme desiccation occurs at 30 °C and 50% of relative humidity, during several days (Molina-Romero et al. 2017c; Pazos-Rojas et al. 2018). PGPB with the capability to tolerate extreme desiccation could be used to formulate more stable inoculants and such bacteria once associated to seeds can remain dormant but ready to colonize plant roots and rhizosphere when favourable conditions of rehydration occur (Molina-Romero et al. 2017b; Pazos-Rojas et al. 2018; Streeter 2003). Based on data of bacterial survival ratio to air desiccation (BSRad), five levels of bacterial tolerance have been proposed (Pazos-Rojas et al. 2018): highly tolerant bacteria (BSRad > 80), tolerant ($60 <$ BSRad ≤ 80), middle tolerant ($40 <$ BSRad ≤ 60), low tolerant ($20 <$ BSRad ≤ 40) and very-low-tolerant (BSRad ≤ 20). Highly tolerant bacteria are desired for the formulation of bacterial inoculants of second generation, because they maintain their ability to promote plant growth even when they experienced desiccation stress (Molina-Romero et al. 2017b; Pazos-Rojas et al. 2018). Second-generation inoculants containing desiccation-tolerant bacteria were observed more efficient in environments with low water availability (Alatorre-Cruz et al. 2015; Molina-Romero et al. 2017b; Pazos-Rojas et al. 2018; Vilchez and Manzanera 2011).

(3) Choosing bacterial species for the mixture. Using the interaction matrix of PGPB (Table 2.5) and the results of tolerance to desiccation, we can select the appropriate compatible bacteria to formulate the inoculants of the second generation. Bacterial mixture formulations theoretically could be more successful than mono-inoculants (Baez-Rogelio et al. 2017). But even a careful selection of compatible bacteria for the hypothetical formulation do not warranty that selected bacteria will be compatible in association to plants or that mixed bacteria will promote the plant growth. In fact, some bacterial mixtures selected at this step could be ineffective to promote plant growth. Experiments of population dynamics of bacteria associated to plants and effectivity studies of these formulations should be carried out to verify the effectiveness of the designed formulations.

(4) Looking for selective media for members of the consortia. A very critical step for the design of bacterial inoculants of second generation is to confirm that bacteria really coexist associated with plants. For this, we require effective selective media to capture every single strain from a mixture discriminating the growth of the other strains sampled from the same environment. Bacterial

capability to growth on different conditions has been explored for each bacterial strain from the mixture (Alatorre-Cruz et al. 2015; Molina-Romero et al. 2017b; Morales-García et al. 2013), for example, use of different carbon and nitrogen sources, tolerance to heavy metals, abilities to grow in microaerophilic semi-solid media without nitrogen source (nitrogen fixation), resistance to different antibiotics, tolerance to salinity, peculiar growth colonies on media, etc. Culture media allow the monitoring of different inoculated strains both for seed adhesion and for bacterial colonization studies (Molina-Romero et al. 2017b; Rodríguez-Andrade et al. 2015). However, the identity of isolated bacteria should be corroborated using molecular tools, for example, by 16S rDNA sequence or by the characteristic restriction pattern of the gene 16S rDNA.

(5) Exploring multi-inoculation in different scenarios. In our experience, all bacteria selected for multispecies inoculants have shown the capability to adhere to seeds and colonize the rhizosphere of plants suggesting its compatibility in association to seeds or rhizosphere (Alatorre-Cruz et al. 2015; Molina-Romero et al. 2017b; Morales-García et al. 2013). However, levels of colonization could be different depending on plant species. For example, the formulation EMMIM-1 which contains 6 bacterial species (Table 2.6), all of them are capable to colonize maize in high numbers (Morales-García et al. 2013). Seed inoculation of tomato and bean with EMMIM-1 formulation resulted in a different *B. unamae* MTl-641, a bacterium part of EMMIM-1 formulation was able to colonize in high numbers the rhizosphere of tomato (Log CFU/gV around 6.8 (± 0.35)) but the numbers observed for colonization of bean rhizosphere (Log CFU/gV around 3.83 (± 0.33)) were lower than those for tomato. Despite these differences, the inoculant multispecies EMMIM-1 was able to increase the plant growth of both plant species suggesting that other strains rather than *B. unamae* MTl-641 promote the growth in bean plants (non-published results).

2.8 Comparing Plant Growth Promotion of Mono and Multi-inoculation

Effects of the bacterial–mixture inoculation on the plant growth promotion should be tested and compared against mono-inoculation. Inoculation of bacterial mixtures had resulted in higher increments of the plant growth than mono-inoculants maybe because there are synergic effects of bacteria.

In our laboratory, better growth-promoting effects were observed on plants inoculated with the multi-species formulations compared with mono-inoculated or non-inoculated plants (Figs. 2.1, 2.4, 2.5, 2.6, and 2.7) (Alatorre-Cruz et al. 2015; Morales-García et al. 2013; Molina-Romero et al. 2017b). An additional feature of multi-strain formulations compared to mono-species formulations is the higher

Table 2.6 Mixed bacterial formulations for enhancement of plant growth

Bacterial species in the formulation	Name of the formulation	Plant model explored	References
Gluconacetobacter diazotrophicus, *Herbaspirillum seropedicae*, *H. rubrisubalbicans*, *A. Amazonense*, *Burkholderia tropica*	No name	Sugarcane	Oliveira et al. (2009)
Acinetobacter sp. EMMS02, *Azospirillum brasilense* Sp7, *Pseudomonas putid*a KT2440, *Sphingomonas* sp. OF178	EMMIM-5	Blue maize	Molina-Romero et al. (2017b)
Ensifer sp. NYM3, *Acinetobacter* sp. P16 *Flavobacterium* sp. KYM3	No name	Cucumber	Wang et al. (2018)
Gluconacetobacter diazotrophicus PAl 5, *Burkholderia unamae* MTl-641, *Sphingomonas* sp. OF178, *Pseudomonas putida* KT2440, *Brdyrhizobium* sp. MS22, *Azospirillum brasilense* Sp7	EMMIM-1	Several maize varieties, potato, tomato, bean	Morales-García et al. (2013)
Microbacterium sp. UAPS01200, *Microbacterium* sp. UAPS01203, *Pseudomonas putida* KT2440	EMMIM-JMAC	*Echinocactus plathyacanthus*	Alatorre-Cruz et al. (2015)
Sphingomonas sp. OF178, *Pseudomonas putida* KT2440, *Gluconacetobacter diazotrophicus* PAl 5	EMMIM-2	Potato	Olovaldo et al. (2017)

Fig. 2.4 Comparison of plants of tomato inoculated with a multi-species formulation (EMMIM-1) versus non-inoculated plants

Fig. 2.5 Bacterial mixture (EMMIM-5) improving the growth of maize plants (Blue maize CAP15-1 TLAX) in comparison with mono-inoculated plants (*Azospirillum brasilense* Sp7), 45 days after the inoculation of germinated seeds

consistency of results obtained in the field crops. In fact, the bacteria of these polymicrobial formulations exert synergic effects on the inoculated plants. Similar results were obtained by a multi-species inoculant applied on corn and rice (Figs. 2.6 and 2.7). The multi-species inoculants containing plant growth-promoting bacteria have been summarized in Table 2.5. Inoculation of bacterial formulations of the second generation is in development, and we hope that in next few years an increased number of new multispecies formulations will appear in the market (Baez-Rogelio et al. 2017). The most advanced multispecies formulation in the context of technology transference in Mexico is the bacterial mixture named EMMIM-1, it is the closest to appear in the market and experiences of inoculation in fields have demonstrated its effectiveness and better positive results than mono-inoculants. Currently, around 5000 ha of maize have been inoculated with EMMIM-1 in the central region of Mexico, and other cultures also have been inoculated in minor proportion.

Fig. 2.6 Maize plants of Cacahuazintle, an autochthonous variety from Calpulalpan-Tlaxcala, México, at 90 days of growth. Maize inoculated with the multispecies formulation EMMIM-1 (**a**) and not inoculated (**b**). The inoculated maize is greener, with higher height, thicker stems, and larger adventitious roots and has less problem of herbs. Fertilization and form of tillage was the same for both treatments

Fig. 2.7 Rice plants inoculated with EMMIM-1 (right) compared to non-inoculated plants (left). Assays from Morelia, Michoacán, México

2.9 Conclusions

Plant growth-promoting bacteria have different mechanisms to perform phytostimulation. Although the use of PGPR-type bacteria for inoculation of crop plants has been moderated, their application is increasing over the time due to their proved benefits on productivity and reducing the chemical fertilizer, pesticides and herbicides inputs in soil. The co-inoculation of beneficial microorganisms on crop plant is apparently more effective in stimulating plant growth than mono-inoculation, perhaps because of the synergistic effect occurring when microorganisms are co-interacting. Despite promising results of co-inoculation, there are still few formulations containing more than three species of microorganisms in consortium. Therefore, a new challenge is the design and formulation of multi-species inoculants able to exercise a greater plant growth promotion in comparison with those of mono-species inoculants. In our laboratory, we have designed some polymicrobial formulations which are already patented. Due to the characteristics and advantages that they have in relation to conventional inoculants, they have been designated as the second-generation inoculants.

Acknowledgements We thank CONACYT and VIEP-BUAP for the financial support to carry out our research.

Conflict of Interest The author(s) have no conflict of interest.

References

Aeron A, Kumar S, Pandey P, Maheshwari DK (2011) Emerging role of plant growth promoting rhizobacteria in agrobiology. In: Maheshwari DK (ed) Bacteria in agrobiology: crop ecosystems. Springer, Berlin, pp 1–36

Aeron A, Maheshwari DK, Dheeman S, Agarwal M, Dubey RC, Bajpai VK (2017) Plant growth promotion and suppression of charcoal-rot fungus (*Macrophomina phaseolina*) in velvet bean (*Mucuna pruriens* L.) by root nodule bacteria. J Phytopathol 165:463–478. https://doi.org/10.1111/jph.12581

Ahemad M, Khan MS (2012) Alleviation of fungicide-induced phytotoxicity in greengram [*Vigna radiata* (L.) Wilczek] using fungicide-tolerant and plant growth promoting *Pseudomonas* strain. Saudi J Biol Sci 19:451–459. https://doi.org/10.1016/j.sjbs.2012.06.003

Ahmad M, Ahmad I, Hilger TH, Nadeem SM, Akhtar MF, Jamil M et al (2018) Preliminary study on phosphate solubilizing *Bacillus subtilis* strain Q3 and *Paenibacillus* sp. strain Q6 for improving cotton growth under alkaline conditions. PeerJ 6:e5122. https://doi.org/10.7717/peerj.5122

Alatorre-Cruz JM, Bustillos-Cristales M del R, Morales-García YE, Hernández-Tenorio AL, Baez-Rogelio A, Pérez-Santos JLM et al (2015) Multi-species formulation to improve the growth of plants from semi-desertic zones, 1–28

Albareda M, Dardanelli MS, Sousa C, Megías M, Temprano F, Rodríguez-Navarro DN (2006) Factors affecting the attachment of rhizospheric bacteria to bean and soybean roots. FEMS Microbiol Lett 259:67–73

Amann RI, Ludwig W, Schleifer K (1995) Phylogenetic identification and in situ detection of individual microbial cells without cultivation. Microbiol Rev 59:143–169

Annan H, Golding AL, Zhao Y, Dong Z (2012) Choice of hydrogen uptake (Hup) status in legume-rhizobia symbioses. Ecol Evol 2:2285–2290. https://doi.org/10.1002/ece3.325

Atieno M, Herrmann L, Okalebo R, Lesueur D (2012) Efficiency of different formulations of Bradyrhizobium japonicum and effect of co-inoculation of *Bacillus subtilis* with two different strains of *Bradyrhizobium japonicum*. World J Microbiol Biotechnol 28:2541–2550. https://doi.org/10.1007/s11274-012-1062-x

Baez-Rogelio A, Morales-García YE, Quintero-Hernández V, Muñoz-Rojas J (2017) Next generation of microbial inoculants for agriculture and bioremediation. Microb Biotechnol 10:19–21. https://doi.org/10.1111/1751-7915.12448

Bal HB, Das S, Dangar TK, Adhya TK (2013) ACC deaminase and IAA producing growth promoting bacteria from the rhizosphere soil of tropical rice plants. J Basic Microbiol 53:972–984. https://doi.org/10.1002/jobm.201200445

Bardin SD, Huang H-C, Pinto J, Amundsen EJ, Erickson RS (2004) Biological control of *Pythium* damping-off of pea and sugar beet by *Rhizobium leguminosarum* bv. *viceae*. Can J Bot 82:291–296. https://doi.org/10.1139/b04-003

Barea JM, Toro M, Orozco MO, Campos E, Azcón R (2002) The application of isotopic (^{32}P and ^{15}N) dilution techniques to evaluate the interactive effect of phosphate-solubilizing rhizobacteria, mycorrhizal fungi and *Rhizobium* to improve the agronomic efficiency of rock phosphate for legume crops. Nutr Cycl Agroecosystems 63:35–42. https://doi.org/10.1023/A:1020589732436

Barman M, Paul S, Choudhury AG, Roy P, Sen J (2017) Biofertilizer as prospective input for sustainable agriculture in India. Int J Curr Microbiol Appl Sci 6:1177–1186. https://doi.org/10.20546/ijcmas.2017.611.141

Baset Mia MA, Shamsuddin ZH (2010) *Rhizobium* as a crop enhancer and biofertilizer for increased cereal production. Afr J Biotechnol 9:6001–6009

Bashan Y, de-Bashan LE, Prabhu SR, Hernandez JP (2014) Advances in plant growth-promoting bacterial inoculant technology: formulations and practical perspectives (1998–2013). Plant Soil 378:1–33. https://doi.org/10.1007/s11104-013-1956-x

Bashan Y, Huang P, Kloepper JW, de-Bashan L (2017) A proposal for avoiding fresh-weight measurements when reporting the effect of plant growth-promoting (rhizo)bacteria on growth promotion of plants. Biol Fertil Soils 53:1–2. https://doi.org/10.1007/s00374-016-1153-1

Benbrook CM (2016) Trends in glyphosate herbicide use in the United States and globally. Environ Sci Eur 28:3. https://doi.org/10.1186/s12302-016-0070-0

Beneduzi A, Ambrosini A, Passaglia LMP (2012) Plant growth-promoting rhizobacteria (PGPR): their potential as antagonists and biocontrol agents. Genet Mol Biol 35:1044–1051. https://doi.org/10.1590/S1415-47572012000600020

Berendsen RL, Pieterse CMJ, Bakker PAHM (2012) The rhizosphere microbiome and plant health. Trends Plant Sci 17:478–486. https://doi.org/10.1016/j.tplants.2012.04.001

Bhattacharyya PN, Jha DK (2012) Plant growth-promoting rhizobacteria (PGPR): Emergence in agriculture. World J Microbiol Biotechnol 28:1327–1350. https://doi.org/10.1007/s11274-011-0979-9

Böltner D, Godoy P, Muñoz-Rojas J, Duque E, Moreno-Morillas S, Sánchez L et al (2008) Rhizoremediation of lindane by root-colonizing Sphingomonas. Microb Biotechnol 1:87–93

Bottini R, Cassán F, Piccoli P (2004) Gibberellin production by bacteria and its involvement in plant growth promotion and yield increase. Appl Microbiol Biotechnol 65:497–503. https://doi.org/10.1007/s00253-004-1696-1

Broadbent JR, McMahon DJ, Welker DL, Oberg CJ, Moineau S (2003) Biochemistry, genetics, and applications of exopolysaccharide production in *Streptococcus thermophilus*: a review. J Dairy Sci 86:407–423. https://doi.org/10.3168/jds.S0022-0302(03)73619-4

Caballero A, Lázaro JJ, Ramos JL, Esteve-Núñez A (2005) PnrA, a new nitroreductase-family enzyme in the TNT-degrading strain *Pseudomonas putida* JLR11. Environ Microbiol 7:1211–1219. https://doi.org/10.1111/j.1462-2920.2005.00801.x

Caballero-Mellado J, Martínez-Aguilar L, Paredes-Valdez G, Estrada-de los Santos P (2004) *Burkholderia unamae* sp. nov., an N_2-fixing rhizospheric and endophytic species. Int J Syst Evol Microbiol 54:1165–1172. https://doi.org/10.1099/ijs.0.02951-0

Caballero-Mellado J, Onofre-Lemus J, Estrada-de Los Santos P, Martínez-Aguilar L (2007) The tomato rhizosphere, an environment rich in nitrogen-fixing *Burkholderia* species with capabilities of interest for agriculture and Bioremediation. Appl Environ Microbiol 73:5308–5319. https://doi.org/10.1128/AEM.00324-07

Çakmakçi R, Dönmez F, Aydın A, Şahin F (2006) Growth promotion of plants by plant growth-promoting rhizobacteria under greenhouse and two different field soil conditions. Soil Biol Biochem 38:1482–1487. https://doi.org/10.1016/j.soilbio.2005.09.019

Carolan MS (2009) Barriers to the adoption of sustainable agriculture on rented land: an axamination of contesting social fields. Rural Sociol 70:387–413. https://doi.org/10.1526/0036011054831233

Cassán F, Vanderleyden J, Spaepen S (2014) Physiological and agronomical aspects of phytohormone production by model Plant-Growth-Promoting Rhizobacteria (PGPR) belonging to the genus *Azospirillum*. J Plant Growth Regul 33:440–459. https://doi.org/10.1007/s00344-013-9362-4

Castagno LN, Estrella MJ, Sannazzaro AI, Grassano AE, Ruiz OA (2011) Phosphate-solubilization mechanism and in vitro plant growth promotion activity mediated by *Pantoea eucalypti* isolated from *Lotus tenuis* rhizosphere in the Salado River Basin (Argentina). J Appl Microbiol 110:1151–1165. https://doi.org/10.1111/j.1365-2672.2011.04968.x

Cawoy H, Debois D, Franzil L, De Pauw E, Thonart P, Ongena M (2015) Lipopeptides as main ingredients for inhibition of fungal phytopathogens by *Bacillus subtilis/amyloliquefaciens*. Microb Biotechnol 8:281–295. https://doi.org/10.1111/1751-7915.12238

Cazorla FM, Romero D, Pérez-García A, Lugtenberg BJJ, De VA, Bloemberg G (2007) Isolation and characterization of antagonistic *Bacillus subtilis* strains from the avocado rhizoplane displaying biocontrol activity. J Appl Microbiol 103:1950–1959. https://doi.org/10.1111/j.1365-2672.2007.03433.x

Chemier JA, Fowler ZL, Koffas MAG, Leonard E (2009) Trends in microbial synthesis of natural products and biofuels. Adv Enzymol Relat Areas Mol Biol 76:151–217

Chen S-K, Edwards CA, Subler S (2001) Effects of the fungicides benomyl, captan and chlorothalonil on soil microbial activity and nitrogen dynamics in laboratory incubations. Soil Biol Biochem 33:1971–1980. https://doi.org/10.1016/S0038-0717(01)00131-6

Cheng X, Etalo DW, van de Mortel JE, Dekkers E, Nguyen L, Medema MH et al (2017) Genome-wide analysis of bacterial determinants of plant growth promotion and induced systemic resistance by *Pseudomonas fluorescens*. Environ Microbiol 19:4638–4656. https://doi.org/10.1111/1462-2920.13927

Compant S, Duffy B, Nowak J, Cle C, Barka EA (2005) Use of plant growth-promoting bacteria for biocontrol of plant diseases: principles, mechanisms of action, and future prospects. Appl Environ Microbiol 71:4951–4959. https://doi.org/10.1128/aem.71.9.4951

Costacurta A, Vanderleyden J (1995) Synthesis of phytohormones by plant-associated bacteria. Crit Rev Microbiol 21:1–18. https://doi.org/10.3109/10408419509113531

Cummings SP (2009) The application of plant growth promoting rhizobacteria (PGPR) in low input and organic cultivation of graminaceous crops; potential and problems. Environ Biotechnol 5:43–50

Curtis TP, Sloan WT, Scannell JW (2002) Estimating prokaryotic diversity and its limits. Proc Natl Acad Sci 99:10494–10499. https://doi.org/10.1073/pnas.142680199

de Oliveira ALM, de Canuto EL, Urquiaga S, Reis VM, Baldani JI (2006) Yield of micropropagated sugarcane varieties in different soil types following inoculation with diazotrophic bacteria. Plant Soil 284:23–32. https://doi.org/10.1007/s11104-006-0025-0

Defez R, Esposito R, Angelini C, Bianco C (2016) Overproduction of indole-3-acetic acid in free-living rhizobia induces transcriptional changes resembling those occurring in nodule bacteroids. Mol Plant-Microbe Interact 29:484–495. https://doi.org/10.1094/MPMI-01-16-0010-R

Dixon R, Kahn D (2004) Genetic regulation of biological nitrogen fixation. Nat Rev Microbiol 2:621

Dobbelaere S, Croonenborghs A, Thys A, Ptacek D, Vanderleyden J, Dutto P et al (2001) Responses of agronomically important crops to inoculation with *Azospirillum*. Funct Plant Biol 28:871–879

Donham KJ (2016) Health effects of agricultural pesticides. Agric Med. https://doi.org/10.1002/9781118647356.ch6

Elbadry M, Taha RM, Eldougdoug KA, Gamal-Eldin H (2006) Induction of systemic resistance in faba bean (*Vicia faba* L.) to bean yellow mosaic potyvirus (BYMV) via seed bacterization with plant growth promoting rhizobacteria. J Plant Dis Prot 113:247–251. https://doi.org/10.1007/bf03356189

Elkoca E, Kantar F, Sahin F (2007) Influence of nitrogen fixing and phosphorus solubilizing bacteria on the nodulation, plant growth, and yield of chickpea. J Plant Nutr 31:157–171. https://doi.org/10.1080/01904160701742097

Esquivel-Cote R, Gavilanes-Ruiz M, Cruz-Ortega R, Pilar Huante Y (2013) Importancia agrobiotecnológica de la enzima acc desaminasa en rizobacterias, una revisión. Rev Fitotec Mex 36:251–258. https://doi.org/10.1016/j.resmic.2006.11.013

Farag MA, Zhang H, Ryu CM (2013) Dynamic chemical communication between plants and bacteria through airborne signals: induced resistance by bacterial volatiles. J Chem Ecol 39:1007–1018. https://doi.org/10.1007/s10886-013-0317-9

Fernández M, Niqui-Arroyo JL, Conde S, Ramos JL, Duque E (2012) Enhanced tolerance to naphthalene and enhanced rhizoremediation performance for *Pseudomonas putida* KT2440 via the NAH7 catabolic plasmid. Appl Environ Microbiol 78:5104–5110. https://doi.org/10.1128/AEM.00619-12

Fibach-Paldi S, Burdman S, Okon Y (2012) Key physiological properties contributing to rhizosphere adaptation and plant growth promotion abilities of *Azospirillum brasilense*. FEMS Microbiol Lett 326:99–108. https://doi.org/10.1111/j.1574-6968.2011.02407.x

Fuentes-Ramirez LE, Caballero-Mellado J (2006) Bacterial Biofertilizers. In: Siddiqui ZA (ed) PGPR: Biocontrol and Biofertilization. Springer, Netherlands, pp 143–172

Fuentes-Ramirez LE, Jimenez-Salgado T, Abarca-Ocampo IR, Caballero-Mellado J (1993) *Acetobacter diazotrophicus*, an indoleacetic acid producing bacterium isolated from sugarcane cultivars of México. Plant Soil 154:145–150. https://doi.org/10.1007/BF00012519

Ghosh N (2004) Promoting biofertilisers in Indian agriculture. Econ Polit Wkly 39:5617–5625

Gkizi D, Lehmann S, Haridon FL, Serrano M, Paplomatas EJ (2016) The innate immune signaling system as a regulator of disease resistance and induced systemic resistance activity against *Verticillium dahliae*. Mol Plant Microbe Interact 29:313–323. https://doi.org/10.1094/MPMI-11-15-0261-R

Glick BR (2014) Bacteria with ACC deaminase can promote plant growth and help to feed the world. Microbiol Res 169:30–39. https://doi.org/10.1016/j.micres.2013.09.009

Glick BR, Penrose DM, Li J (1998) A model for the lowering of plant ethylene concentrations by plant growth-promoting bacteria. J Theor Biol 190:63–68. https://doi.org/10.1006/jtbi.1997.0532

Glick BR, Cheng Z, Czarny J, Duan J (2007) Promotion of plant growth by ACC deaminase-producing soil bacteria. In: Bakker PAHM, Raaijmakers JM, Bloemberg G, Höfte M, Lemanceau P, Cooke BM (eds) New perspectives and approaches in plant growth-promoting Rhizobacteria research. Springer, Netherlands, pp 329–339

Goswami D, Thakker JN, Dhandhukia PC (2016) Portraying mechanics of plant growth promoting rhizobacteria (PGPR): a review. Cogent Food Agric 2:1–19. https://doi.org/10.1080/23311932.2015.1127500

Gutiérrez-Zamora ML, Martínez-Romero E (2001) Natural endophytic association between *Rhizobium etli* and maize (Zea mays L.). J Biotechnol 91:117–126. https://doi.org/10.1016/S0168-1656(01)00332-7

Guzmán A, Obando M, Rivera D, Bonilla R (2012) Selección y caracterización de rizobacterias promotoras de crecimiento vegetal (RPCV) asociadas al cultivo de algodón (*Gossypium hirsutum*). Rev Colomb Biotecnol 14:182–190

Gyaneshwar P, Naresh Kumar G, Parekh LJ, Poole PS (2002) Role of soil microorganisms in improving P nutrition of plants. Plant Soil 245:83–93. https://doi.org/10.1023/A:1020663916259

Hassen W, Neifar M, Cherif H, Najjari A, Chouchane H, Driouich RC et al (2018) *Pseudomonas rhizophila* S211, a new plant growth-promoting rhizobacterium with potential in pesticide-bioremediation. Front Microbiol 9:34

Hernández-Calderón E, Aviles-Garcia ME, Castulo-Rubio DY, Macías-Rodríguez L, Ramírez VM, Santoyo G et al (2018) Volatile compounds from beneficial or pathogenic bacteria differentially regulate root exudation, transcription of iron transporters, and defense signaling pathways in *Sorghum bicolor*. Plant Mol Biol 96:291–304. https://doi.org/10.1007/s11103-017-0694-5

Herridge DF, Peoples MB, Boddey RM (2008) Global inputs of biological nitrogen fixation in agricultural systems. Plant Soil 311:1–18. https://doi.org/10.1007/s11104-008-9668-3

Huang X, Chaparro JM, Reardon KF, Zhang R, Shen Q, Vivanco JM (2014) Rhizosphere interactions: root exudates, microbes, and microbial communities. Botany 92:267–275. https://doi.org/10.1139/cjb-2013-0225

Idris EE, Iglesias DJ, Talon M, Borriss R (2007) Tryptophan-dependent production of Indole-3-Acetic Acid (IAA) affects level of plant growth promotion by *Bacillus amyloliquefaciens* FZB42. Mol Plant-Microbe Interact 20:619–626. https://doi.org/10.1094/MPMI-20-6-0619

Igbedioh SO (1991) Effects of agricultural pesticides on humans, animals, and higher plants in developing countries. Arch Environ Heal An Int J 46:218–224. https://doi.org/10.1080/00039896.1991.9937452

Jha CK, Saraf M (2015) Plant growth promoting Rhizobacteria (PGPR): a review. J Agric Res Dev 5:108–119

Jimenez-Salgado T, Fuentes-Ramirez LE, Tapia-Hernandez A, Mascarua-Esparza MA, Martinez-Romero E, Caballero-Mellado J (1997) Coffea arabica L., a new host plant for Acetobacter diazotrophicus, and isolation of other nitrogen-fixing acetobacteria. Appl Environ Microbiol 63:3676–3683

Johnsen K, Jacobsen CS, Torsvik V, Sørensen J (2001) Pesticide effects on bacterial diversity in agricultural soils—a review. Biol Fertil Soils 33:443–453. https://doi.org/10.1007/s003740100351

Jorquera MA, Shaharoona B, Nadeem SM, de la Luz MM, Crowley DE (2012) Plant growth-promoting rhizobacteria associated with ancient clones of Creosote Bush (*Larrea tridentata*). Microb Ecol 64:1008–1017. https://doi.org/10.1007/s00248-012-0071-5

Kanchiswamy CN, Malnoy M, Maffei ME (2015) Bioprospecting bacterial and fungal volatiles for sustainable agriculture. Trends Plant Sci 20:206–211. https://doi.org/10.1016/j.tplants.2015.01.004

Kang SC, Ha CG, Lee TG, Maheshwari DK (2002) Solubilization of insoluble inorganic phosphates by a soil-inhabiting fungus *Fomitopsis* sp. PS 102. Curr Sci 82:439–442

Kang SM, Khan AL, You YH, Kim JG, Kamran M, Lee IJ (2014) Gibberellin production by newly isolated strain *Leifsonia soli* SE134 and its potential to promote plant growth. J Microbiol Biotechnol 24:106–112. https://doi.org/10.4014/jmb.1304.04015

Kaschuk G, Hungria M (2017) Diversity and importance of diazotrophic bacteria to agricultural sustainability in the tropics. In: de Azevedo JL, Quecine MC (eds) Diversity and benefits of microorganisms from the tropics. Springer International Publishing, Cham, pp 269–292

Khan MS, Zaidi A, Wani PA (2007) Role of phosphate-solubilizing microorganisms in sustainable agriculture—a review. Agron Sustain Dev 27:29–43. https://doi.org/10.1051/agro:2006011

Khan MS, Zaidi A, Ahmad E (2014) Mechanism of phosphate solubilization and physiological functions of phosphate-solubilizing microorganisms. In: Khan MS, Zaidi A, Musarrat J (eds) Phosphate solubilizing microorganisms: principles and application of microphos technology. Springer International Publishing, Cham, pp 31–62

Kloepper JW, Ryu C-M, Zhang S (2004) Induced systemic resistance and promotion of plant growth by *Bacillus* spp. Phytopathology 94:1259–1266. https://doi.org/10.1094/PHYTO.2004.94.11.1259

Krishnan HB, Kang BR, Krishnan AH, Kim KY, Kim YC (2007) *Rhizobium etli* USDA9032 engineered to produce a phenazine antibiotic inhibits the growth of fungal pathogens but is impaired in symbiotic performance. Appl Environ Microbiol 73:327–330. https://doi.org/10.1128/AEM.02027-06

Kumar P, Dubey RC, Maheshwari DK (2012) *Bacillus* strains isolated from rhizosphere showed plant growth promoting and antagonistic activity against phytopathogens. Microbiol Res 167:493–499. https://doi.org/10.1016/j.micres.2012.05.002

Kumar A, Maurya BR, Raghuwanshi R (2014) Isolation and characterization of PGPR and their effect on growth, yield and nutrient content in wheat (*Triticum aestivum* L.). Biocatal Agric Biotechnol 3:121–128. https://doi.org/10.1016/j.bcab.2014.08.003

Lambrese Y, Guiñez M, Calvente V, Sansone G, Cerutti S, Raba J et al (2018) Production of siderophores by the bacterium *Kosakonia radicincitans* and its application to control of phytopathogenic fungi. Bioresour Technol Reports 3:82–87. https://doi.org/10.1016/j.biteb.2018.06.003

Lin W, Bazylinski DA, Xiao T, Wu LF, Pan Y (2014) Life with compass: diversity and biogeography of magnetotactic bacteria. Environ Microbiol 16:2646–2658. https://doi.org/10.1111/1462-2920.12313

Liu K, Newman M, McInroy JA, Hu C-H, Kloepper JW (2017a) Selection and assessment of plant growth-promoting rhizobacteria for biological control of multiple plant diseases. Phytopathology 107:928–936. https://doi.org/10.1094/PHYTO-02-17-0051-R

Liu S-H, Zeng G-M, Niu Q-Y, Liu Y, Zhou L, Jiang L-H et al (2017b) Bioremediation mechanisms of combined pollution of PAHs and heavy metals by bacteria and fungi: a mini review. Bioresour Technol 224:25–33. https://doi.org/10.1016/j.biortech.2016.11.095

Lugtenberg B, Kamilova F (2009) Plant-growth-promoting Rhizobacteria. Annu Rev Microbiol 63:541–556. https://doi.org/10.1146/annurev.micro.62.081307.162918

Mahajan A, Gupta RD (2009) Bio-fertilizers: their kinds and requirement in India. In: Mahajan A, Gupta RD (eds) Integrated Nutrient Management (INM) in a sustainable rice—wheat cropping system. Springer Netherlands, pp 75–100

Maheshwari DK, Aeron A, Dubey RC, Agarwal M, Dheeman S, Shukla S (2015a) Multifaceted beneficial association with *Pseudomonas* and rhizobia on growth promotion of Mucuna pruriensis L. J Pure Appl Microbiol 8:4657–4667

Maheshwari DK, Dubey RC, Agarwal M, Dheeman S, Aeron A, Bajpai VK (2015b) Carrier based formulations of biocoenotic consortia of disease suppressive *Pseudomonas aeruginosa* KRP1 and Bacillus licheniformis KRB1. Ecol Eng 81:272–277. https://doi.org/10.1016/j.ecoleng.2015.04.066

Majumdar K (2015) Bio-fertilizer use in Indian agriculture. Indian J Res 4:377–381

Malik DK, Sindhu SS (2011) Production of indole acetic acid by *Pseudomonas* sp.: effect of coinoculation with *Mesorhizobium* sp. Cicer on nodulation and plant growth of chickpea (*Cicer arietinum*). Physiol Mol Biol Plants 17:25–32. https://doi.org/10.1007/s12298-010-0041-7

Mazid M, Khan TA (2014) Future of bio-fertilizers in Indian agriculture: an overview. Int J Agric Food Res 3:10–23

Mehnaz S, Kowalik T, Reynolds B, Lazarovits G (2010) Growth promoting effects of corn (Zea mays) bacterial isolates under greenhouse and field conditions. Soil Biol Biochem 42:1848–1856. https://doi.org/10.1016/j.soilbio.2010.07.003

Mitter B, Pfaffenbichler N, Flavell R, Compant S, Antonielli L, Petric A et al (2017) A new approach to modify plant microbiomes and traits by introducing beneficial bacteria at flowering into progeny seeds. Front Microbiol 8:1–10. https://doi.org/10.3389/fmicb.2017.00011

Mohamed R, Groulx E, Defilippi S, Erak T, Tambong JT, Tweddell RJ et al (2017) Physiological and molecular characterization of compost bacteria antagonistic to soil-borne plant pathogens. Can J Microbiol 63:411–426. https://doi.org/10.1139/cjm-2016-0599

Molina-Romero D, Bustillos-Cristales MDR, Rodríguez-Andrade O, Morales-García YE, Santiago-Saenz Y, Castañeda-Lucio M et al (2015) Mecanismos de fitoestimulación por rizobacterias, aislamientos en América y potencial biotecnológico. Biológicas 17:24–34

Molina-Romero D, Baez A, Quintero-Hernández V, Castañeda-Lucio M, Fuentes-Ramírez LE, Bustillos-Cristales MDR et al (2017a) Antagonism assays to identify bacterial strains producing antimicrobial compounds. Protoc PLOS one 12:1–2. https://doi.org/10.17504/protocols.io.j4mcqu6

Molina-Romero D, Baez A, Quintero-Hernández V, Castañeda-Lucio M, Fuentes-Ramírez LE, Bustillos-Cristales MDR et al (2017b) Compatible bacterial mixture, tolerant to desiccation, improves maize plant growth. PLoS One 12:e0187913

Molina-Romero D, Baez A, Quintero-Hernández V, Castañeda-Lucio M, Fuentes-Ramírez LE, Bustillos-Cristales MDR et al (2017c) Selection assay to identify desiccation tolerant bacteria. Protoc PLOS One 12:1–2. https://doi.org/10.17504/protocols.io.j4icque

Molina-Romero D, Morales-García YE, Hernández-Tenorio A-L, Castañeda-Lucio M, Netzahuatl-Muñoz AR, Muñoz-Rojas J (2017d) *Pseudomonas putida* estimula el crecimiento de maíz en función de la temperatura. Rev Iberoam Ciencias 4: 80–88

Morales-García YE, Herrera MDC, Munoz-Rojas J (2007) Chloramphenicol, a classic antibiotic as an alternative for the present. Rev Mex Ciencias Farm 38:58–69

Morales-García YE, Juárez-Hernández D, Fuentes-Ramírez LE, Munive-Hernández A, Muñoz-Rojas J (2013) Formulación de un inoculante multiespecies para potenciar el crecimiento de plantas, 1–36

Morales-García YE, Juárez-Hernández D, Aragón-Hernández C, Mascarua-Esparza MA, Bustillos-Cristales MR, Fuentes-Ramírez LE et al (2011) Growth response of maize plantlets inoculated with *Enterobacter* spp., as a model for alternative agriculture. Rev Argent Microbiol 43:287–293. https://doi.org/10.1590/s0325-75412011000400009

Morgan PW, Drew MC (2006) Ethylene and plant responses to stress. Physiol Plant 100:620–630. https://doi.org/10.1111/j.1399-3054.1997.tb03068.x

Muñoz-Rojas J, Caballero-Mellado J (2003) Population dynamics of gluconacetobacter diazotrophicus in sugarcane cultivars and its effect on plant growth. Microb Ecol 46:454–464

Muñoz-Rojas J, Fuentes-Ramírez LE, Caballero-Mellado J (2005) Antagonism among *Gluconacetobacter diazotrophicus* strains in culture media and in endophytic association. FEMS Microbiol Ecol 54:57–66

Myresiotis CK, Vryzas Z, Papadopoulou-Mourkidou E (2012) Biodegradation of soil-applied pesticides by selected strains of plant growth-promoting rhizobacteria (PGPR) and their effects on bacterial growth. Biodegradation 23:297–310. https://doi.org/10.1007/s10532-011-9509-6

Nadeem SM, Zahir ZA, Naveed M, Arshad M (2007) Preliminary investigations on inducing salt tolerance in maize through inoculation with rhizobacteria containing ACC deaminase activity. Can J Microbiol 53:1141–1149. https://doi.org/10.1139/W07-081

Naureen Z, Rehman NU, Hussain H, Hussain J, Gilani SA, Al Housni SK et al (2017) Exploring the potentials of Lysinibacillus sphaericus ZA9 for plant growth promotion and biocontrol activities against phytopathogenic fungi. Front Microbiol 8:1477

O'Callaghan M (2016) Microbial inoculation of seed for improved crop performance: issues and opportunities. Appl Microbiol Biotechnol 100:5729–5746. https://doi.org/10.1007/s00253-016-7590-9

Okon Y, Labandera-Gonzalez CA (1994) Agronomic applications of *Azospirillum*: an evaluation of 20 years worldwide field inoculation. Soil Biol Biochem 26:1591–1601. https://doi.org/10.1016/0038-0717(94)90311-5

Oliveira ALM, Stoffels M, Schmid M, Reis VM, Baldani JI, Hartmann A (2009) Colonization of sugarcane plantlets by mixed inoculations with diazotrophic bacteria. Eur J Soil Biol 45:106–113. https://doi.org/10.1016/j.ejsobi.2008.09.004

Olovaldo Y, Saenz, Laura Ana, Tenorio H, Morales YE, Ana L et al (2017) Method for obtaining patato plants from the extraction of induced sprouts in controlled conditions. 2015014804:1–25

Onofre-Lemus J, Hernández-Lucas I, Girard L, Caballero-Mellado J (2009) ACC (1-aminocyclopropane-1-carboxylate) deaminase activity, a widespread trait in *Burkholderia* species, and its growth-promoting effect on tomato plants. Appl Environ Microbiol 75:6581–6590. https://doi.org/10.1128/AEM.01240-09

Ortíz-Castro R, Contreras-Cornejo HA, Macías-Rodríguez L, López-Bucio J (2009) The role of microbial signals in plant growth and development. Plant Signal Behav 4:701–712. https://doi.org/10.4161/psb.4.8.9047

Pace NR (1997) A molecular view of microbial diversity and the biosphere. Science (80–) 276:734–740. https://doi.org/10.1126/science.276.5313.734

Palus JA, Borneman J, Ludden PW, Triplett EW (1996) A diazotrophic bacterial endophyte isolated from stems of *Zea mays* L. and *Zea luxurians* Iltis and Doebley. Plant Soil 186:135–142. https://doi.org/10.1007/bf00035067

Pandey P, Kang SC, Maheshwari DK (2005) Isolation of endophytic plant growth promoting *Burkholderia* sp. MSSP from root nodules of Mimosa pudica. Curr Sci 89:177–180

Park HB, Lee B, Kloepper JW, Ryu CM (2013) One shot-two pathogens blocked: exposure of Arabidopsis to hexadecane, a long chain volatile organic compound, confers induced resistance against both *Pectobacterium carotovorum* and *Pseudomonas syringae*. Plant Signal Behav 8. https://doi.org/10.4161/psb.24619

Park MG, Blitzer EJ, Gibbs J, Losey JE, Danforth BN (2015) Negative effects of pesticides on wild bee communities can be buffered by landscape context. Proc R Soc B Biol Sci 282

Paterson J, Jahanshah G, Li Y, Wang Q, Mehnaz S, Gross H (2017) The contribution of genome mining strategies to the understanding of active principles of PGPR strains. FEMS Microbiol Ecol 93:1–12. https://doi.org/10.1093/femsec/fiw249

Patten CL, Glick BR (1996) Bacterial biosynthesis of indole-3-acetic acid. Can J Microbiol 42:207–220. https://doi.org/10.1139/m96-032

Patten CL, Glick BR (2002) Role of *Pseudomonas putida* indoleacetic acid in development of the host plant root system. Appl Environ Microbiol 68:3795–3801. https://doi.org/10.1128/AEM.68.8.3795

Pazos-Rojas LA, Rodríguez-Andrade O, Muñoz-Arenas LC, Morales-García YE, Corral-Lugo A, Quintero-Hernández V et al (2018) Desiccation-tolerant rhizobacteria maintain their plant growth- promoting capability after experiencing extreme water stress. SciFed J Appl Microbiol 2:1–13

Pérez-Flores P, Valencia-Cantero E, Altamirano-Hernández J, Pelagio-Flores R, López-Bucio J, García-Juárez P et al (2017) *Bacillus methylotrophicus* M4-96 isolated from maize (*Zea mays*) rhizoplane increases growth and auxin content in *Arabidopsis thaliana* via emission of volatiles. Protoplasma 254:2201–2213. https://doi.org/10.1007/s00709-017-1109-9

Philippot L, Raaijmakers JM, Lemanceau P, van der Putten WH (2013) Going back to the roots: the microbial ecology of the rhizosphere. Nat Rev Microbiol 11:789

Pieterse CMJ, Zamioudis C, Berendsen RL, Weller DM, Van Wees SCM, Bakker PAHM (2014) Induced systemic resistance by beneficial microbes. Annu Rev Phytopathol 52:347–375. https://doi.org/10.1146/annurev-phyto-082712-102340

Prashar P, Kapoor N, Sachdeva S (2014) Rhizosphere: its structure, bacterial diversity and significance. Rev Environ Sci Bio/Technology 13:63–77. https://doi.org/10.1007/s11157-013-9317-z

Príncipe A, Fernandez M, Torasso M, Godino A, Fischer S (2018) Effectiveness of tailocins produced by *Pseudomonas fluorescens* SF4c in controlling the bacterial-spot disease in tomatoes caused by *Xanthomonas vesicatoria*. Microbiol Res 212–213:94–102. https://doi.org/10.1016/j.micres.2018.05.010

Rajput MS, Iyer B, Pandya M, Jog R (2015) Derepression of mineral phosphate solubilization phenotype by insertional inactivation of IclR in *Klebsiella pneumoniae*. PLoS One 10:1–15. https://doi.org/10.1371/journal.pone.0138235

Ramamoorthy V, Viswanathan R, Raguchander T, Prakasam V, Samiyappan R (2001) Induction of systemic resistance by plant growth promoting rhizobacteria in crop plants against pests and diseases. Crop Prot 20:1–11. https://doi.org/10.1016/S0261-2194(00)00056-9

Ramos Martin JL, Muñoz-Rojas J (2006) Microorganismos recombinantes que contienen un gen de resistencia a estrés salino y sus aplicaciones. 1–31

Reis VM, Estrada de los Santos P, Vogel J, Stoffels M, Guyon S, Mavingui P et al (2004) *Burkholderia tropica* sp. nov., a novel nitrogen-fixing, plant-associated bacterium. Int J Syst Evol Microbiol 54:2155–2162. https://doi.org/10.1099/ijs.0.02879-0

Rodríguez H, Fraga R, Gonzalez T, Bashan Y (2006) Genetics of phosphate solubilization and its potential applications for improving plant growth-promoting bacteria. Plant Soil 287:15–21. https://doi.org/10.1007/s11104-006-9056-9

Rodríguez-Andrade O, Fuentes-Ramírez LE, Morales-García YE, Molina-Romero D, Bustillos-Cristales MR, Martínez-Contreras RD et al (2015) The decrease in the population of *Gluconacetobacter diazotrophicus* in sugarcane after nitrogen fertilization is related to plant physiology in split root experiments. Rev Argentina Microbiol 47:335–343. https://doi.org/10.1016/j.ram.2015.09.004

Rojas-Solís D, Zetter-Salmón E, Contreras-Pérez M, Rocha-Granados MDC, Macías-Rodríguez L, Santoyo G (2018) *Pseudomonas stutzeri* E25 and *Stenotrophomonas maltophilia* CR71 endophytes produce antifungal volatile organic compounds and exhibit additive plant growth-promoting effects. Biocatal Agric Biotechnol 13:46–52. https://doi.org/10.1016/j.bcab.2017.11.007

Rosenblueth M, Martínez L, Silva J, Martínez-Romero E (2004) *Klebsiella variicola*, a novel species with clinical and plant-associated isolates. Syst Appl Microbiol 27:27–35. https://doi.org/10.1078/0723-2020-00261

Russo A, Vettori L, Felici C, Fiaschi G, Morini S, Toffanin A (2008) Enhanced micropropagation response and biocontrol effect of *Azospirillum brasilense* Sp245 on *Prunus cerasifera* L. clone Mr.S 2/5 plants. J Biotechnol 134:312–319. https://doi.org/10.1016/j.jbiotec.2008.01.020

Saikia SP, Jain V (2007) Biological nitrogen fixation with non-legumes: an achievable target or a dogma? Curr Sci 92:317–322

Santi C, Bogusz D, Franche C (2013) Biological nitrogen fixation in non-legume plants. Ann Bot 111:743–767. https://doi.org/10.1093/aob/mct048

Santoro M, Cappellari L, Giordano W, Banchio E (2015) Production of volatile organic compounds in PGPR. In: Cassán FD, Okon Y, Creus CM (eds) Handbook for Azospirillum: technical issues and protocols. Springer International Publishing, Cham, pp 307–317

Saravanan VS, Madhaiyan M, Osborne J, Thangaraju M, Sa TM (2008) Ecological occurrence of *Gluconacetobacter diazotrophicus* and nitrogen-fixing Acetobacteraceae members: their possible role in plant growth promotion. Microb Ecol 55:130–140. https://doi.org/10.1007/s00248-007-9258-6

Schneider K, Müller A, Schramm U, Kklipp W (1991) Demonstration of a molybdenum- and vanadium-independent nitrogenase in a nifHDK-deletion mutant of *Rhodobacter capsulatus*. Eur J Biochem 195:653–661. https://doi.org/10.1111/j.1432-1033.1991.tb15750.x

Sevilla M, Burris RH, Gunapala N, Kennedy C (2001) Comparison of benefit to sugarcane plant growth and $^{15}N_2$ incorporation following inoculation of sterile plants with *Acetobacter diazotrophicus* wild-type and nif- mutant strains. Mol Plant-Microbe Interact 14:358–366. https://doi.org/10.1094/MPMI.2001.14.3.358

Shahzad SM, Arif MS, Riaz M, Iqbal Z, Ashraf M (2013) PGPR with varied ACC-deaminase activity induced different growth and yield response in maize (Zea mays L.) under fertilized conditions. Eur J Soil Biol 57:27–34. https://doi.org/10.1016/j.ejsobi.2013.04.002

Sharifi R, Ryu C-M (2018) Revisiting bacterial volatile-mediated plant growth promotion: lessons from the past and objectives for the future. Ann Bot 122:349–358

Sharma CK, Vishnoi VK, Dubey RC, Maheshwari DK (2018) A twin rhizospheric bacterial consortium induces systemic resistance to a phytopathogen *Macrophomina phaseolina* in mung bean. Rhizosphere 5:71–75. https://doi.org/10.1016/j.rhisph.2018.01.001

Shen D (1997) Microbial diversity and application of microbial products for agricultural purposes in China. Agric Ecosyst Environ 62:237–245. https://doi.org/10.1016/S0167-8809(96)01132-2

Simms EL, Taylor DL (2002) Partner choice in nitrogen-fixation mutualisms of legumes and rhizobia. Integr Comp Biol 42:369–380. https://doi.org/10.1093/icb/42.2.369

Singh A, Jain A, Sarma BK, Upadhyay RS, Singh HB (2014a) Rhizosphere competent microbial consortium mediates rapid changes in phenolic profiles in chickpea during *Sclerotium rolfsii* infection. Microbiol Res 169:353–360. https://doi.org/10.1016/j.micres.2013.09.014

Singh S, Singh BK, Yadav SM, Gupta AK (2014b) Potential of biofertilizers in crop production in Indian agriculture. Am J Plant Nutr Fertil Technol 4:33–40. https://doi.org/10.3923/ajpnft.2014.33.40

Singh RP, Shelke GM, Kumar A, Jha PN (2015) Biochemistry and genetics of ACC deaminase: a weapon to "stress ethylene" produced in plants. Front Microbiol 6:1–14. https://doi.org/10.3389/fmicb.2015.00937

Sivasakthi S, Usharani G, Saranraj P (2014) Biocontrol potentiality of plant growth promoting bacteria (PGPR)—Pseudomonas fluorescens and Bacillus subtilis: A review. African J Agric Res 9:1265–1277. https://doi.org/10.5897/AJAR2013.7914

Smith MD, Hartnett DC, Rice CW (2000) Effects of long-term fungicide applications on microbial properties in tallgrass prairie soil. Soil Biol Biochem 32:935–946. https://doi.org/10.1016/S0038-0717(99)00223-0

Streeter JG (2003) Effect of trehalose on survival of *Bradyrhizobium japonicum* during desiccation. J Appl Microbiol 95:484–491. https://doi.org/10.1046/j.1365-2672.2003.02017.x

Strom SL (2008) Microbial ecology of ocean biogeochemistry: a community perspective. Science (80–) 320:1043–1045. https://doi.org/10.1126/science.1153527

Su P, Tan X, Li C, Zhang D, Cheng J, Zhang S et al (2017) Photosynthetic bacterium *Rhodopseudomonas palustris* GJ-22 induces systemic resistance against viruses. Microb Biotechnol 10:612–624. https://doi.org/10.1111/1751-7915.12704

Sundaramoorthy S, Raguchander T, Ragupathi N, Samiyappan R (2012) Combinatorial effect of endophytic and plant growth promoting rhizobacteria against wilt disease of *Capsicum annum* L. caused by *Fusarium solani*. Biol Control 60:59–67. https://doi.org/10.1016/j.biocontrol.2011.10.002

Tahir HAS, Gu Q, Wu H, Raza W, Hanif A, Wu L et al (2017) Plant growth promotion by volatile organic compounds produced by *Bacillus subtilis* SYST2. Front Microbiol 8:171

Tapia-Hernández A, Mascarúa-Esparza MA, Caballero-Mellado J (1990) Production of bacteriocins and siderophore-like activity by *Azospirillum brasilense*. Microbios 64:73–83

Tchakounté Tchuisseu VG, Berger B, Patz S, Fankem H, Ruppel S (2018) Community structure and plant growth-promoting potential of cultivable bacteria isolated from Cameroon soil. Microbiol Res 214:47–59. https://doi.org/10.1016/j.micres.2018.05.008

Tenorio-Salgado S, Tinoco R, Vazquez-Duhalt R, Caballero-Mellado J, Perez-Rueda E (2013) Identification of volatile compounds produced by the bacterium *Burkholderia tropica* that inhibit the growth of fungal pathogens. J Bioeng 4:236–243. https://doi.org/10.4161/bioe.23808

Tortora ML, Díaz-Ricci JC, Pedraza RO (2011) *Azospirillum brasilense* siderophores with antifungal activity against Colletotrichum acutatum. Arch Microbiol 193:275–286. https://doi.org/10.1007/s00203-010-0672-7

van Loon LC (2007) Plant responses to plant growth-promoting rhizobacteria. In: Bakker PAHM, Raaijmakers JM, Bloemberg G, Höfte M, Lemanceau P, Cooke BM (eds) New perspectives and approaches in plant growth-promoting rhizobacteria research. Springer, Netherlands, pp 243–254

van Loon LC, Rep M, Pieterse CMJ (2006) Significance of inducible defense-related proteins in infected plants. Annu Rev Phytopathol 44:135–162. https://doi.org/10.1146/annurev.phyto.44.070505.143425

Vandamme P, Goris J, Chen W-M, de Vos P, Willems A (2002) *Burkholderia tuberum* sp. nov. and *Burkholderia phymatum* sp. nov., nodulate the roots of tropical legumes. Syst Appl Microbiol 25:507–512. https://doi.org/10.1078/07232020260517634

Vilchez S, Manzanera M (2011) Biotechnological uses of desiccation-tolerant microorganisms for the rhizoremediation of soils subjected to seasonal drought. Appl Microbiol Biotechnol 91:1297. https://doi.org/10.1007/s00253-011-3461-6

Vivas A, Marulanda A, Ruiz-Lozano JM, Barea JM, Azcón R (2003) Influence of a *Bacillus* sp. on physiological activities of two arbuscular mycorrhizal fungi and on plant responses to PEG-induced drought stress. Mycorrhiza 13:249–256. https://doi.org/10.1007/s00572-003-0223-z

Vyas P, Gulati A (2009) Organic acid production in vitro and plant growth promotion in maize under controlled environment by phosphate-solubilizing fluorescent *Pseudomonas*. BMC Microbiol 9. https://doi.org/10.1186/1471-2180-9-174

Wang J, Li Q, Xu S, Zhao W, Lei Y, Song C et al (2018) Traits-based integration of multi-species inoculants facilitates shifts of indigenous soil bacterial community. Front Microbiol 9:1–13. https://doi.org/10.3389/fmicb.2018.01692

War Nongkhla FM, Joshi SR (2014) Epiphytic and endophytic bacteria that promote growth of ethnomedicinal plants in the subtropical forests of Meghalaya, India. Rev Biol Trop 62:1295. https://doi.org/10.15517/rbt.v62i4.12138

Zahir ZA, Munir A, Asghar HN, Shaharoona B, Arshad M (2008) Effectiveness of rhizobacteria containing ACC deaminase for growth promotion of peas (*Pisum sativum*) under drought conditions. J Microbiol Biotechnol 18:958–963

Zavarzin GA (2008) A planet of bacteria. Her Russ Acad Sci 78:144–151. https://doi.org/10.1134/S1019331608020056

Zhang S (2018) Interkingdom microbial consortia mechanisms to guide biotechnological applications. Microb Biotechnol Early 1–15. https://doi.org/10.1111/1751-7915.13300

Zhang H, Kim M-S, Krishnamachari V, Payton P, Sun Y, Grimson M et al (2007) Rhizobacterial volatile emissions regulate auxin homeostasis and cell expansion in Arabidopsis. Planta 226:839. https://doi.org/10.1007/s00425-007-0530-2

Zhang H, Sun Y, Xie X, Kim MS, Dowd SE, Paré PW (2009) A soil bacterium regulates plant acquisition of iron via deficiency-inducible mechanisms. Plant J 58:568–577. https://doi.org/10.1111/j.1365-313X.2009.03803.x

Zoppellari F, Chitarra W, Lovisolo C, Spanna F, Bardi L (2014) Improvement of drought tolerance in maize (*Zea mays* L.) by selected rhizospheric microorganisms. Ital J Agrometeorol 19:5–18

Chapter 3
Does PGPR and Mycorrhizae Enhance Nutrient Use Efficiency and Efficacy in Relation to Crop Productivity?

Mahipal Choudhary, Vijay Singh Meena, Ram Prakash Yadav, Manoj Parihar, Arunav Pattanayak, S. C. Panday, P. K. Mishra, J. K. Bisht, M. R. Yadav, Mahaveer Nogia, S. K. Samal, Prakash Chand Ghasal, Jairam Choudhary and Mukesh Choudhary

Abstract With the increasing world's population, higher demand for sustainable food production so as to meet the requirement. It has increased tremendously due to excessive use of agrochemicals. Since, the imbalanced application of agrochemicals in agricultural field leads to soil and environmental degradation. Nowadays, the

M. Choudhary (✉) · V. S. Meena · R. P. Yadav · M. Parihar · A. Pattanayak · S. C. Panday · P. K. Mishra · J. K. Bisht
ICAR–Vivekananda Institute of Hill Agriculture, Almora 263601, Uttarakhand, India
e-mail: mahipal.choudhary@icar.gov.in

V. S. Meena
e-mail: vijayssac.bhu@gmail.com

R. P. Yadav
e-mail: rams34052@gmail.com

M. Parihar
e-mail: manojbhu7@gmail.com

A. Pattanayak
e-mail: director.vpkas@icar.gov.in

S. C. Panday
e-mail: sureshpanday39@yahoo.com

P. K. Mishra
e-mail: misrapank12@gmail.com

J. K. Bisht
e-mail: bishtjk@hotmail.com

M. R. Yadav
Rajasthan Agricultural Research Institute, Durgapura Jaipur, 302018, Rajasthan, India
e-mail: mryadavrari@gmail.com

M. Nogia
ICAR-NBSS&LUP, Reginal Centre, Udaipur 313001, India
e-mail: mahi.nogiya3@gmail.com

D. K. Maheshwari and S. Dheeman (eds.), *Field Crops: Sustainable Management by PGPR*, Sustainable Development and Biodiversity 23,
https://doi.org/10.1007/978-3-030-30926-8_3

scientific community has shifted their focus on alternative eco-friendly management approach. The plant growth-promoting rhizobacteria (PGPR) and mycorrhizae has huge potential to substitute agrochemicals. These efficient eco-friendly microbes have different plant growth-promoting (PGP) activities; hence PGPR and mycorrhizae are gaining importance for restoring soil sustainability and agricultural productivity. Application of these efficient microbes in the soil–plant–environment system will be suitable strategies for improving the soil and crop productivity.

Keywords PGPR · Biological nitrogen fixation · Siderophore · Agronomic efficiency

3.1 Introduction

Nowadays, the sustainable crop production along with enhanced crop productivity relics a key global challenge for different communities (i.e., policy makers, business, and researchers) (Wezel et al. 2014). It is undoubtedly hard to believe that food demand is going to be double in 2050 with the rapid population explosion (Henry et al. 2018). It is also that, being fully dependent on artificial as well as traditional inputs (Kumar et al. 2017). Application of efficient microbes for multi-nutrient solubilization enhances the environmental sustainability (Choudhary et al. 2018c). Among these microbes, the rhizobacteria and arbuscular mycorrhiza (AM) has tremendous capabilities to be colonized in the rhizosphere and to enhance plant growth, development. Apart from this, they could improve nutrient use efficiency (NUE) by access to various nutrients from soil system (Meena et al. 2017). The AM plays a role in nutrient supply and uptake in between fungi and roots and is ubiquitous in nature (Smith and Read 2008). The sustainable crop productions required need-based crop yield that can be achieved prepared via multi-resistance and improve water use efficiency (WUE) in plants. AM is most heterogonous group of soil microbes, ubiquitously form plant–fungal association with 80% of higher plants (Giovannetti et al. 2006). Other than penetrating plant roots, AM fungal hyphae also grow in soil as extraradical hyphae with higher absorptive surface

S. K. Samal
ICAR-Research Complex for Eastern Region, ICAR Parisar, Patna 800014, Bihar, India
e-mail: saubhagya.ssac@gmail.com

P. C. Ghasal · J. Choudhary
ICAR–Indian Institute of Farming Systems Research Modipuram, Meerut 250110, Uttar Pradesh, India
e-mail: pcghasal@gmail.com

M. Choudhary
ICAR–ICAR-Indian Grassland and Fodder Research Institute, Jhansi 284003, Uttar Pradesh, India
e-mail: selmukesh@gmail.com

increases nutrient acquisition. Besides better nutrient availability, AM fungi provides multi-resistance (Parihar and Rakshit 2016), soil nutrient loss reduction (Parihar et al. 2019), improved soil aggregation (Borie et al. 2008), greater soil carbon storage (Wang et al. 2016), and enhanced overall plant productivity.

Judicious application of organic and inorganic minerals improve soil sustainability and crop productivity (Choudhary et al. 2018a), while the sole use of minerals have an adverse effect on the crop growth and development as compared to integrated manner. Imbalanced application of synthetic agrochemicals (mainly fertilizers, pesticides, fungicide, and weedicides) leads to soil and environmental degradation. In this article, the current status of NUE, mode of action of PGPRs and AM fungi, their potential use and benefits for farmer community where they are facing lots of problems related to the high cost of chemical fertilizer, pesticides, and other growth substances highlighted. These factors adversely affected the soil and environment, enhanced NUE through PGPRs and AM fungi, hence needed to encourage farmers for adoption of such cost-effective, efficient, and sustainable technologies for better crop growth and healthy environment.

3.2 Nutrient Use Efficiency (NUE)

It is a measure of nutrient uptake by crop/plants from the soil system as growth and development per unit input (Ramesh et al. 2014). NUE is dynamic and complex term and different scientific fraternities (agronomists, soil scientists, plant physiologists, and agricultural economists) are measured in various ways depending upon the perspective in which it is computed and considered. Conceptually, the word "efficiency" implies the achievement of an envisioned outcome with a lowest possible use of precious input (Fixen 2005). Therefore, NUW is considered an intended output (economic crop yield) as the numerator and input (applied nutrient) as the denominator. Thus, NUE is a simple measure of how beneficially plants use the mineral nutrients available in the soil environment. The determination of NUE is of great importance to differentiate plant species and their genotypes based on their ability to utilize the applied nutrients for maximum production of economic yield with minimum use of nutrient (Barbieri et al. 2008). Improved NUE can be helpful for spreading out of crop production over marginal and degraded lands which are potentially low with respect to nutrient availability so as to reduce mineral fertilization (Dobermann and Cassman 2005). NUE of any crop plant is product of (a) uptake efficiency (depends on ability of plant roots to acquire the nutrient present in soil, their inflow–outflow rate into plant roots, nutrient influx kinetics, and also related to the quantity of the subjected nutrient applied or present in soil), (b) incorporation efficiency (largely depends on ability of plant to transports the acquired nutrient to shoot and leaves, functions of shoot parameters), and (c) utilization efficiency (remobilization of nutrient from different plant parts, functions of whole plant (Wang et al. (2010). Fageria et al. (2008) outlined that nutrient efficient plant are those which can produce more biomass with a defined

Table 3.1 Commonly used NUE indices and their application

NUE index	Questions addressed	Typical use
PFP_N	How productive is the given cropping system in comparison to its nutrient input?	Using as a long-term indicator of NUE trends at various scales
PNB	How much of applied nutrient is being taken out (Uptake efficiency) by the crops in the system?	Also used as a long-term indicator of trends of NUE and provides more practical information when combined with soil fertility dataset
ARE_N	How much of the nutrient applied did the plant take up (the ability of plants to take up the applied nutrient from soil)?	Mostly used as an index of the potential for nutrient loss from the cropping system and to access the efficiency of management practices towards NUE
IUE	How the crop plants are able to transform the acquired nutrients into economic products?	Used to evaluate nutrient efficient and inefficient genotypes in breeding Programs
PE_N	What is the ability of the plant to transform nutrients acquired from the source into economic yield?	Research evaluating NUE among cultivars and other cultural practices

Accepted and modified from Dobermann (2007) and Dobermann et al. (2005)

amount of applied or absorbed nutrient over other similar growing environments. Therefore, it is very clear from the above findings that NUE can be defined in many ways depending on the perspective and is easily misunderstood and misrepresented. The commonly used measure of NUE is discussed below as well as in Table 3.1.

Partial factor productivity (PFPN): It is the simplest measure to depict nutrient use efficiency described as units of output (crop yield) per unit of input (nutrient applied) (Mosier et al. 2004).

$$\mathrm{PFPN(kg\ kg)} = \frac{\text{Crop Yield (kg)}}{\text{Amount of Nutrient plant tissuse (kg)}}$$

Agronomic efficiency (AE_N): AE_N is the most commonly used index, defined as units increase in economic yield per unit inputs (Fageria et al. 2008). To calculate AEN, one essentially needs data on yield from unfertilized plot (without nutrient input). It is critically required to establish a plot with zero nutrient input on the farm (Tilman et al. 2002).

$$\mathrm{AEN(kg\ kg)} = \frac{\text{Yield of fertillized plot (kg)} - \text{Yield of unfertillized plot (kg)}}{\text{Amount of nutrient applied (kg)}}$$

Physiological Efficiency (PEN): Physiological growth in relation to the increase in nutrient uptake by the crop. Like AEN and AREN, it requires knowledge of without application of the nutrient.

$$\mathrm{PEN(kg\,kg)}=\frac{\mathrm{BYf\,(kg)-BYuf\,(kg)}}{\mathrm{NUf\,(kg)-NUuf\,(kg)}}$$

where BYf = Biomass yield (grain + straw) in fertilized plot, BYuf = Biomass yield (grain + straw) in unfertilized plot, NUf = Nutrient uptake in fertilized plot and NUuf = Nutrient uptake in unfertilized plot.

Apparent Recovery Efficiency (AREN)-: It is one of the more complex forms of NUE expressions, described by the difference in nutrient uptake (above-ground biomass of crops) between the fertilized plot and that of unfertilized plot relative to the quantity of input applied.

$$\mathrm{AREN(\%)}=\frac{\mathrm{NUf\,(kg)-NUuf\,(kg)}}{\mathrm{Amount\,applied\,(kg)}}\times 100$$

where NUf = nutrient uptake in grain plus straw of fertilized plot and NUuf = nutrient uptake in grain plus straw of unfertilized plot.

Internal Utilization Efficiency (IEU): It is a simple indexing measure of quantifying nutrient use efficiency based on crop yield and nutrient uptake. The value of this index is depending upon agro-climatic conditions, crop cultivar, and level of soil–plant management (Witt et al. 2005). Nutrient utilization efficiency is the cross product of physiological efficiency and apparent recovery efficiency of nutrient which is calculated by using the following equation:

$$\mathrm{IUE}\;\left(\mathrm{kgkg^{-1}}\right)\;=\;\mathrm{PE_N}\;\times\;\mathrm{ARE_N}$$

where PEN is physiological efficiency and AREN is apparent recovery efficiency as defined above.

Partial nutrient balance (PNB): It is the simplest form to characterize NUE of crop production system. It is described as nutrient output in relation to nutrient input (Lopez-Bellido and Lopez-Bellido 2001). Over the short term and on individual farms, PNB can show substantial fluctuations due to cash flow and market conditions, especially for P and K. Long-term assessment of PNB over several years is, therefore, more useful (Table 3.2).

3.3 Current Scenario of NUE

Nowadays, an increase in global food demand ~110% is expected by 2050 compared to 2005, requiring a tremendous increase in production (Tilman et al. 2011). Approaches considered economic, social, and environmental dimensions are essential to sustainable agricultural systems and therefore provide an appropriate context for specific NUE indicators. Global phosphorus consumption continuously has been increased since 1960 (Fig. 3.1). Phosphorus is the second important

Table 3.2 Components of NUE and processes that influence genotypic differences in NUE in plants

A. Nutrient acquisition
• 1. Nutrient availability from soil system to roots through diffusion and/or mass flow: soil buffering power, nutrient density, bio–physico–chemical properties of element, tortuosity, soil moisture content, bulk density, and soil temperature
• 2. Root morphological factors: number and length of roots, root hair density and its efficiency, root forage area, and root density
• 3. Rhizosphere engineering
• 4. Physiological: root: shoot ratio, rhizosphere microorganisms dynamics (rhizobia, azotobacter, mycorrhizal fungi), concentration of acquired nutrient, transpiration rate (water Uptake), internal and outflow nutrient rates, nutrient movement within roots and shoots, affinity to uptake (Km), threshold concentration (Cmin)
• 5. Biochemical: secretion of various enzymes (phosphatase), chelating compounds, phytosiderophores, proton exudates, organic acid exudates
B. Nutrient movement in root
• 1. Transfer of acquired nutrient through endodermis and transport within the root
• 2. Compartmentalization/binding of absorbed nutrient within roots
• 3. Rate of nutrient release to root and shoot xylem
C. Nutrient accumulation and remobilization in shoot
• 1. Demand for given nutrient at the cellular level and their storage in vacuoles
• 2. Remobilization of stored nutrient from older to younger leaves and from vegetative to reproductive tissues as and when needed
• 3. Dynamics of chelates during transportation of nutrient through the xylem
D. Nutrient utilization and growth
• 1. Nutrient metabolism at cellular level
• 2. Concentration of nutrient in supporting structure, particularly stem
• 3. Elemental substitution (such as Na for K, Fe for Mn etc.)
• 4. Biochemical processes and factors involved in metabolism (nitrate reductase for N-use efficiency, glutamate dehydrogenase for N metabolism, peroxidase for Fe efficiency, pyruvate kinase for K deficiency, metallothionein for metal toxicities, ascorbic acid oxidase for Cu, carbonic anhydrase for Zn)

Modified from Baligar et al. (2001); Fageria and Baligar (2005)

primary nutrient element after nitrogen. Low P use efficiency (PUE) is a significant challenge for agricultural production on P-deficient soil as well as in acidic and calcareous soils (Shenoy and Kalagudi 2005). Acquisition of soil and fertilizer P by crops depends on soil and plant properties.

On the other hand, K use efficiency is more limited than either N or P. This is partly due to the environmentally benign nature of K where interest in efficiency is driven primarily by agronomic or economic factors. In general, NUE of sulfur is only 8–10% due to immobilization and leaching with water whereas in case of micronutrients NUE is only 2–5% due to fixation in soil.

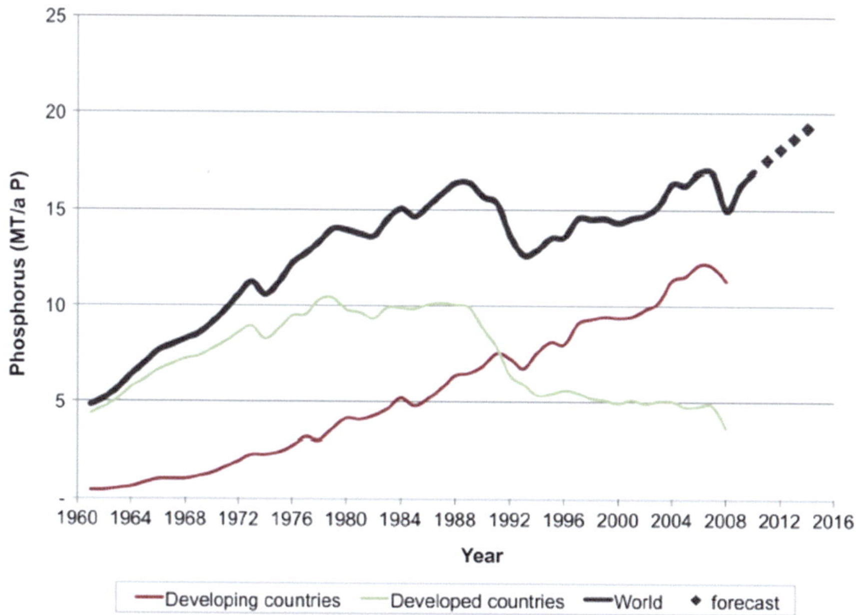

Fig. 3.1 Global phosphorus fertilizer consumption (International fertilizer industry association)

3.4 Why NUE Is Important for Agricultural System?

Nutrient deficiency is critically important and most yield-limiting constraints to crop production in the agro-ecological (Fan et al. 2004). The adoption of intensive chemical fertilizer led nutrient management during the past four to five decades undoubtedly reported to increase global food production around four times but has also increased chemical fertilizer consumption by fourfold which created implications for the environmental safety (Tilman et al. 2001). Overall, the contribution of chemical fertilizers in raising crop yields is $\sim$40% in poor fertile soils when other productions factors (genotypes, irrigation and plant protection measures) are at an adequate level (Rahimizadeh et al. 2010). However, it is estimated that chemical fertilizer inputs in modern agriculture share about 30% of the total cost of production system (Richardson et al. 2011).

Ensuring food and fodder for human and livestock on sustainable systems approach has significant challenges and is highly critical especially in thhhe developing region of the world. Therefore, to combat this challenge, farmers are forced to overuse agrochemicals that leads in the deteriorating soil and environment (Foley et al. 2011). Long-term adoption of nutrient management practices dominated by chemical fertilizers led to declining NUE making fertilizer consumption uneconomical, imparting adverse effects on environment and groundwater quality (Hungate et al. 2003). Furthermore, the ANR of inputs is quite low such as global cereals N recovery efficiency is ~33%. Likewise, the global recovery efficiency of PK is estimated less than 20% and 40%, respectively (Syers et al. 2008). Thus, it is very clear that usually a large quantum of applied nutrient through the precious inorganic fertilizers is lost from the soil–plant system or unavailable to crops. The lower efficiency of applied fertilizer is attributed to leaching and runoff, gaseous losses, fixation by soil, and use of inefficient nutrient absorbing/utilizing plant species or cultivars (Yadav et al. 2017).

Eutrophication, a process of water enrichment with chemicals (NP) can cause excessive algal growth of aquatic algal communities is another consequence associated with lower NUE (Garnett et al. 2009). This can cause a shortage of oxygen, and may produce substances, which are directly toxic to aquatic communities and indirectly to livestock and humans, is now big threat throughout the world (Baligar et al. 2001). Further, ammonia emitted to the atmosphere through volatilization losses from agricultural fields and other anthropogenic activities can return back to earth surface as co-deposition with sulfur dioxide gas (Buresh et al. 2004). Aquatic and forests ecosystems of the world are more prone to N deposition and excess enrichment of N can cause disruption in ecosystem functions and services. Larger N deposition in ecosystems can also lead to global warming due to larger emissions of nitrogen-based greenhouse gases, soil acidification due to excessive aluminum dissolution and reduced carbon stocks in the soil (Galloway et al. 2008). Nitrous oxide (N_2O) generated as a product of denitrification process considered as an important N based GHG gas which is responsible for around five percent of the total global climate change (Shoji et al. 2001).

Inefficient and excessive use of N-based fertilizers leads in environmental problems related due to associated large scale emission of NH_3, N_2, and N_2O to the atmosphere (Rockstorm et al. 2009). Further, improved NUE can play a great role in extending the crop production over marginal and degraded lands, which potentially low with respect to nutrient availability (Fig. 3.2).

3.5 What Are PGPR and Mycorrhizae?

PGPR is a group of rhizobacteria that inhabited in the rhizosphere, the term PGPR refers to rhizobacteria that colonize in the rhizosphere (Vejan et al. 2016). Under the PGPR, wide range of genera of bacterial species (*Pseudomonas*, *Alcaligens*, *Azospirillum*, *Arthrobacter*, *Burkholderia*, *Azotobacter*, *Bacillus*, *Klebsiella*, *Serratia*, *Enterobacter*) reported to improving growth and development (Saharan and Nehra 2011). The example of various mechanisms involved in growth and development by rhizobacteria has been reported by different researchers (Table 3.3).

Mycorrhizae define a mutualistic symbiotic beneficial relationship between the root of a plant (mainly woody plants) and a fungus that colonizes the plant root. In many plants, mycorrhizae are fungi that grow inside the plant's roots, or on the surfaces of the roots (Smith and Read 2008).

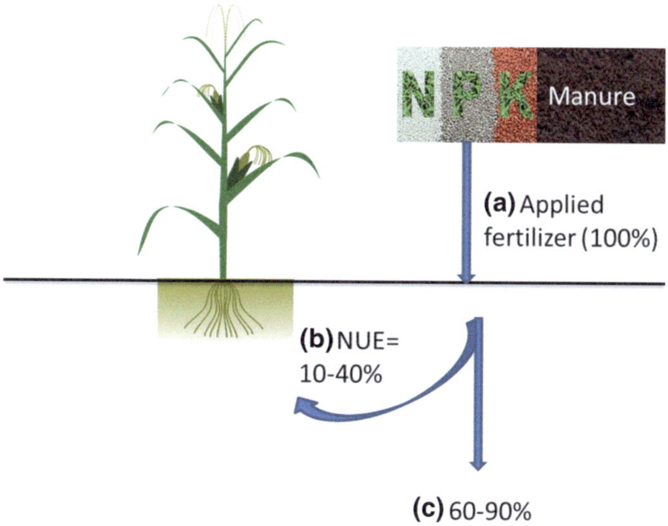

Fig. 3.2 Model for improved plant NUE with inoculants **a** total amount of fertilizer or manure applied to plants, **b** 10–40% of the applied fertilizer or manure is taken up by plants, and **c** 60–90% of the applied fertilizer or manure is lost

3.6 Mode of Action of Plant Growth-Promoting Rhizobacteria (PGPR)

3.6.1 Direct Mechanism

The direct mechanism of efficient PGPR strain entails either providing the plant with PGP traits that are synthesized by the bacterium. Biofertilizers contribute to plant nutrition both by facilitating nutrient uptake and by enhancing mineral availability by direct mechanisms, as fixing atmospheric nitrogen, solubilizing mineral nutrients like phosphorus, potassium, etc., mineralizing organic compounds, producing phytohormones and production of siderophores which sequestering micronutrient (Kaur et al. 2016).

3.6.2 Indirect Mechanism

The indirect mechanism of plant growth occurs has an insignificant role in plant nutrient management practices. Infect, the PGPR prevent deleterious effects of one or more phytopathogenic microbes (Kundan et al. 2015).

Table 3.3 Different plant growth-promoting rhizobacteria (PGPR) tested for various crop types

PGPRs	Host crop	Results of the addition of bacteria to crops	References
Achromobacter xylosoxidans	*Vigna radiata*	Influence plant homeostasis	Ma et al. (2009)
Azorhizobium		Nitrogen fixation	Sabry et al. (1997)
Azospirillum	–	Nitrogen fixation	Sahoo et al. (2014)
Acinetobacter spp.	–	IAA, phosphate solubilization, siderophores	Rokhbakhsh-Zamin et al. (2011)
Bacillus sp. *PSB10*	–	IAA, siderophores, HCN, ammonia	Wani and Khan (2010)
Bacillus	–	Potassium solubilization	Han et al. (2006)
Enterobacter sakazaki 8MR5	–	Inoculation increased growth parameters	Babalola et al. (2003)
Rhizobium Pseudomonas	*Medicago sativa, Trigonella foenum-graecum, Trigonella* sp., *Trifolium* sp., *Vigna radiata*	Biological nitrogen fixation	Maheshwari et al. (2010), Choudhary et al. (2017a)
P. fluorescens C7	*Arabidopsis thaliana* plants	The ironpyoverdin synthesized, it increased the iron level inside the plants and improved their growth	Vansuyt et al. (2007)
Xanthomonas sp. *RJ3, Pseudomonas* sp. *RJ10, Bacillus RJ31, Azomonas* sp. *RJ4*	*Brassica napus*	Stimulated plant growth and increased cadmium accumulation	Sheng and Xia (2006)
Rahnella aquatilis	–	Phosphate solubilization, IAA, ACC Deaminase	Mehnaz et al. (2010)
Bradyrhizobium MRM6	*Vigna radiate* Soybean, wheat	When herbicide-tolerant *Rhizobium* strain MRP1 was used with herbicide, it increased the growth parameters at all tested concentrations of herbicides (quizalafop-p-ethyl and clodinafop)	Ahemad and Khan (2012a)

(continued)

Table 3.3 (continued)

PGPRs	Host crop	Results of the addition of bacteria to crops	References
Pseudomonas sp. PS1	*Vigna radiate*	Significantly increased plant dry weight, nodule numbers, total chlorophyll content, leghaemoglobin, root N, shoot N, root P, shoot P, seed yield, and seed protein	Ahemad and Khan (2012b)
Pseudomonas aeruginosa	*Cicer arietinum*	Positively stimulate potassium and phosphorus uptake	Ahemad and Kibret (2014)
Pseudomonas cepacia	*Cucumis sativus*	Prevent pathogens in *Pythium ultimum*	Montano et al. (2014)
	Gossypium hirsutum	Help fight the *Rhizoctonia solani* virus	
Pseudomonas fluorescens	*Triticum aestivum* *Hordeum vulgare*	Help prevent *Fusarium culmorum*	Santoro et al. (2016)
	Pennisetum glaucum	Significant increase in plant height, ear length, yield	Arora (2003)
	Camellia sinensis	Increased NUE (+7%); nitrogen (+52%); phosphorus (+67%); potassium (+18%)	Thomas et al. (2010)
Pseudomonas sp.	–	Phosphate solubilization, IAA, siderophore, HCN, biocontrol potentials	Tank and Saraf (2009)
Bradyrhizobium sp. *750* *Pseudomonas* sp., *Ochrobactrum cytisi*	*Lupinus luteus*	Increased both biomass and nitrogen content, accumulation of metals (phytostabilization potential)	Dary et al. (2010)
AM Fungi (*Coccus* DIM7, *streptococcus* PIM6) and PSB (*Bacillus* sp. PIS7)	*Zea mays*	Rock phosphate mineralization in soil and improved all growth parameters including shoot (56%), root yield (52%), height (41%), N (80%), and P (91%) uptake by the maize plants as compared to control	Wahid et al. (2016)

(continued)

Table 3.3 (continued)

PGPRs	Host crop	Results of the addition of bacteria to crops	References
Pseudomonas sp. *Paenibacillus Polymyxa*	*Piper nigrum Zea mays*	Significantly increased soil enzyme activities, total productivity, and nutrient uptake. Significantly increased the biomass of plants and elicited induced systemic resistance against bacterial spot pathogen *Xanthomonas axonopodis* pv. *Vesicatoria* untreated plants.	Sharma et al. (2011)
Pseudomonas and *Azospirillum*	*Piper nigrum*	Phosphate solubilization and increases the availability of phosphorus	Ramachandran et al. (2007)
Azotobacter chroococcum	*Brassica juncia*	Stimulated plant growth	Orlandini et al. (2014)
	Triticuma estivum	Phosphate solubilization	Bhattacharyya and Jha (2012)
	Fagopyrum esculentum	Biological nitrogen fixation	
Azospirillum brasilense	*Festuca arundinacea*	Increases biomass and increase plant tolerance to polycyclic aromatic hydrocarbons	Orlandini et al. (2014)
	Saccharum officinarum	Alter plant root architecture by increasing the formation of lateral and adventitious roots and root hairs	
	Camellia sinensis	Increased NUE (+13%); nitrogen (+65%); phosphorus (+25%); potassium (+14%)	Thomas et al. (2010)
AM fungi, *Azospirillum, Pseudomomas* sp.	*Pinus sabiniana, Solanum lycopersicum, Lactuca sativa, Triticum aestivum, Zea mays*	Roots stimulate root colonization, limits soil salinity stress, and affects plant yield,	Kohler et al. (2010),

(continued)

Table 3.3 (continued)

PGPRs	Host crop	Results of the addition of bacteria to crops	References
Acinetobacter sp.	*Oryza sativa*	Significantly increased the growth and yield parameters of rice as well as also increases the solubility of Zn from source of zinc such as ZnO and $ZnCO_3$ with inoculation	Gandhi and Muralidharan (2016)
Bacillus sp.	*Glycine max*	Phosphate solubilization and increases the availability of phosphorus	Wahyudi et al. (2011)
Bacillus polymyxa	*Lycopersicon esculentum*	Proline accumulation improved the physiological and biochemical parameters of plants	Shintu and Jayaram (2015)
Bacillus subtilis	*Brassica juncia*	Facilitate Nickel accumulation,	Prathap and Ranjitha (2015), Oyedele et al. (2014)
	Hordeum vulgare	Prevent powdery mildew	
	Gossypium hirsutum	Prevent from *Meloidogyne incognita* and *M. arenaria*	
Bacillus subtilis Pseudomonas aeruginosa	*Solanum lycopersicum, Abelmoschus esculentus, Amaranthus*	Dry biomass increased 31% for tomato, 36% for okra 83% for African spinach	Adesemoye et al. (2008)
PSB *(P. fluorescens* BAM-4 and *B. cepacia* BAM-12 and AM fungus *(G. etunicatum)*	*Triticum aestivum*	Significantly improved the plant growth and nutrient uptake and showed a significant increase in grain yield of wheat with the increases the availability of P from insoluble P sources.	Minaxi et al. (2013)
Azotobacter and *Bacillus*		Wheat inoculation increased its seeds yield of 30 and 43%, respectively. This increase due to the production of certain	Kloepper et al. (1991)

(continued)

Table 3.3 (continued)

PGPRs	Host crop	Results of the addition of bacteria to crops	References
		growth hormones such as IAA	
Enterobacter strain CIK-521R and *Klebsiella* strain CIK-518	*Zea mays*	Higher tolerance to Cd and thus could be deployed to manage Cd-contaminated soil. Both of the strains could be effective inoculants to get higher maize production in normal as well as in Cd-contaminated soils.	Ahmad et al. (2016)

3.7 Mode of Action by Mycorrhizae

AM fungi are obligate biotrophs depends on the plant for their carbon nutrition with the exchange of different nutrients (Ortas and Rafique 2017), have the ability to improve the availability of nutrients for plants growth and development (Teotia et al. 2017). The greater nutrient acquisitions by mycorrhizae inoculated plants are possibly due to (i) AM fungi-mediated nutrient mobilization and acquisition and (ii) transfer of nutrients from fungal hyphae to plant roots. Mycorrhizal pathways of nutrient uptake in plants root are rapid transit and well-regulated system. AM fungi form a large web of extraradical mycelium (ERM) which can explore a huge volume of soil beyond the nutrient depletion zone and facilitate uptake of inaccessible and mineral bound nutrients. Several studies revealed that mycorrhiza could be engaged in organic matter decomposition and relocate nutrients to host plants (Reynolds et al. 2005). Other than organic matter, AM fungi as alone or in alliance with other microbes are also take part in nutrient mobilization by mineral weathering. Some reports suggest that mycorrhizae produces low molecular organic compounds which solubilize rock bounds nutrients and enhance their absorption (Smith and Read 2008) as well as fungal ability to reduce different nutrient losses (runoff, leaching and volatilization) from the soil system (Parihar et al. 2019).

3.8 Mechanism of PGPR for Increasing Bioavailability of Nutrients and Crop Productivity

3.8.1 Nitrogen Fixation

Nitrogen (N) is one of the most important nutrients from plant nutrition and productivity point of view. Despite large reservoir of nitrogen in atmosphere, plants are unable to take it directly due to the high energy requirement to break the triple bond

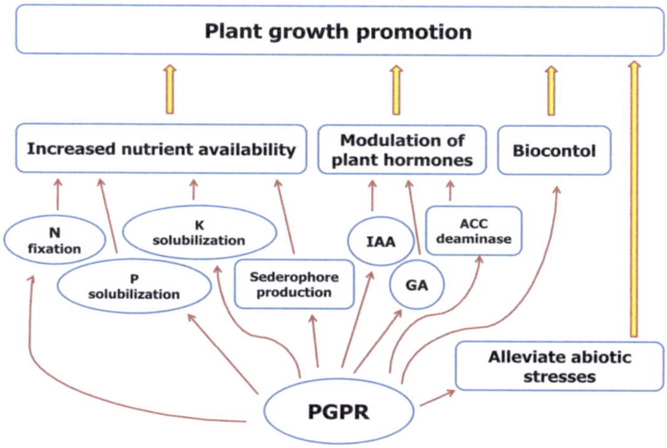

Fig. 3.3 Schematic representation of various mechanisms involved in plant promotion by PGPR

of nitrogen. Plants avail it through biological N_2 fixation (BNF) by changing nitrogen to ammonia by the help of N-fixing microbes with involvement of nitrogenase (Kim and Rees 1994). Two types of BNF are found in nature depending on types of microorganism involved, first, symbiotic nitrogen fixation, in which plant and microbes in symbiotic relationship and fix nitrogen (Ahemad and Khan 2012a), second, associative and nonassociative, non-symbiotic nitrogen fixation, which involves free-living microbes (Choudhary et al. 2017b). Hence, symbiotic N-fixation plays a major role in N nutrition of plant (Fig. 3.3). At the beginning of the process legume plant secret a specific flavonoid compound and rhizobium specific to that compound attracted towards the plant root. The bacteria attach firmly themselves to the root hair that involves lectin fibrils of the host plant. The host plant then senses the nod factor secreted by rhizobium leading to colonized root hairs to curl.

3.8.2 *Phosphorus Solubilization and Mineralization*

Phosphorus (P) is next to nitrogen among the essential nutrients that limit crop growth (Tak et al. 2012) and is present in huge amount in the soil in form of both inorganic and organic P. Besides its abundance in soil, it remains in insoluble form, thus its availability to plant is limited. The plant can absorb P in two of its soluble forms {monobasic ($H_2PO_4^-$) and diabasic ($H_2PO_4^{2-}$) ions}. The insoluble P remains in form of inorganic mineral, e.g., apatite and organic P, largely as inositol phosphate (Glick 2012). In the soil environment a considerable number of beneficial microbes are capable of releasing unavailable soil through P-solubilization (Bhattacharyya and Jha 2012). These groups of microbes are termed as

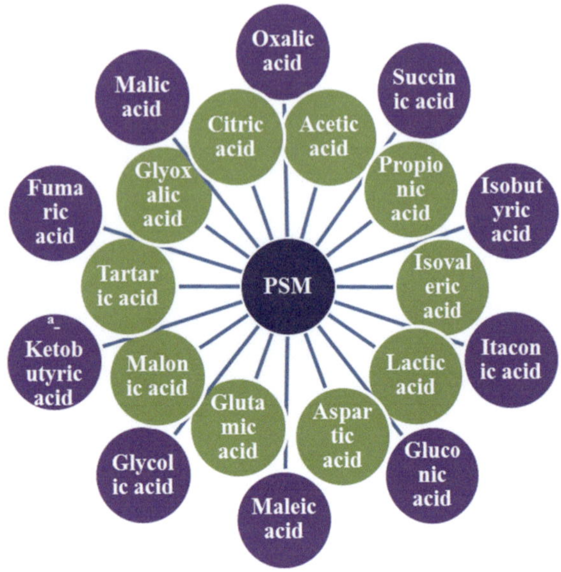

Fig. 3.4 Schematic representation of numerous organic acids released by PSM

P-solubilizing microbes (PSM). Many efficient bacteria (*Azotobacter*, *Bacillus*, *Enterobacter, Beijerinckia, Burkholderia, Erwinia, Microbacterium, Pseudomonas, Flavobacterium, Rhizobium and Serratia*), fungi, Actinobacteria, and algae show P-solubilization (Fig. 3.4). The major soluble form of inorganic phosphate is $H_2PO_4^-$, which found its existence at lower pH.

The existence of each category of the enzyme depends upon soil pH condition. Acid phosphatases are prevalent in acid soils, whereas, alkaline phosphatases are abundantly found in neutral and alkaline soils (Renella et al. 2006). Another enzyme involved in the process of organic P mineralization that is produced by PSM is phytase.

3.8.3 Potassium Solubilization

Potassium (K) is the third major essential plant nutrient affecting plant growth. Soil contains a large amount of K than any other nutrient, but its phytoavailability is limited. Potassium is present in four different forms in the soil such as mineral K, non-exchangeable K, exchangeable K, and solution K. In general, 90–98% of soil K remains in form of mineral K, hence, remains unavailable to plant (Sparks and Huang 1985). Among these microbes, the frequently studied by the researcher are K solubilizing bacteria (KSB), namely, *B. mucilaginosus, B. edaphicus* and *B.circulanscan* are effective in K solubilization (Meena et al. 2014, 2016). The major mechanisms involved are (i) lowering of pH; (ii) chelation of the cations bound to K; (iii) acidolysis (Meena et al. 2014; Maurya et al. 2014).

3.8.4 Siderophore Production

Iron (Fe) is an important essential micronutrient as it acts as a cofactor of many enzymes and is required in many physiological processes of plants such as N_2 fixation, photosynthesis, respiration. However, in aerobic environment, it occurs principally in the form of Fe^{3+} and is very likely to form insoluble hydroxides and oxyhydroxides, thus making it unavailable to both plants and microbes (Rajkumar et al. 2010). Microbes and plants acquire Fe through production of low molecular mass chelators termed as siderophore. In bacteria, synthesis of siderophores is induced by the low level of Fe^{3+}. Siderophore released by microbes form complex with iron (Fe^{3+}), further this complex is reduced to Fe^{2+} on bacterial membrane.

3.8.5 Plant Hormone Production

Plant growth is strongly influenced by the phytohormones through their various activities. Microbes also produce plant hormones; thus, affect plant growth through modulating the phytohormone level. Plant growth promotion by PGPR occurs through bacterial synthesis of hormones like indole-3-acetic acid (IAA), cytokinin, and gibberellins (GA) and breakdown of plant produced. Auxins in plant promote plant growth through their function such as cell division, extension, and differentiation. Microbial produced IAA change plant auxin pool, thus promote plant growth by interfering with above physiological processes.

3.8.6 Alleviate Abiotic Stress

Abiotic stresses are major sources of crop yield reduction. Extensive study has been done regarding the mechanisms of alleviating abiotic stresses in plants by PGPR Moreover, PGPR also promote plant growth when it is used as biofertilizers (Vessey 2003).

3.9 Mechanism of Mycorrhizae for Increasing Bioavailability of Nutrients and Crop Productivity

Like PGPR, mycorrhiza also enhances plan growth through various mechanisms. Between the two common types of mycorrhizae (AM and ECM) found in nature, AM are probably the dominant fungi that are commonly found in agricultural soils (Willis et al. 2013). These fungi form haustoria-like structure called arbuscule by penetrating into root cortical cells; the arbuscular function as a mediator for the

Table 3.4 Function of mycorrhizae in promotion of plant growth under stress conditions

Mycorrhizae	Type of stress	Stress alleviation and plant growth promotion trait	References
Glomus fasciculatum and *G. macrocarpum*	Semi-arid wasteland	Mycorrhizae decreased the alkalinity of rhizosphere and high concentration of P, K, Cu, and Zn was observed in inoculated plants	Giri et al. (2005)
Glomus spp.	Heavy metal	Plant height, basal diameter, seedling biomass, and superoxide-dismutase activity was more in mycorrhizal plants. Significant high lead concentration was observed in mycorrhizal plants roots	Zhang et al. (2010)
Glomus mosseae, Glomus intraradices	Heavy metal	AM fungus increased the infection of sunflower root and also increased the pollution tolerance and yield of sunflower in a degraded soil	Adewole et al. (2010)
Glomus intraradices	Water stress	Mycorrhizae protected the plant from drought. Higher leaf water potential was recorded in inoculated plants and kept the plant protected against oxidative stress	Porcel and Ruiz-Lozano (2004)
Glumos intraradices	Water stress	Drought caused a negative impact on sorghum length, shoot dry matter, 1000 kernel weight and yield. Mycorrhizae inoculation dilutes the negative impact of stress and enhanced yield. Grain yield increased 17.5% due to inoculation compared to drought	Alizadeh et al. (2011)
Glomus mosseae	*Salinity stress*	Plant salt tolerance increased in mycorrhizal plants mainly due to elevated levels of superoxide-dismutase, catalase, ascorbate peroxidase, and peroxidase which degraded reactive oxygen species and alleviated membrane damage	He et al. (2007)
Glomus etunicatum	Salinity Stress	Mycorrhizal inoculated plants grow better in saline conditions than uninoculated plants. Fresh and dry weight, root colonization, and proline contents were more in salt pretreated fungus that than non-salt-pretreated fungus	Sharifi et al. (2007)

interchange of metabolites in between fungus and host cytoplasm. The proliferated hypae of mycorrhizal increases the access of roots to a greater soil volume that facilitates host plant to acquire more nutrients than by non-mycorrhizal plants (Guo et al. 2010). Mycorrhizae can enhance the phytoavailability of slowly diffusing ions like phosphate (McArther and Knowles 1993). In addition to improving P nutrition, AM fungi are also capable of increasing availability of nutrients like N, K, Mg, Cu, and Zn, specifically where they exist in less soluble forms in soil (Meding and Zasoski 2008). The mycorrhizal mycelium also contributes to forms stable soil aggregates (Singh 2012) (Table 3.4).

3.10 Future Perspective

In the current scenario, due to abundant use of synthetic chemicals on crops, the sustainability of agriculture systems has distorted; the cost of cultivation has increased at a high rate; the income of farmers stagnated; and the provision of food security and safety has become a frightening challenge (Panday et al. 2018). Imbalanced application of agrochemicals leads to soil degradation (Choudhary et al. 2018b). For these reasons, PGPR that make use of microbes to improve soil sustainability has been recognized (Suhag 2016). PGPR add nutrients through the natural processes of N-fixation, nutrient solubilization, and PGP traits.

3.11 Conclusions

To know about the role of NUE by efficient PGPR aims to increase food security, and sustainable plant productivity, while maintaining environmental quality. However, to achieve this, basic and strategic studies must be undertaken to improve our knowledge of microbial interactions in the rhizosphere. Considering the good impact of PGPR exerts a positive influence on crop productivity and ecosystem functioning, encouragement should be given to its implementation in sustainable agriculture.

Acknowledgments We are thankful to the editors and anonymous reviewers for their productive comments, which help us to improve the manuscript.

Conflict of Interest: The author(s) have no conflict of interest.

References

Adesemoye AO, Obini M, Ugoji EO (2008) Comparison of plant growth promotion with *Pseudomonas aerugenosa* and *Bacillus subtilis* in three vegetables. Brazilian J Microbiol 39:423–426

Adewole MB, Awotoye OO, Ohiembor MO, Salami AO (2010) Influence of mycorrhizal fungi on phytoremediating potential and yield of sunflower in Cd and Pb polluted soils. J Agric Sci 55:17–28

Ahemad M, Khan MS (2012a) Evaluation of plant-growth promoting activities of rhizobacterium *Pseudomonas putida* under herbicide stress. Ann Microbiol 62:1531–1540

Ahemad M, Khan MS (2012b) Effect of fungicides on plant growth promoting activities of phosphate solubilizing Pseudomonas *putida* isolated from mustard (*Brassica compestris*) rhizosphere. Chemosphere 86:945–950

Ahemad M, Kibret M (2014) Mechanisms and applications of plant growth promoting rhizobacteria: current perspective. J King Saud Uni Sci 26:1–20

Ahmad I, Akhtar MJ, Asghar HN, Ghafoor U, Shahid M (2016) Differential effects of plant growth-promoting rhizobacteria on maize growth and cadmium uptake. J Plant Growth Regul 35:303–315

Alizadeh O, Zare M, Nasr AH (2011) Evaluation effect of Mycorrhiza inoculate under drought stress condition on grain yield of sorghum (*Sorghum bicolor*). Adv Environ Biol 5:2361–2364

Arora DR (2003) Text book of microbiology. CBS Publisher, New Delhi, pp 4–48

Babalola OO, Osir EO, Sanni A, Odhaimbo GD, Bulimo WD (2003) Amplification of 1-aminocyclopropane-1-carboxylic (ACC) deaminase from plant growth promoting rhizobacteria in Striga -infested soils. African J Biotechnol 2:157–160

Baligar VC, Fageria NK, He Z (2001) Nutrient use efficiency in plants. Commun Soil Sci Plant Anal 31:921–950

Barbieri P, Echeverría HE, Saínz Rozas HR, Andrade FH (2008) Nitrogen use efficiency in maize as affected by nitrogen availability and row spacing. Agron J 100:1094–1100

Bhattacharyya PN, Jha DK (2012) Plant growth-promoting rhizobacteria (PGPR): Emergence in agriculture. World J Microbiol Biotechnol 28:1327–1350

Borie F, Rubio R, Morales A (2008) Arbuscular mycorrhizal fungi and soil aggregation. In: *segundo simposio internacional suelos, ecología y medioambiente universidad de la frontera* (15)

Buresh RJ, Rowe EC, Livesley SJ, Cadisch G, Mafongoya P (2004) Opportunities for capture of deep soil nutrients. In: Below-ground interactions in tropical agroecosystems: concepts and models with multiple plant components, pp. 109–125

Choudhary M, Panday SC, Meena VS (2017a) Azolla's cultivation and its uses in mountain ecosystem. Indian Farm 67(09):09–12

Choudhary M, Patel BA, Meena VS, Yadav RP, Ghasal PC (2017b) Seed bio-priming of green gram with *Rhizobium* and levels of nitrogen and sulphur fertilization under sustainable agriculture. Legume Res LR-3837:1–6

Choudhary M, Panday SC, Meena VS, Singh S, Yadav RP, Mahanta D, Mondal T, Mishra PK, Bisht JK, Pattanayak A (2018a) Long-term effects of organic manure and inorganic fertilization on sustainability and chemical soil quality indicators of soybean-wheat cropping system in the Indian mid-Himalayas. Agric Ecosyst Environ 257:38–46

Choudhary M, Panday SC, Meena VS, Yadav RP, Singh S, Mahanta D, Pattanayak A, Bisht JK (2018b) Effect of long-term fertilization and manure on soil organic carbon fraction and micronutrient status after harvest of wheat under soybean-wheat cropping system. In: Tripathi AK et al (eds) Abstracts proceeding international conference on "Sustainability of smallholder agriculture in develop-ing countries under changing climatic scenario". CSAUAT, Kanpur (UP), India, p 98

Choudhary M, Ghasal PC, Yadav RP, Meena VS, Mondal T, Bisht JK (2018c) Towards Plant-Beneficiary Rhizobacteria and Agricultural Sustainability. In: Meena V. (eds.) Role of Rhizospheric Microbes in Soil. Springer, Singapore. https://doi.org/10.1007/978-981-13-0044-8_1

Dary M, Chamber-Pérez MA, Palomares AJ, Pajuelo E (2010) In situ phytostabilisation of heavy metal polluted soils using Lupinus luteus inoculated with metal resistant plant-growth promoting rhizobacteria. J Hazard Mater 177:323–330

Dobermann A (2005) Nitrogen use efficiency-state of the art. In: Proceedings of IFA International Workshop on Enhanced-efficiency Fertilizers, Frankfurt, Germany, June 28–30(2005):1–18

Dobermann A (2007) Nutrient use efficiency—measurement and management. IFA international workshop on fertilizer best management practices. Belgium, Brussels, pp 1–28

Dobermann A, Cassman KG (2005) Cereal area and nitrogen use efficiency are drivers of future nitrogen fertilizer consumption. Sci China 48:745–758

Fageria NK, Baligar VC (2005) Enhancing nitrogen use efficiency in crop plants. Adv Agron 88:97–185

Fageria NK, Baligar VC, Li YC (2008) The role of nutrient efficient plants in improving crop yields in twenty first century. J Plant Nutr 31:1121–1157

Fan X, Li F, Liu F, Kumar D (2004) Fertilization with a new type of coated urea: Evaluation for nitrogen efficiency and yield in winter wheat. J Plant Nutr 27:853–865

Fixen PE (2005) Understanding and improving nutrient use efficiency as an application of information technology. In: Proceedings of the symposium on information technology in soil fertility and fertilizer management, a satellite symposium at the XV international plant nutrient colloquium, 14–16 September 2005, Beijing, China

Foley JA (2011) Can we feed the world & sustain the planet? Sci Am 305(5):60–65

Galloway JN, Townsend AR, Erisman JW et al (2008) Transformation of the nitrogen cycle: recent trends, questions and potential solutions. Science 320:889–892

Gandhi A, Muralidharan G (2016) Assessment of zinc solubilizing potentiality of *Acinetobacter* sp. Isolated from rice rhizosphere. Euro J Soil Biol 76:1–8

Garnett T, Conn V, Kaiser BN (2009) Root based approaches to improving nitrogen use efficiency in plants. Plant, Cell Environ 32(9):1272–1283

Giovannetti M, Avio L, Fortuna P, Pellegrino E, Sbrana C, Strani P (2006) At the root of the wood wide web. Plant Signal Behavior 1:1–5

Giri B, Kapoor R, Mukerji KG (2005) Effect of the arbuscular mycorrhizae *Glomus fasciculatum* and *G. macrocarpum* on the growth and nutrient content of *Cassia siamea* in a semi-arid Indian wasteland soil. New For 29:63–73

Glick BR (2012) Plant growth-promoting bacteria: mechanisms and applications. Scientifica. http://dx.doi.org/10.6064/2012/963401, 963401

Guo Y, Ni Y, Huang J (2010) Effects of rhizobium, arbuscular mycorrhiza and lime on nodulation, growth and nutrient uptake of lucerne in acid purplish soil in China. Trop Grasslands 44:109–114

Han HS, Supanjani S, Lee KD (2006) Effect of co-inoculation with phosphate and potassium solubilizing bacteria on mineral uptake and growth of pepper and cucumber. Plant Soil Environ 52:130–136

He Z, He C, Zhang Z, Zou Z, Wang H (2007) Changes of antioxidative enzymes and cell membrane osmosis in tomato colonized by arbuscular mycorrhizae under NaCl stress. Colloids Surf B Biointer 59:128–133

Henry RC, Engstro¨m K, Olin S, Alexander P, Arneth A, Rounsevell MDA (2018) Food supply and bioenergy production within the global cropland planetary boundary. PLoS ONE 13(3): e0194695. https://doi.org/10.1371/journal.pone.0194695

Hungate BA, Dukes JS, Shaw MR, Luo Y, Field CB (2003) Nitrogen and climate change. Science 302:1512–1513

Kaur H, Kaur J, Gera R (2016) Plant growth promoting rhizobacteria: a boon to agriculture. Int. J Cell Sci Biotechnol 5:17–22

Kim J, Rees DC (1994) Nitrogenase and biological nitrogen fixation. Biochemistry 33:389–397

Kloepper JW, Zablowicz RM, Tipping B, Lifshitz R (1991) Plant growth mediated by bacterial rhizosphere colonizers. In: Gregan B (ed) Keister DL. The rhizosphere and plant growth, BARC Symp, pp 315–326

Kohler J, Caravaca F, Roldan A (2010) An AM fungus and a PGPR intensify the adverse effects of salinity on the stability of rhizosphere soil aggregates of *Lactuca sativa*. Soil Biol Biochem 42:429–434

Kumar A, Maurya BR, Raghuwanshi R, Meena VS, Islam MT (2017) Co-inoculation with Enterobacter and Rhizobacteria on yield and nutrient uptake by wheat (*Triticum aestivum* L.) in the alluvial soil under Indo-Gangetic plain of India. J Plant Growth Regul. https://doi.org/10.1007/s00344-0169663-5

Kundan R, Pant G, Jadon N, Agrawal PK (2015) Plant growth promoting Rhizobacteria: mechanism and current prospective. J Fertil Pestic 6:2

Lopez-Bellido RJ, Lopez-Bellido L (2001) Efficiency of nitrogen in wheat under Mediterranean condition: effect of tillage, crop rotation and N fertilization. Field Crop Res 71:31–64

Ma Y, Rajkumar M, Freitas H (2009) Inoculation of plant growth promoting bacterium *Achromobacter xylosoxidans* strain Ax10 for the improvement of copper phytoextraction by *Brassica juncea*. J Environ Manag 90:831–837

Maheshwari DK, Kumar S, Kumar B, Pandey P (2010) Co-inoculation of urea and DAP tolerant *Sinorhizobium meliloti* and *Pseudomonas aeruginosa* as integrated approach for growth enhancement of *Brassica juncea*. Indian J Microbiol 50(4):425–431

Maurya BR, Meena VS, Meena OP (2014) Influence of Inceptisol and Alfisol's potassium solubilizing bacteria (KSB) isolates on release of K from Waste mica. Vegetos 27:181–187

McArther DAJ, Knowles NR (1993) Influence of VAM and phosphorus nutrition on growth, development and mineral nutrition of potato. Plant Physiol 102:771–782

Meding SM, Zasoski RJ (2008) Hyphal-mediated transfer of nitrate, arsenic, cesium, rubidium, and strontium between arbuscular mycorrhizal forbs and grasses from California oak woodland. Soil Biol Biochem 40:126–134

Meena VS, Maurya BR, Verma JP (2014) Does a rhizospheric microorganism enhance K+ availability in agricultural soils? Microbiol Res 169:337–347

Meena VS, Maurya BR, Verma JP, Meena RS (2016) Potassium solubilizing microorganisms for sustainable agriculture. Springer

Meena VS, Meena SK, Verma JP, Kumar A, Aeron A, Mishra PK, Bisht JK, Pattanayak A, Naveed M, Dotaniya ML (2017) Plant beneficial rhizospheric microorganism (PBRM) strategies to improve nutrients use efficiency: a review. Ecol Eng 107:8–32

Mehnaz S, Baig DN, Lazarovits G (2010) Genetic and phenotypic diversity of plant growth promoting rhizobacteria isolated from sugarcane plants growing in Pakistan. J Microbiol Biotechnol 20:1614–1623

Minaxi J, Saxena S, Chandra S, Nain L (2013) Synergistic effect of phosphate solubilizing rhizobacteria and arbuscular mycorrhiza on growth and yield of wheat plants. J Soil Sci Plant Nut 13:511–525

Montano FP, Villegas CA, Bellogia RA, Cerro PD, Espuny MR, Guerrero IJ (2014) Plant growth promotation in cereals and leguminous agricultural important plants from microorganisms capacities to crop production. Microbiol Res 169(5–6):325–336

Mosier AR, Syers JK, Freney JR (2004) Agriculture and the nitrogen cycle. Assessing the impacts of fertilizer use on food production and the environment, Scope-65. Island Press, London

Orlandini V, Emiliani G, Fondi M, Maida E, Perrin E, Fani R, (2014) Network analysis of plasmidomes: The *Azospirillum brasilense* Sp245 Case. Hindawi Publishing Corporation, pp. 1–14

Ortaş I, Rafique M (2017) The mechanisms of nutrient uptake by arbuscular mycorrhizae. In Mycorrhiza-Nutrient Uptake, Biocontrol, Ecorestoration (1–19). Springer, Cham

Oyedele OA, Samuel T (2014) Antifungal activities of *Bacillus subtilis* isolated from some condiments and soil. Afr J Microbiol Res 8(18):1841–1849

Panday SC, Choudhary M, Singh S, Meena VS, Mahanta D, Yadav RP, Pattanayak A, Bisht JK (2018) Increasing farmer's income and water use efficiency as affected by long-term fertilization under a rainfed and supplementary irrigation in a soybean wheat cropping system of Indian mid-Himalaya. Field Crop Res 219:214–221

Parihar M, Rakshit A (2016) Arbuscular Mycorrhiza: a versatile component for alleviation of salt stress. Nat Environ Pollut Technol 15(2):417

Parihar M, Meena VS, Mishra PK. Rakshit M, Choudhary M, Yadav RP, Rana K, Bisht JK (2019). Arbuscular mycorrhiza: a viable strategy for soil nutrient loss reduction. Archives of Microbiology, pp 1–13. https://doi.org/10.1007/s00203-019-01653-9

Porcel R, Ruiz-Lozano JM (2004) Arbuscular mycorrhizal influence on leaf water potential, solute accumulation, and oxidative stress in soybean plants subjected to drought stress. J Exp Bot 55:1743–1750

Prathap M, Ranjitha KBD (2015) A Critical review on plant growth promoting rhizobacteria. J Plant Pathol Microbiol 6(4):1–4

Rahimizadeh M, Kashani A, Zare-Feizabadi A, Koocheki AR, Nassiri-Mahallati M (2010) Nitrogen use efficiency of wheat as affected by preceding crop, application rate of nitrogen and crop residues. Aust J Crop Sci 4:363–368

Rajkumar M, Ae N, Prasad MNV, Freitas H (2010) Potential of siderophore-producing bacteria for improving heavy metal phytoextraction. Trends Biotechnol 28:142–149

Ramachandran K, Srinivasan V, Hamza S, Anandaraj M (2007) "Phosphate solubilizing bacteria isolated from the rhizosphere soil and its growth promotion on black pepper (*Piper nigrum* L.) cuttings," in Proceedings of the First International Meeting on Microbial Phosphate Solubilization, Vol. 102, eds E. Velázquez and C. Rodríguez-Barrueco (Dordrecht: Springer), 325–331

Ramesh A, Sharma SK, Sharma MP, Yadav N, Joshi OP (2014) Inoculation of Zinc solubilizing *Bacillus aryabhattai* strains for improved growth, mobilization and biofortification of zinc in soybean and wheat cultivated in Vertisols of central India. Appl Soil Ecol 73:87–96

Renella G, Egamberdiyeva D, Landi L, Mench M, Nannipieri P (2006) Microbial activity and hydrolase activities during decomposition of root exudates released by an artificial root surface in Cd-contaminated soils. Soil Biol Biochem 38:702–708

Reynolds HL, Hartley AE, Vogelsang KM, Bever JD, Schultz PA (2005) Arbuscular mycorrhizal fungi do not enhance nitrogen acquisition and growth of old-field perennials under low nitrogen supply in glasshouse culture. New Phytol 167(3):869–880

Richardson AE, Lynch JP, Ryan PR (2011) Plant and microbial strategies to improve the phosphorus efficiency of agriculture. Plant Soil 349:121–156

Rockstorm J, Steffen W, Noone K, Persson A, Chapin AS III et al (2009) A safe operating space for humanity. Science 461:472–475

Rokhbakhsh-Zamin F, Sachdev D, Kazemi-Pour N, Engineer A, Pardesi KR, Zinjarde S, Dhakephalkar PK, Chopade BA (2011) Characterization of plant-growth-promoting traits of *Acinetobacter* species isolated from rhizosphere of Pennisetum glaucum. J Microbiol Biotechnol 21:556–566

Sabry SRS, Saleh SA, Batchelor CA (1997) Endophytic establishment of Azorhizobium caulinodans in wheat. Proc Biol Sci 264:341–346

Saharan BS, Nehra V (2011) Plant growth promoting Rhizobacteria: a critical review. Life Sci Med Res 21:1–30

Sahoo RK, Ansari MW, Pradhan M (2014) Phenotypic and molecular characterization of native *Azospirillum* strains from rice fields to improve crop productivity. Protoplasma 251:943–953

Santoro MV, Bogino PC, Nocelli N, Cappellari LR, Giordano WF, Banchio E (2016) Analysis of plant growth promoting effects of Fluorescent *Pseudomonas strains* isolated from *Mentha piperita* Rhizosphere and effects of their volatile organic compounds on essential oil composition. Front Microbiol 7(1085):1–17

Sharifi M, Ghorbanli M, Ebrahimzadeh H (2007) Improved growth of salinity-stressed soybean after inoculation with salt pre-treated mycorrhizal fungi. J Plant Physiol 164:1144–1151

Sharma SK, Johri BN, Ramesh A, Joshi OP, Prasad SVS (2011) Selection of plant growth promoting *Pseudomonas* spp. that enhanced productivity of soybean-wheat cropping system in central India. J Microbiol Biotechnol 21:1127–1142

Sheng XF, Xia JJ (2006) Improvement of rape (*Brassica napus*) plant growth and cadmium uptake by Cadmium resistant bacteria. Chemophore 64:1036–1042

Shenoy VV, Kalagudi GM (2005) Enhancing plant phosphorus use efficiency for sustainable cropping. Biotechnol Adv 23:501–513

Shintu PV, Jayaram KM (2015) Phosphate solubilising bacteria (*Bacillus polymyxa*)—and effective approach to mitigate drought in tomato (*Lycopersicon esculentum*). Trop Plant Res 2:17–22

Shoji S, Delgado J, Mosier A, Miura Y (2001) Use of controlled release fertilizers and nitrification inhibitors to increase nitrogen use efficiency and to conserve air andwater quality. Comm Soil Sci Plant Anal 32(7–8):1051–1070

Singh PK (2012) Role of glomalin related soil protein produced by arbuscular mycorrhizal fungi: a review. Agric Sci Res J 2:119–125

Smith SE, Read DJ (2008) Mycorrhizal symbiosis, 3rd edn. Academic Press, London

Sparks DL, Huang PM (1985) Physical chemistry of soil potassium. Potassium in agriculture. 201– 276

Suhag M (2016) Potential of biofertilizers to replace chemical fertilizers. International Advanced Research Journal in Science, Engineering and Technology 3:163–167

Syers JK, Johnson AE, Curtin D (2008) Efficiency of soil and fertilizer phosphorus use: Reconciling changing concepts of soil phosphorus behaviour with agronomic information. FAO: Food and Agriculture Organization of the United Nations, Rome. Fert Plant Nutr Bull 18

Tak HI, Ahmad F, Babalola OO, Inam A (2012) Growth, photosynthesis and yield of chickpea as influenced by urban wastewater and different levels of phosphorus. Int J Plant Res 2:6–13

Tank N, Saraf M (2009) Enhancement of plant growth and decontamination of nickel-spiked soil using PGPR. J Basic Microbiol 49:195–204

Teotia P, Kumar M, Prasad R, Kumar V, Tuteja N, Varma A (2017) Mobilization of Micronutrients by Mycorrhizal Fungi. In Mycorrhiza-Function, Diversity, State of the Art (9–26). Springer, Cham

Thomas J, Ajay D, Kumar R, Mandal AK (2010) Influence of beneficial microorganisms during in vivo acclimatization of in vitro-derived tea (*Camellia sinensis*) plants. Plant Cell, Tissue Organ Cult 101:365–370

Tilman D, Cassman K, Matson P (2002) Agricultural sustainability and intensive production practices. Nature 418:671–677

Tilman D, Christian B, Jason H, Belinda LB (2011) Global food demand and the sustainable intensification of agriculture. Proc Natl Acad Sci 108(50):20260–20264

Tilman D, Fargione J, Wolff B (2001) Forecasting agriculturally driven global environmental change. Science 292:281–284

Vansuyt G, Robin A, Briat JF, Curie C, Lemanceau P (2007) Iron acquisition from Fe pyoverdine by *Arabidopsis thaliana*. Mol Plant Microbes Interact 20:441–447

Vejan P, Abdullah R, Khadiran T, Ismail S, Boyce AN (2016) Role of plant growth promoting Rhizobacteria in agricultural sustainability—A review. Molecules 21(573):1–17

Vessey JK (2003) Plant growth promoting rhizobacteria as biofertilizers. Plant Soil 255:571–586

Wahid F, Sharif M, Steinkellner S, Khan MA, Marwat KB, Khan SA (2016) Inoculation of arbuscular mycorrhizal fungi and phosphate solubilizing bacteria in the presence of rock phosphate improves phosphorus uptake and growth of maize. Pak J Bot 48:739–747

Wahyudi AT, Astuti RP, Widyawati A, Meryandini A, Nawangsih AA (2011) Characterization of *Bacillus* sp. strains isolated from rhizosphere of soybean plants for their use as potential plant growth for promoting Rhizobacteria. J Microbiol Antimicrob 3:34–40

Wang X, Shen J, Liao H (2010) Acquisition or utilization, which is more critical for enhancing phosphorus efficiency in modern crops? Plant Sci 179:302–306

Wang ZG, Bi YL, Jiang B, Zhakypbek Y, Peng SP, Liu WW, Liu H (2016) Arbuscular mycorrhizal fungi enhance soil carbon sequestration in the coalfields, northwest China. Scientific reports 6:336–343

Wani PA, Khan MS (2010) *Bacillus* species enhance growth parameters of chickpea (*Cicer arietinum* L.) in chromium stressed soils. Food Chem Toxicol 48:3262–3267

Wezel A, Casagrande M, Celette F, Vian JF, Ferrer A, Peigné J (2014) Agroecological practices for sustainable agriculture. a review. Agron Sust Dev 34:1–20. https://doi.org/10.1007/s13593-013-0180-7

Willis A, Rodriguesb BF, Harrisa PJC (2013) The ecology of arbuscular mycorrhizal fungi. Crit Rev Plant Sci 32:1–20

Witt C, Fairhurst TH, Griffiths W (2005) Proceedings of 5th national ISP seminar, Johor, Bahru, Malaysia, 27–28 June 2005. Incorporated Society of Planters, pp 1–22

Yadav MR, Kumar R, Parihar CM, Yadav RK, Jat SL, Ram H, Meena RK, Singh M, Verma AP, Kumar U, Ghosh A (2017) Strategies for improving nitrogen use efficiency: A review. Agric Rev 38(1):29–40

Zhang HH, Tang M, Chen H, Zheng C, Niu Z (2010) Effect of inoculation with AM fungi on lead uptake, translocation and stress alleviation of *Zea mays* L. seedlings planting in soil with increasing lead concentrations. Eur J Soil Biol 46:306–311

Chapter 4
In Sustainable Agriculture: Assessment of Plant Growth Promoting Rhizobacteria in Cucurbitaceous Vegetable Crops

Musa Seymen, Ertan Sait Kurtar, Atilla Dursun and Önder Türkmen

Abstract One of the most important vegetable families commonly grown in around of the world is Cucurbitaceae for economic value, nutrition, consumer's preference, general adaptability and extent of cultivation. Plant growth promoting rhizobacteria (PGPR) mostly associated with the plant rhizosphere have been established as beneficial for plant growth, yield and crop quality. They are important to promote the circulation of plant nutrients and reduce the need for chemical fertilizers and interest in eco-friendly, sustainable and organic agricultural practices as well. Use of PGPR's containing beneficial microorganisms in lieu of inorganic chemicals are positively known to affect plant growth and may help to sustain environmental health and soil productivity, even in biotic and abiotic stress conditions. PGPR's also have potential bio-control agents against to a wide range of bacterial and fungal pathogens in agriculture. The effects of PGPR's on physiological mechanisms, plant growth, yield and yield components, uptake of mineral elements and contents, biotic and abiotic stress conditions in Cucurbits vegetables and future perspectives have been discussed in the review.

Keywords PGPR · Cucurbits · Growth · Yield · Biotic · Abiotic · Minerals

M. Seymen (✉) · E. S. Kurtar · Ö. Türkmen
Department of Horticulture, Faculty of Agriculture, Selçuk University, Konya, Turkey
e-mail: mseymen@selcuk.edu.tr

E. S. Kurtar
e-mail: ertansaitkurtar@selcuk.edu.tr

Ö. Türkmen
e-mail: turkmenonder@hotmail.com

A. Dursun
Department of Horticulture, Faculty of Agriculture, Atatürk University, Erzurum, Turkey
e-mail: atilladursun@atauni.edu.tr

D. K. Maheshwari and S. Dheeman (eds.), *Field Crops: Sustainable Management by PGPR*, Sustainable Development and Biodiversity 23,
https://doi.org/10.1007/978-3-030-30926-8_4

4.1 Introduction

Cucurbitaceae is the largest family of summer vegetable group with about 125 genera and 960 species (Rahmatullah 2012). Cucurbitaceae family has economically important species grown all over the world such as watermelon (*Citrullus lanatus*), melon (*Cucumis melo*), cucumber (*Cucumis sativus*), summer squash (*Cucurbita pepo* L.), winter squash (*Cucurbita maxima* Duch.) and pumpkin (*Cucurbita moschata* Duch.), besides, bitter gourd (*Momordica charantia* L.), ridge gourd (*Luffa actangua* L. Roxb), and snake gourd (*Trichosanthes anguina*) are the members of family as a minor group. Watermelons, melons, cucumber, and squashes are used either fresh or cooked for consumption, while mature seeds of some species are used directly or indirectly in human nutrition. Cucurbitaceae family has an important share (about 25%) in world vegetable production with a total 255 million tons (117 million tons of watermelon, 80 million tons of cucumber, 31 million tons of melon and 26 million tons of pumpkin) production value (FAO 2016). 1.1 billion tons of vegetables are nearly produced in the world.

The cultivation techniques of vegetable species are more difficult than many herbal products due to higher water content, sensitive root, and morphological structure. Thus, vegetables are more affected by biotic and abiotic stress conditions. For these reasons, it is known that chemical pesticide and fertilizers are widely used in vegetable production at a high rate. Chemical fertilization, which has been unconsciously made in agriculture for many years, causes the accumulation of toxic substances, the depletion of organic carbon, the degradation of microflora and fauna, and consequently the reduction of soil fertility (Sairam and Reddy 2013). On the other hand, pesticides used for plant disease and pest control have negative effects on human health and ecology, along with, enhanced the systematic resistance of pest.

In countries with low-income levels, the use of pesticides is limited due to high pesticide prices it is, therefore, control of pest and diseases is a big problem (Dardanelli et al. 2010). Otherwise, it has become a necessity to increase the productivity in the unit area to meet the nutritional needs of the growing world population. To increase efficiency 1- better agricultural land management, 2- use of more chemicals, including fertilizers, 3- use of pesticides and herbicides with safely and effectively, 4- more agricultural mechanization, 5- widespread use of transgenic crops, and 6- use of plant growth promoting rhizobacteria (PGPR) (Glick 2014) have been examined by various workers.

PGPRs are important microorganisms to increase productivity in the sense of sustainability and ecological practices in agriculture and to reduce the use of chemical fertilizers and pesticides. Rhizobacteria positively interacts with plant roots and play an important role in the growth of the plants (Agrawal et al. 2014). Bacteria enhance the circulation of nutrients and provide limited use of chemical nutrients (Dursun et al. 2010; Seymen et al. 2013a, b). PGPRs greatly affect soil properties and play an important role in crop production in inefficient and poor-quality soils (Gouda et al. 2018). The use of PGPR has become an important application for an increase in soil quality and yield in many parts of the world (Gabriela et al. 2015).

Beneficial, bacteria are also used as symbiotic (*Rhizobium, Bradyrhizobium, Mesorhizobium*) and non-symbiotic (*Pseudomonas*, *Bacillus*, *Klebsiella*, *Azotobacter*, *Acotobacter*, *Azospirillum*, *Azomonas*) agents to increase in plant growth (Saharan and Nehra 2011) in biotic stress conditions such as herbicides (Ahemad and Khan 2010), insecticides (Ahemad and Khan 2011), fungicides (El-Sharkawy et al. 2015; García-Gutiérrez et al. 2012; Salman et al. 2017) as well as abiotic stress conditions such as drought (El-Meihy 2016; Kang et al. 2014; Wang et al. 2012), salinity (Yıldırım et al. 2006), heavy metal (Tóth et al. 2013), soil alkalinity (Esitken et al. 2016; Ipek et al. 2014, 2017), low temperature (Selvakumar et al. 2008) and groundwater.

In this review, the effects of PGPR applications as plant growth regulators were evaluated in order to the agricultural sustainability and eco-friendly production in the Cucurbitaceae family, which contains the most cultivated vegetable species in the world. Likewise, the recent studies have been collected and future perspectives have also been discussed.

4.2 What Are the PGPR's?

The plants have been in constant cross-talk with the microorganisms (bacteria and fungi) found in the soil during development and growth periods. The microorganisms the plant species, provides important benefits to plant development through different mechanisms. PGPR's are able to colonize root surface, survive, reproducible and compete with other organisms throughout the development process in the plant besides, having positive effects on plant growth (Kloepper 1994). If the bacteria show a positive effect on plant development about 2–5% when inoculated into another soil flora, it is evaluated as PGPR (Kloepper and Schroth 1978). PGPR's can penetrate different parts of the plant such as stomata, lenticels, nodules, wounds, and cracks (Fig. 4.1).

The most important features of PGPR's are the fixation of free N in the atmosphere, solubilize organic phosphorus, production of some secondary metabolites (plant hormone, siderophore and antibiotics etc.), increase in the systemic resistance, competition of nutrient and colonization and suppress disease factor (İmriz et al. 2014). PGPR's are evaluated into two groups as directly and indirectly, according to their action mechanism (Gouda et al. 2018; Gupta et al. 2015) (Fig. 4.2).

Fig. 4.1 The penetration ways of PGPR's in pumpkin plant (*Cucurbita pepo* L.)

4.2.1 Direct Mechanisms

PGPR's promote plant growth directly by providing nutrients to plants, N fixation, mineralization of organic compounds, resolution of mineral nutrients, and phytohormone production (Bhardwaj et al. 2014). These mechanisms vary according to the plant species, but different bacteria may show different effects in the rhizosphere. In this case, determination of PGPR strains is one of the most important issues in terms of plant species.

4.2.1.1 Nitrogen Fixation

Nitrogen (N_2) is one of the essential nutrients for plant growth and yield (Gupta et al. 2015; İmriz et al. 2014). It is the basic component of nitrogen nucleotides,

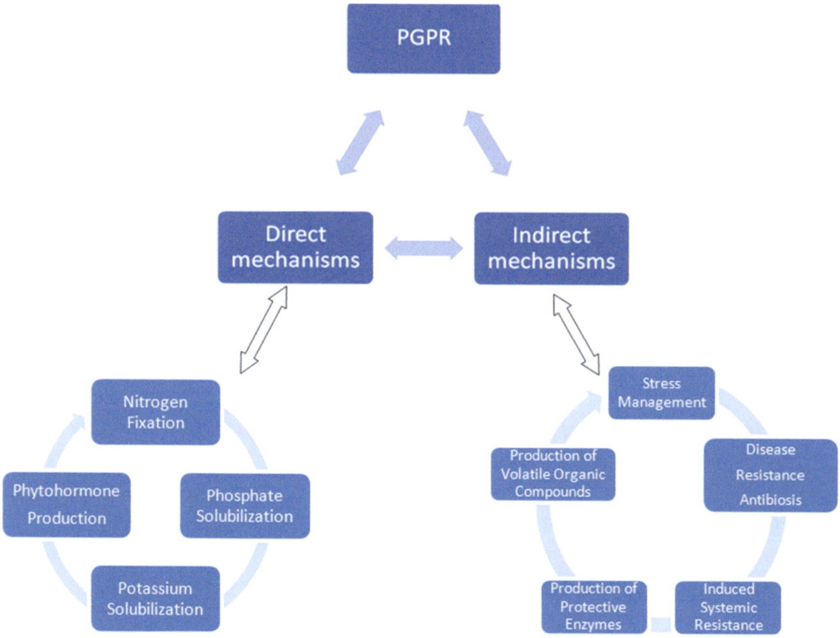

Fig. 4.2 The action mechanisms of PGPR's

membrane lipids and amino acids (Marschner 1995). Biological nitrogen fixation is the use of fixed nitrogen, which accounts for about two-thirds of the world's atmosphere (Shridhar 2012). The most studied N fixative PGPR's are *Rhizobium* sp., *Azoarcus* sp., *Pantoea agglomerans, Beijerinckia* sp. and *Klebsiella pneumoniae* (Ahemad and Kibret 2014). Rhizobacteria have positive effects on nodule production when they are applied as a mixture. Inoculation of PGPRs to plants or growing areas provides the nitrogen requirement of the soil and plant, as well as to prevent disease and to promote plant development (Damam et al. 2016).

4.2.1.2 Phosphate Solubilization

Phosphorus plays an important role in plant development for almost all metabolic processes such as energy transfer, signaling between plant receptors, respiration, macromolecular biosynthesis, and photosynthesis (Anand et al. 2016). However, 95–99% of the phosphorus is immobilized or precipitated, and is difficult to be taken up by plants. Plants can only utilize phosphorus from the soil as monobasic (H_2PO_4) and dibasic (HPO_{4-2}) forms (Bhattacharyya and Jha 2012; Gouda et al. 2018). One of the most important features of PGPRs is to solubilize insoluble phosphorus through the low molecular weight organic acids they synthesize (Bahadır et al. 2018; Sharma et al. 2013) of bacterial origin.

The most common of these organic acids which can solubilize the phosphorus produced by PGPR's, which contribute to the formation of organic acids in the soil, are lactic acid, gluconic acid, acetic acid, formic acid, oxalic acid, tartaric acid, fumaric acid, and succinic acid. Most common genera such as *Bacillus polymyxa*, *B. megatarium*, *B. circulans*, *B. subtilis*, *B. firmus*, *Pseudomonas striata*, *P. rathonia*, *Rhizobium leguminosarum* and *R. meliloti* are PGPR bacteria that can phosphorylate by producing organic acid (İmriz et al. 2014). Studies have shown that such bacteria increase phosphorus solubility when applied alone or as a mixture (consortia) (Zaidi et al. 2009).

4.2.1.3 Potassium Solubilization

Potassium is the third important plant nutrient element require for plant growth. The soluble potassium concentration in the soil is usually very low, and more than about 90% of potassium is in the silicates and rock form (Parmar and Sindhu 2013). In addition, irregular and unbalanced potassium fertilization has negative effect on plant development. Potassium deficiency causes slow root development of the plant, negatively affecting on plant growth, and causes small seed formation and loss of yield.

To ensure healthy plant development, it is necessary to find an alternative endemic source of potassium that provides potassium source from the soil (Kumar and Dubey 2012). Rhizobacterium which promotes plant development helps to dissolve potassium rocks by organic acid production and release and transform them into a useful form for plants (Han and Lee 2006).

Bacterium species such as *Acidothiobacillus* sp., *Ferrooxidans* sp., *B. edaphicus*, *Pseudomonas* sp., *B. mucilaginosus*, *Burkholderia* sp. and *Paenibacillus* sp. showed significant effects on potassium solubility (Liu et al. 2012). Therefore, the use of PGPRs in integrated nutrient management has an important role in the restriction of chemicals (Gupta et al. 2015; Setiawati and Mutmainnah 2016) led to sustainable agricultural production.

4.2.1.4 Phytohormone Production

In plant growth, some phytohormones such as indole-3-acetic acid (IAA), cytokinin and gibberellins have important role in the root development, expansion of root surface area, seed germination, chlorophyll accumulation, leaf growth, seed germination, dormancy, initiation of enzyme function, senescence of leaves and fruits etc. (İmriz et al. 2014; Sureshbabu et al. 2016). It is well known that PGPR's have important role in the synthesis of these phytohormones (Gouda et al. 2018). It was determined that 80% of the microorganisms isolated from the soil could synthesize IAA (Patten and Glick 2002). Indole-pyruvic acid and indole-3-acetic aldehyde mediated IAA synthesis were determined in *Erwinia herbicola*, in some saprophytic *Pseudomonas* and *Agrobacterium* species, *Bardyrhizobium*, *Rhizobium*,

Azospirillum, Klebsiella, and Enterobacter. In addition, tryptophan-derived IAA was found in Cyanobacterium, while tryptophan-independent IAA production demonstrated in Azospirila and cyanobacteria (Ahemad and Kibret 2014).

4.2.2 Indirect Mechanisms

PGPRs produce repressive substances to prevent or mitigate the harmful effects of phytopathogens on plants and increase in natural resistance (Singh and Jha 2015). This mechanism can also be described as a process that helps plants to actively grow under environmental stress (abiotic stress) or to protect plants from infections (biotic stress) (Akhgar et al. 2014).

4.2.2.1 Stress Management

Stress is one of the most important factors that imparts negative effects on plant growth (Foyer et al. 2016). In all stress conditions, reactive oxygen species (ROS) such as H_2O_2, O_2- and OH- are increased. Producing excessive ROS in plants causes oxidative damage to plants by oxidizing photosynthetic pigments, membrane lipids, proteins, and nucleic acids (Gouda et al. 2018). Plants accumulate metabolites to avoid stress such as poly-sugars, proline, glycine-betaine, abscisic acid, and upregulation in the synthesis of enzymatic and non-enzymatic antioxidants, as superoxide dismutase (SOD), catalase (CAT), ascorbate peroxidase (APX), glutathione reductase, ascorbic acid, α- tocopherol, and glutathione (Agami et al. 2016). Under abiotic stress conditions that occur in different environmental conditions, PGPR applications provide positive effects on plant growth by providing metabolite balance in plants. Beneficial effects of PGPR's have also been observed in abiotic stress conditions, as for drought (El-Meihy 2016; Kang et al. 2014; Wang et al. 2012), salinity (Yıldırım et al. 2006), heavy metal (Tóth et al. 2013), soil alkalinity (Esitken et al. 2016; Ipek et al. 2014, 2017), low temperature, etc. (Selvakumar et al. 2008).

Different pathogens such as bacteria, viruses, nematodes, fungi, insects, and viroids are important biotic factors agricultural production and adversely affected on plant growth (Haggag et al. 2015). Various genera of PGPR such as *B. amyloliquefaciens* strain HYD-B17, *P. polymyxa* strain B2, B3, B4, *B. licheniformis* strain HYTAPB18, *P. favisporus* strain BKB30, *B. thuringiensis* strain HYDGRFB19 and *B. subtilis* strain RMPB44 gave positive response against to biotic agents (Gouda et al. 2018).

4.2.2.2 Disease Resistance Antibiosis

The use of PGPRs against plant pathogens in agriculture means the application and restriction of the use of agricultural chemicals. *Bacillus* spp. and *Pseudomonas* sp. microorganisms prevent pathogen damage by releasing antibiotics (Gouda et al. 2018).

The effects of microorganisms that produce antibiotics are the most studied topics in recent years. Some antifungal antibiotics (phenazines, phenazine-1-carboxylic acid, phenazine-1-carboxamide, pyrrolnitrin, pyoluteorin, 2,4 diacetyl phloroglucinol, rhamnolipids, oomycin A, cepaciamide A, ecomycins, viscosinamide, butyrolactones, N-butylbenzene sulfonamide, pyocyanin) and bacterial antibiotics (pseudomonic acid andazomycin), antitumor antibiotics (FR901463 and cepafungins) and antiviral antibiotics (Karalicine) are mostly produced by *Pseudomonas* species (Ramadan et al. 2016). Alcohols, aldehydes, ketones, sulfides and hydrogen cyanide are called volatile antibiotics, whereas, polyketides, cyclic lipopeptides, amino polyols, phenylpyrrole, and heterocyclic nitrogenous compounds are non-volatile antibiotics in nature (Fouzia et al. 2015).

4.2.2.3 Induced Systemic Resistance

When the plants are exposed to attack by pathogens, PGPRs are used to stimulate their defense mechanisms (Pieterse et al. 2009). This mechanism called Induced Systemic Resistance (ISR) provides resistance to pathogens the plant is exposed in future (Van Loon 2007). Gram-positive (*B. pumilus)*, Gram-negative (*P. fluorescens, P. putida, P. aeruginosa*), and enterobacteria (*Serratia marcesens, S. plymuthica*) provide resistance in ISR mechanism (Jourdan et al. 2009).

4.2.2.4 Production of Protective Enzymes

Ethylene is a very important plant signal hormone that shows a rapid increase in abiotic and biotic stress conditions, which play a regulatory role in seed germination, root growth, root nodulation, flowering and fruit ripening (İmriz et al. 2014). 1-aminocyclopropane-1-carboxylate (ACC) deaminase, also produced by PGPRs, promotes plant growth and development by regulating ethylene production in the plant (Yang et al. 2009). After the ethylene signal ACC is synthesized in the plant, it is taken up by the bacteria from the plant roots and hydrolyzed to the enzyme and 2-oxobutanoate. Thus, the level of ACC concentration in the plant is reduced and excessive ethylene formation is prevented and finally plant growth is promoted (Ahemad and Kibret 2014). In recent studies, ACC-deaminase activity has been determined in many bacterial strains belonging to the genus *Acinetobacter, Achromobacter, Agrobacterium, Alcalienes, Azospirillum, Bacillus, Burkholderia, Enterobacter, Pseudomonas, Ralstonia* and *Rhizobium* (Shaharoona et al. 2006; Zahir et al. 2009).

4.2.2.5 Production of Volatile Organic Compounds (VOCs)

VOCs increase in plant growth, providing systemic resistance to plants against phytopathogens by reducing the effects of fungal pathogens, bacteria and nematodes (Reza et al. 2016). Some bacterial species such as in *Pseudomonas*, *Bacillus, Arthrobacter, Stenotrophomonas* and *Serratia* induce plant growth promoting VOCs. 2, 3-Butanediol and acetone produced by *Bacillus* spp. stimulate plant growth by suppressing fungal development (Santoro et al. 2016). VOCs obtained directly or indirectly from PGPRs have positive effects on disease resistance, abiotic stress tolerance and plant biomass (Gouda et al. 2018).

However, the quantity and identity of the VOCs emitted vary among species, some soil microorganisms produced benzene, cyclohexane, methyl, 2,6,10-trimethyl, 2-(benzyloxy) ethanamine, dodecane, decane, 1-(N-phenylcarbamyl)- 2- morpholinocyclohexene, 1-chlorooctadecane, benzene (1-methylnonadecyl), tetradecane, 11-decyldocosane and dotriacontane having common characteristics of VOCs (Kanchiswamy et al. 2015).

4.3 The Effects of PGPR's on Growth and Yield in Cucurbits

Although the functions of PGPRs have not yet been fully described, but have lots of unique advantages on plant production and agricultural systems. Many studies have been conducted on the effects of PGPR's in cucurbit crops and it has been exactly revealed that PGPRs have positive effects on seed and seedling quality, plant growth, yield components, etc. But, a majority of field research examined under controlled conditions is scarce. Besides, the effectiveness of PGPR's depends on plant genotype, rhizobacteria strain, inoculation doses and ecology in relation to the plant growth and yield components.

4.3.1 Seed and Seedling Quality

As in all plants, a healthy seed germinates to produce seedling of plant production in cucurbit crops. It is well known that some soil-borne fungal diseases caused by *Fusarium* spp., *Didymella* spp. and *Phytophthora* spp. caused poor seedling growth and seedling loss at the beginning, thus, yield and quality are significantly reduced in Cucurbits. Seed and seedling applications of root-colonizing rhizobacteria are used for both controlling plant disease and also promote seedling growth without using any chemical preparation in watermelon, melon, cucumber, and squash. In this regard, PGPRs are known as eco-friendly practices in today's agriculture.

PGPR's can be applied to the seeds by seed soaking, alginate-encapsulation and seed coating (Bashan 1998; Reed and Glick 2005). The rhizobacteria are applied to seedlings by "roots dipped" or foliar spraying methods. Seeds were inoculated some PGPR strains and the symptoms of *Fusarium* spp., *Didymella bryoniae* and *Myrothecium* were significantly reduced, thus the seed germination and seedling vigor reported to increase in watermelon (Lokesh et al. 2007). The plants raised by seed treatments of some isolates (GBO3 and INR7) were effective on *Colletotrichum orbiculare*, *P. syringae* pv. *lachrymans* and *Erwinia tracheiphila* in *C. sativus* L. under greenhouse conditions (Raupach and Kloepper 1998). PGPR treatments provide considerable growth of seedling, stem diameter, leaf number, cotyledon diameter, shoot and root weight were significantly increased with treatments of *B. pumilus* and *Alcaligenes piechaudii* in cucumber (Kidoglu et al. 2007; Yıldırım et al. 2015). Similarly, Kokalis-Burelle et al. (2003) reported that PGPR's increased seedling quality (shoot weight, shoot length, and stem diameter) in muskmelon and watermelon. *Azotobacter chroococcum*, *B. megaterium,* and *B. mucilaginosus* generated germination and growth rate, the biomass of shoots, the number of lateral roots, and the hair root area in cucumber seedlings (Sokolova et al. 2011).

Root length was significantly increased in cucumber seedlings by bacterial inoculation under in vitro conditions (Utkhede and Koch 1999). *Gigaspora margarita, B. subtilis*, *Thermomonospora* sp. and *Thermoactinomyces* sp. considerably interrupted damping-off in 2–3 weeks old cucumber seedlings (Kabayashi 1989).

Coinoculation of seeds with soaking methods *A. brasilense* strains enhanced germination value, vigor index and the endogenous IAA content in cucumber seed and seedlings (Mangmang et al. 2015). The high germination vigor and emergence rate have been associated with phytohormones, (IAA and GA) synthesized by PGPR's (Kang et al. 2015).

4.3.2 Plant Growth and Yield

PGPR's promote plant growth and yield by involvement of different mechanism such as biological nitrogen fixation, phosphate solubilization, siderophores and exopolysaccharides production, phyto stimulation by IAA and GA production, bio-control for plant diseases, supply of some mineral elements and production of antibiotic in Cucurbit crops (Noumavo et al. 2016; Vikram et al. 2007). Phytohormones produced by some PGPR's are recognized as unique signaling substances for plant growth and strongly enhance to root growth, root surface area, and branching and plant habitus (Cassán et al. 2009; Levanony and Bashan 1989; Vessey 2003). In addition, recent studies stated that few proteins responsible for growth and photosynthesis activated by root inoculation of some PGPR's (Yaoyao et al. 2017).

In watermelon, continuous cropping system resulted to low yield and poor quality due to high incidence of soil-borne diseases (especially *Fusarium oxysporum* f. sp. *niveum*) and limited plant growth (Adhikari et al. 2017; Ling et al. 2014). These handicaps were effectively eliminated by application of bio-organic fertilizers and functional microbes' combination (Zhao et al. 2018). Likewise, *Fusarium* wilt was considerably reduced by using organic fertilizer *Paenibacillus polymyxa* SQR-21 based (Ling et al. 2012). In a greenhouse experiment, P. *polymyxa* and *Sinomonas atrocyanea* enhanced total chlorophyll content, plant height, fresh weight and dry weight of three watermelon varieties (Adhikari et al. 2017). Earlier, the positive effects of PGPR treatments on shoot weight, shoot length and stem diameter were reported in muskmelon and watermelon transplant (Kokalis-Burelle et al. 2003).

The beneficial effects of rhizobacteria on cucumber growth and yield have been also observed by various researchers (Kang et al. 2014; Sturz et al. 2000). The application of *Pantoea agglomerans* FF increased in fruit number and fruit weight per plant, plant length, fruit width, fruit length and dry matter. The highest average fruit weight was obtained from *Bacillus megatorium*-GC subgroup A. MFD-2 in cucumber (Dursun et al. 2010). The combination of biogas slurry and *P. fluorescens* resulted in higher shoot fresh weight, shoot dry weight, root fresh weight, root dry weight and total yield (Ahamd et al. 2015). When the PGPR or compost tea was used as a bio-fertilizer, yield and quality were significantly increased under the sandy soil (Abou-El-Hassan et al. 2014). Farrag et al. (2015) reported that the inoculation of *Azospirillum* and *Trichoderma* promote vegetative growth, leaves dry matter percentage, early emergence and number of female flowers, and total yield (cv. hybrid Prince). Diverse genera such as *Rhodobacter sphaeroides, Lactobacillus plantarum,* and *Saccharomyces cerevisiae* have substantial increase in the shoot length, root length, shoot fresh weight, shoot dry weight, and chlorophyll content, via secretion of optimum IAA and/or organic acids and low ABA level in cucumber (Kang et al. 2015). *Pseudomonas putida* strain P13 and *Pantoea agglomerans* strain P5 have positive effects on cucumber yield and yield components (Isfahani and Besharati 2012). *R. sphaeroides*, *L. plantarum*, and *S. cerevisiae* increased in the quantities of amino acids perhaps responsible for photosynthesis and nitrogen fixation in cucumber (Kang et al. 2010; Kang et al. 2014; Kang et al. 2015). Earlier, Rhizobacteria strain FE-43 and N-17/3 inoculations had shown the positive effects on yield and yield components (Seymen et al. 2010). In soilless culture, Utkhede and Koch (1999) reported that the strains BACT-0 of *B. subtilis* and *P. aphanidermatum* promote growth affected by fruit yield, and fruit number in commercial greenhouse conditions. Furthermore, root volume and plant growth increased in by using AMF and charcoal compost (Kabayashi 1989).

In *Cucumis melo* L., *G. mossae* and *B. cepacia* either separately or in combination enhance the growth and yield (Zulkarami et al. 2012). As an eco-friendly application, the bio-fertilizers were found to be effective on growth and yield in soilless squash production (Dasgan et al. 2010). The application of *G. intraradices* improved growth and yield in Zucchini squash (*C. pepo* L.) (Colla et al. 2008). The combination of *Azotobacter chroococum* and mineral N-fertilizer was found to be a profitable practice in sandy soil (Refai et al. 2010). Bio-fertilizer Halex-2 significantly enhanced fruit yield due to female flower formation at a higher rate

(Abd-El-Fattah and Sorial 2000). Root and shoot dry weight, number of leaves, shoot length, stem diameter and number of ramifications significantly increased by application of three nitrogen (N) sources and PGPR in squash (Tchiaze et al. 2016).

4.3.3 Fruit Quality

The majority of the researches on PGPR application concentrated on plant growth and yield, but the knowledge about fruit quality parameters is still not sufficient and need to investigate in Cucurbit crops.

According to the findings of Dursun et al. (2010), PGPR's (*Pantoea agglomerans, Acinetobacter baumannii,* and *B. megaterium*) increased mineral contents (N, P, Mg, Ca, Na, K, Cu, Mn, Fe, and Zn) and dry matter in fruits, but TSSC (Total Soluble Solid Content) were not changed in cucumber. Likewise, FE-43 and N-17/3 PGPR strains had positive effects on water-soluble dry matter and pH content in cucumber fruit (Seymen et al. 2010). Soluble sugar, amino acids, and soluble proteins were considerably enhanced by the inoculation of *G. versiforme*, *G. mosseae,* and *G. intraradices* (Lu et al. 2006). Biogas slurry + *P. fluorescens* and some PGPR treatments produced high-quality cucumber fruits (Ahamd et al. 2015; Elwan and Abd 2015). Several other studies of plant growth and productivity have been sumarised-up in the Table 4.1 which illustrate the significanc of PGPRs in enahncement of yield and yield components of cucurbit crops.

4.4 The Role of PGPR's on the Uptake of Mineral Elements in Cucurbits

Rhizobacteria promotes plant growth by increasing the supply or availability of primary nutrients to the host plant (Isfahani and Besharati 2012: Seymen et al. 2014, 2015a). PGPR's, chemical or organic substances fertilizers either separately or combination with different doses are used for mineral resources in Cucurbit crops. The efficiency of PGPR's on mineral elements uptake and availability depended on many factors such as bacterial strains, application doses, plant species and soil fertility status.

In *C. sativus* L., the rock materials (K and P) and bacterial strains (*B. megaterium* var. *phosphaticum* and *B. mucilaginosus)* consistently increased in availability and uptake of minerals (Han and Lee 2006). Foliar spraying of rhizobacteria had positive effects on N, P and K percentages in cucumber leaves (Farrag et al 2015). *R. sphaeroides*, *L. plantarum*, and *S. cerevisiae* reported IAA-producing microorganisms and enhanced mineral nutrient uptake by plant roots such as potassium, magnesium, phosphate, and calcium (Kang et al. 2015; Nimnoi et al. 2014). Furthermore, some bacteria (*L. plantarum* and *S. cerevisiae*) produce

Table 4.1 The effects of PGPR's on plant growth, yield and yield components of cucurbit crops

Species	PGPR's	Comments	References
	G. versiforme, *G. mosseae* and *G. intraradices*	Soluble sugar, Amino acids, Soluble proteins	Lu et al. (2006)
	N-52/1, N-17/3, FE-43, F-21/3, 637 Ca, MfdCa1	Yield, Fruit per plant, Fruit weights, Dry matter pH content	Seymen et al. (2010)
	Burkholderia sp. KCTC11096BP	Shoot length, Shoot fresh weight, Shoot dry weight, Root fresh weight, Root dry weight, Chlorophyll (SPAD)	Kang et al. (2010)
	P. putida strain P1 3, *Pantoea agglomerans* strain P5	Yield, Length of plant, Chlorophyll content Shoots dry weight, Fresh weight, Dry weight, Fresh weight	Isfahani and Besharati (2012)
	B.s subtilis BA-142, *B. megaeorium-GC subgroup* A. MFD-2, *A. baumannii* CD-1 *P. agglomerans* FF	Fruit number, Fruit weight, Plant length, Fruit width, Fruit length, Dry matter pH	Dursun et al. (2010)
Cucumber	*B. pumilis, Alcaligenes piechaudii*	Seedling growth and quality	Yıldırım et al. (2015)
	P. fluorescens	Growth, Yield, Fruit quality, Shoot fresh weight, Shoot dry weight, Root fresh weight, Root dry weight	Ahamd et al. (2015)
	Azospirillum *Trichoderma*	Growth, Leaves dry matter Number of female, Fruit length, Fruit diameter, Early and total yield	Farrag et al. (2015)
	B. megaterium var. *phosphaticum* *B. mucilaginosus*	Plant growth, Dry weight	Han and Lee (2006)
	Rhodobacter sphaeroides, Lactobacillus plantarum, Saccharomyces cerevisiae	Amino acids, Shoot length, Root length, Fresh weight, Dry weight, Chlorophyll (SPAD)	Kang et al. (2015)
	P. fluorescens DF57	Root dry weight, Root length	Ravnskov and Jakobsen (1999)
Watermelon	*Paenibacillus polymyxa* (SN-22), *Sinomonas atrocyanea* (NSB-27)	Total chlorophyll content, Plant height, Total fresh weight, Total dry weight	Adhikari et al. (2017)
Watermelon and Melon	*B. subtilis* *amyloliquefaciens B. subtilis* *B. pumilus*	Shoot weight, Shoot length Stem diameter	Kokalis-Burelle et al. (2003)
	A. chroococum strains (5 mutants and wild-type strain)	Stem length, Flowering, Fruits, Early and total yield	Refai et al. (2010)

(continued)

Table 4.1 (continued)

Species	PGPR's	Comments	References
Squash	*B. subtilis, B. amyloliquefaciens, B. pumilus,* *B. licheniformis, S. cerevisiae*	Root dry weight, Shoot dry weight, Total plant dry weight, Number of leaves, Shoot length, Stem diameter, Number of ramifications	Tchiaze et al. (2016)
	Paenibacillus polymyxa *B. megaterium*	Early fruit yield, Total yield, Good quality of fruits	Moussa (2006)
	A. brasilense AC1 *B. subtilis* AC2	Fruit yield Fruit weight Growth	Elwan and Abd (2015)

organic acids and increase in the availability of mineral elements, mainly phosphorus (Wang et al. 2014).

In *Cucurbita pepo* L., the treatments of rhizobacteria significantly reduced P- and N-fertilizer without any reduction in squash yield (Moussa 2006; Elwan and Abd 2015), importantly for sustainable plant production. The combinations of PGPR and different N sources (NO_3^-, NH_4^+, and NO_3NH_4) reported as efficient bio-fertilizers to significantly improve the minerals uptake in squash (Tchiaze et al. 2016). G. intraradices caused higher K and lower Na concentration in leaf tissue in squash (Colla et al. 2008).

4.5 Abiotic Stress Conditions and PGPR's in Cucurbits

The soil has a complex and dynamic system that promote plant growth. Some stress factors should be carefully examined in order to maintain plant growth in soil, regularly (Nadeem et al. 2014). While the most important abiotic factors affecting on plant growth are drought, salinity, and heavy metal stresses, some stress elements such as high temperature, soil alkalinity, high groundwater and low temperature also have negative effects on plant growth. In sustainable agriculture, abiotic stress factors are considered to be the main causes of yield loss. On the other hand, the effect of stress factors leads to yield losses of up to 50–82%, varying with soil structure, plant species and varieties (Christensen et al. 2007).

These stresses include plant growth hormone and nutritional imbalance among general effects, physiological disorders such as epinasty, abscess and aging, and predisposition to diseases (El-Iklil et al. 2000; Zhu et al. 1997). Drought and salinity stress directly affect plant growth, leading to increase in ethylene release in the plant and limitation of root and shoot development (Glick et al. 2007). Drought and salinity are typically encountered in arid and semi-arid regions of the world and effective about 7.6 million km^2 areas.

Agricultural areas used in the world are affected by drought stress (26%), mineral material stress (20%), cold and frost stress (15%) and other stress factors (29%) (Kalefetoğlu and Ekmekçi 2005). Unfortunately, it is also estimated that this area will be doubled and water resources will decrease by 30% up to 2050 (Falkenmark 2013). For this reason, some solutions are being searched to increase in plant development and productivity in stress conditions for sustainability of agriculture. It is known that rhizobacteria (PGPR), which support plant growth under stress conditions, play an important role in the growth and metabolism of plants (Kang et al. 2014). PGPR's produce induced systemic tolerance for the development of plants under abiotic stress conditions via generating physical and chemical changes in plants (Yang et al. 2009). It has also been reported that PGPR's are more effective against to abiotic stress when applied in combination with mycorrhiza (Mayak et al. 2004).

4.5.1 *Drought Stress*

Drought is examined under four main headings; meteorological, agricultural, hydrological and socioeconomic drought. Droughts begin as meteorological, develop as agricultural and hydrological and become visible as socio-economical (Örs and Ekinci 2015). In particular, drought stress is an important factor that negatively effects on agricultural productivity in arid and semi-arid regions and limited plant growth (Seymen et al. 2016; Yavuz et al. 2015a; Yavuz et al. 2015b). Ethylene, which released in arid conditions from plant tissue, has the property of restricting plant growth, inhibiting photosynthesis and changing chlorophyll content. Some chemicals such as amino ethoxy vinyl glycine (AVG), cobalt ion (Co^{2+}) and silver ion (Ag^{+}) can be applied to decrease the ethylene level and to protect against stress. However, these chemicals are not preferred because they are both expensive and adversely affect human health (Mckeon et al. 1995).

Besides, these types of chemicals are not recommended because they have a permanent effect on the soil and they will cause environmental pollution and other stress factors in the future. As an alternative method, eco-friendly production techniques and practices should be developed to ensure sustainability in agriculture. While plants have developed some specific mechanisms to combat such stress conditions, it is known that some useful microorganisms in root regions play an important role in reducing stress intensity (Nadeem et al. 2014). It is possible to explain the mechanisms by which PGPRs develop against abiotic stress factors; 1- produces some phytohormones such as ABA, GA and IAA, 2- reduce the level of ethylene in the roots with ACC deaminase, 3- creates systemic resistance with bacterial compounds, 4- enhance bacterial exopolysaccharides (Kim et al. 2013; Timmusk et al. 2014; Yang et al. 2009). In sustainable agriculture, PGPR's, an environmentally friendly practice, have been realized to promote plant development with nodules formed in the root zone of the plant in case of drought stress (El-Meihy 2016; Wang et al. 2012).

Cucumber, melon, watermelon, and squash need a lot of water in economic cultivation for both open field and protected cultivation. The drought stress during the cultivation season causes great yield loss in an economic sense. In particular, melon, watermelon, and squash are grown in open field conditions. These species have medium root structure and they are moderate resistant vegetable species to drought stress. In fact, some species and varieties are grown by limited irrigation in arid and semi-arid regions with their deep root structure (Sensoy et al. 2007; Yavuz et al. 2015a; Yavuz et al. 2015b). However, PGPR applications have an important role in the cultivation of Cucurbit species due to increasing in the agricultural land affected by drought and correspondingly more economical use of irrigation water to improve plant development and productivity.

In a study, the effects of *B. cereus* AR 156, *B. subtilis* Sin 21 and *Serratia* sp. XY 21 bacterial strains were investigated in cucumber. As a result of the study, the bacterial strains increased the electrical conductivity and the root ratio by 40% and 50%, respectively. In addition, PGPR inoculations decreased the effects of drought stress by increased MDA, SOD, proline, and ascorbate peroxidase contents (Wang et al. 2012). Likewise, the effectiveness of *Burkholdera cepacia* SE4, *Promicromonospora* sp. SE188 and *A. colcoaceticus* SE370 were examined and used for reducing the effects of drought and salinity stress, consequently, bacteria strains increased in biomass and chlorophyll contents in drought stress conditions. PGPR practices have led to increases in water potential by reducing the electrolytic exudation. In addition to reducing oxidative stress, PGPR applications have also affected on phosphorus and potassium uptake, positively. It has also been reported that PGPR practices increase productivity in stress conditions and PGPR's are an important application for sustainable agriculture (Kang et al. 2014). In squash (*C. pepo* L.), *A. chroococcum* ML1, *B. circulans* ML2, *B. megaterium* ML3 and *P. fluorescence* ML4 enhance the plant growth and produce a higher yield in drought stress. Bacteria have reduced the effect of drought stress by causing enzyme activities, IAA, and GA3 increases. It is clarified that the abscisic acid secreted under stress conditions was found to be less in bacterial applications. Researchers have reported that PGPR inoculations constitute better results on plant growth and yield than that of humic acid and chemical fertilizer applications (El-Meihy 2016) (Table 4.2).

4.5.2 Salinity Stress

There are about 1.5 billion hectares of agricultural land in the world, and 77 million hectares (5%) of this area are considered as inefficient agricultural land due to the high salt content (Abdel Latef and Chaoxing 2011). Besides, 20% of world agricultural land is negatively affected by salinity due to the using salty irrigation water (Wu et al. 2010). The salinity problem is increasing in regularly all over the world. When soil salinity increases, osmotic stress is triggered in the plant. Thus, the high

Table 4.2 The effects of PGPRs on the mineral elements uptake of cucurbit crops

Species	PGPR's	Comments	References
	N-52/1, N-17/3, FE-43, F-21/3, 637 Ca, MfdCa1	N, P, K, Mg, Ca, Na, Cu, Fe, S and B	Seymen et al. (2015b)
	B.subtilis BA-142, *B. megaeorium-GC subgroup A.* MFD-2, *A. baumannii* CD-1 *Pantoea agglomerans* FF	N, P, Mg, Ca, Na, K, Cu, Mn, Fe, and Zn	Dursun et al. (2010)
	P. fluorescens	N, P, and K	Ahamd et al. (2015)
Cucumber	*Azospirillum* *Trichoderma*	N, P, and K	Farrag et al. (2015)
	B. megaterium var. *phosphaticum* *B. mucilaginosus*	N, P, and K	Han and Lee (2006)
	Rhodobacter sphaeroides, Lactobacillus plantarum, Saccharomyces cerevisiae	Ca, Mg, P, and K	Kang et al. (2015)
	A. brasilense	Fe	Pii et al. (2015), (2016)
	P. fluorescens DF57	P	Ravnskov and Jakobsen (1999)
	B. subtilis, B. amyloliquefaciens, B. pumilus, *B. licheniformis, Saccharomyces cerevisiae*	N, P, K and Mn	Tchiaze et al. (2016)
Squash	*Paenibacillus polymyxa* *B. megaterium*	N, P, and K	Moussa (2006)
	A. brasilense AC1 *B. subtilis* AC2	N and P	Elwan and Abd (2015)

concentration of Na^+ and Cl^- in the soil causes nutrient imbalance, which reduces nutrient uptake and ion toxicity (Daei et al. 2009).

Soil salinity induces drought stress, ion toxicity, ion imbalance and degradation of soil structure and pH, hereby, inhibit plant growth (Gopal et al. 2012). While some plants try to counter the stress with the mechanisms of genetic structure in order to sustain its development in saline soil, some sensitive species are adversely affected. Some of the plant growth promoting rhizobacteria and fungi in the rhizosphere are reducing plant pathogens or affecting the production of phytohormones, increasing in plant nutrient uptake from the soil and increasing in plant growth, indirectly (Grichko and Glick 2001). Many studies have reported that PGPR's have positive effects on plant growth under salt stress conditions (Kang et al. 2014; Palacio-Rodríguez et al. 2017; Yıldırım et al. 2006).

Cucumber, melon, watermelon, and squash are economically important species of the Cucurbitaceae family and known as sensitive or moderately susceptible species to salinity (Francois 1985). Today, cucumber, melon, watermelon, and pumpkin irrigated with drip irrigation system for both open field and protected areas. The results of salinity of the irrigation waters, salinity increases in agricultural areas day by day and the cultivation of cucumber, melon, watermelon, and squash is negatively affected. For this reason, PGPR practices have been important in the cultivation of cucumber, melon, watermelon, and pumpkin in salty soils.

Some PGPR inoculation *(Burkholdera cepacia* SE4, *Promicromonospora* sp. SE188 and *Acinetobacter colcoaceticus* SE370) produced the higher yield under drought and salty soil conditions in cucumber (Kang et al. 2014). In squash, AgBlend, SoilBuilder, Yield Shield, Plant Shield, Inoculoid, and Equity bacterial strains considerably increased in plant fresh weight under salt stress compared to control lots. On the other hand, bacterial applications showed positive effects on potassium uptake. In 100 mM salt application, SoilBuilder, Yield Shield, and Equity bacteria inoculations increased in sodium concentration while other applications reduced. The most important effect of the bacteria is to increase in the K^+/Na^+ ratio which is an important parameter for plant development. Researchers have reported that bacterial applications have positive effects on plant growth and are feasible under salt stress conditions (Yıldırım et al. 2006). In another study conducted in watermelon and cucumber under salt stress conditions, from the 38 isolates, *Bacillus* sp. (LBEndo1) and *Pseudomonas lini* (KBEcto4) strains have shown important results on stimulation of root and shoot growth. It was determined that the bacteria increased in IAA and phosphorus uptake. Palacio-Rodríguez et al. (2017) demonstrated that the bacteria provide tolerance to salt stress and promote plant growth (Table 4.3).

4.5.3 Heavy Metals Stress

Some nutrient sources are needed for plant development at microelement level. However, when it is above the desired levels, it causes a toxic effect on the plant and limited the plant growth. Besides, mineral toxicity like the other stress factors has a negative effect on growth and economic yield of plants economically by accelerating the synthesis of ethylene (Safronova et al. 2006).

Fields contaminated with mineral toxicity is increasing in as a result of misapplications on fertilization made by human beings or in naturally. Agriculture is restricted in such areas and the consumption of crops yielded these areas also negatively affects human health. Accumulation of heavy metals such as Cd, Cr, and Pb in the human body leads to serious systemic health problems (Oliver 1997). One of the most important factors of heavy metal accumulation in the soil is the irrigation with dirty waste city or industrial water, because of these reasons, Cd, Zn, Cr, Ni, Pb, and Mn accumulate in the surface soil (Sharma et al. 2007). Many strategies have been conducted to counteract toxic substances in the soil in order to make the reclamation of agricultural areas. The use of microorganisms to increase in plant productivity and resistance to heavy metal pollution in the soil is one of these strategies (Gopal et al. 2012). Although heavy metal stress is not a very

Table 4.3 Effect of different PGPR's on cucurbit species in abiotic stress conditions

Plant species	PGPR	Comments (enhancements)	References
Cucumber (in drought stress)	*B. cereus* AR 156 *B. subtilis* Sin 21 *Serratia* sp. XY 21	40% electrical conductivity 50% root Prolin content Ascorbate peroksidaz MDA SOD	Wang et al. (2012)
Cucumber (in Al stress)	*A. chroococcum* *B. megaterium*	Shoot dry weight Root dry weight Root/shoot ratio Chlorophyll content Fe, Mn, Zn and Mg uptake	Tóth et al. (2013)
Cucumber (in drought and salinity stress)	*Burkholdera cepacia* SE4 *Promicromonospora* sp. SE188 *A. colcoaceticus* SE370	Shoot dry weigh Chlorophyll content K and P uptake Salicylic acid Gibberellin	Kang et al. (2014)
Squash (in salinity stress)	AgBlend SoilBuilder Yield Shield Plant Shield Inoculoid Equity	Biomas K uptake K^+/Na^+	Yıldırım et al. (2006)
Squash (in drought stress)	*A. chroococcum* ML1 *B. circulans* ML2 *B. megaterium* ML3 *P. fluorescence* ML4	Plant growth Enzyme activity IAA GA_3 Higher yield Low Abscisic acid	El-Meihy (2016)
Cucumber and Watermelon (in salinity stress)	*Bacillus* sp. (LBEndo1) *P. lini.* (KBEcto4)	Plant growth IAA P uptake	Palacio-Rodríguez et al. (2017)
Squash (in cold stress)	*S. marcescens* SRM	Shoot and root length Shoot and root dry weight N, P, K uptake	Selvakumar et al. (2008)

common problem in today's vegetable industry (Savvas et al. 2010), approximately 7% of cadmium was determined in eggplant fruits in Japan (Arao et al. 2008). This level is highly over the limits established by the international health organization

(Savvas et al. 2010). It is inevitable that these kinds of problems increase with the use of non-agricultural wastewater for plant production areas where water-scarce. Cucurbitaceae family vegetables are not so much affected by heavy metal toxicity by reason of the rootstocks used with some grafting techniques (Savvas et al. 2010). Although the grafting technique is widely used in watermelon cultivation, it is not preferred much in cucumber, melon and zucchini species; therefore, cucurbit production areas are in danger of being faced with heavy metal toxicity. To overcome this problem, it is important to develop PGPRs that reduce the efficacy of heavy metal toxicity and promote plant growth in such soil.

A. chroococcum and *B. megaterium* bacteria used against aluminum (Al) stress decreased Al contents in cucumber leaves. It has also been reported that bacteria inoculations reduced the negative effects of Al on the roots and caused to higher chlorophyll content in shoots. Al applications have been shown to reduce the root/shoot rate while bacterial applications have removed this negative effect. The use of biological fertilizers instead of chemical fertilizers has emerged as an alternative way. Biological fertilizers also positively affected on plant growth in the case of heavy metal stress. However, researchers think that the interaction of these bacteria with other bacteria and their effects may be different in the main growing sites of the plant under heavy metal stress conditions (Tóth et al. 2013) (Table 4.3).

4.6 Biotic Stress Conditions and PGPR's in Cucurbits

Since the beginning of agriculture, synthetic pesticides have been widely used for plant diseases and pests. The pesticides have caused adverse effects on human health and the environment as well as the systemic resistance of the pests. In addition, countries with low-income levels have limited use due to high pesticide prices (Dardanelli et al. 2010). PGPR's have been used as an alternative to chemicals against soil-borne diseases through human health and environmentalist approaches in sustainable agriculture (Kloepper et al. 1993). When PGPR's were started to be used in agriculture, they decreased in plant growth and productivity as well as the effect of diseases and pests (Gerhardson 2002; Reddy 2016).

In general, bacteria used as biopesticides belong to the *Agrobacterium*, *Bacillus* and *Pseudomonas* (Fravel 2005). In insect control, *Bacillus thuringiensis* constitute 70% of the used bacteria (Sanchis and Bourguet 2008). *Bacillus* such as *B. subtilis*, *B. licheniformis*, and *B. pumilus* represent approximately half of the bacterial biocontrol agents present in the market (Ongena and Jacques 2007).

Many soil pathogens are one of the most important factors limiting productivity. Soil pathogens are more dangerous than other diseases and pests, and their control is more difficult. These pathogens can cause serious damage in different parts of the world. For example, fungal and bacterial pathogens caused yield losses of 7–15% in products such as wheat, soybean, corn, which are the most cultivated in the world between 2001 and 2003 (Oerke 2005). It is very difficult to predict the yield losses because of the difficulty of diagnosing the pathogen (Dardanelli et al. 2010).

PGPR's penetrate the rhizosphere of plants, and they destroy the negative effects of nematodes. PGPR-nematode interactions have been studied to control nematodes. It is important to select bacteria that reduce or eliminate the effect of nematodes in plants (Reddy 2016). Beneficial bacteria are reduced to nematode populations by colonizing the plant's rhizosphere.

Agrobacterium, Alcaligenes, Bacillus, Clostridium, Desulfovibrio, Pseudomonas, Serratia and *Streptomyces* are the important bacteria that reduce or eliminate the population of nematode (Reddy 2016). PGPR's also provide systemic resistance to nematode pests. Although the use of PGPR against nematodes is limited, successful results have been obtained from studies on the patato cystic nematode (Sikora 1992).

The use of PGPR against to insects has been a subject of limited study. Bacterial pathogens of insects and the majority of related taxa are found in the Bacillaceae, Pseudomonadaceae, Enterobacteriaceae, Streptococcaceae and Micrococcaceae families (Reddy 2016). Most of these bacteria are toxic and very few have a lethal effect, and often expose insects to environmental stress and slow their growth and development.

Cucurbitaceae is a large family that is consumed as vegetables and has economic importance in the world. Ecological and biotic factors caused the biggest yield losses in cucurbit crops. In cucurbit production, pests and diseases are the preliminary important factors which have negative effects on productivity. In squash *Didymella bryoniae* (Black Rot), *Meloidogyne javanica*, and Root Rot/Wilt, *Macrophomina phaseolina, Fusarium oxysporum, F. solani* (Root-Knot Nematode), in cucumber, *Pythium* sp. (Damping-Off), *Colletotrichum orbiculare, C. lagenarium (*Anthracnose*), F. oxysporum* f. sp. *cucumerinum (*Fusarium Wilt*), Erwinia Tracheiphila (*Bacterial Wilt*), P. syringae* pv. *Lachrymans (*Angular Leaf Spot*)*, Cucumber Mosaic Virus (CMV), *Meloidogyne* spp. (Root-Knot Nematodes), *Acalymma vittatum*; Spotted Cucumber Beetle, *Diabrotica undecimpunctata (*Striped Cucumber Beetle*)*, in watermelon *F. oxysporum* f. sp. *niveum (*Fusarium Wilt*)*, Watermelon Mosaic Virus, *Meloidogyne javanica* and Wilt, *Fusarium solani* Disease Complex (Root-Knot Nematode), in melon *Fusarium oxysporum* f. sp. *melonis (*Fusarium Wilt*)* are the important pests and diseases (Reddy 2016). The successful reports have been reported in cucumber (*Pseudomonas* spp., *Bacillus* spp., *Flavomonas oryzihabitans, S. marcescens, Klebsiella pneumonia, Stenotrophomonas maltophilia, Flavomonas oryzihabitans*), in watermelon (*Pseudomonas* spp., *P. polymyxa,* and *Trichoderma harzianum*), in melon (*Pseudomonas* spp. and *Bacillus* spp.) and in squash (*Bacillus* spp.) (Table 4.4).

4.7 The Future Perspectives of PGPR's in Cucurbits

In today's agricultural perspective, increasing the productivity of the unit area as well as cultivating healthy crop is one of the most important issues. For this reason, PGPR's applications are a current and important issue and are the priority areas in

Table 4.4 Effect of different PGPRs on cucurbit species in biotic stress conditions

Plant species	PGPR	Diseases	References
Cucumber	*P.* WB1, Wb15, Wb52	*Pythium ultinum* *Rhizoctonia solani*	Arndt et al. (1998)
Cucumber	*B. subtilis,* *P. fluorescens,* *A. chroococcum* sp.	*Cucumber mosaic cucumovirus*	El-Borollosy and Oraby (2012)
Cucumber	*B. pumilus* INR-7 *Flavomonas oryzihabitans* INR-5	*Diabrotica undecimpunctata howardi* Barber	Zehnder et al. (1997)
Cucumber	*B. subtilis,* *B. amyloliquefaciens,* *P. aeruginosa,*	*Pythium aphanidermatum*	Elazzazy et al. (2012)
Cucumber	*B. polymixa,* *P. fluorescens,*	Squash mosaic virus	Firmansyah (2017)
Cucumber	*P. putida* BTP1, M3	*P. aphanidermatum*	Ongena et al. (2000)
Cucumber	*P. putida* strain 18/1 K, *S. marcescens* strain 62	*F. oxysporum f.* sp. *cucumerinum*	Gül et al. (2013)
Cucumber	*A. brasilense SBR,* *A. chroococcum ZCR,* *Klebsiella pneumoneae KPR,*	*F. oxysporum f.* sp. *lycopersici,* *R. solani,* *Pythium* sp.,	Hassouna et al. (1998)
Cucumber	New strains of *P. stutzeri,* *B. subtilis,* *Stenotrophomonas maltophilia,* *B. amyloliquefaciens.*	*P. crown*	Islam et al. (2016)
Cucumber	*S. marcescens* (90-166), *P. fluorescens* (89B61)	*Colletotrichum orbiculare*	Jeun et al. (2004)
Cucumber	*P. fluorescens* (isolate 9A-14), *Pseudomonas* sp. (isolate 8D-45), *B. subtilis* (isolate 8B-1).	*P. ultimum*	Khabbaz and Abbasi (2013)
Cucumber	*P. putida* strain 89B-61, *S. marcesens* strain 90-166, *F. oryzihabitans* strain INR-5, *B. pumilis* strain INR-7.	*Erwinia tracheiphila*	Zehnder et al. (2001)
Cucumber	*P. putida* 89B-27 *S. marcescens* 90-166	*F. oxysporum f.* sp. *cucumerinum*	Liu et al. (1995)
Cucumber	*P. fluorescens* *P. corrugata* *S. plymuthica*	*P. aphanidermatum*	McCullagh et al. (1996)
Cucumber	*P. fluorescens,* *B. subtilis,* *Rhizobium* sp., *Trichoderma harzianum* *T. viride*	*F. oxysporum f.* sp. *cucumerinum*	El-Sharkawy et al. (2015)

(continued)

Table 4.4 (continued)

Plant species	PGPR	Diseases	References
Cucumber	*B. pumilus* INR7, *B. subtilis* GB03, *Curtobacterium flaccumfaciens* ME1	*Colletotrichum orbiculare, P. syringae pv. lachrymans, Erwinia tracheiphila*	Raupach and Kloepper (1998)
Cucumber	*B. subtilis* B4	*Colletotrichum orbiculare*	Park et al. (2013)
Watermelon	*P. aeruginosa* 231-1	*Didymella bryoniae*	Nga et al. (2010)
Watermelon	*P. aeruginosa* 231-1	*F. oxysporum f.* sp. *niveum* *Didymella bryoniae*	Nga et al. (2013)
Watermelon	*P. fluorescens*	*F. oxysporum f.* sp. *niveum*	Salman et al. (2017)
Watermelon	*P. polymyxa* *Trichoderma harzianum*	*F. oxysporum f.* sp. *niveum.*	Wu et al. (2009)
Watermelon	*P. fluorescens* WR8-3, WR9-11 *P. putida* WR9-16	*Didymella byroniae*	Lee et al. (2001)
Melon	*P. fluorescens*	C. mosaic virus	Al-ani and Adhab (2012)
Melon	*Pseudomonas* spp. *Bacillus* spp.	Powdery mildew	García-Gutiérrez et al. (2009)
Melon	*B. subtilis* UMAF6614, UMAF6639 *B. cereus* UMAF8564, *P. fluorescens* strains, UMAF6031, UMAF6033	Powdery mildew	García-Gutiérrez et al. (2012)
Melon	*Bacillus* sp. RAB9	*Acidovorax citrulli*	Medeiros et al. (2009)
Squash	*B. macauensis* (SE52, SE76, INR7, IN937a, and IN937b)	*P. capsici*	Zhang et al. (2010)
Squash	*B.* IN937b	*P. capsici*	Mo (2013)
Pumpkin	*B. subtilis,* *B. pumilus,*	Watermelon mosaic potyvirus	Elbeshehy et al. (2015)
Watermelon and Melon	*C. flaccumfaciens,* *Microbacterium oxydans,* *P. oryzihabitans, P. fluorescens*	*A. citrulli*	Horuz and Aysan (2016)

new agricultural research which are increasing our knowledge of plant-bacterial interactions that take place in the root zone of the plant will allow us to develop molecular and biotechnological approaches to achieve integrated management of microorganisms in the soil (Gouda et al. 2018). PGPR's are also utilized as an important key factor for agricultural sustainability and eco-friendly vegetable production, worldwide. In this sense, PGPR's, PGPR strains and their combinations (multi-strain or combined inoculants) with or without commercial fertilizers have shown great advantages for both farmers and researchers. In spite of PGPR's have multi-way advantages, their performances are not clear under open field conditions. Because of numerous researches about PGPR's were examined under laboratory or controlled greenhouse. Hence, PGPR's should be experienced in large areas with multiple effects to determine the compatibility and interaction among the PGPR's, plants and the other microbial organisms. Thus, appropriate combinations and multiple applications of PGPR's will be clarified and recommended for local production in larger areas. Because of, appropriate PGPR strain for each different production area and extreme conditions should be determined due to complex environmental interaction. Additionally, it will become more important in the context of global climate change.

The cost of PGPR's in larger areas is the other important point for widespread applications. Within sustainable agriculture and eco-friendly production, PGPR's should be available easily and cheaply. Molecular biology and biotechnology may be a solution to generating low-priced and specialized PGPR strains with a different formulation. Nevertheless, in environmentalist approach, transgenic technology has a series problem such as bio-safety and phytotoxicity.

In respect to Cucurbit crops, application methods of PGPRs (such as seeds, roots, leaves), application periods (such as seedlings, flowering) and put on the market are important issues. PGPRs, which give positive results in many studies, have not been actualized on as bio-fertilizer. A number of studies have been conducted on the effects of PGPRs on plant growth, flowering, and fruit quality. When the studies are examined, most of the works about yield and quality are concentrated on cucumber and squash in cucurbit crops. Limited works have been realized on the other important cucurbits such as melon and watermelon. It is important to conduct PGPR studies on these species which are widely cultivated on open field conditions in the world.

A number of studies have been conducted on the effects of PGPRs on nitrogen and phosphorus uptake from soil. However, there is not much work on potassium uptake and mechanisms in the soil. It is obviously sad that this nutrient element which is important for plant development will be subject to the future-day's research. Investigation of the effects of PGPRs on nutrient uptake in watermelon and melon is one of the important issues.

Understanding the mechanisms of abiotic stresses expressed by multiple genes and causing adverse effects on plants is an important issue. Molecular and biotechnological approaches have been included in current work in order to reveal the genetic structure of these mechanisms. It will be possible to develop more effective PGPR's for abiotic stress conditions in future studies by the introduction of genetic expressions of bacteria. Limited works have been conducted on PGPR application in cucumber, watermelon, and squash, on the other hand, there is not any study in melon. Moreover, determination of bacteria that increase in plant growth against salt stress is an important topic for cucurbit species grown under protected areas. In addition, studies on other abiotic stress conditions are very limited, and it is thought that working with other stresses will be increased in day by day. When PGPR's were started to be used in agriculture, they increased in plant growth and productivity while decreased in the effect of diseases and pests

(Gerhardson 2002; Reddy 2016). In cucurbits, there has been a lot of work on this subject, especially in the cucumber.

Cucurbits are more susceptible to phytopathogens than many plant species due to having more sensitive root structure and grown under optimal conditions where diseases and pests develop. Apart from fungal pathogens, insect damage is one of the other important points should be worked on.

4.8 Conclusion

The most important factors affecting on agricultural cultivation is ecological conditions. To feed the population of the world, the effects of bacteria on increasing in plant growth and yield, and also the relationship between bacteria and plants should be examined. The yield of the plants should be increased in the feeding of the population. Biotic and abiotic agricultural areas are getting an increase in the world and chemical fertilizers applied the area may cause economically important problems in agriculture area and health problems in the world. Because of the reason, uses of PGPR's containing beneficial microorganisms instead of inorganic chemicals are positively known to effect on plant growth in terms of supplying of plant nutrients and may help to sustain human and environmental health and soil productivity. A number of inoculated bacterial species mostly associated with the plant rhizosphere in laboratory conditions have been tested and determined to be beneficial for plant growth, yield and crop quality so far. But, it is necessary to know PGPR's effects on field conditions.

The effects of PGPR's on yield, plant growth, quality, uptake of mineral elements, biotic and abiotic stresses in Cucurbits especially cucumber and squash have been studied so far. However, not much is known about promoting effects of PGPR's on yield, growth and nutrient contents of melon and watermelon vegetable species. The species should be evaluated in terms of effecting of PGPR's.

Nowadays, biotic and abiotic stress conditions in agricultural area have restricted the agricultural management. While there are adequate studies on biotic stress conditions, there is limited study on abiotic stress conditions in cucurbits. Especially, it is very important to know the effects of PGPR's on arid and salinity soil conditions.

References

Abdel Latef AAH, Chaoxing H (2011) Effect of arbuscular mycorrhizal fungi on growth, mineral nutrition, antioxidant enzymes activity and fruit yield of tomato grown under salinity stress. Sci Hortic 127:228–233. https://doi.org/10.1016/j.scienta.2010.09.020

Abd-El-Fattah MA, Sorial ME (2000) Sex expression and productivity responses of summer squash to biofertilizer application under different nitrogen levels. Zagzig J Agric Res 27:255–281

Abou-El-Hassan S, Abdrabbo MA, Desoky AH (2014) Enhancing organic production of cucumber by using plant growth promoting rhizobacteria and compost tea under sandy soil condition. Research J Agric Bio Sci 10:162–169

Adhikari M, Yadav DR, Kim SW et al (2017) Biological control of bacterial fruit blotch of watermelon pathogen (Acidovorax citrulli) with rhizosphere associated bacteria. Plant Patholog J 33:170–183. https://doi.org/10.5423/PPJ.OA.09.2016.0187

Agami RA, Medani RA, Abd El-Mola IA, Taha RS (2016) Exogenous application with plant growth promoting rhizobacteria (PGPR) or proline induces stress tolerance in basil plants (*Ocimum basilicum* L.) exposed to water stress. Int J Environ Agric Res 2:78–92

Agrawal PK, Agrawal S, Kundan R, Bhatt M (2014) Application and perspective of plant growth promoting rhizobacteria in development of sustainable agriculture. Int J Cur Res 6:9044–9051

Ahamd M, Zeshan MSH, Nasim M (2015) Improving the productivity of cucumber through combined application of organic fertilizers and *Pseudomonas fluorescens*. Pak J Agri Sci 52:1011–1016

Ahemad M, Khan MS (2010) Growth promotion and protection of lentil (*Lens esculenta*) against herbicide stress by Rhizobium species. Ann Microbiol 60:735–745. https://doi.org/10.1007/s13213-010-0124-2

Ahemad M, Khan MS (2011) Insecticide-tolerant and plant growth promoting *Bradyrhizobium* sp. (vigna) improves the growth and yield of green gram [*Vigna radiata* (L.) Wilczek] in insecticide stressed soils. Symbiosis 54:17–27. https://doi.org/10.1007/s13199-011-0122-6

Ahemad M, Kibret M (2014) Mechanisms and applications of plant growth promoting rhizobacteria: current perspective. J King Saud Univ Sci 26:1–20. https://doi.org/10.1016/j.jksus.2013.05.001

Akhgar R, Arzanlou M, Bakker PAHM, Hamidpour M (2014) Characterization of 1-aminocyclopropane-1-carboxylate (ACC) deaminase-containing *Pseudomonas sp.* in the rhizosphere of salt-stressed canola. Pedosphere 24:161–468. https://doi.org/10.1016/S1002-0160(14)60032-1

Al-Ani RA, Adhab MA (2012) Protection of melon plants against Cucumber mosaic virus infection using *Pseudomonas fluorescens* biofertilizer. Afr J Biotechnol 101:16579–16585. https://doi.org/10.5897/AJB12.2308

Anand K, Kumari B, Mallick MA (2016) Phosphate solubilizing microbes: an effective and alternative approach as bio-fertilizers. Int J Pharm Sci 8:37–40

Arao T, Takeda H, Nishihara E (2008) Reduction of cadmium translocation from roots to shoots in eggplant (*Solanum melongena*) by grafting onto Solanum torvum rootstock. Soil Sci Plant Nutr 54:555–559. https://doi.org/10.1111/j.1747-0765.2008.00269.x

Arndt W, Kolle C, Buchenauer H (1998) Effectiveness of fluorescent pseudomonads on cucumber and tomato plants under practical conditions and preliminary studies on the mode of action of the antagonists. J Plant Dis Prot 105:198–215

Bahadır PS, Liaqat F, Eltem R (2018) Plant growth pFromoting properties of phosphate solubilizing Bacillus species isolated from the Aegean Region of Turkey. Turk J Bot 42:183–196. https://doi.org/10.3906/bot-1706-51

Bashan Y (1998) Inoculants of plant growth-promoting bacteria for use in agriculture. Biotechnol Adv 16:729–770. https://doi.org/10.1016/S0734-9750(98)00003-2

Bhardwaj D, Ansari MW, Sahoo RK, Tuteja N (2014) Biofertilizers function as key player in sustainable agriculture by improving soil fertility, plant tolerance and crop productivity. Microb Cell Factories 66:1–10. https://doi.org/10.1186/1475-2859-13-66

Bhattacharyya PN, Jha DK (2012) Plant growth-promoting rhizobacteria (PGPR): emergence in agriculture. World J Microbiol Biotechnol 28:1327–1350. https://doi.org/10.1007/s11274-011-0979-9

Cassán F, Perrig D, Sgroy V, Masciarelli O, Penna C, Luna V (2009) *Azospirillum brasilense* Az39 and *Bradyrhizobium japonicum* E109, inoculated singly or in combination, pro-mote seed germination and early seedling growth in corn (*Zea mays* L.) and soybean (*Glycine max* L.). Eur J Soil Biol 45:28–35. https://doi.org/10.1016/j.ejsobi.2008.08.005

Christensen JH, Hewitson B, Busuioc A et al (2007) Regional climate projections. In: Solomon S, Qin D, Manning M, Chen Z, Marquis M, Averyt KB, Tignor M, Miller HL (eds) Climate change 2007: the physical science basis. Contribution of working group I to the fourth assessment report of the intergovernmental panel on climate change. Cambridge University Press, Cambridge, UK and New York, NY, pp 848–940

Colla G, Rouphael Y, Cardarelli M, Tullio M, Rivera CM, Rea E (2008) Alleviation of salt stress by arbuscular mycorrhizal in zucchini plants grown at low and high phosphorus concentration. Biol Fertil Soils 44:501–509. https://doi.org/10.1007/s00374-007-0232-8

Daei G, Ardekani MR, Rejali F, Teimuri S, Miransari M (2009) Alleviation of salinity stress on wheat yield, yield components, and nutrient uptake using Arbuscular mycorrhizal fungi under field conditions. J Plant Physiol 166:617–625. https://doi.org/10.1016/j.jplph.2008.09.013

Damam M, Kaloori K, Gaddam B, Kausar R (2016) Plant growth promoting substances (phytohormones) produced by rhizobacterial strains isolated from the rhizosphere of medicinal plants. Int J Pharm Sci Rev 37:130–136

Dardanelli MS, Carletti SM, Paulucci NS et al (2010) Benefits of plant growth-promoting rhizobacteria and rhizobia in agriculture. In: Maheshwari DK (ed) Plant growth and health promoting bacteria. Springer, Berlin, pp 1–20

Dasgan HY, Aydoner G, Akyol M (2010) Use of some microorganisms as bio-fertilizers in soilless grown squash for saving chemical nutrients. In: Procedings of the XXVIIIth IHC – IS on Greenhouse, pp 155–162

Dursun A, Ekinci M, Donmez MF (2010) Effects of foliar application of plant growth promoting bacterium on chemical contents, yield and growth of tomato (Lycopersicon esculentum L.) and cucumber (*Cucumis sativus* L.). Pak J Bot 42:3349–3356

Elazzazy AM, Almaghrabi OA, Moussa TA, Abdelmoneim TS (2012) Evaluation of some plant growth promoting rhizobacteria (pgpr) to control Pythium aphanidermatum in cucumber plants. Life Sci J 9:3147–3153

Elbeshehy EKF, Youssef SA, Elazzazy AM (2015) Resistance induction in pumpkin *Cucurbita maxima* L. against Watermelon mosaic potyvirus by plant growth-promoting rhizobacteria. Biocontrol Sci Tech 25:525–542. https://doi.org/10.1080/09583157.2014.994198

El-Borollosy AM, Oraby MM (2012) Induced systemic resistance against Cucumber mosaic cucumovirus and promotion of cucumber growth by some plant growth-promoting rhizobacteria. Ann Agric Sci 57:91–97. https://doi.org/10.1016/j.aoas.2012.08.001

El-Iklil Y, Karrou M, Benichou M (2000) Salt stress effect on epinasty in relation to ethylene production and water relations in tomato. Agron 20:399–406. https://hal.archives-ouvertes.fr/hal-00886055

El-Meihy RM (2016) Evaluation of pgpr as osmoprotective agents for squash (*Cucurbita pepo* L.) growth under drought stress. Middle East J 5:583–595

El-Sharkawy EES, Abdalla MY, El-Shemy AO (2015) Biological control and induction of systemic resistance against cucumber fusarium wilt by plant growth promoting rhizo-organisms. Egypt J Biol Pest Control 25:407–413

Elwan MWM, Abd EA (2015) Effects of plant growth promoting rhizobacteria on summer squash growth, yield, nutrients uptake and availability under nitrogen and phosphorus fertilization levels. Arab Univ J Agricul Sci 23:497–513

Esitken A, Ipek M, Arikan Ş, Aras S, Sahin M, Pirlak L, Donmez F, Turan M (2016) Effects of plant growth promoting rhizobacteria on Fe nutrition and FC-R activity of peach under calcareous soil conditions. J Plant Nutr Soil Sci (in process)

Falkenmark M (2013) Growing water scarcity in agriculture: future challenge to global water security. Philos Trans R Soc A 371:20120410. https://doi.org/10.1098/rsta.2012.0410

FAO (2016) Food and Agriculture Organization of the United Nations. http://www.fao.org/faostat/en/#data/TP. Accessed 14 Mar 2018

Farrag DK, Omara AA, El-Said MN (2015) Significance of foliar spray with some growth promoting rhizobacteria and some natural biostimulants on yield and quality of cucumber plant. Egypt J Hort 42:321–333

Firmansyah D (2017) Use of chitosan and plant growth promoting rhizobacteria to control squash mosaic virus on cucumber Plants. Institut Pertanian Bogor, pp-44

Fouzia A, Allaoua S, Hafsa C, Mostefa G (2015) Plant growth promoting and antagonistic traits of indigenous fluorescent *Pseudomonas* spp. Isolated from wheat rhizosphere and a thalamus endosphere. Eur Sci J 11:129–148

Foyer CH, Rasool B, Davey JW, Hancock RD (2016) Cross-tolerance to biotic and abiotic stresses in plants: a focus on resistance to aphid infestation. J Exp Bot 7:2025–2037. https://doi.org/10.1093/jxb/erw079

Francois LE (1985) Salinity effects on germination, growth, and yield of two squash cultivars. Horticience 20:1102–1104

Fravel DR (2005) Commercialization and implementation of biocontrol. Annu Rev Phytopathol 43:337–359. https://doi.org/10.1146/annurev.phyto.43.032904.092924

Gabriela F, Casanovas EM, Quillehauquy V, Yommi AK, Goni MG, Roura SI, Barassi CA (2015) Azospirillum inoculation effects on growth, product quality and storage life of lettuce plants grown under salt stress. Sci Hortic 195:154–162. https://doi.org/10.1016/j.scienta.2015.09.015

García-Gutiérrez L, Romero D, Zeriouh H et al (2012) Isolation and selection of plant growth-promoting rhizobacteria as inducers of systemic resistance in melon. Plant Soil 358:201–212. https://doi.org/10.1007/s11104-012-1173-z

García-Gutiérrez L, Romero D, Zeriouh H, Cazorla FM, Vicente AD, Pérez-García A (2009) Induction of systemic resistance by PGPR, a suitable means to consider for managing of cucurbit powdery mildew. Multitrophic Interact SoilIOBC/wprs Bull 42:83–86

Gerhardson B (2002) Biological substitutes for pesticides. Trends Biotechnol 20:338–343. https://doi.org/10.1016/S0167-7799(02)02021-8

Glick BR (2014) Bacteria with ACC deaminase can promote plant growth and help to feed the world. Microbiol Res 169:30–39. https://doi.org/10.1016/j.micres.2013.09.009

Glick BR, Cheng Z, Czarny J, Duan J (2007) Promotion of plant growth by ACC deaminase-producing soil bacteria. Eur J Plant Patholo 119:329–339. https://doi.org/10.1007/s10658-007-9162-4

Gopal S, Chandrasekaran M, Shagol C, Kim KY, Sa TM (2012) Spore associated bacteria (SAB) of arbuscular mycorrhizal fungi (AMF) and plant growth promoting rhizobacteria (PGPR) increase nutrient uptake and plant growth under stress conditions. Korean J Soil Sci Fert 45:582–592. https://doi.org/10.7745/KJSSF.2012.45.4.582

Gouda S, Kerry RG, Das G, Paramithiotis S, Shin HS, Patra JK (2018) Revitalization of plant growth promoting rhizobacteria for sustainable development in agriculture. Microbiol Res 206:131–140. https://doi.org/10.1016/j.micres.2017.08.016

Grichko VP, Glick BR (2001) Amelioration of flooding stress by ACC deaminase-containing plant growth-promoting bacteria. Plant Physiol Biochem 39:11–17. https://doi.org/10.1016/S0981-9428(00)01212-2

Gül A, Özaktan H, Kıdoğlu F, Tüzel Y (2013) Rhizobacteria promoted yield of cucumber plants grown in perlite under Fusarium wilt stress. Sci Hortic 153:22–25. https://doi.org/10.1016/j.scienta.2013.01.004

Gupta G, Parihar SS, Ahirwar NK, Snehi SK, Singh V (2015) Plant growth promoting rhizobacteria (PGPR): current and future prospects for development of sustainable agriculture. J Microb Biochem Technol 7:096–102. https://doi.org/10.4172/1948-5948.1000188

Haggag WM, Abouziena HF, Abd-El-Kreem F, Habbasha S (2015) Agriculture biotechnology for management of multiple biotic and abiotic environmental stress in crops. J Chem Pharm 7:882–889

Han HS, Lee KD (2006) Effect of co-inoculation with phosphate and potassium solubilizing bacteria on mineral uptake and growth of pepper and cucumber. Plant Soil Environ 52:130–136

Hassouna MG, El-Saedy MAM, Saleh HM (1998) Biocontrol of soil-borne plant pathogens attacking cucumber (Cucumis sativus) by Rhizobacteria in a semiarid environment. Arid Land Res Manag 12:345–357. https://doi.org/10.1080/15324989809381523

Horuz S, Aysa Y (2016) Biological control of watermelon seedling blight caused by acidovorax citrulli using antagonistic bacteria. Plant Protect Sci 1–9. https://doi.org/10.17221/168/2016-PPS

İmriz G, Özdemir F, Topal BE, Taş MN, Yakışır E, Okur O (2014) Bitkisel üretimde bitki gelişimini teşvik eden rizobakteri (pgpr)'ler ve etki mekanizmaları. Elektronik Mikrobiyoloji Dergisi 12:1–19

Ipek M, Aras S, Arikan Ş, Eşitken A, Pirlak L, Donmez F, Turan M (2017) Root plant growth promoting rhizobacteria inoculations increases ferric chelate reductase (FC-R) activity and Fe nutrition in pear under calcareous soil conditions. Sci Hortic 219:144–151. https://doi.org/10.1016/j.scienta.2017.02.043

Ipek M, Pirlak L, Esitken A, Figen Dönmez M, Turan M, Sahin F (2014) Plant growth-promoting rhizobacteria (PGPR) increase yield, growth and nutrition of strawberry under high-calcareous soil conditions. J Plant Nutr 37:990–1001. https://doi.org/10.1080/01904167.2014.881857

Isfahani FM, Besharati H (2012) Effect of biofertilizers on yield and yield components of cucumber. J Biol Earth Sci 2:83–92

Islam S, Akanda AM, Prova A, Islam MT, Hossain MM (2016) Isolation and identification of plant growth promoting rhizobacteria from cucumber rhizosphere and their effect on plant growth promotion and disease suppression. Front microbiol 6:1360. https://doi.org/10.3389/fmicb.2015.01360

Jeun YC, Park KS, Kim CH, Fowler WD, Kloepper JW (2004) Cytological observations of cucumber plants during induced resistance elicited by rhizobacteria. Biol Control 29(1):34–42. https://doi.org/10.1016/S1049-9644(03)00082-3

Jourdan E, Henry G, Duby F, Dommes J, Barthelemy JP, Thonart P, Ongena M (2009) Insights into the defenserelated events occurring in plant cells following perception of surfactin-type lipopeptide from *Bacillus subtilis*. Mol Plant Microbe Interact 22:456–468. https://doi.org/10.1094/MPMI-22-4-0456

Kabayashi N (1989) Suppression of *Rhizoctonia* and *Pythium* damping-off of cucumber by microorganisms in charcoal and VAM Fungi. In: Hattori H et al (eds) Research advances in microbial ecology. Japan Scientifi c Press, Japan, pp 242–246

Kalefetoğlu T, Ekmekçi Y (2005) The effects of drought on plants and tolerance mechanisms. Gazi Üniversitesi Fen Bilimleri Dergisi 18:723–740

Kanchiswamy CN, Malnoy M, Maffei ME (2015) Chemical diversity of microbial volatiles and their potential for plant growth and productivity. Front Plant Sci 6:151. https://doi.org/10.3389/fpls.2015.00151

Kang SM, Hamayun M, Joo GJ et al (2010) Effect of *Burkholderia* sp. KCTC 11096BP on some physiochemical attributes of cucumber. Eur J Soil Biology 46:264–268. https://doi.org/10.1016/j.ejsobi.2010.03.002

Kang SM, Radhakrishnan R, You YH et al (2015) Cucumber performance is improved by inoculation with plant growth-promoting microorganisms. Acta Agric Scand, Sect B Soil Plant Sci 65:36–44. https://doi.org/10.1080/09064710.2014.960889

Kang SM, Khan AL, Waqas M et al (2014) Plant growth-promoting rhizobacteria reduce adverse effects of salinity and osmotic stress by regulating phytohormones and antioxidants in *Cucumis sativus*. J Plant Int 9:673–682. https://doi.org/10.1080/17429145.2014.894587

Khabbaz SE, Abbasi PA (2013) Isolation, characterization, and formulation of antagonistic bacteria for the management of seedlings damping-off and root rot disease of cucumber. Can J Microbiol 60:25–33. https://doi.org/10.1139/cjm-2013-0675

Kidoglu F, Gül A, Ozaktan H, Tüzel Y (2007) Effect of rhizobacteria on plant growth of different vegetables. In: ISHS Acta Horticulturae 801: International Symposium on High Technology for Greenhouse System Management: Greensys 2007

Kim YC, Glick B, Bashan Y, Ryu CM (2013) Enhancement of plant droughttolerance by microbes. In: Aroca R (ed) Plant responses to drought stress. Springer, Berlin

Kloepper JW (1994) Plant growth-promoting rhizobacteria (other systems). In: Okon Y (ed) Azospirillum/plant associations. CRC Press, Boca Raton, FL, USA, pp 111–118

Kloepper JW, Schroth M (1978) Plant growth-promoting rhizobacteria on radishes. In Proceedings of the International Conference on Plant Pathogenic Bacter, vol 2, pp 879–882

Kloepper JW, Tuzun S, Liu L, Wei G (1993) Plant growth- promoting rhizobacteria as inducers of systemic disease resistance. In: Lumsden RD, Waughn JL (eds) Pest management: biologically based technologies. American Chemical Society Books, Washington, DC, pp 156–165

Kokalis-Burelle N, Vavrina CS, Reddy MS, Kloepper JW (2003) Amendment of muskmelon and watermelon transplant media with plant growth-promoting rhizobacteria: effects on seedling quality, disease, and nematode resistance. HortTechnology 13:476–482

Kumar P, Dubey RC (2012) Plant Growth Promoting Rhizobacteria for biocontrol of phytopathogens and yield enhancement of Phaseolus vulgaris. J Curr Perspect App Microbiol 1:6–38

Lee YH, Lee WH, Lee DK, Shim HK (2001) Factors relating to induced systemic resistance in watermelon by plant growth-promoting *Pseudomonas* sp. Plant Pathol J 17:174–179

Levanony H, Bashan Y (1989) Localization of specific antigens of Azospirillum brasilense Cd in its exopolysaccharide by immune-gold staining. Curr Microbiol 18:145–149

Ling N, Deng K, Song Y et al (2014) Variation of rhizosphere bacterial community in watermelon continuous mono-cropping soil by long-term application of a novel bioorganic fertilizer. Microbiol Res 169:570–578. https://doi.org/10.1016/j.micres.2013.10.004

Ling N, Zhang W, Tan S, Huang Q, Shen Q (2012) Effect of the nursery application of bioorganic fertilizer on spatial distribution of *Fusarium oxysporum* f. sp. *niveum* and its antagonistic bacterium in the rhizosphere of watermelon. Appl Soil Ecol 59:13–19. https://doi.org/10.1016/j.apsoil.2012.05.001

Liu D, Lian B, Dong H (2012) Isolation of *Paenibacillus* sp. and assessment of its potential for enhancing mineral weathering. J Geomicrobiol 29:413–421. https://doi.org/10.1080/01490451.2011.576602

Liu L, Kloepper JW, Tuzun S (1995) Induction of systemic resistance in cucumber against Fusarium wilt by plant growth-promoting rhizobacteria. Phytopathology 85:695–698

Lokesh S, Bharath BG, Raghavendra VB, Govindappa M (2007) Importance of plant growth-promoting rhizobacteria in enhancing the seed germination and growth of watermelon attacked by fungal pathogens. Acta Agron Hung 55:243–249. https://doi.org/10.1556/AAgr.55.2007.2.12

Lu G, Chen G, Qi G, Gao Z (2006) Effects of arbuscular mycorrhizal fungi on the growth and fruit quality of plastic greenhouse *Cucumis sativus* L. J Appl Ecol 17:2352–2356

Mangmang JS, Deaker R, Rogers G (2015) Early seedling growth response of lettuce, tomato and cucumber to *Azospirillum brasilense* inoculated by soaking and drenching. Hort Sci 42: 37–46. https://doi.org/10.17221/159/2014-HORTSCI

Marschner H (1995) Mineral Nutrition of Higher Plants. Academic Press, San Diego, CA

Mayak S, Tirosh T, Glick BR (2004) Plant growth-promoting bacteria that confer resistance to water stress in tomatoes and peppers. Plant Sci 166:525–530. https://doi.org/10.1016/j.plantsci.2003.10.025

McCullagh M, Utkhede R, Menzies JG, Punja ZK, Paulitz TC (1996) Evaluation of plant growth-promoting rhizobacteria for biological control of Pythium root rot of cucumbers grown in rockwool and effects on yield. Eur J Plant Pathol 102:747–755

McKeon TA, Fernandez-Maculet JC, Yang SF (1995) Biosynthesis and metabolism of ethylene. In: Davies PJ (ed) Plant hormones physiology, biochemistry and molecular biology. Kluwer Academic Publishers, Dordrecht, Netherlands, pp 118–139

Medeiros FH, Moraes IS, da Silva Neto EB, Silveira EB, Rosa de Lima RM (2009) Management of melon bacterial blotch by plant beneficial bacteria. Phytoparasitica 37:453–460. https://doi.org/10.1007/s12600-009-0063-2

Mo X (2013) Interactions between plant growth-promoting rhizobacteria (PGPR) and squash plants—the role of PGPR-mediated suppression of phytophthora blight. (Doctoral dissertation, University of Florida, pp 187

Moussa SAM (2006) Interactıon effect between phosphate dissolving micro-organisms and boron on squash (*Cucurbita pepo L.*) growth, endogenous phyto. J Biol Chem Environ Sci 1:751–774

Nadeem SM, Ahmad M, Zahir ZA, Javaid A, Ashraf M (2014) The role of mycorrhizae and plant growth promoting rhizobacteria (PGPR) in improving crop productivity under stressful environments. Biotechnol Adv 32:429–448. https://doi.org/10.1016/j.biotechadv.2013.12.005

Nga NT, Tien DT, Linh VT et al (2013) Control of plant diseases by the endophytic rhizobacterial strain *Pseudomonas aeruginosa* 23 1-1. Recent advances in biofertilisers and biofungicides (PGPR) for sustainable agriculture, pp 8–18

Nga NTT, Giau NT, Long NT et al (2010) Rhizobacterially induced protection of watermelon against Didymella bryoniae. J App Microbiol 109:567–582. https://doi.org/10.1111/j.1365-2672.2010.04685.x

Nimnoi P, Pongsilp N, Lumyong S (2014) Co-inoculation of soybean (*Glycine max*) with actinomycetes and Bradyrhizobium japonicum enhances plant growth, nitrogenase activity and plant nutrition. J Plant Nut 37:432–446. https://doi.org/10.1080/01904167.2013.864308

Noumavo PA, Agbodjato NA, Baba-Moussa F, Adjanohoun A, Baba-Moussa L (2016) Plant growth promoting rhizobacteria: beneficial effects for healthy and sustainable agriculture. Afric J Biotech 15:1452–1463. https://doi.org/10.5897/AJB2016.15397

Oerke EC (2005) Crop losses to pests. J Agric Sci 144:31–43. https://doi.org/10.1017/S0021859605005708

Oliver MA (1997) Soil and human health: a review. Eur J Soil Sci 48:573–592. https://doi.org/10.1111/j.1365-2389.1997.tb00558.x

Ongena M, Jacques P (2007) Bacillus lipopeptides: versatile weapons for plant disease biocontrol. Trends Microbiol 16:115–125. https://doi.org/10.1016/j.tim.2007.12.009

Ongena MARC, Daayf F, Jacques P et al (2000) Systemic induction of phytoalexins in cucumber in response to treatments with fluorescent pseudomonads. Plant Pathol 49:523–530. https://doi.org/10.1046/j.1365-3059.2000.00468.x

Örs S, Ekinci M (2015) Kuraklık stresi ve bitki fizyolojisi. Derim 32:237–250

Palacio-Rodríguez R, Coria-Arellano JL, López-Bucio J et al (2017) Halophilic rhizobacteria from Distichlis spicata promote growth and improve salt tolerance in heterologous plant hosts. Symbiosis 73:179–189. https://doi.org/10.1007/s13199-017-0481-8

Park K, Park JW, Lee SW, Balaraju K (2013) Disease suppression and growth promotion in cucumbers induced by integrating PGPR agent Bacillus subtilis strain B4 and chemical elicitor ASM. Crop Prot 54:199–205. https://doi.org/10.1016/j.cropro.2013.08.017

Parmar P, Sindhu SS (2013) Potassium solubilisation by Rhizosphere Bacteria: influence of nutritional and environmental conditions. J Microb Res 3:25–31. https://doi.org/10.5923/j.microbiology.20130301.04

Patten CL, Glick BR (2002) Role of *Pseudomonas putida* indoleacetic acid in development of the host plant root system. Appl Environ Microbiol 68:3795–3801. https://doi.org/10.1128/AEM.68.8.3795-3801.2002

Pieterse CMJ, Leon-Reyes A, Van der Ent S, Van Wees SCM (2009) Networking by small-molecule hormones in plant immunity. Nat Chem Biol 5:308–316. https://doi.org/10.1038/nchembio.164

Pii Y, Marastoni L, Springeth C et al (2016) Modulation of Fe acquisition process by Azospirillum brasilense in cucumber plants. Environ Exp Bot 130:216–225. https://doi.org/10.1016/j.envexpbot.2016.06.011

Pii Y, Penn A, Terzano R, Crecchio C, Mimmo T, Cesco S (2015) Plant-microorganism-soil interactions influence the Fe availability in the rhizosphere of cucumber plants. Plant Physiol Biochem 87:45–52. https://doi.org/10.1016/j.plaphy.2014.12.014

Rahmatullah M, Biswas A, Haq WM, Seraj S, Jahan R (2012) An ethnomedicinal survey of cucurbitaceae family plants used in the folk medicinal practices of Bangladesh. Chron Young Sci 3(3):212

Ramadan EM, AbdelHafez AA, Hassan EA, Saber FM (2016) Plant growth promoting rhizobacteria and their potential for biocontrol of phytopathogens. Afr J Microbiol Res 10:486–504. https://doi.org/10.5897/AJMR2015.7714

Raupach GS, Kloepper JW (1998) Mixtures of plant growth-promoting rhizobacteria enhance biological control of multiple cucumber pathogens. Phytopathology 88:1158–1164. https://doi.org/10.1094/PHYTO.1998.88.11.1158

Ravnskov S, Jakobsen I (1999) Effects of Pseudomonas fluorescens DF57 on growth and P uptake of two arbuscular mycorrhizal fungi in symbiosis with cucumber. Mycorrhiza 8:329–334

Raza W, Ling N, Yang L, Huang Q, Shen Q (2016) Response of tomato wilt pathogen *Ralstonia solanacearum* to the volatile organic compounds produced by a biocontrol strain *Bacillus amyloliquefaciens* SQR-9. Sci Rep 6:24856. https://doi.org/10.1038/srep24856

Reddy PP (2016) Plant growth promoting rhizobacteria for horticultural crop protection. Springer. Indian Institute of Horticultural Research, Bangalore, Karnataka, India, pp 310

Reed ML, Glick BR (2005) Growth of canola (*Brassica napus*) in the presence of plant growth-promoting bacteria and either copper or polycyclic aromatic hydrocarbons. Can J Microbiol 51:1061–1069. https://doi.org/10.1139/w05-094

Refai EF, Foly H, Dakhly OF (2010) Growth and yield of zucchini type summer squash (*Cucurbita pepo* L.) fertilized by combined Azotobacter chroococum mutants and mineral n-fertilization. Egypt J Agric Res 88:241–255. https://doi.org/10.1139/w05-094

Safronova VI, Stepanok VV, Engqvist GL, Alekseyev YV, Belimov AA (2006) Root-associated bacteria containing 1- minocyclopropane- 1-carboxylate deaminase improve growth and nutrient uptake by pea genotypes cultivated in cadmium supplemented soil. Biol Fert Soils 42:267–272. https://doi.org/10.1007/s00374-005-0024-y

Saharan BS, Nehra V (2011) Plant growth promoting rhizobacteria: a critical review. Life Sci Med Res 21:30

Sairam KVSS, Reddy NSR (2013) Global success of prathista industries organic agri-inputs for sustainable agriculture. In: Recent advances in biofertilizers and biofungicides (PGPR) for sustainable agriculture. Proceedings of 3rd Asian conference on plant growth-promoting rhizobacteria (PGPR) and other microbials, manila, philippines, 21–24 April, pp 8–12

Salman M, Shahin N, Abu-Khalaf N et al (2017) Antagonistic activity of *Pseudomonas fluorescens* against *Fusarium oxysporum* f. sp. Nievum isolated from soil samples in palestine. J Plant Stud 6:2. https://doi.org/10.5539/jps.v6n2p1

Sanchis V, Bourguet D (2008) Bacillus thuringiensis: applications in agriculture and insect resistance management: a review. Agron Sustain Dev 28:11–20. https://doi.org/10.1051/agro:2007054

Santoro MV, Bogino PC, Nocelli N, Cappellari LR, Giordano WF, Banchio E (2016) Analysis of plant growth promoting effects of Fluorescent pseudomonas strains isolated from Mentha piperita Rhizosphere and effects of their volatile organic compounds on essential oil composition. Front Microbiol 1085:1–17. https://doi.org/10.3389/fmicb.2016.01085

Savvas D, Colla G, Rouphael Y, Schwarz D (2010) Amelioration of heavy metal and nutrient stress in fruit vegetables by grafting. Sci Hortic 127:156–161. https://doi.org/10.1016/j.scienta.2010.09.011

Selvakumar G, Mohan M, Kundu S et al (2008) Cold tolerance and plant growth promotion potential of Serratia marcescens strain SRM (MTCC 8708) isolated from flowers of summer squash (*Cucurbita pepo*). Lett Appl Microbiol 46:171–175. https://doi.org/10.1111/j.1472-765X.2007.02282.x

Sensoy S, Ertek A, Gedik I, Kucukyumuk C (2007) Irrigation frequency and amount affect yield and quality of field-grown melon (*Cucumis melo* L.). Agri Wat Manag 88:269–274. https://doi.org/10.1016/j.agwat.2006.10.015

Setiawati TC, Mutmainnah L (2016) Solubilization of Potassium containing mineral by microorganisms from sugarcane Rhizosphere. Agri Sci Procedia 9:108–117. https://doi.org/10.1016/j.aaspro.2016.02.134

Seymen M, Eyice R, Türkmen Ö, Paksoy M, Dönmez FM, Dursun A (2015a) Bazı Bakteri Aşılamalarının Hıyarın (*Cucumis sativus* L.) Besin Elementi İçeriğine Etkileri. Selçuk Tarım ve Gıda Bilimleri Dergisi 27:1–7

Seymen M, Turkmen O, Dursun A, Donmez MF, Paksoy M (2010) Effects of bacterium inoculation on yield and yield components of cucumber (*Cucumis sativus*). Bull UASVM Hortic 67:274–277

Seymen M, Türkmen Ö, Dursun A, Paksoy M (2014) Effects of bacteria inoculation on yield, yield components and mineral contents of tomato. Selcuk J Agr Food Sci 28:52–57

Seymen M, Türkmen Ö, Dursun A, Paksoy M, Dönmez MF (2013b) Effects of bacteria inoculation on yield, yield components and mineral composition in eggplant (*Solanum melongena* L.). In: ICOEST Conference 2013 (Special Issue-1) pp, 403–413. Journal of Selcuk University Natural and Applied Science

Seymen M, Türkmen Ö, Paksoy M (2013a) Effects of plant growth-promoting rhizobacteria (PGPR) on yield, yield components and mineral contents of pepper under greenhouse conditions. J Ecosys Ecolo Sci 3:645–650

Seymen M, Türkmen Ö, Paksoy M (2015a) Bacteria inoculation affect on yield, yield components and mineral contents of *(Capsicum annum L.)* bell pepper. In: 7th ICGBEEAH2015 July 10–11, pp 87–94. ISSN 1947-8321

Seymen M, Yavuz D, Yavuz N, Türkmen Ö (2016) Effect on yield and yield components of different irrigation levels in edible seed pumpkin growing. Int J Biol Biomol Agric Food Biotechnol Eng 10:214–219

Shaharoona B, Arshad M, Zahir ZA (2006) Effect of plant growth promoting rhizobacteria containing ACC-deaminase on maize (*Zea mays* L.) growth under exenic conditions and on nodulation in mung bean. Lett Appl Microbiol 42:155–159. https://doi.org/10.1111/j.1472-765X.2005.01827.x

Sharma A, Johri BN, Sharma AK, Glick BR (2013) Plant growth-promoting bacterium Pseudomonas sp. strain GRP3 influences iron acquisition in mung bean (Vignaradiata L. Wilzeck). Soil Biol Biochem 35:887–894. https://doi.org/10.1016/S0038-0717(03)00119-6

Sharma RK, Agrawal M, Marshall F (2007) Heavy metal contamination of soil and vegetables in suburban areas of Varanasi, India. Ecotoxicol Environ Saf 66:258–266. https://doi.org/10.1016/j.ecoenv.2005.11.007

Shridhar BS (2012) Review: nitrogen fixing microorganisms. Int J Microb Res 3:46–52

Sikora RA (1992) Management of the antagonistic potential in agricultural ecosystems for the biological control of plant parasitic nematodes. Ann Rev Phytopatho 30:245–270. https://doi.org/10.1146/annurev.py.30.090192.001333

Singh RP, Jha PN (2015) Molecular identification and characterization of Rhizospheric bacteria for plant growth promoting ability. Int J Curr Biotechnol 3:12–18

Sokolova MG, Akimova GP, Vaishlia OB (2011) Effect of phytohormones synthesized by rhizosphere bacteria on plants. Prikl Biokhim Mikrobiol 47:302–307. https://doi.org/10.1134/S0003683811030148

Sturz AV, Christie BR, Novak J (2000) Bacterial endophytes: potential role in developing sustainable system of crop production. Crit Rev Plant Sci 19:1–30. https://doi.org/10.1080/07352680091139169

Sureshbabu K, Amaresan N, Kumar K (2016) Amazing multiple function properties of plant growth promoting rhizobacteria in the rhizosphere soil. Int J Curr Microbiol Appl Sci 5:661–683. http://dx.doi.org/10.20546/ijcmas.2016.502.074

Tchiaze AI, Taffouo VD, Fankem H et al (2016) Influence of nitrogen sources and plant growth-promoting rhizobacteria Inoculation on growth, crude fiber and nutrient uptake in squash. Not Bot Horti Agrobot Cluj-Napoca 44:53. https://doi.org/10.15835/nbha44110169

Timmusk S, Islam A, El Abd D et al (2014) Drought-tolerance of wheat improved by rhizosphere bacteria from harshenvironments: enhanced biomass production and reduced emissions of stressvolatiles. PLoS One 9:1–13. https://doi.org/10.1371/journal.pone.0096086

Tóth B, Lévai L, Kovács B, Varga M, Veres S (2013) Compensation effect of bacterium containing biofertilizer on the growth of *Cucumis sativus* L. under Al-stress conditions. Acta Biol Hung 64:60–70. https://doi.org/10.1556/ABiol.64.2013.1.6

Utkhede RS, Koch CA (1999) Rhizobacterial growth and yield promotion of cucumber plants inoculated with Pythium aphanidermatum. Can J Plant Path 21:265–271. https://doi.org/10.1080/07060669909501189

Van Loon LC (2007) Plant responses to plant growth-promoting rhizobacteria. Eur J Plant Pathol 119:243–254. https://doi.org/10.1007/s10658-007-9165-1

Vessey JK (2003) Plant growth promoting rhizobacteria as biofertilizers. Plant Soil 255:571–586

Vikram A, Hamzehzarghani H, Alagawadi AR, Krishnaraj PU, Chandrashekar BS (2007) Production of plant growth promoting substances by phosphate solubilizing bacteria isolated from vertisols. J Plant Sci 2:326–333

Wang CJ, Yang W, Wang C et al (2012) Induction of drought tolerance in cucumber plants by a consortium of three plant growth-promoting rhizobacterium strains. PLoS One 7:e52565. https://doi.org/10.1371/journal.pone.0052565

Wang T, Liu MQ, Li HX (2014) Inoculation of phosphate-solubilizing bacteria Bacillus thuringiensis B1 increases available phosphorus and growth of peanut in acidic soil. Acta Agr Scand B-S P 64:252–259. https://doi.org/10.1080/09064710.2014.905624

Wu HS, Yang XN, Fan JQ et al (2009) Suppression of Fusarium wilt of watermelon by a bio-organic fertilizer containing combinations of antagonistic microorganisms. BioControl 54:287–300. https://doi.org/10.1007/s10526-008-9168-7

Wu QS, Zou YN, Liu W, Ye XF, Zai HF, Zhao LJ (2010) Alleviation of salt stress in citrus seedlings inoculated with mycorrhiza: changes in leaf antioxidant defense systems. Plant Soil Environ 56:470–475

Yang JW, Kloepper JW, Ryu CM (2009) Rhizosphere bacteria help plants tolerate abiotic stress. Trends Plant Sci 14:1–4. https://doi.org/10.1016/j.tplants.2008.10.004

Yaoyao E, Yuan J, Yang F et al (2017) PGPR strain *Paenibacillus polymyxa*SQR 21 potentially benefits watermelon growth by re shaping root protein expression. AMB Expr 7:104. https://doi.org/10.1186/s13568-017-0403-4

Yavuz D, Seymen M, Yavuz N, Türkmen Ö (2015a) Effects of irrigation interval and quantity on the yield and quality of confectionary pumpkin grown under field conditions. Agri Wat Manag 159:290–298. https://doi.org/10.1016/j.agwat.2015.06.025

Yavuz D, Yavuz N, Seymen M, Türkmen Ö (2015b) Evapotranspiration, crop coefficient and seed yield of drip irrigated pumpkin under semi-arid conditions. Sci Hortic 197:33–40. https://doi.org/10.1016/j.scienta.2015.11.010

Yıldırım E, Ekinci M, Dursun A, Karagöz K (2015) Plant growth-promoting rhizobacteria improved seedling growth and quality of cucumber (*Cucumis sativus* L.). In: International Conference on Chemical, Food and Environment Engineering (ICCFEE'15) Jan 11–12, Dubai (UAE), pp 6–8

Yıldırım E, Taylor AG, Spittler TD (2006) Ameliorative effects of biological treatments on growth of squash plants under salt stress. Sci Hortic 111:1–6. https://doi.org/10.1016/j.scienta.2006.08.003

Zahir ZA, Ghani U, Naveed M, Nadeem SM, Asghar HN (2009) Comperative effectiveness of rhizobacteria containing ACC-deaminase for growth promotion of pea (*Pisum sativum*) under drought conditions. J Microbiol Biotechnol 18:958–963

Zaidi A, Khan MS, Ahemad M, Oves M (2009) Plant growth promotion by phosphate solubilizing bacteria. Acta Microbiol Immunol Hung 56:263–284. https://doi.org/10.1556/AMicr.56.2009.3.6

Zehnder G, Kloepper J, Tuzun S et al (1997) Insect feeding on cucumber mediated by rhizobacteria-induced plant resistance. Entomol Exp Appl 83:81–85. https://doi.org/10.1046/j.1570-7458.1997.00159.x

Zehnder GW, Murphy JF, Sikora EJ, Kloepper JW (2001) Application of rhizobacteria for induced resistance. Eur J Plant Patholog 107:39–50

Zhang S, White TL, Martinez MC, McInroy JA, Kloepper JW, Klassen W (2010) Evaluation of plant growth-promoting rhizobacteria for control of Phytophthora blight on squash under greenhouse conditions. Biolog Cont 53:129–135. https://doi.org/10.1016/j.biocontrol.2009.10.015

Zhao J, Liu J, Liang H et al (2018) Manipulation of the rhizosphere microbial community through application of a new bioorganic fertilizer improves watermelon quality and health. PLoS One 13:e0192967. https://doi.org/10.1371/journal.pone.0192967
Zhu JK, Hasegawa PM, Bressan RA (1997) Molecular aspects of osmotic stress in plants. Crit Rev Plant Sci 16:253–277. https://doi.org/10.1080/07352689709701950
Zulkarami B, Tajul MI, Fariz A et al (2012) Effects of bacteria and arbuscular mycorhizae inoculation at different electrical conductivity level on growth and yield of rockmelon (*Cucumis melo*) under soilless culture. Austr J Crop Sci 6:1494–1501

Chapter 5
Harnessing Beneficial *Bacillus* in Productivity Improvement of Food Security Crops of Himalayan Agro-Climatic Zones

Shrivardhan Dheeman, Dinesh Kumar Maheshwari, Ramesh Chand Dubey, Sandeep Kumar, Nitin Baliyan and Sandhya Dhiman

Abstract Food security is a burning problem before scientists, which mainly concern with the scarcity and accessible food for all. Scientists are ever-involved to find-out the solutions to overcome on these hurdles, majorly by boosting soil fertility. Over the last 20 to 30 years, the role of soil microbiome in the improvement of soil fertility and crop productivity by involving role of the microbial life has been emerged sustain and facilitated soil nutrients availability to the plants. The microbial life as the great engines enriching soil and helping to facilitate the breakdown of organic matter, that can be recycled into new life, if only they could be unlocked from that organic matter and also, from the mineral matter. The microbes, especially endospore forming *Bacilli* are incredibly important in the

S. Dheeman (✉) · D. K. Maheshwari · R. C. Dubey · S. Kumar · N. Baliyan · S. Dhiman
Department of Botany and Microbiology, Gurukula Kangri Vishwavidyalaya, Haridwar 249404, Uttarakahnd, India
e-mail: svdheeman@gmail.com

D. K. Maheshwari
e-mail: maheshwaridk@gmail.com

R. C. Dubey
e-mail: profrcdubey@gmail.com

S. Kumar
e-mail: sandeepchokar@gmail.com

N. Baliyan
e-mail: baliyan.nitin394@gmail.com

S. Dhiman
e-mail: sandhyadhiman93@gmail.com

S. Dheeman
Department of Microbiology, Sardar Bhagwan Singh University, Dehradun 248161, Uttarakhand, India

D. K. Maheshwari and S. Dheeman (eds.), *Field Crops: Sustainable Management by PGPR*, Sustainable Development and Biodiversity 23,
https://doi.org/10.1007/978-3-030-30926-8_5

rhizosphere. *Bacillus* is a versatile candidate providing its services in plant growth and health promotion tirelessly. This aerobic microorganism is powered with super abilities to produce endospore resting fruiting bodies, whenever it feels unhappy under adversity. *Bacillus* is also known for its several characteristics, to help crop plants by virtue of producing metabolites similar to plant in the form of phytohormones. Besides, it is able to mineralize rocks and minerals into bio-available form and can metabolize organic matter into much simpler forms. Plant pumps huge amount of sugar and other organic matter in the soil through roots, which attract bacteria to fasten their growth on spent of these supplements. In addition, *Bacillus* is known to produce sort of secretions of which assist in root colonization due to rhizo-competence on root surface and sometimes invade in root cells. In this chapter, we have reviewed the benefits of *Bacillus* species to the food security crops in term of raising productivity and yield.

Keywords *Bacillus* · Food security · *Eleusine coracona* · *Amaranth* · Buckwheat

5.1 Introduction

Mankind is practicing agriculture since time immemorial for their survival and proliferation on earth. However, due to ever increasing population and depleting amount of food raised an alarming situation. This situation can be revised looking in thorough history of agricultural revolutions. Presently, the agricultural fields being fertilized abundantly with pesticides and agrochemicals that were began in nineteenth and twentieth centuries. Initially, the application of agrochemical had shown quantum jump in crop productivity but later, with the rapid technological evolution and un-precedent application of increased dose of fertilizers declined in fertility indices of soil. This not only increased the soil toxicity, but also increased the financial un-sustainability of farmers especially in Asia. Hence, application of chemicals is no more a solution for sustainable crop production. It is accounted that around 870 million populations are under hunger-stricken slab due to severe scarcities of accessible food and FAO in 2012 putted several reports addressed the need to increase the agriculture production by 60% to plateauing food production corresponding to population blast, which is roughly estimated to reach about 9 billion by 2050.

In the present scenario, farmers have blind reliance on the use of chemical fertilizers and pesticides, which not only depleting nutrition index in crops but also affecting microscopic life of soil. Across the globe, governments are now discouraging farmers for the use the chemical fertilizers and pesticides in agriculture production. To sort this matter, farmers have to raise trend of organic crop production and their alternatives specifically, the use of beneficial microorganisms. Actually, earlier, it was not believed that microbial lives in the soil can naturally regulates the ecology of soil and provide plenty and sustained nutrients for crop health and growth. But, now at this diaspora, to obviate these problems, it has become an indispensable need to administer such methods of farming which are ecologically compatible, holistic and organic in nature. Sustainable approaches in agricultural sector are of utmost importance to improve the food security and nutrition problem around the globe (Arora 2018).

Looking to the history, the very first microbe as the fertilizer was symbiotic nitrogen fixing strain of rhizobia that was patented in the year 1896 under the brand name "Nitragin" for 17 different leguminous plants. Evolution in the knowledge about the role of microbes in agriculture resulted over 2000 tons of rhizobia based inoculants production in the recent scenario which is increasing exponentially and regressed in market size to be reached around 13.25% (annual growth rate) by the year 2020. On the other hand, the microbe-based inoculants are lower in cost, protect indigenous flora and fauna and other natural resources, have less or no harmful byproducts, improve soil organic matter and above all, maintain sustainability of ecosystems (Arora 2018). At present we know that biological alternatives are the only ways to save planet Earth from the abusive chemical fertilizers entering in our food chain and causing serious health hazards to mankind.

Food security and climate change are emerging issues all over the world and consequently resulting serious influence on crop production in general and pressure for higher production of major cereal crops in particular (Asseng et al. 2013; Springmann et al. 2016). The major crops such as rice, wheat, and maize are cultivating to mitigate food demand since a decade besides, over-looked cultivation of other crops such as millets bearing rich amount of nutrients, i.e. essential amino acids, vitamins, minerals etc. is an another cause of nutrient deficiency and malnutrition especially in children (Gundersen and Ziliak 2015).

Cultivation of major crops due to their high demand (domestic as well as industrial) and good economic value, the other nutritional crops has attracted less attention after evergreen revolution (Swaminathan et al. 1998). The marginalized and poor farmers of India, particularly in remote area(s) of hill districts lies in the provinces of Kumaun and Garhwal of Indian Himalaya are deprived of procuring better seeds and practices for raising the crop yield.

In the geographical point of view, the mighty Himalaya is a scenic beauty of diverse ecosystems and eco-regions, including different forest types and eco-agriculture regions at distinct elevation bands. Majorly, the Himalayan subtropical covers approximately 3,000 km pine forests. Moreover, it expenses into broadleaf forests covering an area of 38,000 km^2. It is also coincide with Western Himalaya covering subalpine conifer forests and appear as temperate forests. The temperate forests in the eastern Himalayas also include broadleaf forests, subalpine conifer forests, Northern Triangle temperate forests, and Northeastern Himalayan subalpine conifer forests. Thus, Himalaya itself enunciates its wealth in biodiversity to emerge as hotspots on the world's chart and offer a haven for endemic plants and animals. On the other hand, The Himalayan terrains also have significant agro-climatic zone suitable for the cultivation of range of crops and cultivars in relatively less area available for the cultivation of agricultural crops.

Biogeography of Uttarakhand Hills in India differs from plain especially to the concern of topography, elevation, geographical features, diversity of habitats for flora and fauna, ethnic diversity, land use system and socio-economic conditions. Mehta et al. (2010) stated the productivity of cereals, oilseed crop, millets and vegetables is major hindrance in the food security of the region. Bungla et al. (2012) stated, among the diverse food and staple crops grown in Uttarakhand region *Mandua* (local name) or *Ragi* (commercial name) may act as an alternative of wheat. Mehta et al. (2010) underpin several non-conventional but ethnic crops in food security. According to National Food Security Mission (NFSM 2012) and clause of Food Security Bill (NFSB 2013), several native crops of Himalayan provinces comes under food security and, therefore, essential to be considered for increase their production and productivity using microbial life back into course of fertilization and plant growth improvement.

To mitigate the hurdles of productivity with the benign use of microbes in general and aerobic endospore forming *Bacillus* in particular improvements in soil nutritional indices and soil fertility is needed, thereby stated that there are certain crops which are nutritionally rich but neglected by the farmers. Such crops are defined as "Underutilized crops" which include grain amaranth, barley, ragi, sorghum, buckwheat, barnyard millet, etc., growing in agricultural belts of mighty Himalaya covering the Garhwal ranges of Himalaya. This review is focused to assign the introduction to inoculants of crops, economic importance, productivity status, and current advancement of use of *Bacillus* for raising these valuable crops in a package. These further extend the horizons to address the hurdles of cultivating "Food Security" on higher altitude of agricultural lands. The review is outlined to explore the current prospects and scenario of using beneficial soil inhabiting "*Bacilli*" to the food security crops for enhancement of production enhancement.

5.2 Need of Crop Productivity Improvement

As earlier, the sound problems of agriculture in the means yet to meet the challenges of food requirements of the burgeoning population and plateauing productivity of agricultural lands persuade via adopting "Ever-sustained Green Revolution". The major crop such as rice and wheat were largely and adaptively cultivated on the call of green revolution (Swaminathan 2006) but now, found no more sustainable to mitigate food demand and affordable agriculture, besides hitting back the fertility of soil. Since, these crops are staple food, scientists coming across the world have paid ultimate attention to enhance the production of rice and wheat. Magkos et al. (2003) stated that most of these crops possess less amounts of essential nutrients such as amino acids, vitamins, minerals, etc., thus do not meet out the complete nutrition requirement for a healthy life.

On the other hand, Blum and Bartha (1980) mentioned that the chemicals such as pesticides and fertilizers got accumulated in the ecosystem and consequently enter in the food chain, when used in high doses in order to raise disease free crops.

The pesticides once escaped in the environment influence soil and climatic factors according to their half-life, amounts applied and physicochemical reactions with the existing minerals of the soil (Navarro et al. 2007). Pesticide residues have been detected from various leafy vegetables including spinach, fenugreek, mustard and cabbage ($\sim$21.5 ppm), tomato ($\sim$17.5 ppm), cauliflower ($\sim$1.70 ppm), etc., which unfortunately are above the maximum acceptable daily intake (ADI) as prescribed by WHO in India (Bakore et al. 2002). The subsequent investigation revealed the presence of significant amounts of pesticide residues in the ground water resources and trophic levels at a magnitude (Maloney 2001). Chauhan and Singhal (2006) examined these residual chemicals and reported mutations in chromosomes of humans, animals and thereby inducing carcinoma of lung and liver. The harmful effect of such chemicals in ground water, and agricultural products have been discussed in various studies elaborating agriculture-environment relationship (Eser and Geçit 2010; Meena et al. 2016a, b; Chandra et al. 2018). Majorly, under-utilized crops including amaranth (*Amaranthus hypochondriacus*), barley (*Hordeum vulgare*), buckwheat (*Fagopyrum esculentum*), finger millet (*Eleusine coracana*), and barnyard millet (*Echinochloa frumentacea*) enriched with plenty essential nutrients so as to complete a balanced food and can be used to accomplish the escalating food requirement (Mayes et al. 2011).

5.2.1 Solution(s) for Enhancement of Crop Productivity

The harmful effect of agrochemicals as a concern among the scientific community and also provoked to search other avenue so as to earn better productivity. One of the eco-friendly and cost effective approaches is the use of beneficial group of bacteria such as plant growth promoting rhizobacteria (PGPR) as a bio-fertilizer and biocontrol agents to reduce crop losses and enhance crop productivity and soil fertility. Along with plant growth promotion they reforest eroded areas, restored the contaminated sites and thereby cause positive effect on degraded ecosystem (Gupta et al. 2015). These PGPR seemed as successful bacteria in getting established in soil ecosystem due to their high adaptability in a wide variety of environments, faster growth rate and biochemical versatility to metabolize a wide range of natural and xenobiotic compounds. Several worker have earlier studied that in the rhizosphere ecology, plant growth promoting rhizobacteria (PGPR) enhance plant growth by a wide variety of functions like biological nitrogen fixation, phytohormone production, phosphate solubilization, siderophore production, 1-Aminocyclopropane-1-carboxylate deaminase production (ACC), exhibiting antifungal activity due to the secretion of chitinolytic enzymes and production of volatile organic compounds (VOCs), promoting beneficial plant-microbe symbiosis, interference with pathogen toxin production, etc. (Nelson 2004; Babalola 2010; Bhattacharya and Jha 2012; Kundan et al. 2015; Prasad et al. 2015; Maheshwari 2010, 2011a, b, c, 2015; Goswami et al. 2016). Successful studies using PGPR on the growth enhancement of various crops have been achieved in laboratory and field conditions (Gray and

Smith. 2005; Agarwal et al. 2017). In the recent scenario, harnessing the benefits of PGPR exhibited quite interesting to unlock new horizons of sustainable agriculture (Maheshwari et al. 2013).

To overcome these situations, there is a need to develop eco-friendly and cost-effective microbial inoculants which may be withstand in the extreme environmental conditions of Himalayas. In Himalayan hills, soil temperature remains low (6–25 °C), during different cropping seasons. Few other stresses such as water-deficit and salinity stress also impose serious impact on morphology and physiology of plants in the mighty Himalaya. PGPB are known to improve crop health even at low soil temperature (Mishra et al. 2012) and the survival of the endospore forming *Bacilli* at different temperatures makes them suitable candidates for enhancement of vegetative and reproductive plant growth even in adverse conditions of Himalayan hills. Various workers across the globe stated that these bacteria are well known for their beneficial effect on different host plant (Schroth and Kloepper 1978; Glick 1995; Gray and Smith 2005; Yadav et al. 2014; Maheshwari et al. 2015a, b; Agarwal et al. 2017; Gouda et al. 2017).

As stated, PGPR ameliorate plant health and productivity by enhancing the nutrient status of soil and host plants subsequently (Dey et al. 2004). The bioavailability of nutrients, their increased uptake may significantly enhance the nutrient use efficiency of plants. Increased solubilization of nutrients (macro and micro) by PGPBs enhanced their uptake and accumulation in grain amaranth (Parmar and Patel 2009). Nutrient availability has been influenced by solubilization, chelation and oxidation–reduction reaction in soil (de Santiago et al. 2011).

The plant growth significantly influenced by the micronutrients along with the macronutrients supported metabolic and enzymatic activities in plant. Application of PGPR has been reported as an effective approach under field conditions in the management of plant diseases (Siddiqui and Akhtar 2009; Beneduzi et al. 2012; Kumar et al. 2012a, b). PGPR such as *Bacillus*, has been reported for crop production of few food security crops such as *Eleusine coracana*, *Fagopyrum esculentum* Moench. (Agarwal et al. 2017), *Amaranthus hypocondriacus* (Pandey et al. 2018).

5.3 Food Security Crops

Cultivation of food security crops offering a solution to maintain a sustainable production of indigenous crops that are overlooked in recent scenario or suffering with productivity loss. Most of the nutritional crop varieties are not been cultivated widely in entire globe and thus, meets shortage of access to the consumers, and hence been considering under food security. It is an urgency to mitigate the health problems of human race with organic crop production and improve financial state of farmers especially in developing countries of Asia to revitalize agricultural sustainability. Few of them are below:

5.3.1 *Buckwheat (*Fagopyrum esculemtum* Moench)*

Buckwheat belongs to the dicot family Polygonaceae also known as 'Japanese buckwheat' and 'silverhull buckwheat'. It is widely cultivated in the Himalayas as staple grain crop. Buckwheat is not related to wheat, despite of its name. The name 'buckwheat' or 'beech wheat' understands its triangular seeds shape. It is a translation of Middle Dutch boecweite. Beside this, buckwheat is also related to knotweed, sorrel, and rhubarb. The seeds of buckwheat and its floor are consumed due to its richness with carbohydrates, protein and fiber, therefore, it is also referred to as a pseudocereal.

Buckwheat contains 9–13% crude fat and 2–2.7% ash (Thakur et al. 2016), 57.9–103.4 mg/g N methionine content, 62.2–79.2 mg/g tryptophan etc. (Dogra 2010). It can be cultivated in moist soil to semi-dry soil as its branching root system has profused penetration power to reaches deeply into the soil (Stone 1906). It produces white and pink flowers as per plant variety (Li and Zhang 2001). Buckwheat plant has more complete adaptations of environment due to their ability of producing suckers (vegetative growth) but, sometimes reduce productivity (Stone 1906). The seed hull density is less than that of water, making the hull easy to remove (Li and Zhang 2001).

5.3.1.1 Economic and Nutritional Importance

The buckwheat is a short-term crop having importance due to nutritional profile that contains high amount of protein, amino acids vitamins, starch, minerals, dietary fiber and a plenty source of gluten-free diet for celiac disease patients, Sedej et al. (2011), Wronkowska et al. (2010). It is sometimes also used as silage or manure..

Buckwheat is a trustworthy cover crop during summer as it quickly establishes defence mechanisms to cope temperature changes (Björk et al. 2008). The harvesting period is observed from 10–12 weeks especially at high latitude to grow it up to 30–50 inches (75–125 cm) tall.

Being an important functional food, cultivated various polyphenols (Luthar 1992), proteins with high biological value and balanced amino acid composition (Eggum et al. 1980, Ikeda et al. 1986, Kayashita et al. 1999, Liu et al. 2001 and Tomotake et al. 2000), relatively high fiber content (Bonafaccia et al. 2003 and Steadman et al. 2001), retrograded starch in groat products (Skrabanja and Kreft 1998; Skrabanja et al. 2001; Kreft and Skrabanja 2002), high contents of available zinc, copper, and manganese (Ikeda et al. 1990, Ikeda and Yamaguchi 1993 and Ikeda and Yamashita 1994) and dietary selenium (Stibilj et al. 2004) proved its significance in modern world.

On the other hand, it also contains some anti-nutritional factors. Allergenic reactions occurred due to ingesting buckwheat and its products or by exposure to buckwheat dust. The hypersensitive symptoms involve asthma and asthmatic attacks, urticaria, and gastrointestinal disorders (Li and Zhang 2001 and Wieslander and Norback 2001). As buckwheat does not contain gluten, it is a common supplement for patients with coeliac disease. Buckwheat intolerance is rare among patients with gluten intolerance alone, but more common in those with coeliac disease combined with other food allergies. Allergic reactions are caused by ingestion of allergenic buckwheat proteins (Wieslander and Norback 2001). Ingestion of the entire plant can cause serious photosensitization. Phototoxity of fagopyrin is connected with sensitivity to ultraviolet rays (Li & Zhang 2001). Polyphenols in buckwheat may also inhibit the action of certain enzymes (Eggum et al. 1980).

Basically, flour and groats dishes are generally used to prepare from buckwheat seeds. The buckwheat noodles are more often popular in Asian countries like China and Japan. The buckwheat is rich with plenty amount of rutin as one of the important cause for the desired production. The tartary buckwheat (*Fagopyrum tataricum* Gaertn.) is another variety which has higher amount of rutin than other variety of buckwheat (Suzuki et al. 2002). Few other commercially important products are prepared by buckwheat includes floral honey (Nagai et al. 2011), green tea (Paradkar and Irudayaraj 2002), buckwheat sprouts, vinegar, beer, and spirit irrespective to use as vegetable in some Asian countries such as India, Pakistan, China, and Japan.

In summarizing as nutritious crop, it consists of 25% starch, 75% amylose and amylopectin. It is riched with crude proteins of 18% and the high concentration of all essential amino acids especially lysine, threonine, tryptophan, and the sulfur-containing amino acids. Minerals like iron, zinc, selenium are importantly found in seeds additionally it has antioxidants properties with due to rutin average 10–200 ppm and fagopyrin range 0.4–0.6 mg/g of flour. Aromatic compound like salicylaldehyde has also identified as a characteristic component of buckwheat aroma.

On Hindu fasting days in northern states of India people eat items made of buckwheat flour except eating cereals such as wheat or rice being prohibited. Since, buckwheat is not a cereal, it is considered acceptable for consumption. Buckwheat (Kuttu/Ugal) is an important crop and is commonly cultivated for international trade. Alike, consumption increase the economic value in India and requisite the increase production and productivity. This crop is extensively grown in hilly region of Uttarakhand, especially in province of Garhwal in India. Taking into consideration of such properties present research has foremost emphasizes to develop some bioinoculant, could be used organically grown buckwheat crop having significant in productivity and production.

5.3.1.2 Productivity Enhancement: Status of Use of PGPR

Buckwheat is widely cultivated in several Asian countries in certain European countries. Therefore, buckwheat emerged as important food crop but, productivity observed declined sharply in the twentieth century especially in hilly regions of India beside, its popularity as vegetable and pseudocereals. It is also noted that

Table 5.1 Effect of PGPR and fungal pathogens on plant growth promoting parameters of *Fagopyrum esculentum* (60 DAS) (Adopted from Agarwal et al. 2017)

Treatment	Seed Germination (%)	Length (cm)		Fresh wt. (mg)		Dry wt. (mg)		Disease Index (DI)
		Root	Shoot	Root	Shoot	Root	Shoot	
B. pumilus MSUA3	93.5	4.550**	13.480**	0.258**	2.087**	0.034**	0.212**	0
R. solani + *B. pumilus* MSUA3	72.0	3.897**	11.206**	0.216**	1.947**	0.028**	0.182**	20.0
F. oxysporum + *B. pumilus* MSUA3	61.7	2.583**	7.917**	0.194**	1.413**	0.019**	0.136**	20.0
F. oxysporum	10	1.140**	3.820**	0.044**	0.867**	0.003**	0.017**	90.0
R. solani	12	1.118**	3.112**	0.054**	0.756**	0.004**	0.014**	90.0
Control	65.0	2.613	10.153	0.192	1.890	0.023	0.165	

Values are mean of ten replicates; Analysis of Variance (ANOVA): [a] = significant at 0.01 level of lowest standard deviation (LSD) as compared to control; [**] = Significant at 0.01 level of LSD as compared to control; [*] = Significant at 0.05 level of LSD as compared to control; [ns] = Not Significant at 0.05 level of LSD as compared to control. None of the value in ANOVA were observed significant at 0.01 level (a), at 0.05 level (*) and not significant (ns)

people colonized in hilly regions with the adoption of fertilizer in farming experienced decreased in net productivity of other staples also. Irrespective to this, when this is compared with the other cereals, seed yield of *F. esculentum* is often relatively low without adding fertilizers (Dogra 2010). Low productivity is mainly due to abortive fertilization and several other physiological factors such as self-incompatibility based dimorphic heterostylism, underdeveloped female reproductive organs, and kernel development in the early stage of embryo development (Kumar et al. 2002). The fungal diseases in the buckwheat are also observed as detrimental factor for productivity loss beside, buckwheat itself has significant metabolite with the ability of biocontrol of fungal pathogens. However, not reviewed extensively after Sidorova (1963).

If the crop becomes susceptible, the standing crop suffers from 50% loss due to infection of fungal pathogens, *Fusarium oxysporum* and *Rhizoctonia solani* causing wilting and root rot diseases respectively (Nyvall 1989; Huang et al. 2015). To combat these pathogens in an eco-friendly manner beneficial *Bacillus* spp. of native soil was devised to mitigate the harmful effect of both the pathogens (Agarwal et al. 2017). Biocontrol agents in carrier-based inoculant (vermiculite) were observed suitable in enhancing the yield (Table 5.1; Fig. 5.1).

5.3.2 *Ragi (*Eleusine coracana)

Eleusine coracana L. (Moench.) (Commercial name: Ragi; Common/local name: Mandua) belongs to family Poaceae. Ragi is a significant food crop and world-wide known with several names (Table 5.2). Ragi (Finger millet) is important as food grain and source of straw. It is very nutritious abundant protein, minerals, fibers,

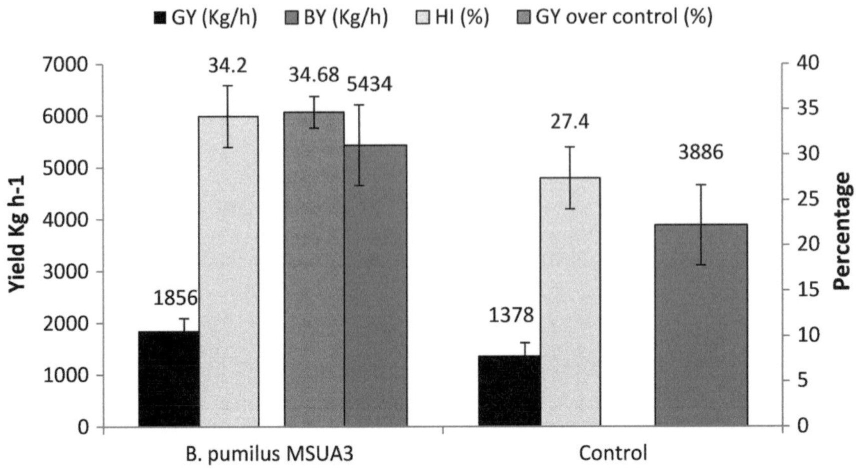

Fig. 5.1 Yield and biocontrol parameters of *Fagopyrum esculetnum* at harvesting (Abbreviation: GY = Grain yield, BY = Biological yield, HI = Harvest Index; error bar on column represented standard deviation at 0.5%) (Adopted from Agarwal et al. 2017)

Table 5.2 Common name of Ragi in different of the globe

Countries/Languages	Common name
UK	Finger millet, African millet, Ragi, Koracan
France	Eleusine cultivee, Coracan, Koracan
Germany	Fingerhirshe
India	Ragi (Kannada, Telugu), Manduva (North India), Nachani (Marathi)
Kenya	Wimbi, Ugimbi
Nepal	Koddo
Sri Lanka	Kurakkan
Zimbabwe	Rapoko, Zviyo, Njera, Poho

and vitamins along with 8–10 time more carbohydrate than in rice and wheat. Finger millet's has slower digestibility and regarded as storage food. It has unique malting properties so why it is used in variety of food processing like brewing porridge, bread, biscuits, and pudding production.

Among the diverse food and staple crops in Uttarakhand, Himalayan region, the crop is grown for alternative of wheat, locally called Mandua or Ragi (commercial name) and is widely grown in Himalayan region to south coastal region of India. It is invariably adaptable to costal land and higher elevations of Himalaya up to 2,300 m above mean sea level (MSL).

5.3.2.1 Economic and Nutritional Importance

Finger millet (*Eleusine coracana* L.) is an important millet crop cultivated extensively in Asia and Africa and produced as low as ranked sixth after wheat, rice, maize, sorghum, and bajra in India. Anatomically small round-shaped seed are coated with hull. The naked seed has brick red-colored seed coat. It is used as complete meal for chapati and biscuits preparation. As stated earlier it is rich in nutrition with high calcium (344 mg/100 g), phytates (0.48%), polyphenols (0.61%), trypsin and dietary fibers (Brajdes and Vizireanu 2012). The entire seed is edible with its seed coat as rich source of phytochemicals, dietary fiber and polyphenols (0.2–3.0%) (Hadimani and Malleshi 1993; Ramachandra et al. 1977). Earlier Bravo (1998) has reported that phytates, polyphenols, and tannins provides antioxidant activity as a factor to affect health, aging, and metabolism of consumers (Bravo 1998). As per above, it is submitted as same that it is beneficial as antidiabetic, antioxidant and hypocholesterolaemic (Devi et al. 2014). The mixture of benzoic acid and cinnamic acid derivatives as millet polyphenols exhibit enzymatic activities and anti-cataractogenic properties. Finger millet consumption as regular in snacks helps to control diabetes (Devi et al. 2014).

Probably tradition of ragi's cultivation first reached in the south coastal regions of India and subsequently in Himalayan regions which is a hotspot of diversity and distribution (Hilu and Wet De 1976). It (Chandra et al. 2016).

Among the food grain (Maize, Jower, Ragi, Bajra, Small Millets and Barley) its cultivation decreased from 45.9 million hectares in 1970–71 to 24.1 million hectares in 2014–15 and ranked among food security crop as millet crops in the world. Food grain production is increased from 50.83 million tonnes in 2015–16. It is expected to attain higher by 1.14 million tonnes over the production of 252.02 million tonnes during 2018–19 (Economic Survey 2018–19).

5.3.2.2 Productivity Enhancement: Status of Use of PGPR

This crop has low productivity in Himalayan range for fulfillment of food demand in sub-terrain of mountains and Indo-Gangetic population. Thus, it is essential to elicit this hurdle and boost up the production of *E. coracana* using *Bacillus* as beneficial PGPR and enhance the productivity in term of grain yield, biological yield, and harvest index.

Accordingly, soil and climatic condition of Himalayan region as well as in south India, Ragi is season bound crop and the best season for sowing is December–January and June–July. Ragi is adapted to a wide range of environmental adversities, tolerance to a significant level of salinity, relatively resistant to water (Dida et al. 2007). It is a stable diet in many villages across south India, especially in north Karnataka besides Maharashtra and North West zones of Tamil Nadu, India.

Several workers have reported use of diverse *Bacillus* for productivity enhancement of other millet crop belong to family Poeacae such as, *Pennisetum glaucum* (Pearl millet) (Raj et al. 2003) and *Setaria italica* (foxtail millet)

Table 5.3 Yield parameters of *E. coracana* at harvesting (120 DAS). (Unpublished data)

Treatment	1000 seeds weight (gm.)	Shelling percentage	Grain yield (Kg/ hectare)	Biological yield (Kg/ hectare)	Harvest Index (%)	% Grain yield rise over control
T1	1.520 ns	34.36 ns	1074 ns	8430 ns	12.74 ns	8.19 ns
T2	2.056**	49.55**	1481**	10740**	13.78**	33.42**
T3	2.096*	50.22*	1511*	9510*	15.88*	34.74*
T4	1.710*	35.22*	1083*	8540*	12.68*	8.95*
T5	2.124**	51.14**	1522**	10870**	14.00**	35.21**
T6	2.187*	52.35*	1587*	9646*	16.45*	37.87*
Control	1.300	37.82	986	7972	12.36	

Abbreviations: T1 = MSTA8 bacterized seed; T2 = MSTD26 bacterized seed; T3 = MSTA8 + MSTD26 bacterized seed; T4 = bioinoculant of MSTA8; T5 = bioinoculant of MSTD26; T6 = bioinoculant of consortia of MSTA8 and MSTD26; C = Control (non-bacterized seed). Values are mean of ten replicates; Analysis of Variance (ANOVA): [a] = significant at 0.01 level of lowest standard deviation (LSD) as compared to control;[**] = Significant at 0.01 level of LSD as compared to control; [*] = Significant at 0.05 level of LSD as compared to control; [ns] = Not Significant at 0.05 level of LSD as compared to control

(Khatri et al. 2016). Veer and Goel (2015) have reported increase in plant growth parameters including fresh/dry weight and length of root/shoot but, available literature exhibited the limited information on its cultivation during this era of food security to raise crop productivity using microbial inoculant. However, few publications reported yield improvement using chemical fertilizers as an alternative way to raise crop productivity (Raman et al. 2004; Rangaraj et al. 2007; Bama and Ramakrishnan 2010).

From the author's lab *Bacillus pumilus* MSTA8, *Bacillus amyloliquefaciens* MSTD26 and their consortia (MSTA8 + MSTD26) increased vegetative parameters such as plant length, fresh weight and dry weight with respect to controls. Effects of bioinoculant formulation on harvesting parameters and yield of ragi were recorded higher to that of non-bacterized seeds. During our investigations, an increase in seed yield by 37.87% with 16.45 harvest index over control at 120 DAS was reported (Table 5.3).

5.3.3 *Amaranth* (**Amaranthus hypochondriacus**)

Among the different underutilized crops grain amaranth (*Amaranthus hypochondriacus*) is one of the important nutritionally rich pseudocereals crop which belongs to the order Caryophillales, Sub-family Amaranthoideae, family Amaranthaceae, Amicarelli and Camaggio (2012), Østerberg et al. (2017), Niranjana and Kumar (2017). The genus *Amaranthus* embraces approximately 60 species and most of them are weeds. Among all, widespread species are *A. hypochondriacus, A.*

cruentus, A. spinosus, A. hybridus, A. tricolor, etc. Amaranth plant has an outward appearance which is having branched inflorescence bearing thousands of minute flowers to set seeds with variable colors for instance cream, golden, pink, and black. Different species of amaranth are used either as leafy vegetable, grain, animal feed, or as ornamentals in different countries. The seeds of amaranth are lenticular and relatively small (Pandey et al. 2013). Grain amaranth is grown worldwide in many countries including Africa, Argentina, China, India, Pakistan, Peru, and USA. However, much variation in the productivity is recorded among the major countries growing amaranth (Chan et al. 2016).

5.3.3.1 Economic Importance

Amaranthus hypochondriacus has gained maximum attention due to its high nutritive value as reported by Raina and Dutta (1992). A very high content of carbohydrates (62–67.9%), fiber (4.9–5%), fat (6.1–7.3%), and protein (15–16.6%) was observed in this crop. Dodok et al. (1997) observed higher (5.95 g/16 gN) lysine content in the amaranth grain in comparison to wheat flour (2.9 g/16 gN). Gorinstein et al. (2002) acquainted grain amaranth as a substitute of conventional cereals due to the presence of high essential and other nutrients. Amino acids (Lysine 5.95%, Histidine 2.5–3%, Tryptophan 0.9–1.8% etc.) in the seeds of amaranth were reported by Písaříková et al. (2005) and Mlakar et al. (2009). Thus, the protein and high essential amino acid contents found in this crop proved to be useful to lessen the scarcity of nutritionally rich food to human beings (De La Rosa et al. 2009; Barrio and Añón 2010). Due to the high nutritive value, nurturing of this nutritive crop suggested to improve food security and to augment farm incomes (Mburu et al. 2012). Sanz-Penella et al. (2013) noted the nutritive value of amaranth seeds as prosperous resource of macro and micronutrients such as iron, calcium, magnesium, zinc, as well as vitamins including riboflavin, ascorbic acid, niacin and thiamine. The amaranth is rich in protein, essential amino acids, vitamins, micronutrients and useful as a medicinal plant to cure various diseases. Recently, Perales-Sánchez et al. (2014) reviewed the antioxidant, anti-allergic properties, plasma cholesterol level decreasing and immunomodulatory abilities of grain amaranth. Several studies have put forward to advocate the quality of amaranth proteins comparable to the optimum protein reference pattern in human diet, and almost reaching the requirements according to the norms given by FAO/WHO (Rastogi and Shukla 2013; Shevkani et al. 2014a, b). Amaranth has high medicinal value, provides vitamin C, purifies blood, soothes cough, and improves general health (Miraj 2016; Shirani et al. 2017). Additionally, being a gluten-free grain, it is suitable for diets of celiac disease patients (Lamacchia et al. 2014; Taylor and Awika 2017). Different compounds such as alkaloids, triterpanoids, anthocyanins, and amaranthine (Castrillón-Arbeláez and Frier 2016; Shirani et al. 2017) are responsible for its beneficial effects. Globulins, albumins, and glutelins are the proteins present in the amaranth seeds with essential amino acids which are deficient in traditional cereals and legumes (Awasthi et al. 2011; Jimenez-Aguilar and

Grusak 2017). Lysine-Methionine richness dwelling gene (*AmA*1) was reported from *Amaranthus* which was successfully used in the development of genetically modified sweet potatoes for enhanced protein quality with balanced amino acid composition.

5.3.3.2 Productivity Enhancement: Status of Use of PGPR

In India, it is cultivated in different regions of Uttarakhand, Himachal Pradesh, and southern parts of Kerala and Tamil Nadu. In most of the areas it is grown as traditional crop but has gained recognition among the farmers in past few decades. Its low cultivation cost, short growth cycle and good adaptability to varied agro-climatic conditions proved its suitability among the farmers. Amaranth seeds are recommended as a health food for feeding children to increase their immunity (Kanensi et al. 2013; Dhangrah et al. 2015). It has significant influence to overcome various ailments/diseases including malnutrition. Amaranth has been called as poor farmer's crop but now attracted attention worldwide due to its high protein and amino acid contents in comparison to maize and wheat (Das 2015).

In India, amaranth is used as an important ingredient in the meal of the people from the entire Himalayan region, and to some extent in the states of Gujarat, Maharashtra, Karnataka, and Eastern parts of Uttar Pradesh. In northern parts, it is regarded as "Holy food" and consumed during ritual fasts where the other cereals and pulses are restricted.

In Uttarakhand, the crop is mostly cultivated in hilly areas. However, its cultivation area (6072 ha) is much lower in comparison to rice (2,73,686 ha), wheat (3,79,196 ha) and maize (28,283 ha) (https://www.agriculture.uk.gov.in). Other than the less market value and low consumption, frequent attack of fungal diseases, pests and insects are responsible for its lower productivity (Kagali et al. 2013; Awurum and Uchegbu. 2013). Since, organic farming in agriculture policy recommends not to use harmful chemicals for raising crop productivity further, due to their toxicity besides deleterious to beneficial microorganisms, undesirable to soil and environment (Glenn et al. 2013; Meena et al. 2016a, b; Mallick et al. 2018). Hence, disease containment through an eco-friendly biocontrol approach, using natural antagonistic microflora, is becoming an inevitable component in the management strategy of plant diseases.

Among the major insects, *Spodoptera litura, Helicoverpa armigera, Hymenia recurvalis, Sylepta derogate, Hypoxilus nubilosus, Epilachnae laterii,* etc., were found most harmful (Sharma 2009; Aderolu et al. 2013). Prior to this study, the lepidopteran insects were reported to cause significant damage in amaranth crop and further two groups of lepidopterans reported to affect and reduce maximum productivity in several countries (Clarke-Harris et al. 2004). The first group of the lepidopteran includes the leafwebbers or webworms which folds their larvae in order to produce web or glue on amaranth leaves and feed within the leaves (Storer et al. 2010) whereas, *Udea ferrugalis, Pterolophia basalis, Herpetogrammabi punctalis* and *Achyrarantalis* considered the second lepidopteran group, belongs to the family Crambidae reported as the infrequent pests of amaranth (Othim et al. 2018). Their larvae feed on amaranth but unlike webworms they do not glue or fold amaranth leaves. Aderolu et al. (2013) and Mureithi et al. (2017) reported *Spodoptera litura* and *Spodoptera furgiperda* as the major representative of the lepidopteran's group. On the other hand, the great loss in the crop productivity was observed by James et al. (2010) due the frequent attack of other insects. García et al. (2011) reported 92% damage to the *A. hypochondriacus* due to attack of borers and 45% decline was due to white grubs thus; the overall yield was reduced by the other general insects of foliage and soil.

In the accompaniment with insects, different microorganisms, for instance, various fungi have been reported to influence amaranth productivity by causing extensive damage to the crop through tissue discoloration, decay of stems and branches of mature plants. Since decades, *Rhizoctonia solani* has been reported as an eminent causal agent of amaranth (Kamala et al. 1996) and exhibited huge number of symptoms such as light and dark brown spots in stem, necrotic lesions, blister like lesions on the leaves and sheaths. Smitha (2000) observed leaf blight symptoms on the foliage of amaranth with small irregular whitish cream spots on leaves, which enlarged under high humidity. Several other fungi such as *F. oxysporum, F. subglutinans, F. sambucinum, Alternaria solani, Alternaria tenuissima,* etc., were reported as the major pathogens of amaranth (Blodgett et al. 2004). Recently, various workers observed foliar blight disease in amaranth caused by *R. solani* owing to diminished productivity of amaranth in our country (Nair and Anith. 2009).

On the other hand, some other fungi were described as a factor behind the diminished productivity of amaranth. For instance, *Pythium myrlotylum* accounted for causing damping off in different amaranth species including *A. hypochondriacus* (Sealy et al. 1988). Mihail and Champaco (1993) observed death and decay of *A. hypochondriacus* and *A. hybridus* due to the infection of *M. phaseolina* causing green black spots on stems. It was stated that severe pre and post emergence of damping off caused by *P. aphanidermatum* in *A. hypochondriacus* and *A. hybridus* were most common and prevalent diseases affects seed germination and seedling survival. Discoloration of branches, stems, and root collars due to *Fusarium oxysporum, Fusarium subqlultinans, A. tenuissima* were observed antecedently by Blodgett et al. (1998). Further, Blodgett and Swart (2002) noticed dark brown to black, necrotic lesions on the leaves of *A. hybridus* accounted *Alternaria tenuissima* as causative agent. Subsequently, Mandal and Das (2002) observed severe damage as a result of leaf spot disease caused by *A. amaranthi*. It was noted that phytotoxin produced by the *F. oxysporum,* toxic to the seedling and callus growth and thereby caused stem decay and root rot of amaranth (Chen and Swart 2001). Further, Priyadarsini (2003) stated that the symptoms of leaf blight disease of amaranth began as small irregular whitish cream spots on the foliage that enlarged under high humidity. In the later stage, the spots became translucent with irregular brown margins and shot hole was observed in severely infected leaves which finally led to defoliation. On the other hand, Talukder et al. (2012) described defoliation and withering of whole plant of red amaranth due to white rust caused by *Albugo occidenta*. The agrochemicals were used to restrain these pests and fungi with instant effect, however, their arbitrary use has caused a lot of side effects on human beings (Lawrence et al. 2004, Chaturvedi et al. 2013), soil and environment (Aktar et al. 2009; Tiryaki and Temur 2010) and animals (Odukkathil and Vasudevan. 2013).

5.4 *Bacillus*: The Protagonist of Crop Productivity

Specially aerobic endospore-forming bacteria *Bacillus* (AEFB'B) remain viable in drastic environmental conditions due to dormant body (endospore) which, in favorable condition become viable and showed their finest rhizospheric competence rhizosphere (Kadyan et al. 2013). There are several free-living, associative and symbiotic bacterial genera effectively involved in plant growth promotion. The other important and widely accepted genera are *Acetobacter, Acidovorax, Azorhizobium, Azotobacter, Azospirillum, Beijrenckia, Bradyrhizobium, Burkholderia, Ensifer, Enterobacter, Erwinia, Mesorhizobium, Paenibacillus, Pseudomonas, Rhizobium, Serratia,* etc., have been exploited for growth promotion and suppression of crop diseases (Maheshwari 2016). *Bacillus* such as *Bacillus subtilis*, *B. licheniformis*, *B. polymyxa*, *B. pumilus*, *B. amyloliquifaciens,* etc., are closely related species having biocontrol potential against different fungal pathogens including *Fusarium oxysporum, Rhizoctonia solani, Sclerotium rolfsii, Macrophomina phaseolina* (Dubey and Gupta 2012; Agarwal et al. 2017). *Bacillus* spp. aggressively colonizes the root system, and exhibited biocontrol and plant growth promoting properties.

Among the PGPR, *Bacillus* genera has predictably immense potential application in term of functionality to raise productivity of crop under field trials (Maheshwari et al. 2013). On the other hand, many regions of Uttarakhand state in India are becoming food insecure today with net sown area, per capita food availability and across to food especially in the hilly districts have been declined. The availability of pulses and cereals has significantly narrowed (Chopra and Pasi 2002). A large number of crops have been introduced in this region by early settlers or civilians those who created a huge diversity of crop for human consumption. In Garhwal Himalaya, average 78–85% population solely depends upon agriculture for their livelihood and 80% of area consists of hills, whereas, remain plain is maximum urban. So, the cereals, pulses, oil-seeds crops and millets grown have low productivity (NFSM 2012). In fact, agricultural production is also affected with climatic and seasonal changes, as in monsoon, flood and landslide are common in Himalaya and in winter snow fall proved deleterious one for various crops (Sharma and Ambili 2009).

Among all endospore-forming *Bacillus* spp. offer several advantages in comparison to that of other beneficial bacteria. These *Bacilli* tolerate adverse conditions including cold temperature and low pH and other abiotic stresses exist in of the high-altitude regions. Diversified populations of aerobic endospore forming species of *Bacillus* occur in soil, and agricultural fields contributed to crop productivity (Dheeman et al. 2017; Agarwal et al. 2017). Physiological traits, such as multi-layered cell wall, stress resistant endospores formation, secretion of peptide or non-peptide secondary metabolites and extracellular enzymes are ubiquitous to these *Bacilli* make them most suitable to restore benefits to the plants in adverse environmental conditions for extended periods of time. Exploiting these abilities, the bacteria exert and inhabit diverse niches in agro-ecosystems and compete with other microorganisms during plant-microbe interactions. The colonization niches for the bacteria being reproducibly stable are also likely to be used in precision management of agro-ecosystems (Timmusk et al. 2014).

Numerous *Bacillus* species express in promotion of plant growth and suppress soil-borne plant pathogens by three ecological pathways viz., providing nutrients for plant growth promotion (Chauhan et al. 2017), antagonism against fungi, and bacteria, and stimulation of host defence mechanisms (Sharma et al. 2018). Recently, a number of *Bacillus* species have been developed commercially as plant growth promoters and biocontrol agents for their use in agriculture (Muis and Quimio 2016).

Drastic changes in Himalayan region are a problem to sustain crop productivity. The more frequent and intensity of climate cannot be ruled out. Hence, there is always a necessity to overcome these abiotic factors using a technology that address issues of farmers in general and seed industry in particular. One of the possibilities is by using the cold tolerant microbes endowed with the ability to synthesize several cryoprotectant compounds for instance, glycine betaine, glycerol, mannitol, sorbitol, glucose, and fructose to overcome the ill effects of cold temperature induced stress (Angelidis and Smith. 2003). Cold tolerant PGPB were suggested as growth promoter because of their ability to solubilize nutrients which enhanced their bioavailability to the plants (Katiyar and Goel. 2003; Trivedi and Sa. 2008). Rinu and Pandey (2009) conducted field based experiment to evaluate the growth promotion abilities of *Bacillus* for cultivars, i.e., lentil at Himalayan terrain under cold conditions. An increase in protein concentration and yield improvement was reported due to the effect posed by combination of PGPB. Plant growth promoting abilities of *Bacillus megaterium* isolated from the cold region of Himalayan terrains was found pragmatic in the form of bioformulation (Trivedi and Pandey. 2008). Kumar et al. (2013) conducted green-house experiment at 20–25 °C by inoculating phytase producing psychrotolerant *Bacillus* sp. in solubilization of phosphorus in the soil. A significant enhancement in biomass of *Brassica juncea* was reported in such condition. Sati et al. (2013) advocated the use of psychrotolerant PGPBs for the growth promotion and productivity enhancement of crops especially in the low temperature environments. Growth promotion of *Cicer arietinum* (L.), *Vigna mungo* (L.) Hepper, *Vigna radiata* (L.) Wilczek, *Cajanus cajan* (L.) Mill sp. and *Eleusine coracona* (L.) Gaertn. at 28 °C was due to the increased uptake of N and P by psychrotolerant *Pseudomonas jesenii* strain MP1 (Kumar et al. 2014). Kaur et al. (2015) observed increased seedling parameters of i.e. lentil with increase in productivity by the inoculation of psychrotolerant *Rhizobium* and other PGPRs

Several other workers have also observed biocontrol activities of *Bacillus* spp. against many common fungal phytopathogens (Kadaikunnan et al. 2015; Zohora et al. 2016; Abdallah et al. 2018). Diverse populations provide better resources for the improvement of plant growth promotion and biocontrol abilities, as different strains possess varied mode of action and survival in extreme environmental conditions. Since, climatic conditions may affect their potential, it is imperative to isolate native *Bacilli* strains from the soils having PGP activity that could influence the crop.

Bacillus spp. considered as the safe microorganisms that hold remarkable abilities for synthesizing a vast array of beneficial substances for the suppression of pathogens (Stein 2005). Majority of antifungal metabolites produced by *Bacillus* spp. could be peptides (Bacillomycin, Iturins, Mycosubtilin, Fengycin, Surfactins, etc.) or non-peptides (Zwittermicin A, Kanosamine, etc.) (Fernando et al. 2005). Largely these are secretary compounds released by PGPBs in soil or any other habitant (Podile and Kishore 2007).

Bacillus species such as *B. amyloliquifaciens, B. subtilis, B. cereus, B. pumilus, B. mycoides* and *B. sphaericus* have been reported to elicit significant reduction in the severity of various diseases of diversified host (Kloepper et al. 2004; Choudhary et al. 2008). Shanmugam et al. (2013) evaluated chitinase producing *B. subtilis* for disease management of ginger because of reduction in the incidence of rhizome rot caused by *F. oxysporum* and *F. solani.* Tan et al. (2013) observed *B. amyloliquifaciens* as antagonist toward *Ralstonia solanacearum.* Consortium of *P. aeruginosa* KRP1 and *B. licheniformis* KRB1 were reported suppressive for the fungal phytopathogens *F. oxysporum* and *S. sclerotiorum* causing disease in *Brassica campestris* (Maheshwari et al. 2015a, b). Pane and Zaccardelli (2015) reported suppression of early blight disease of *Solanum lycopersicum* L. caused by *Alternaria alternata* by *Bacillus* spp. *Bacillus subtilis* treated seeds of *Solanum lycopersicum* showed suppression of *R. solani* with 80% reduction in disease incidence (Zohora et al. 2016). Recently, Rais et al. (2018) evaluated *Bacillus* spp. for blast disease suppression on rice crop. *Bacillus subtilis* XF-1 also reported to inhibit the growth of *Plasmodiophora brassicae* causing infection in cruciferous crops (Liu et al. 2018).

This implies that PGPBs competence strongly enhanced quality enhancement and crop growth. Supanjani et al. (2006) applied *Bacillus megaterium* var. *phosphaticum* and *Bacillus mucilaginosus* in nutrient limited stressed soil where the strains increased bioavailability of minerals, their uptake and subsequently enhanced growth of pepper and cucumber. Hafeez et al. (2006) suggested *Bacillus pumilus* as a bioinoculant to promote the crop yield in wheat. Beneduzi et al. (2008) reported *Bacillus* isolate SVPR30 as an efficient bioinoculant for growth enhancement of the rice. Furthermore, Singh et al. (2008) reported increase in the dry weight of root with the *Bacillus* bacterization of seeds of *Pinus roxburghii* in Himalayan terrains. On the other hand, Zongzheng et al. (2010) evaluated growth promoter effect of *Bacillus subtilis* SY1. A significant increase in seedling parameters such as sprout tendency, germination percentage, sprout index and vigor index were observed. *Bacillus* isolates exhibited seedling length, fresh weight and dry weight in cow pea (Thomas et al. 2010). Agrawal and Agrawal (2013) reported growth promotion of tomato by PGP *Bacillus* sp. Mehta et al. (2015) supported the fact of planting value parameter enhancement by the treatment of P-solubilizing *Bacillus circulans* CB7 on the growth enhancement of tomato. Significant increase in early vegetative plant growth parameters along with increase in nitrogen, potassium, and phosphorus was observed.

5.5 *Bacillus*: In Action as PGPR

5.5.1 Bacillus *as Mineral Solubilizers*

Evidences of phosphorus solubilizing microorganism (PSM) were dates back to 1903 (Wani et al. 2008). Khan et al. (2010) stated that efficiency of P fertilizer throughout the world is around 10–25% and bio-available P in soil reaching the level of 1.0 mg kg–1 soil (Goldstein 1994). Further, several contributions were made to establish the concept of phosphorus (P) solubilization as major growth-limiting factor for the plants (Richardson 2001; Ezawa et al. 2002; Saharan and Nehra 2011; Pingale and Virkar 2013). He et al. (2004) observed heterotrophic microorganisms secreted organic acids responsible for inorganic P solubilization into phosphatic minerals and/or chelate cationic partners. An advance finding in the same manner on the metabolic processes that diverse group from autotrophs to heterotrophs and diazotrophs to phototrophs have been reported to secrete enzyme phosphatases (Gupta et al. 2007). The production of microbial metabolites resulted in decrease in soil pH, probably played a major role in P solubilization (Chen et al. 2010). Yadav and Verma (2012) applied P as fertilizer which become immobile pools via precipitation in soils. It has also been noted that PGPR as PSB inoculants enhances the nutritional uptake of plant and influence growth factors under stressed conditions (Nadeem et al. 2014a, b). In addition, Nenwani et al. (2010) proposed proton extrusion and organic acid production as a microbial processes/mechanisms for phosphate solubilization.

In between lot of concept notes were postulated on P-solubilizing microorganisms. Dubey et al. (2014) reported that the *Bacillus subtilis* BSK17 solubilized inorganic phosphate in soil resulting its easy uptake and enhanced yield of the *Cicer arietinum.* Inoculation of the *Bacilli* compensated the nutrient deficiency and improved the overall plant growth and development (Schoebitz et al. 2013; Sandilya et al. 2018). Due to the treatment of *B. thuringiensis,* an 11% increase in phosphorus availability in soil and 34% in wheat plant occurred (Delfim et al. 2018).

Organic phosphate being another form of phosphate in soil constitutes 30–50% of total phosphorus available amounts in soil. It was stated that organic phosphorus is largely in the form of inositol phosphate (soil phytate), the most stable form of organic phosphate synthesised by microorganisms and plants (Dalai 1977).

Myo-, scyllo-, chiro- and neo-inositol phosphates were widely spread phytases, of which most myo-inositol has been reported as most common and distributed in various plant tissues. Phytase produced by microorganisms able to change-over soil phytate into phosphate for the plant's uptake (Patel and Kanungo 2010). Earlier, Idriss et al. (2002) reported *B. amyloliquefaciens* in the degradation of extracellular phytate (myo-inositol hexaklsphosphate). Inoculation of the culture filtrate of these *Bacilli* containing phytase at enhanced root weight, shoot weight and root length of maize. Various studies on enhancement of plant growth by the phytase producing *Bacillus* strains have been described by researchers (Kumar et al. 2012a, b; Balaban et al. 2016).

5.5.2 **Bacillus** *as Iron Scavengers*

Iron is the fourth most essential element of the earth's crust (Pasek and Lauretta 2005). dates back in 1952, "Neilands" isolating and purifying siderophores in crystalline form and honored as the "Father of ferruginous facts". The term siderophore stands for "iron carriers" or "iron bearers" in Greek.

Iron has been proved essential nutrient for virtually all microorganisms to control enzymatic reactions as necessary cofactor. More than 500 siderophores have been reported (Boukhalfa and Crumbliss 2002; Hider and Kong 2010). Kraemer (2004) stated most, but not all, siderophores are hexadentate ligands forming 1:1 complexes with Fe^{3+}. Miethke and Marahiel (2007) observed carboxylate siderophore produced by fungi in acidic environment has low capacity to form stable Fe^{2+} complexes. Also, Miethke and Marahiel (2007) examined siderophore complexes varies based on their structure and ligand type varies their stability of Fe^{3+} influence pH and affect chelation efficiency thereby. On the other hand, Raymond et al. (2003) observed higher affinity for Fe^{3+} in catecholates siderophore.

Earlier, Liles et al. (2000) discovered *Legionella pneumophila* produced a non-classical siderophore, legiobactin, conserved among the members of legionellae. Like other PGPR, rhizobia utilize a large spectrum of these molecule to overcome iron starvation (Plessner et al.1993) and produce strain specific siderophores (Arora et al.2001; Deshwal et al. 2003). Sharma and Johri (2003) characterized bacterial siderophores of low-molecular-weight with high Fe^{3+} chelating affinities meant for easy transport across the membrane. Further, their role to inhibit the growth of plant pathogens was also reported by siderophore producing microorganism. Siderophore production has been reported in *Pseudomonads* spp. (Haas and Défago 2005), *Bradyrhizobium* (Deshwal et al. 2003), *Rhizobia* (Arora et al. 2001). Pandey et al. (2005) have been reported bacteria produce hydroxamate-type and catecholate-types siderophores invariably and observed that *Pseudomonas aeruginosa* GRC_1, excel in PGP traits and also have prolific production ability of hydroxamate siderophore in iron-deficient conditions. The purified siderophore appeared to be of pyoverdin type with typical amino acid composition. On the other hand, siderophore production by root nodulating rhizobia

(Carson et al. 1992; Arora et al. 2001). Various strains of PGPR have been reported to produce a wide range of siderophores such as rhizobactin, citrate, hydroxymate, catechol, anthranalate under low iron (deficient) conditions (Guerinot et al.1990; Sridevi and Mallaiah 2008).

5.5.3 **Bacillus** ***as Plant Growth Regulators***

Several decades before, the term "Auxin" was introduced into the identification of scientific community (Went and Thimann 1937). IAA significantly influenced plant growth and development (Went and Thimann 1937). Several workers observed mechanism of plant hormones secreted by PGPR in plant growth and development (Kumar et al. 2009; Maheshwari et al. 2012; Reetha et al. 2014). The indole acetic acid (IAA) enhance root proliferation and increase nutrient uptake by involvement of most common IAA biosynthesis by Indole-3-Acetamide (IAM) and Indole-3-Pyruvate (IPyA) (Barazani and Friedman 1999). Representatives of *B. subtilis* and *B. amyloliquefaciens* group secreted IAA-like substance in reasonable amount when fed with tryptophan (Idris et al. 2007). Lambrecht et al. (2000) studied the involvement of IAM pathway in plant gall size, whereas the IPyA pathway determined epiphytic fitness in plants. On the other hand, Ali et al. (2009) reported auxin-producing rhizobacteria exert positive effect in *Triticum aestivum.* Whereas, Grunewald et al. (2009) explained auxin responses during microbe-induced de novo organ formation which modify their host's auxin transport. Hussain and Hasnain (2009) examined rhizobacterial extract of *B. subtilis* BC1 and *P. aeruginosa* E2 strain in enhancement of cell division, fresh weight and cotyledons size in cucumber. Morrone et al. (2009) demonstrated that *B. japonicum* encodes separate diterpenoid from plant and fungi *ent*-copalyl diphosphate and *ent*-kaurene involved in biosynthesis of gibberellin. Earlier, Bottini et al. (2004) reviewed the involvement of gibberellins-mediated symbiotic and soil-endophytic microorganisms in PGP activities. Cytokinin is dominantly produced by rhizobacteria (Arkhipova et al. 2007). On the other hand, Liu et al. (2013) observed cytokinin producing PGPR inoculation in seedlings alleviated drought stress in arid environments.

5.5.4 **Bacillus** ***in Stress Management***

Certain enzyme containing microbes have dual role to be played in plant–bacteria interactional studies. The mechanism of PGP involve 1-aminocyclopropane-1-carboxylate deaminase (ACCD) production. PGPRs played significant role in the regulation of the plant hormone, ethylene, thus influence the growth and development of plants (Davies 2010). Belimov et al. (2009) demonstrated bacteria-mediated ACCD in alleviation of stress-induced ethylene-mediated adverse impact in plants, abiotic and biotic factors. Accelerated ethylene production under high and chilling temperatures has been reported by researchers in both plants and rhizospheric microbial species (Wright and Osborne 1974). Penrose and Glick (2003) proposed the standard methodology to characterize rhizospheric bacteria for the production of ACC-deaminase in vitro. Thereafter, several advancement were made in of rhizosphere microbiology, especially to establish the characteristically ACCD active in PGPRs. A psychrotolerant ACCD producing bacterium *P. putida* UW4 reported to promote plant growth at low temperature under salt stress conditions in canola (Cheng et al. 2007). PGPR containing ACCD boost plant growth particularly under stressed conditions due to regulation of accelerated ethylene production in response to a multitude of stresses such as salinity, temperature, and drought (Marshall et al. 2012), water logging, pathogenicity (Lulai and Suttle 2004), and contamination (Arteca and Arteca 2007). Various workers observed PGPR having ACCD caused marked improvement in root growth and biomass production under stressed conditions (Belimov et al. 2001, 2005; Safronova et al. 2006; Zahir et al. 2009; Nadeem et al. 2014a, b). The decreased ACC level resulted in low endogenous ethylene concentration reduced the harmful effects of stress-induced ethylene so as to allow plants to develop a better root system. PGPR containing ACCD mitigates the ACC-imposed effect in the similar way as did the chemical inhibitor such as carbon-di-oxide (Shaharoona et al. 2007; Nadeem et al. 2010).

5.5.5 **Bacillus** ***in Biocontrol of Phytopathogens***

Among diverse group of microorganisms, about 350 bacterial species studied to date, are important producers of volatile substances (Wenke et al. 2010). Beauchamp and Drury (1991) reported that rhizobia are relatively less efficient in HCN production than that of other rhizosphere bacteria but its production is common by fluorescent group of pseudomonads (Sacherer et al. 1994; Bagnasco et al. 1998; Rodríguez and Fraga 1999; Siddiqui et al. 2006; Ahmad et al. 2008). Cyanide as one among secondary metabolite produced from glycine by HCN synthase reported by Castric and Deal (1994). Secondary volatile metabolite produced by majority of gram-negative bacteria have deleterious effect on growth of phytopathogens (Knowles and Bunch 1986). Cyanides play important role in the suppression of root rot in tobacco caused by *Thilaviopsis basicola* (Ahl et al. 1986). HCN inhibits the enzyme Cytochrome C oxidase and other metalloenzymes (Voisard et al. 1989) of the pathogens thus helps plants against the attack soil-borne diseases (Blumer and Haas 2000). Long back, Glick et al. (1998) observed that certain bacteria along with fungi were also reported that inhibits by the action of HCN because HCN negative microorganisms do not resist the HCN activity. The volatile cyanogens production in liquid cultures proved inhibitory to various fungal genera (O'Sullivan and O'Gara 1992; Dowling and O'Gara 1994). Haas and Défago (2005) elaborated understanding of action of HCN in plant disease control.

Kremer and Souissi (2001) demonstrated that the production of HCN by rhizobacteria was similar to its exogenous concentrations, inhibiting the seedling growth, suggesting that HCN produced in rhizosphere of proved potential and environmentally compatible mechanism for biocontrol. Dubey et al. (2014) found plant growth-spromoting *Bacillus subtilis* BSK17 strain able to produce HCN and other inhibitory metabolites to act against *Fusarium oxysporum*.

5.6 Biofilm Formation by *Bacilli*

The root exudation process includes the secretion of ions, free oxygen and water, enzymes, mucilage, and a universal array of carbon-containing primary and secondary metabolites helps in maintaining rhizospheric ecology. In rhizosphere microbes in root-association forms biofilms, basically depends on the availability of root exudates nutritional factor to favor cell aggregation (Seneviratne et al. 2011). Biofilm formation is central factor triggering root colonization (Ortíz-Castro et al. 2009; Bais et al. 2006). Chemotaxis behaviors of rhizobacteria (in particular flagellated bacteria) for root exudates have also been observed (Tan et al. 2013; Yuan et al. 2015;). Root exudates are less diverse but often compose a larger proportion of the exudates by mass. They include low-molecular weight compounds such as amino acids, organic acids, sugars, phenolics secretion of low- and high-molecular weight root exudates into the soil changes its environment. The polysaccharides, mucilage, and proteins are comes under the category of high-molecular root exudates. Mucilages are released from the root cap, these are primary cell between epidermal and sloughed root cap (including root hairs). Few rhizobacteria also produce microbial mucilages. Further, autolysis of root cells is another reason for the production of lysates as root exudates in the soil. Collectively, lysate, plant and microbial mucilages associate organic and mineral matter (Rovira 1969).

A marked concept about root exudates was put forwarded by Bais et al. (2004) that plant root cells has functional way to secrete chemicals which also includes activity and turnover of microbes and thus, influence nutrient availability (Leff et al. 2015) in the rhizosphere (Magiorakos et al. 2012). On the other hand, chemicals and polysaccharides secretions of root cells are responsible for initial attachment of other low-molecular organic compounds such as ethylene, sugars, amino acids, vitamins, polysaccharides, and enzymes present in soil. Hence, these nutritional resources influence microbial population structure and play a vital role in niche stabilization and rhizospheric homeostasis. During microbial interactions, the root cells release exudates either passively (diffusates) and actively (secretions). Whereas, the low-molecular weight organic compounds released in a passive process along with their own steep concentration-gradient that usually exists between the cytoplasm of intact cells and the external solution (Kuzyakov 2002).

Exopolysaccharide (EPS) of bacterial origin modulate the chemical and physical properties of biofilms on abiotic surfaces (Friedman and Kolter 2004). The bacterial EPS production merit as important factor for effective colonization on root surfaces.

More than 90% of the EPS volume consists of water (Schmitt & Flemming 1999) and proteins and polysaccharides which enhance bacterial adhesion (Sutherland 2001) while lipopolysaccharides (Prakash et al. 2003), uronic acids (Kavita et al. 2014) and biosurfactants (Banat et al. 2010) depress bacterial adhesion. Sutherland (2001) noted that structure of the EPS has a marked effect on the biofilm formation. Biofilm EPS is not generally uniform but vary spatially and temporally. Honma et al. (2007) described the inhibitory role of glycosylated surface-glycoprotein in *Tannerella forsythia* biofilms. Li et al. (2008) stated that EPS play vital role in the biofilm formation and structure of membrane-aerated biofilms (MABs). Production of a novel glucose, galactose, and mannose-rich polymer that contributes to cell–cell interactions necessary for pellicle and biofilm formation and its stability (Armitano et al. 2014).

Several workers have studied the role of biofilms in various systems of industrial, ecological and environmental backgrounds. However, two major tools in the past few decades have emerged to enhance our understanding of biofilm and its accomplishment. Foremost is the use of scanning electron microscopy (SEM) to elucidate ultrastructure of biofilm (West et al. 2014), and second, exploration of the genes governs metabolite production that involve in cell adhesion and biofilm formation (Mielich-Süss and Lopez 2015).

Although, *Bacillus* and *Paenibacillus* species among *Bacilli* showed pragmatic biofilm formation (Branda et al. 2006; Bais et al. 2004). But, *B. subtilis* forms adhering biofilms on inert surfaces under the control of a variety of transcription factors (Cho et al. 2003; Kinsinger et al. 2003; Bais et al. 2004). Zeriouh et al. (2014) reported that the surfactin production by *Bacillus subtilis* not only necessary for biofilm formation but also exhibits antimicrobial activities. The antimicrobial activity of surfactin seems to be responsible for the inhibition of several phytopathogens. The role of biofilm in biocontrol initiation has also been highlighted by Haggag and Timmusk (2008) who observed the biofilm-forming *P. polymyxa* in controlling crown root rot disease caused by *Aspergillus niger* and suggested that the superior biofilm former offers significantly better disease protection.

Biofilms hold tremendous practical significance, exhibiting both beneficial and detrimental activities (Zhang et al. 2003; Ribeiro et al. 2015). The *in vitro* production of biofilm formation of PGPR gave better crop yields through a range of plant growth mechanisms bio fertilizers through improved N_2 fixation and by solubilization of micro- and macronutrient and their uptake. Further, enhanced growth observed due to the production of beneficial metabolites. Their performance during host-plant bacteria relationship exhibited significant and sustainable enhancement of productivity and yield both in green houses and field practices.

5.7 Conclusion

Scientists in general and biologists in particular focus their attention for sustainable crop production because, food security as a challenge of now cope food security problems. Different species of *Bacillus* are really versatile with that rendered enhancement of plant growth and health promotion powered with super abilities to produce resting fruiting bodies in the form of endospores. During adverse situations, several characteristics these organisms and because of provide all benefits to the productivity enhancement in field crops. Thus, assistance of such group of bacteria is fitted for luxuriant growth and food production enhancement. However, a proper evaluation be made about their pathogenicity before releasing for commercial practices.

Acknowledgements DKM is thankful to Uttarakhand Council of Science & Technology (UCOST), Dehradun, India for the financial assistance in the form of R&D project UCS&T/R&D-6/17-18/14281

References

Abdallah DB, Tounsi S, Gharsallah H, Hammami A, Frikha-Gargouri O (2018) Lipopeptides from *Bacillus amyloliquefaciens* strain 32a as promising biocontrol compounds against the plant pathogen *Agrobacterium tumefaciens*. Environ Sci Poll Res 25(36):36518–36529

Aderolu IA, Omooloye AA, Okelana FA (2013) Occurrence, abundance and control of the major insect pests associated with Amaranths in Ibadan, Nigeria. Entomol Ornithol Herpetol 2 (112):2161-0983

Agarwal M, Dheeman S, Dubey RC, Kumar P, Maheshwari DK, Bajpai VK (2017) Differential antagonistic responses of *Bacillus pumilus* MSUA3 against *Rhizoctonia solani* and *Fusarium oxysporum* causing fungal diseases in *Fagopyrum esculentum* Moench. Microbiol Res 205:40–47

Agrawal DP, Agrawal S. (2013) Characterization of *Bacillus* sp. strains isolated from rhizosphere of tomato plants (*Lycopersicon esculentum*) for their use as potential plant growth promoting rhizobacteria. Int J Curr Microbiol App Sci 2(10):406–17

Ahl P, Voisard C, Défago G (1986) Iron bound-siderophores, cyanic acid, and antibiotics involved in suppression of Thielaviopsis basicola by a Pseudomonas fluorescens strain. J Phytopathol 116(2):121–134

Ahmad F, Ahmad I, Khan MS (2008) Screening of free-living rhizospheric bacteria for their multiple plant growth promoting activities. Microbiol Res 163(2):173–181

Aktar W, Sengupta D, Chowdhury A (2009) Impact of pesticides use in agriculture: their benefits and hazards. Interdiscipline Toxicol 2(1):1–12

Ali B, Sabri AN, Ljung K, Hasnain S (2009) Auxin production by plant associated bacteria: impact on endogenous IAA content and growth of *Triticum aestivum* L. Lett Appl Microbiol 48 (5):542–547

Amicarelli V, Camaggio G (2012) Amaranthus: a crop to rediscover. In: Forum Ware International, vol. 2, International Society of Commodity Science and Technology (IGWT), Vienna, pp 4–11

Angelidis AS, Smith GM (2003) Three transporters mediate uptake of glycine betaine and carnitine by *Listeria monocytogenes* in response to hyperosmotic stress. Appl Environ Microbial 69(2):1013–1022

Arkhipova TN, Prinsen E, Veselov SU, Martinenko EV, Melentiev AI, Kudoyarova GR (2007) Cytokinin producing bacteria enhance plant growth in drying soil. Plant Soil 292(1–2):305–315

Armitano J, Méjean V, Jourlin-Castelli C (2014) G ram-negative bacteria can also form pellicles. Environ Microbiol Rep 6(6):534–544

Arora NK (2018) Environmental Sustainability—necessary for survival. Environ Sustain 1:2

Arora NK, Kang SC, Maheshwari DK (2001) Isolation of siderophore-producing strains of *Rhizobium meliloti* and their biocontrol potential against *Macrophomina phaseolina* that causes charcoal rot of groundnut. Curr Sci 673–677

Arteca RN, Arteca JM (2007) Heavy-metal-induced ethylene production in *Arabidopsis* thaliana. J Plant Physiol 164(11):1480–1488

Asseng S, Ewert F, Rosenzweig C, Jones JW, Hatfield JL, Ruane AC, Boote KJ, Thorburn PJ, Rötter RP, Cammarano D, Brisson N (2013) Uncertainty in simulating wheat yields under climate change. Nature Clim Chan 3(9):827

Awasthi CP, Kumar A, Singh N, Thakur R (2011) Biochemical composition of grain amaranth genotypes of Himachal pradesh. Indian J Agric Biochem 24(2):141–144

Awurum AN, Uchegbu PC (2013) Development of wet rot disease of *Amaranthus cruentus* L. caused by *Choanephora cucurbitarum* (Berk. and Rav.) Thax. in response to phytochemical treatments and inoculation methods. Adv Med Plant Res 1(3):66–71

Babalola OO (2010) Beneficial bacteria of agricultural importance. Biotechnol Lett 32(11):1559–1570

Bagnasco P, De La Fuente L, Gualtieri G, Noya F, Arias A (1998) Fluorescent *Pseudomonas* spp. as biocontrol agents against forage legume root pathogenic fungi. Soil Biol Biochem 30 (10–11):1317–1322

Bais HP, Fall R, Vivanco JM (2004) Biocontrol of *Bacillus subtilis* against infection of *Arabidopsis* roots by *Pseudomonas syringae* is facilitated by biofilm formation and surfactin production. Plant Physiol 134(1):307–319

Bais HP, Weir TL, Perry LG, Gilroy S, Vivanco JM (2006) The role of root exudates in rhizosphere interactions with plants and other organisms. Ann Rev Plant Biol 57:233–266

Bakore N, John PJ, Bhatnagar P (2002) Evaluation of organochlorine insecticide residue levels in locally marketed vegetables of Jaipur City, Rajasthan, India. J Environ Biol 23(3):247–252

Balaban NP, Suleimanova AD, Valeeva LR, Chastukhina IB, Rudakova NL, Sharipova MR, Shakirov EV (2016) Microbial phytases and phytate: exploring opportunities for sustainable phosphorus management in agriculture. Am J Mol Biol 7(01):11

Bama ME, Ramakrishnan K (2010) Effects of combined inoculation of *Azospirillum* and AM fungi on the growth and yield of finger millet (*Eleusine coracana* Gaertn) Var. Co 12. J Exp Sci 8:10–11

Banat IM, Franzetti A, Gandolfi I, Bestetti G, Martinotti MG, Fracchia L, Smyth TJ, Marchant R (2010) Microbial biosurfactants production, applications and future potential. Appl Microbial Biotechnol 87(2):427–444

Barazani OZ, Friedman J (1999) Is IAA the major root growth factor secreted from plant-growth-mediating bacteria? J Chem Ecol 25(10):2397–2406

Barrio DA, Añón MC (2010) Potential antitumor properties of a protein isolate obtained from the seeds of *Amaranthus mantegazzianus*. Euro J Nutr 49(2):73–82

Beauchamp EG, Drury CF (1991) Ammonium fixation, release, nitrification, and immobilization in high-and low-fixing soils. Soil Sci Soc Am J 55(1):125–129

Belimov AA, Dodd IC, Hontzeas N, Theobald JC, Safronova VI, Davies WJ (2009) Rhizosphere bacteria containing 1-aminocyclopropane-1-carboxylate deaminase increase yield of plants grown in drying soil via both local and systemic hormone signalling. New Phytol 181(2):413–423

Belimov AA, Hontzeas N, Safronova VI, Demchinskaya SV, Piluzza G, Bullitta S, Glick B R (2005) Cadmium-tolerant plant growth-promoting bacteria associated with the roots of Indian mustard (*Brassica juncea* L. Czern.). Soil Biol Biochem 37(2):241–250

Belimov AA, Safronova VI, Sergeyeva TA, Egorova TN, Matveyeva VA, Tsyganov VE, Borisov AY, Tikhonovich IA, Kluge C, Preisfeld A, Dietz KJ (2001) Characterization of plant growth promoting rhizobacteria isolated from polluted soils and containing 1-aminocyclopropane-1-carboxylate deaminase. Can J Microbiol 47(7):642–652

Beneduzi A, Peres D, Vargas LK, Bodanese-Zanettini MH, Passaglia LM (2008) Evaluation of genetic diversity and plant growth promoting activities of nitrogen-fixing bacilli isolated from rice fields in South Brazil. Appl Soil Ecol 39(3):311–320

Beneduzi A, Ambrosini A, Passaglia LM (2012) Plant growth-promoting rhizobacteria (PGPR): their potential as antagonists and biocontrol agents. Genet Mol Biol 35(4):1044–1051

Bhattacharya PN, Jha DK (2012) Plant growth promoting rhizobacteria (PGPR): Emergence in agriculture. World J Microbiol Biotechnol 28(4):1327–1350

Björk RG, Björkman MP, Andersson MX, Klemedtsson L (2008) Temporal variation in soil microbial communities in Alpine tundra. Soil Biol Biochem 40(1):266–268

Blodgett JT, Swart WJ (2002) Infection, colonization, and disease of *Amaranthus hybridus* leaves by the *Alternaria tenuissima* group. Plant Dis 86(11):1199–1205

Blodgett JT, Swart WJ, Louw SV (2004) Identification of fungi and fungal pathogens associated with *Hypolixus haerens* and decayed and cankered stems of *Amaranthus hybridus*. Plant Dis 88 (4):333–337

Blodgett JT, Swart WJ, Louw SV (1998) First report of *Fusarium sambucinum*, *F. oxysporum,* and *F. subglutinans* associated with stem decay of *Amaranthus hybridus* in South Africa. Plant Dis 82(9):1062–1062

Blum JE, Bartha R (1980) Effect of salinity on methylation of mercury. Bull Environ Cont Toxicol 25(1):404–408

Blumer C, Haas D (2000) Mechanism, regulation, and ecological role of bacterial cyanide biosynthesis. Arch Microbiol 173(3):170–177

Bonafaccia G, Marocchini M, Kreft I (2003) Composition and technological properties of the flour and bran from common and tartary buckwheat. Food Chem 80(1):9–15

Bottini R, Cassán F, Piccoli P (2004) Gibberellin production by bacteria and its involvement in plant growth promotion and yield increase. Appl Microbiol Biotechnol 65(5):497–503

Boukhalfa H, Crumbliss AL (2002) Chemical aspects of siderophore mediated iron transport. Biometals 15(4):325–339

Brajdes C, Vizireanu C (2012) Sprouted buckwheat an important vegetable source of antioxidants. The Annals of the University of Dunarea de Jos of Galati. Fascicle VI. Food Technol 36(1):53

Branda SS, Chu F, Kearns DB, Losick R, Kolter R (2006) A major protein component of the *Bacillus subtilis* biofilm matrix. Mol Microbiol 59(4):1229–1238

Bravo L (1998) Polyphenols: chemistry, dietary sources, metabolism, and nutritional significance. Nutr Rev 56(11):317–333

Bungla PS, Tewari LM, Bhatt ID, Rawal RS, Kishor K (2012) The traditional food system of Kuloor watershed of Kumaon Himalaya, India. Afr J Food Sci 6(23):546–553

Carson KC, Holliday S, Glenn AR, Dilworth MJ (1992) Siderophore and organic acid production in root nodule bacteria. Arch Microbiol 157(3):264–271

Castric PA, Deal CD (1994) Differentiation of *Pseudomonas aeruginosa* pili based on sequence and B-cell epitope analyses. Infect Immun 62(2):371–376

Castrillón-Arbeláez PA, Frier JPD (2016) Secondary Metabolism in *Amaranthus* spp. A genomic approach to understand its diversity and responsiveness to stress in marginally studied crops with high agronomic potential. In: Abiotic and biotic stress in plants-recent advances and future perspectives. http://dx.doi.org/10.5772/61820

Chan CL, Gan RY, Corke H (2016) The phenolic composition and antioxidant capacity of soluble and bound extracts in selected dietary spices and medicinal herbs. Int J Food Sci Technol 51 (3):565–573

Chandra S, Askari K, Kumari M (2018) Optimization of indole acetic acid production by isolated bacteria from *Stevia rebaudiana* rhizosphere and its effects on plant growth. J Gene Eng Biotechnol 16(2):581–586

Chandra D, Chandra S, Sharma AK (2016) Review of Finger millet (*Eleusine coracana* (L.) Gaertn): a power house of health benefiting nutrients. Food Sci Hum Welln 5(3):149–155

Chaturvedi M, Sharma C, Tiwari M (2013) Effects of pesticides on human beings and farm animals: a case study. Res J Chem Environ Sci 1(3):14–19

Chauhan AK, Maheshwari DK, Dheeman S, Bajpai VK (2017) Termitarium-inhabiting *Bacillus* spp. enhanced plant growth and bioactive component in turmeric (*Curcuma longa* L.). Curr Microbiol 74(2):184–92

Chauhan RS, Singhal L (2006) Harmful effects of pesticides and their control through cowpathy. Int J Cow Sci 2(1):61–70

Chen F, Wang M, Zheng Y, Luo J, Yang X, Wang X (2010) Quantitative changes of plant defense enzymes and phytohormone in biocontrol of cucumber *Fusarium* wilt by *Bacillus subtilis* B579. World J Microbiol Biotechnol 26(4):675–684

Chen WQ, Swart WJ (2001) Genetic variation among *Fusarium oxysporum* isolates associated with root rot of *Amaranthus hybridus* in South Africa. Plant Dis 85(10):1076–1080

Cheng Z, Park E, Glick BR (2007) 1-Aminocyclopropane-1-carboxylate deaminase from *Pseudomonas putida* UW4 facilitates the growth of canola in the presence of salt. Can J Microbiol 53(7):912–918

Cho HS, Mason K, Ramyar KX, Stanley AM, Gabelli SB, Denney DW Jr, Leahy DJ (2003) Structure of the extracellular region of HER2 alone and in complex with the Herceptin Fab. Nature 421(6924):756

Chopra R, Pasi S (2002) Where are the empty thalis in Uttarakhand. District Level Food Insecurity Analysis of Uttarakhand. People's Science Institute, Dehradun

Choudhary DK, Johri BN, Prakash A (2008) Volatiles as priming agents that initiate plant growth and defence responses. Curr Sci 10:595–604

Clarke-Harris D, Fleischer SJ, Fuller C, Bolton J (2004) Evaluation of the efficacy of new chemistries for controlling major *lepidoptera* pests on vegetable amaranth in Jamaica. CARDI Rev 4:12–19

Dalai RC (1977) Soil organic phosphorus, vol 29. Advances in Agronomy, Academic Press, pp 83–117

Das S (2015) *Amaranthus parganensis* (Amaranthaceae), a new species from West Bengal, India. Novon 23(4):406–410

Davies PJ (2010) The plant hormones: their nature, occurrence, and functions. In: Plant hormones. Springer, Dordrecht, pp 1–15

De La Rosa AB, Fomsgaard IS, Laursen B, Mortensen AG, Olvera-Martínez L, Silva-Sánchez C, Mendoza-Herrera A, González-Castañeda J, De León-Rodríguez A (2009) Amaranth (*Amaranthus hypochondriacus*) as an alternative crop for sustainable food production: Phenolic acids and flavonoids with potential impact on its nutraceutical quality. J Cereal Sci 49 (1):117–121

de Santiago A, Quintero JM, Avilés M, Delgado A (2011) Effect of *Trichoderma asperellum* strain T34 on iron, copper, manganese, and zinc uptake by wheat grown on a calcareous medium. Plant Soil 342(1–2):97–104

Delfim J, Schoebitz M, Paulino L, Hirzel J, Zagal E (2018) Phosphorus availability in wheat, in volcanic soils inoculated with phosphate-solubilizing *Bacillus thuringiensis*. Sustainability 10 (1):144

Deshwal VK, Dubey RC, Maheshwari DK (2003) Isolation of plant growth-promoting strains of *Bradyrhizobium* (Arachis) sp. with biocontrol potential against *Macrophomina phaseolina* causing charcoal rot of peanut. Curr Sci 443–448

Devi PB, Vijayabharathi R, Sathyabama S, Malleshi NG, Priyadarisini VB (2014) Health benefits of finger millet (*Eleusine coracana* L.) polyphenols and dietary fiber: a review. J Food Sci Technol 51(6):1021–40

Dey RKKP, Pal KK, Bhatt DM, Chauhan SM (2004) Growth promotion and yield enhancement of peanut (*Arachis hypogaea* L.) by application of plant growth-promoting rhizobacteria. Microbiol Res 159(4):371–394

Dhangrah VK, Mandal J, Bhat JS (2015) Heritable variation and predicted selection response of green yield and its component traits in vegetable amaranth. Int J Bio-res Environ Agric Sci 1 (4):146–152

Dheeman S, Maheshwari DK, Agarwal M, Dubey RC, Aeron A, Kim K, Bajpai VK (2017) Polyphasic and functional diversity of high altitude culturable *Bacillus* from rhizosphere of *Eleusine coracana* (L.) Gaertn. Appl Soil Ecol 110:127–136

Dida MM, Ramakrishnan S, Bennetzen JL, Gale MD, Devos KM (2007) The genetic map of finger millet, *Eleusine coracana*. Theor Appl Genet 114(2):321–332

Dodok L, Modhir AA, Buchtova V, Halasova G, Polaček I (1997) Importance and utilization of amaranth in food industry. Part 2. Composition of amino acids and fatty acids. Food/ Nahrung 41(2):108–110

Dogra D (2010) Biochemical evaluation of buck wheat *Fagopyrum esculentum* Moench genotypes. PhD thesis. CSK Himachal Pradesh Krishi Vishwavidyalaya HP, India

Dowling DN, O'Gara F (1994) Metabolites of *Pseudomonas* involved in the biocontrol of plant disease. Trend Biotechnol 12(4):133–141

Dubey P, Gupta GP (2012) Synergistic and antagonistic interactions among endophytic bacterial isolates of *Vigna mungo* (L.) Hepper. J Curr Perspect Appl Microbiol 1(2):1–12

Dubey RC, Khare S, Kumar P, Maheshwari DK (2014) Combined effect of chemical fertilisers and rhizosphere-competent *Bacillus subtilis* BSK17 on yield of *Cicer arietinum*. Arch Phytopathol Plant Protec 47(19):2305–2318

Eggum BO, Kreft I, Javornik B (1980) Chemical composition and protein quality of buckwheat (*Fagopyrum esculentum* Moench). Plant Food Hum Nutr 30(3–4):175–179

Eser D, Geçit HH (2010) Ekoloji. AÜ Ziraat Fakültesi 1584

Ezawa T, Smith SE, Smith FA (2002) P metabolism and transport in AM fungi. Plant Soil 244(1–2):221–230

Fernando WD, Nakkeeran S, Zhang Y (2005) Biosynthesis of antibiotics by PGPR and its relation in biocontrol of plant diseases. In: PGPR: biocontrol and biofertilization. Springer, Dordrecht, pp. 67–109

Friedman L, Kolter R (2004) Genes involved in matrix formation in *Pseudomonas aeruginosa* PA14 biofilms. Mol Microbiol 51(3):675–690

García AA, Huato MÁD, Lara MH, Sáenz-de-Cabezón FJ, Pérez-Moreno I, Marco-Mancebón V, López-Olguín JF (2011) Insect occurrence and losses due to phytophagous species in the amaranth *Amaranthus hypocondriacus* L. crop in Puebla, Mexico. African J Agric Res 6 (27):5924–5929

Glenn EP, Anday T, Chaturvedi R, Martinez-Garcia R, Pearlstein S, Soliz D et al (2013) Three halophytes for saline-water agriculture: An oilseed, a forage and a grain crop. Environ Exp Bot 92:110–121

Glick BR (1995) The enhancement of plant growth by free-living bacteria. Can J Microbiol 41 (2):109–117

Glick BR, Penrose DM, Li J (1998) A model for the lowering of plant ethylene concentrations by plant growth-promoting bacteria. J Theor Biol 190(1):63–68

Goldstein AH (1994) Involvement of the quinoprotein glucose dehydrogenase in the solubilization of exogenous phosphates by gram-negative bacteria. In: Phosphate in microorganisms: cellular and molecular biology. ASM Press, Washington, DC, pp 197–203

Gorinstein S, Pawelzik E, Delgado-Licon E, Haruenkit R, Weisz M, Trakhtenberg S (2002) Characterisation of pseudocereal and cereal proteins by protein and amino acid analyses. J Sci Food Agri 82(8):886–891

Goswami D, Thakker JN, Dhandhukia PC (2016) Portraying mechanics of plant growth promoting rhizobacteria (PGPR): A review. Cogent Food Agric 2(1):1127500

Gouda S, Nayak S, Bishwakarma S, Kerry RG, Das G, Patra JK (2017) role of microbial technology in agricultural sustainability. Microbial Biotechnology. Springer, Singapore, pp 181–202

Gray EJ, Smith DL (2005) Intracellular and extracellular PGPR: commonalities and distinctions in the plant-bacterium signaling processes. Soil Biol Biochem 37:395–412

Grunewald W, Van Noorden G, Van Isterdael G, Beeckman T, Gheysen G, Mathesius U (2009) Manipulation of auxin transport in plant roots during *Rhizobium* symbiosis and nematode parasitism. Plant Cell 21(9):2553–2562

Guerinot ML, Meidl EJ, Plessner O (1990) Citrate as a siderophore in *Bradyrhizobium japonicum*. J Bacteriol 172(6):3298–3303

Gundersen C, Ziliak JP (2015) Food insecurity and health outcomes. Health Aff 34(11):1830–1839

Gupta G, Parihar SS, Ahirwar NK, Snehi SK, Singh V (2015) Plant growth promoting rhizobacteria (PGPR): current and future prospects for development of sustainable agriculture. J Microb Biochem Technol 7(2):096–102

Gupta RP, Kalia A, Kapoor S (2007) Bioinoculants: a step towards sustainable agriculture. New India Publishing, New Delhi, India

Haas D, Défago G (2005) Biological control of soil-borne pathogens by fluorescent pseudomonads. Nature Rev Microbiol 3(4):307–319

Hadimani NA, Malleshi NG (1993) Studies on milling, physico-chemical properties, nutrient composition and dietary fibre content of millets. J Food Sci Technol (India) 30(1):17–20

Hafeez FY, Yasmin S, Ariani D, Zafar Y, Malik KA (2006) Plant growth-promoting bacteria as biofertilizer. Agron Sustain Dev 26(2):143–150

Haggag WM, Timmusk S (2008) Colonization of peanut roots by biofilm-forming Paenibacillus polymyxa initiates biocontrol against crown rot disease. J Appl Microbiol 104(4):961–969

He Z, Griffin TS, Honeycutt CW (2004) Enzymatic hydrolysis of organic phosphorus in swine manure and soil. J Environ Qual 33(1):367–372

Hider RC, Kong X (2010) Chemistry and biology of siderophores. Nat Prod Rep 27(5):637–657

Hilu KW, De Wet JMJ (1976) Domestication of *Eleusine coracana*. Econ Bot 30(3):199–208

Honma K, Inagaki S, Okuda K, Kuramitsu HK, Sharma A (2007) Role of a Tannerella forsythia exopolysaccharide synthesis operon in biofilm development. Microb Pathog 42(4):156–166

Huang X, Yang S, Gong J, Zhao Y, Feng Q, Gong H et al (2015) Genomic analysis of hybrid rice varieties reveals numerous superior alleles that contribute to heterosis. Nat Comm 6:6258

Hussain A, Hasnain S (2009) Cytokinin production by some bacteria: its impact on cell division in cucumber cotyledons. Afr J Microbiol Res 3(11):704–712

Idris EE, Iglesias DJ, Talon M, Borriss R (2007) Tryptophan-dependent production of indole-3-acetic acid (IAA) affects level of plant growth promotion by *Bacillus amyloliquefaciens* FZB42. Mol Plant Microb Interact 20(6):619–626

Idriss EE, Makarewicz O, Farouk A, Rosner K, Greiner R, Bochow H et al (2002) Extracellular phytase activity of *Bacillus amyloliquefaciens* FZB45 contributes to its plant-growth-promoting effecta. Microbiology 148(7):2097–2109

Ikeda K, Matsuda Y, Katsumaru A, Teranishi M, Yamamoto T, Kishidai M (1986) Factors affecting protein digestibility in soybean foods. Electrophoresis 10(100):1–3

Ikeda S, Yamaguchi Y (1993) Zinc contents in various samples and products of buckwheat. Buckwheat Newslett 1:1–3

Ikeda S, Yamashita Y (1994) Buckwheat as a dietary source of zinc, copper and manganese. Fagopyrum 14:29–34

Ikeda S, Edotani M, Naito S (1990) Zinc in buckwheat. Fagopyrum 10:51–56

James B, Atcha-Ahowe C, Godonou I, Baimey H, Goergen H, Sikirou R, Toko M (2010) Integrated pest management in vegetable production: A guide for extension workers in West African International Institute Tropical Agriculture (IITA), Ibadan, Nigeria, p 120

Jimenez-Aguilar DM, Grusak MA (2017) Minerals, vitamin C, phenolics, flavonoids and antioxidant activity of *Amaranthus* leafy vegetables. J Food Compos Anal 58:33–39
Kadaikunnan S, Rejiniemon TS, Khaled JM, Alharbi NS, Mothana R (2015) In-vitro antibacterial, antifungal, antioxidant and functional properties of *Bacillus amyloliquefaciens*. Annal Clin Microbiol Antimicrob 14(1):9–10
Kadyan S, Panghal M, Kumar S, Singh K, Yadav JP (2013) Assessment of functional and genetic diversity of aerobic endospore forming *Bacilli* from rhizospheric soil of *Phyllanthus amarus* L. World J Microbiol Biotechnol 29(9):1597–1610
Kagali RN, Kioko EN, Osiemo Z, Muya S, Wachera C (2013) Insect abundance and diversity on cultivated *Amaranthus* spp. (Amaranthacea) in Meru County, Kenya. American Int J Contemp Res 3(7):111–113
Kamala N, Gokulapalan C, Nair MC (1996) A new foliar blight of *Amaranthus* caused by *R. solani.* Indian Phytopathol 49:407
Kanensi OJ, Ochola S, Gikonyo NK, Makokha A (2013) Effect of steeping and germination on the diastatic activity and sugar content in amaranth grains and viscosity of porridge. J Agric Food Technol 3(1):1–7
Katiyar V, Goel R (2003) Solubilization of inorganic phosphate and plant growth promotion by cold tolerant mutants of *Pseudomonas fluorescens*. Microbiol Res 158(2):163–168
Kaur J, Khanna V, Kumari P, Sharma R (2015) Influence of psychrotolerant plant growth-promoting rhizobacteria (PGPR) as coinoculants with *Rhizobium* on growth parameters and yield of lentil (*Lens culinaris* Medikus). Afr J Microbiol Res 9(4):258–264
Kavita K, Singh VK, Mishra A, Jha B (2014) Characterisation and anti-biofilm activity of extracellular polymeric substances from *Oceanobacillus iheyensis*. Carbo Polym 101:29–35
Kayashita J, Shimaoka I, Nakajoh M, Kishida N, Kato N (1999) Consumption of a buckwheat protein extract retards 7, 12-dimethylbenz [α] anthracene-induced mammary carcinogenesis in rats. Biosci Biotechnol Biochem 63(10):1837–1839
Khan MS, Zaidi A, Ahemad M, Oves M, Wani PA (2010) Plant growth promotion by phosphate solubilizing fungi–current perspective. Arch Agron Soil Sci 56(1):73–98
Khatri D, Durgapal A, Joshi PK (2016) Biofertilization enhances productivity and nutrient uptake of foxtail millet plants. J Crop Improv 30(1):32–46
Kinsinger RF, Shirk MC, Fall R (2003) Rapid surface motility in *Bacillus subtilis* is dependent on extracellular surfactin and potassium ion. J Bacteriol 185(18):5627–5631
Kloepper JW, Ryu CM, Zhang S (2004) Induced systemic resistance and promotion of plant growth by *Bacillus* spp. Phytopathology 94(11):1259–1266
Knowles CJ, Bunch AW (1986) Microbial cyanide metabolism. Adv Microb Physiol 27(27):73–111
Kraemer SM (2004) Iron oxide dissolution and solubility in the presence of siderophores. Aqua Sci 66(1):3–18
Kreft I, Skrabanja V (2002) Nutritional properties of starch in buckwheat noodles. J Nutr Sci Vitaminol 48(1):47–50
Kremer RJ, Souissi T (2001) Cyanide production by rhizobacteria and potential for suppression of weed seedling growth. Curr Microbiol 43(3):182–186
Kumar S, Suyal DC, Dhauni N, Bhoriyal M, Goel R (2014) Relative plant growth promoting potential of Himalayan Psychrotolerant *Pseudomonas jesenii* strain MP1 against native *Cicer arietinum* (L.)., *Vigna mungo* (L.) Hepper; *Vigna radiata* (L.) Wilczek., *Cajanus cajan* (L.) Millsp. and *Eleusine coracana* (L.) Gaertn. African J Microbiol Res 8(50):3931–43
Kumar V, Singh P, Jorquera MA, Sangwan P, Kumar P, Verma AK, Agrawal S (2013) Isolation of phytase-producing bacteria from Himalayan soils and their effect on growth and phosphorus uptake of Indian mustard (*Brassica juncea*). World J Microbiol Biotechnol 29(8):1361–1369

Kumar A, Kumar A, Devi S, Patil S, Payal C, Negi S (2012a) Isolation, screening and characterization of bacteria from Rhizospheric soils for different plant growth promotion (PGP) activities: an in vitro study. Rec Res Sci Technol 4(1):1

Kumar A, Palni LMS, Sood A, Sharma M, Palni UT, Gupta AK (2002) Heat-shock induced somatic embryogenesis in callus cultures of gladiolus in the presence of high sucrose. J Horti Sci Biotechnol 77(1):73–78

Kumar A, Sharma S, Mishra S (2009) Effect of alkalinity on growth performance of Jatropha curcas inoculated with PGPR and AM fungi. J Phytol 1(3):1

Kumar P, Dubey RC, Maheshwari DK (2012b) *Bacillus* strains isolated from rhizosphere showed plant growth promoting and antagonistic activity against phytopathogens. Microbiol Res 167 (8):493–499

Kundan R, Pant G, Jadon N, Agrawal PK (2015) Plant growth promoting rhizobacteria: mechanism and current prospective. J Fertil Pest 6:155. https://doi.org/10.4172/2155-6202.1000155

Kuzyakov Y (2002) Factors affecting rhizosphere priming effects. J Plant Nutr Soil Sci 165 (4):382–396

Lamacchia C, Camarca A, Picascia S, Di Luccia A, Gianfrani C (2014) Cereal-based gluten-free food: how to reconcile nutritional and technological properties of wheat proteins with safety for celiac disease patients. Nutrients 6(2):575–590

Lambrecht M, Okon Y, Broek AV, Vanderleyden J (2000) Indole-3-acetic acid: a reciprocal signalling molecule in bacteria–plant interactions. Trend Microbiol 8(7):298–300

Lawrence G, Richards CA, Cheshire L (2004) The environmental enigma: why do producers professing stewardship continue to practice poor natural resource management? J Environ Pol Plan 6(3–4):251–270

Leff JW, Jones SE, Prober SM, Barberán A, Borer ET, Firn JL et al (2015) Consistent responses of soil microbial communities to elevated nutrient inputs in grasslands across the globe. Proc Natl Acad Sci 112(35):10967–10972

Li SQ, Zhang QH (2001) Advances in the development of functional foods from buckwheat. Crit Rev Food Sci Nutr 41(6):451–464

Li T, Bai R, Liu J (2008) Distribution and composition of extracellular polymeric substances in membrane-aerated biofilm. J Biotechnol 135(1):52–57

Liles MR, Scheel TA, Cianciotto NP (2000) Discovery of a nonclassical siderophore, legiobactin, produced by strains of *Legionella pneumophila*. J Bacteriol 182(3):749–757

Liu C, Yang Z, He P, Munir S, Wu Y, Ho H, He Y (2018) Deciphering the bacterial and fungal communities in clubroot-affected cabbage rhizosphere treated with *Bacillus subtilis* XF-1. Agric Ecosyst Environ 256:12–22

Liu F, Xing S, Ma H, Du Z, Ma B (2013) Cytokinin-producing, plant growth-promoting rhizobacteria that confer resistance to drought stress in *Platycladus orientalis* container seedlings. Appl Microbiol Biotechnol 97(20):9155–9164

Liu Z, Ishikawa W, Huang X, Tomotake H, Kayashita J, Watanabe H, Kato N (2001) Research Communication: A buckwheat protein product suppresses 1, 2-Dimethylhydrazine–induced colon carcinogenesis in rats by reducing cell proliferation. J Nutr 131(6):1850–1853

Lulai EC, Suttle JC (2004) The involvement of ethylene in wound-induced suberization of potato tuber (*Solanum tuberosum* L.): a critical assessment. Postharvest Biol Technol 34(1):105–112

Luthar Z (1992) Polyphenol classification and tannin content of buckwheat seeds (*Fagopyrum esculentum* Moench). Fagopyrum 12:36–42

Magiorakos AP, Srinivasan A, Carey RB, Carmeli Y, Falagas ME, Giske CG et al (2012) Multidrug-resistant, extensively drug-resistant and pandrug-resistant bacteria: an international expert proposal for interim standard definitions for acquired resistance. Clin Microbiol Infect 18(3):268–281

Magkos F, Arvaniti F, Zampelas A (2003) Organic food: nutritious food or food for thought? A review of the evidence. Int J Food Sci and Nutr 54(5):357–371

Maheshwari DK, Dubey RC, Agarwal M, Dheeman S, Aeron A, Bajpai VK (2015a) Carrier based formulations of biocoenotic consortia of disease suppressive *Pseudomonas aeruginosa* KRP1 and *Bacillus licheniformis* KRB1. Ecol Eng 81:272–277

Maheshwari DK (2010) Plant health and growth promoting bacteria. Microbiology Monograph. Springer, Heidelberg, Germany

Maheshwari DK (2011a) Bacteria in agrobiology: crop ecosystem. Springer, Heidelberg, Germany

Maheshwari DK (2011b) Bacteria in agrobiology: plant growth responses. Springer, Heidelberg, Germany

Maheshwari DK (2011c) Bacteria in agrobiology: plant nutrient management. Springer, Heidelberg, Germany

Maheshwari DK (2015) Halophiles- biodiversity and sustainable exploitation. Springer, Gewerbestrasse, Switzerland

Maheshwari DK (2016) Application of plant microbes in enhancing agricultural productivity: Success stories and bottlenecks–Presidential Address. In: Maheshwari DK (ed) Abstract Book, 103rd Indian Science Congress. Mysore, University of Mysore, Mysore, pp 5–38

Maheshwari DK, Dheeman S, Agarwal M (2015b) Phytohormone-producing PGPR for sustainable agriculture. In: Maheshwari DK (ed) Bacterial metabolites in sustainable agroecosystem. Springer, Cham, pp 159–182

Maheshwari DK, Dubey RC, Aeron A, Kumar B, Kumar S, Tewari S, Arora NK (2012) Integrated approach for disease management and growth enhancement of *Sesamum indicum* L. utilizing *Azotobacter chroococcum* TRA2 and chemical fertilizer. World J Microbiol Biotechnol 28 (10):3015–3024

Maheshwari DK, Saraf M, Aeron A (2013) Bacteria in Agrobiology: crop productivity. Springer Science & Business Media, Springer, Heidelberg, Germany

Mallick I, Bhattacharyya C, Mukherji S, Dey D, Sarkar SC, Mukhopadhyay UK, Ghosh A (2018) Effective rhizoinoculation and biofilm formation by arsenic immobilizing halophilic plant growth promoting bacteria (PGPB) isolated from mangrove rhizosphere: a step towards arsenic rhizoremediation. Sci Total Environ 610:1239–1250

Maloney SE (2001) Pesticide degradation. British mycological society symposium series 23:188–223

Mandal N, Das PK (2002) Intra-and interspecific genetic diversity in grain *Amaranthus* using random amplified polymorphic DNA markers. Plant Tissue Cult 12(1):49–56

Marshall VM, Lewis MM, Ostendorf B (2012) Buffel grass (*Cenchrus ciliaris*) as an invader and threat to biodiversity in arid environments: a review. J Arid Environ 78:1–12

Mayes S, Massawe FJ, Alderson PG, Roberts JA, Azam-Ali SN, Hermann M (2011) The potential for underutilized crops to improve security of food production. J Exp Bot 63:1075–1079

Mburu MW, Gikonyo NK, Kenji GM, Mwasaru AM (2012) Nutritional and functional properties of a complementary food based on Kenyan amaranth grain (*Amaranthus cruentus*). Afr J Food Agric Nutr Dev 12(2):5959–5977

Meena SK, Rakshit A, Meena VS (2016a) Effect of seed bio-priming and N doses under varied soil type on nitrogen use efficiency (NUE) of wheat (*Triticum aestivum* L.) under greenhouse conditions. Biocatal Agric Biotechnol 6:68–75

Meena VS, Bahadur I, Maurya BR, Kumar A, Meena RK, Meena SK, Verma JP (2016b) Potassium-solubilizing microorganism in evergreen agriculture: an overview. Potassium solubilizing microorganisms for sustainable agriculture. Springer, New Delhi, pp 1–20

Mehta P, Walia A, Kulshrestha S, Chauhan A, Shirkot CK (2015) Efficiency of plant growth-promoting P-solubilizing Bacillus circulans CB7 for enhancement of tomato growth under net house conditions. J Basic Microbiol 55(1):33–44

Mehta PS, Negi KS, Ojha SN (2010) Native plant genetic resources and traditional foods of Uttarakhand Himalaya for sustainable food security and livelihood. Indian J Natur Prod Resour 1(1):89–96

Mielich-Süss B, Lopez D (2015) Molecular mechanisms involved in *Bacillus subtilis* biofilm formation. Environ Microbiol 17(3):555–565

Miethke M, Marahiel MA (2007) Siderophore-based iron acquisition and pathogen control. Microbiol Mol Biol Rev 71(3):413–451

Mihail JD and Champaco ER (1993) Diseases of *Amaranthus* spp. caused by *Pythium aphanidermatum* and *Macrophomina phaseolina*. Can J Botany 71(9):1219–1223

Miraj S (2016) A Medicinal plant with antioxidant activity in Iranian folk medicine: *Amaranthus*. Der Pharm Lett 8(14):27–30

Mishra PK, Bisht SC, Mishra S, Selvakumar G, Bisht JK, Gupta HS (2012) Coinoculation of *Rhizobium leguminosarum*-PR1 with a cold tolerant *Pseudomonas* sp. improves iron acquisition, nutrient uptake and growth of field pea (*Pisum sativum* L.). J Plant Nutri 35 (2):243–256

Mlakar SG, Turinek M, Jakop M, Bavec M, Bavec F (2009) Nutrition value and use of grain amaranth: potential future application in bread making. Agricultura 6(4):1

Morrone D, Chambers J, Lowry L, Kim G, Anterola A, Bender K, Peters RJ (2009) Gibberellin biosynthesis in bacteria: Separate ent-copalyl diphosphate and ent-kaurene synthases in Bradyrhizobium japonicum. FEBS Lett 583(2):475–480

Muis A, Quimio AJ (2016) Biological control of banded leaf and sheath blight disease (*Rhizoctonia solani* Kuhn) in corn with formulated *Bacillus subtilis* BR23. Indo J Agric Sci 7 (1):1–7

Mureithi DM, Komi FK, Ekesi S, Meyhöfer R (2017) Important arthropod pests on leafy Amaranth (*Amaranthus viridis*, *A. tricolor* and A. blitum) and broad-leafed African nightshade (*Solanum scabrum*) with a special focus on host-plant ranges. Afr J Horti Sci 11:1–17

Nadeem SM, Ahmad M, Zahir ZA, Javaid A, Ashraf M (2014a) The role of mycorrhizae and plant growth promoting rhizobacteria (PGPR) in improving crop productivity under stressful environments. Biotechnol Advances 32(2):429–448

Nadeem SM, Ahmad M, Zahir ZA, Javaid A, Ashraf M (2014b) The role of mycorrhizae and plant growth promoting rhizobacteria (PGPR) in improving crop productivity under stressful environments. Biotechnol Adv 32(2):429–448

Nadeem SM, Zahir ZA, Naveed M, Asghar HN, Arshad M (2010) Rhizobacteria capable of producing ACC-deaminase may mitigate salt stress in wheat. Soil Sci Soc Am J 74(2):533–542

Nagai T, Sakai M, Inoue R, Inoue H, Suzuki N (2011) Antioxidative activities of some commercially honeys, royal jelly, and propolis. Food Chem 75(2):237–240

Nair CB, Anith NK (2009) Efficacy of acibenzolar-S-methyl and rhizobacteria for the management of foliar blight disease of amaranth. Int J Trop Agric 47(1):43–47

Navarro S, Vela N, Navarro G (2007) An overview on the environmental behaviour of pesticide residues in soils. Spanish J Agric Res 3:357–375

Nelson LM (2004) Plant growth promoting rhizobacteria (PGPR): prospects for new inoculants. Plant Manag Netw. https://doi.org/10.1094/cm-2004-0301-05-rv

Nenwani V, Doshi P, Saha T, Rajkumar S (2010) Isolation and characterization of a fungal isolate for phosphate solubilization and plant growth promoting activity. J Yeast Fungal Res 1(1):009–014

NFSB (2013) National food security bill. The Gazette of India: extraordinary, New Delhi, the 10th September, 2013/Bhadra 19, 1935 (Saka) http://indiacode.nic.in/acts-in-pdf/202013.pdf

NFSM (2012) National food security mission, 12th Five Year Plan. NFSM-Coarse cereals. http://www.nfsm.gov.in/CoarseCereals.aspx

Niranjana M, Kumar JSA (2017) Grain amaranth (*Amaranthus* sp.)-an underutilized crop species for nutritional security and climate resilience. Mysore J Agric Sci 51(1):12–20

Nyvall RF (1989) Diseases of cotton. In: Field crop diseases handbook Springer, Boston, MA, pp 171–210

Odukkathil G, Vasudevan N (2013) Toxicity and bioremediation of pesticides in agricultural soil. Rev Env Sci Biotech 12(4):421–444

Ortíz-Castro R, Contreras-Cornejo HA, Macías-Rodríguez L, López-Bucio J (2009) The role of microbial signals in plant growth and development. Plant Signal Behavior 4(8):701–712

Østerberg JT, Xiang W, Olsen LI, Edenbrandt AK, Vedel SE, Christiansen A et al (2017) Accelerating the domestication of new crops: feasibility and approaches. Trend Plant Sci 22 (5):373–384

O'sullivan DJ, O'Gara F (1992) Traits of fluorescent *Pseudomonas* spp. involved in suppression of plant root pathogens. Microbiol Rev 56(4):662–676

Othim STO, Kahuthia-Gathu R, Akutse KS, Foba CN, Fiaboe KKM (2018) Seasonal occurrence of amaranth *Lepidopteran defoliators* and effect of attractants and amaranth lines in their management. J Appl Entomol 2018:12513. https://doi.org/10.1111/jen.12513

Pandey C, Dheeman S, Negi YK, Maheshwari DK (2018) Differential response of native *Bacillus* spp. isolates from agricultural and forest soils in growth promotion of *Amaranthus hypochondriacus*. Biotechnol Res 4(1):54–61

Pandey N, Gupta B, Pathak GC (2013) Foliar application of Zn at flowering stage improves plant's performance, yield and yield attributes of black gram. Indian J Exp Biol 51:548–555

Pandey P, Kang SC, Maheshwari DK (2005) Isolation of endophytic plant growth promoting *Burkholderia* sp. MSSP from root nodules of *Mimosa pudica*. Curr Sci 177–180

Pane C, Zaccardelli M (2015) Evaluation of *Bacillus* strains isolated from solanaceous phylloplane for biocontrol of Alternaria early blight of tomato. Biol Cont 84:11–18

Paradkar MM, Irudayaraj J (2002) Rapid determination of caffeine content in soft drinks using FTIR–ATR spectroscopy. Food Chem 78(2):261–266

Parmar JK, Patel JJ (2009) Effect of organic and inorganic nitrogen and biofertilizer on nutrient content and uptake by amaranthus (*Amaranthus hypochondriacus* L.). Asian J Soil Sci 4 (1):135–138

Pasek MA, Lauretta DS (2005) Aqueous corrosion of phosphide minerals from iron meteorites: a highly reactive source of prebiotic phosphorus on the surface of the early Earth. Astrobiology 5 (4):515–535

Patel DK, Kanungo VK (2010) Phytoremediation potential of duckweed (lemnaminor l: a tiny aquatic plant) in the removal of pollutants from domestic wastewater with special reference to nutrients. Bioscan 5(3):355–358

Penrose DM, Glick BR (2003) Methods for isolating and characterizing ACC deaminase-containing plant growth-promoting rhizobacteria. Physiol Plant 118(1):10–15

Perales-Sánchez JX, Reyes-Moreno C, Gómez-Favela MA, Milán-Carrillo J, Cuevas-Rodríguez EO, Valdez-Ortiz A, Gutiérrez-Dorado R (2014) Increasing the antioxidant activity, total phenolic and flavonoid contents by optimizing the germination conditions of amaranth seeds. Plant Food Hum Nutr 69(3):196–202

Pingale SS, Virkar PS (2013) Study of influence of phosphate dissolving micro-organisms on yield and phosphate uptake by crops. Euro J Exp Biol 3(2):191–193

Písaříková B, Kráčmar S, Herzig I (2005) Amino acid contents and biological value of protein in various amaranth species. Czech J Anim Sci 50(4):169–174

Plessner O, Klapatch T, Guerinot ML (1993) Siderophore utilization by *Bradyrhizobium japonicum*. Appl Environ Microbiol 59(5):1688–1690

Podile AR, Kishore GK. (2007) Plant growth-promoting rhizobacteria. In: Plant-associated bacteria. Springer, Dordrecht, pp 195–230

Prakash B, Veeregowda BM, Krishnappa G (2003) Biofilms: a survival strategy of bacteria. Curr Sci 1299–1307

Prasad R, Kumar M, Varma A (2015) Role of PGPR in soil fertility and plant health. In: Plant-growth-promoting rhizobacteria (PGPR) and medicinal plants. Springer International Publishing, pp 247–260

Priyadarsini P (2003) Ecofriendly management of *Rhizoctonia* leaf blight of amaranthus. MSc (Ag.) thesis, Kerala Agricultural University, Thrissur, India

Raina A, Datta A (1992) Molecular cloning of a gene encoding a seed-specific protein with nutritionally balanced amino acid composition from *Amaranthus*. Proc Natl Acad Sci 89 (24):11774–11778

Rais A, Shakeel M, Malik K, Hafeez FY, Yasmin H, Mumtaz S, Hassan MN (2018) Antagonistic *Bacillus* spp. reduce blast incidence on rice and increase grain yield under field conditions. Microbiol Res 208:54–62

Raj SN, Deepak SA, Basavaraju P, Shetty HS, Reddy MS, Kloepper JW (2003) Comparative performance of formulations of plant growth promoting rhizobacteria in growth promotion and suppression of downy mildew in pearl millet. Crop Protec 22(4):579–588

Ramachandra G, Virupaksha TK, Shadaksharaswamy M (1977) Relation between tannin levels and in vitro protein digestibility in finger millet (*Eleusine coracana* Gaertn.). J Agric Food Chem 25(5):1101–1104

Raman R, Kuppuswamy G, Krishnamoorthy R (2004) Response of organic mulching practices on weed anagement and yield of cotton. In: 6th IFAOM-Asia scientific conference, p 476

Rangaraj T, Somasundaram EM, Amanullah M, Thirumurugan V, Ramesh S, Ravi S (2007) Effect of Agro-industrial wastes on soil properties and yield of irrigated finger millet (*Eleusine coracana* L. Gaertn) in coastal soil. Res J Agric Biol Sci 3(3):153–156

Rastogi A, Shukla S (2013) Amaranth: a new millennium crop of nutraceutical values. Crit Rev Food Sci Nutr 53(2):109–125

Raymond KN, Dertz EA, Kim SS (2003) Enterobactin: an archetype for microbial iron transport. Proc Natl Acad Sci 100(7):3584–3588

Reetha S, Bhuvaneswari G, Thamizhiniyan P, Mycin TR (2014) Isolation of indole acetic acid (IAA) producing rhizobacteria of *Pseudomonas fluorescens* and *Bacillus subtilis* and enhance growth of onion (*Allim cepa.* L). Int J Curr Microbiol Appl Sci 3(2):568–574

Ribeiro APD, Andrade MC, Bagnato VS, Vergani CE, Primo FL, Tedesco AC, Pavarina AC (2015) Antimicrobial photodynamic therapy against pathogenic bacterial suspensions and biofilms using chloro-aluminum phthalocyanine encapsulated in nanoemulsions. Laser Med Sci 30(2):549–559

Richardson AE (2001) Prospects for using soil microorganisms to improve the acquisition of phosphorus by plants. Func Plant Biol 28(9):897–906

Rinu K, Pandey A (2009) *Bacillus subtilis* NRRL B-30408 inoculation enhances the symbiotic efficiency of *Lens esculenta* Moench at a Himalayan location. J Plant Nutr Soil Sci 172 (1):134–139

Rodríguez H, Fraga R (1999) Phosphate solubilizing bacteria and their role in plant growth promotion. Biotechnol Adv 17(4–5):319-339

Rovira AD (1969) Plant root exudates. Botanic Rev 35(1):35–57

Sacherer P, Défago G, Haas D (1994) Extracellular protease and phospholipase C are controlled by the global regulatory gene *gac*A in the biocontrol strain *Pseudomonas fluorescens* CHA0. FEMS Microbiol Lett 116(2):155–160

Safronova VI, Stepanok VV, Engqvist GL, Alekseyev YV, Belimov AA (2006) Root-associated bacteria containing 1-aminocyclopropane-1-carboxylate deaminase improve growth and nutrient uptake by pea genotypes cultivated in cadmium supplemented soil. Biol Fertil Soils 42(3):267–272

Saharan BS, Nehra V (2011) Plant growth promoting rhizobacteria: a critical review. Life Sci Med Res 21(1):30

Sandilya SP, Bhuyan PM, Gogoi DK, Kardong D (2018) Phosphorus solubilization and plant growth promotion ability of rhizobacteria of *R. communis* L growing in Assam, India. Proc Natl Acad Sci India B: Biol Sci 88(3):959–966

Sanz-Penella JM, Wronkowska M, Soral-Smietana M, Haros M (2013) Effect of whole amaranth flour on bread properties and nutritive value. LWT-Food Sci Technol 50(2):679–685

Sati P, Dhakar K, Pandey A (2013) Microbial diversity in soil under potato cultivation from cold desert Himalaya, India. ISRN Biodivers 3:2013

Schmitt J, Flemming HC (1999) Water binding in biofilms. Water Sci Technol 39(7):77–82

Schoebitz M, López MD, Roldán A (2013) Bioencapsulation of microbial inoculants for better soil–plant fertilization. A review. Agron Sustain Dev 33(4):751–765

Schroth MN, Kloepper JW (1978) Plant growth promoting rhizobacteria on radish. In: Proceedings of the fourth conference plant pathogenic bacteria. Ed Station de Pathogenic Vegetable ET (Phytobacteriologic) INRA Angers, pp 876–882

Sealy RL, Kenerley CM, McWilliams EL (1988) Evaluation of *Amaranthus* accessions for resistance to damping-off by *Pythiurn myriotylutn*. Plant Dis 72:985–989

Sedej I, Sakač M, Mandić A, Mišan A, Tumbas V, Hadnađev M (2011) Assessment of antioxidant activity and rheological properties of wheat and buckwheat milling fractions. J Cereal Sci 54 (3):347–353

Seneviratne G, Jayasekara APDA, De Silva MSDL, Abeysekera UP (2011) Developed microbial biofilms can restore deteriorated conventional agricultural soils. Soil Biol Biochem 43 (5):1059–1062

Shaharoona B, Jamro GM, Zahir ZA, Arshad M, Memon KS (2007) Effectiveness of various *Pseudomonas* spp. and *Burkholderia caryophylli* containing ACC-Deaminase for improving growth and yield of wheat (*Triticum aestivum* I.). J Microbiol Biotechnol 17(8):1300

Shanmugam V, Thakur H, Gupta S (2013) Use of chitinolytic *Bacillus atrophaeus* strain S2BC-2 antagonistic to *Fusarium* spp. for control of rhizome rot of ginger. Ann Microbiol 63(3):989–96

Sharma CK, Vishnoi VK, Dubey RC, Maheshwari DK (2018) A twin rhizospheric bacterial consortium induces systemic resistance to a phytopathogen *Macrophomina phaseolina* in mung bean. Rhizosphere 5:71–75

Sharma A, Johri BN (2003) Growth promoting influence of siderophore-producing *Pseudomonas strains* GRP3A and PRS sub 9 in maize (*Zea mays* L.) under iron limiting conditions. Microbiol Res 158(3):243

Sharma BR, Ambili G (2009) Impact of climate change on water resources and potential adaptations for agriculture in the Indus-Gangetic basin. In: 60th IEC meeting & 5th Asian regional conference on improving in efficiency of irrigation projects through technology upgradation and better operation & maintenance, pp 6–11

Sharma N (2009) Phosphate solubilizing potential of plant growth promoting rhizobacteria (PGPR) isolated from tomato seedlings. Ph.D. Thesis, Dr. Yashwant Singh Parmar University of Horticulture and Forestry, Nauni, Solan – 173230 H.P. India https://krishikosh.egranth.ac.in/displaybitstream?handle=1/96993 (accessed on 07.09.2019)

Shevkani K, Singh N, Rana JC, Kaur A (2014) Relationship between physicochemical and functional properties of amaranth (*Amaranthus hypochondriacus*) protein isolates. Int J Food Sci Tech 49(2):541–550

Shirani M, Raeisi R, Heidari-Soureshjani S, Asadi-Samani M, Luther T (2017) A review for discovering hepatoprotective herbal drugs with least side effects on kidney. J Nephropharmacol 6(2):38–48

Siddiqui IA, Shaukat SS, Sheikh IH, Khan A (2006) Role of cyanide production by *Pseudomonas fluorescens* CHA0 in the suppression of root-knot nematode, *Meloidogyne javanica* in tomato. World J Microbiol Biotechnol 22(6):641–650

Siddiqui ZA, Akhtar MS (2009) Effects of antagonistic fungi, plant growth-promoting rhizobacteria, and arbuscular mycorrhizal fungi alone and in combination on the reproduction of *Meloidogyne incognita* and growth of tomato. J Gen Plant Pathol 75(2):144

Sidorova SF (1963) Review of fungal diseases of buckwheat in the USSR Trudy. Vses Inst Zashch Rast 19:25–31

Singh N, Pandey P, Dubey RC, Maheshwari DK (2008) Biological control of root rot fungus *Macrophomina phaseolina* and growth enhancement of *Pinus roxburghii* (Sarg.) by rhizosphere competent *Bacillus subtilis* BN1. World J Microbiol Biotechnol 24(9):1669

Skrabanja V, Kreft I (1998) Resistant starch formation following autoclaving of buckwheat (*Fagopyrum esculentum* Moench) groats: An in vitro study. J Agric Food Chem 46(5):2020–2023

Skrabanja V, Liljeberg Elmståhl HG, Kreft I, Björck IM (2001) Nutritional properties of starch in buckwheat products: studies in vitro and in vivo. J Agric Food Chem 49(1):490–496

Smitha KP (2000) Management of foliar blight of Amaranthus (*Amaranthus tricolor* L.) caused by *Rhizoctonia solani* Kuhn using microbial antagonists. MSc (Ag.) Thesis. Thiruvananthapuram. p 74

Springmann M, Godfray HC, Rayner M, Scarborough P (2016) Analysis and valuation of the health and climate change cobenefits of dietary change. Proc Natl Acad Sci 113(15):4146–4151

Sridevi M, Mallaiah KV (2008) Production of hydroxamate-type of siderophores by *Rhizobium* strains from *Sesbania sesban* (L.). Merr. Int J Soil Sci 3:24–28

Steadman KJ, Burgoon MS, Lewis BA, Edwardson SE, Obendorf RL (2001) Buckwheat seed milling fractions: description, macronutrient composition and dietary fibre. J Cereal Sci 33 (3):271–278

Stein T (2005) *Bacillus subtilis* antibiotics: structures, syntheses and specific functions. Mol Microbiol 56(4):845–857

Stibilj V, Kreft I, Smrkolj P, Osvald J (2004) Enhanced selenium content in buckwheat (*Fagopyrum esculentum* Moench) and pumpkin (*Cucurbita pepo* L.) seeds by foliar fertilisation. Euro Food Res Technol 219(2):142–144

Stone JL (1906) Buckwheat, vol. 238. Cornell University

Storer NP, Babcock JM, Schlenz M, Meade T, Thompson GD, Bing JW, Huckaba RM (2010) Discovery and characterization of field resistance to Bt maize: *Spodoptera frugiperda* (*Lepidoptera*: Noctuidae) in Puerto Rico. J Econ Entomol 103(4):1031–1038

Supanjani HH, Jung JS, Lee KD (2006) Rock phosphate-potassium and rock-solubilising bacteria as alternative, sustainable fertilizers. Agron Sustain Dev 26(4):233–240

Sutherland IW (2001) Biofilm exopolysaccharides: a strong and sticky framework. Microbiology 147(1):3–9

Suzuki T, Honda Y, Funatsuki W, Nakatsuka K (2002) Purification and characterization of flavonol 3-glucosidase, and its activity during ripening in tartary buckwheat seeds. Plant Sci 163(3):417–423

Swaminathan JM, Smith SF, Sadeh NM (1998) Modeling supply chain dynamics: a multiagent approach. Decis Sci 29(3):607–632

Swaminathan MS (2006) An evergreen revolution. Crop Sci 46(5):2293–2303

Talukder MMR, Riazuddin M, Rahman MM, Uddin MS, Khan MSI (2012) Efficacy of fungicodes to control white rust (*Albugo occidentalis*) of red amaranth (*Amaranthus* sp.). Bangladesh Phytopathol Soc 28:1–2

Tan S, Yang C, Mei X, Shen S, Raza W, Shen Q, Xu Y (2013) The effect of organic acids from tomato root exudates on rhizosphere colonization of *Bacillus amyloliquefaciens* T-5. Appl Soil Ecol 64:15–22

Taylor J, Awika J (2017) Gluten-free ancient grains: cereals, pseudocereals, and legumes: sustainable, Nutritious, and health-promoting foods for the 21st century. Woodhead Publishing

Thakur R, Kumar S, Awasthi CP (2016) Mineral composition and protein fractionation of tartary buckwheat grains grown in cold dry desert of Himachal Pradesh. J Hill Agricul 7(1):125–128

Thomas L, Gupta A, Gopal M, George P, Thomas GV (2010) Plant growth promoting potential of *Bacillus* spp. isolated from rhizosphere of cocoa (*Theobroma cacao* L.). J Plant Crop 38(2):97–104

Timmusk S, El-Daim IAA, Copolovici L, Tanilas T, Kännaste A, Behers L et al (2014) Drought-tolerance of wheat improved by rhizosphere bacteria from harsh environments: enhanced biomass production and reduced emissions of stress volatiles. PLoS One 9(5):e96086

Tiryaki O, Temur C (2010) The fate of pesticide in the environment. J Biol Environ Sci 4(10):29–38

Tomotake H, Shimaoka I, Kayashita J, Yokoyama F, Nakajoh M, Kato N (2000) A buckwheat protein product suppresses gallstone formation and plasma cholesterol more strongly than soy protein isolate in hamsters. J Nutr 130(7):1670–1674

Trivedi P, Pandey A (2008) Plant growth promotion abilities and formulation of *Bacillus megaterium* strain B 388 (MTCC6521) isolated from a temperate Himalayan location. Indian J Microbiol 48(3):342–347

Trivedi P, Sa T (2008) *Pseudomonas corrugata* (NRRL B-30409) mutants increased phosphate solubilization, organic acid production, and plant growth at lower temperatures. Curr Microbiol 56(2):140–144

Veer SA, Goel R (2015) Plant growth promoting efficiency of *chryseobacterium* sp. PSR 10 on finger millet (*Eleusine coracana*). J Global Biosci 4(6):2569–2575

Voisard C, Keel C, Haas D, Dèfago G (1989) Cyanide production by *Pseudomonas fluorescens* helps suppress black root rot of tobacco under gnotobiotic conditions. EMBO J 8(2):351–358

Wani PA, Khan MS, Zaidi A (2008) Chromium-reducing and plant growth-promoting *Mesorhizobium* improves chickpea growth in chromium-amended soil. Biotechnol Lett 30 (1):159–163

Wenke K, Kai M, Piechulla B (2010) Belowground volatiles facilitate interactions between plant roots and soil organisms. Planta 231(3):499–506

Went FW, Thimann KV (1937) Phytohormones, New York, p 294

West S, Horn H, Hijnen WAM, Castillo C, Wagner M (2014) Confocal laser scanning microscopy as a tool to validate the efficiency of membrane cleaning procedures to remove biofilms. Sep Purif Technol 122:402–411

Wieslander G, Norback D (2001) Buckwheat consumption and its medical and pharmacological effects–a review of the literature. In: Proceedings of the 8th international symposium on Buckwheat–ISB, September 2001, pp 608–612

Wright M, Osborne DJ (1974) Abscission in *Phaseolus vulgaris* the positional differentiation and ethylene-induced expansion growth of specialised cells. Planta 120(2):163–170

Wronkowska M, Zielińska D, Szawara-Nowak D, Troszyńska A, Soral-Śmietana M (2010) Antioxidative and reducing capacity, macroelements content and sensorial properties of buckwheat-enhanced gluten-free bread. Int J Food Sci Technol 45(10):1993–2000

Yadav BK, Verma A (2012) Phosphate solubilization and mobilization in soil through microorganisms under arid ecosystems. In: The functioning of ecosystems. InTech

Yadav J, Verma JP, Jaiswal DK, Kumar A (2014) Evaluation of PGPR and different concentration of phosphorus level on plant growth, yield and nutrient content of rice (*Oryza sativa*). Ecol Eng 62:123–128

Yuan J, Zhang N, Huang Q, Raza W, Li R, Vivanco JM, Shen Q (2015) Organic acids from root exudates of banana help root colonization of PGPR strain *Bacillus amyloliquefaciens* NJN-6. Scientific Rep 5:13438

Zahir ZA, Ghani U, Naveed M, Nadeem SM, Asghar HN (2009) Comparative effectiveness of *Pseudomonas* and *Serratia* sp. containing ACC-deaminase for improving growth and yield of wheat (*Triticum aestivum* L.) under salt-stressed conditions. Arch Microbiol 191(5):415–424

Zeriouh H, de Vicente A, Pérez-García A, Romero D (2014) Surfactin triggers biofilm formation of *Bacillus subtilis* in melon phylloplane and contributes to the biocontrol activity. Environ Microbiol 16(7):2196–2211

Zhang YQ, Ren SX, Li HL, Wang YX, Fu G, Yang J et al (2003) Genome-based analysis of virulence genes in a non-biofilm-forming Staphylococcus epidermidis strain (ATCC 12228). Mol Microbiol 49(6):1577–1593

Zohora US, Ano T, Rahman MS (2016) Biocontrol of *Rhizoctonia solani* K1 by iturin A producer *Bacillus subtilis* RB14 seed treatment in tomato plants. Adv Microbiol 6(06):424

Zongzheng Y, Xin L, Zhong L, Jinzhao P, Jin Q, Wenyan Y (2010) Effect of *Bacillus subtilis* SY1 on antifungal activity and plant growth. Int J Agric Biol Eng 2(4):55–61

Chapter 6
Utilization of Endophytic Bacteria Isolated from Legume Root Nodules for Plant Growth Promotion

Winston Franz Ríos-Ruiz, Renzo Alfredo Valdez-Nuñez, Eulogio J. Bedmar and Antonio Castellano-Hinojosa

Abstract For decades, rhizobia were described as the single inhabitant of legume nodules. However, other bacteria, which are not typical rhizobia, are often found within nodules and suggested to affect the behaviour and fitness of the host plant. Here, we highlight their diversity, role in the promotion of legumes growth and in the recuperation of degraded soils. Studies have shown the capacity of Non-Nodular Endophytic Bacteria (NNEB) to stimulate plant growth by direct mechanisms including production of phytohormones such as auxins, facilitation of acquisition of plant resources/nutrients that plants lack such as fixed nitrogen, iron, phosphorous, besides, production of the enzyme 1-aminocyclopropane-1-carboxylate (ACC deaminase) involved in reduction of plant ethylene, etc. Among the indirect mechanisms, those associated to the enhancement of plant growth are production of siderophores, antibiotics and lytic enzymes. Finally, we described greenhouse and field experiments that successfully used NNEB to both increase the growth and yield of legumes and to recover degraded soils.

Keywords Legume · Plant-growth promotion · Nodule · Endophytic bacteria · Soil

---

W. F. Ríos-Ruiz · R. A. Valdez-Nuñez
Universidad Nacional de San Martín–Tarapoto, Tarapoto, Perú
e-mail: wrios@unsm.edu.pe

R. A. Valdez-Nuñez
e-mail: revanu@gmail.com

E. J. Bedmar · A. Castellano-Hinojosa (✉)
Department of Soil Microbiology and Symbiotic Systems,
Estación Experimental del Zaidín, CSIC, Granada, Spain
e-mail: ach@ugr.es

E. J. Bedmar
e-mail: eulogio.bedmar@eez.csic.es

A. Castellano-Hinojosa
Pharmacy Faculty, Department of Microbiology, University of Granada, Granada, Spain

D. K. Maheshwari and S. Dheeman (eds.), *Field Crops: Sustainable Management by PGPR*, Sustainable Development and Biodiversity 23,
https://doi.org/10.1007/978-3-030-30926-8_6

6.1 The Bacteria Inside the Legume Root Nodules

Legumes (family Fabaceae or Leguminosae) constitute the third largest family of angiosperms plants with a critical importance in human and animal feeding as well as for benefits of agricultural systems. Leguminous and actinorhizal plants are unique among the living beings because they have the characteristic to establish N_2-fixing symbiosis with soil bacteria, collectively known as rhizobia; as a consequence of the plant-bacteria interaction, roots, and occasionally stems, nodules are formed where N_2-fixation occurs (Gresshoff and Ferguson 2017; Lace and Ott 2018). Briefly, nodulation is a complex and specific molecular 'conversation' between the host legume and the rhizobia where the former produce flavonoids in their radical exudates which are detected by rhizobia (Hayashi et al. 2014). Then, flavonoids induce a cascade of nodulation genes in the rhizobia that start the synthesis of nodulation factors, the formation of an infection thread to deliver rhizobia within the epidermal cells of the plant host and, finally the transformation of vegetative rhizobial cells into specialized, N_2-fixing bacteroids (Oldroyd and Downie 2008; Downie 2014).

6.2 The Diversity of Nodule-Forming Rhizobial Bacteria

The rhizobia are a polyphyletic group of bacteria which, in addition to their symbiotic N_2-fixing lifestyle with legumes, are also able to colonize plant roots and even to maintain an endophytic role in non-legume plants (Bhattacharjee et al. 2008; Poole et al. 2018; Schneijderberg et al. 2018). Classical nodule-forming, N_2-fixing, symbiotic rhizobia belong to the Alphaproteobacteria and Betaproteobacteria classes, albeit members of the Gammaproteobacteria and Actinobacteria have been also reported as symbiotic bacteria (Martínez-Hidalgo and Hirsch 2017 and references therein). Currently, classical rhizobia include the genera *Rhizobium*, *Ensifer* (formerly *Sinorhizobium*), *Neorhizobium*, *Pararhizobium* and *Allorhizobium* within the family Rhizobiaceae, and genera *Mesorhizobium*, *Bradyrhizobium* and *Azorhizobium* of the Phyllobacteriaceae, Nitrobacteriaceae (Bradyrhizobiaceae) and Hyphomicrobiaceae families, respectively. Comprehensive reviews on the current classification of bacteria able to establish N_2-fixing legume symbioses, including both classical and non-classical rhizobia, have been published (Mousavi et al. 2015; Peix et al. 2015; Shamsheldin et al. 2017; Martínez-Hidalgo and Hirsch 2017; Velázquez et al. 2017).

The ability for nodulation of non-classical rhizobia, those which do not belong to the above-mentioned genera, has been related to their coexistence with classical symbiotic rhizobia leading to horizontal transfer of nodulation genes (Martínez-Hidalgo and Hirsch 2017). This attribute constitutes the main driver of evolution for the acquisition of new symbiotic abilities for bacteria residing within the nodule (Estrela et al. 2016; Zgadzaj et al. 2016). A list of valid species of rhizobia is constantly updated and recorded in the list of Prokaryotic names with standing in nomenclature (https://www.bacterio.cict.fr).

6.3 Non-nodulating Bacteria from Legume Root Nodules

In the interior of the legume nodules, in addition to the symbiotic rhizobia, the presence of a vast diversity of bacteria unable to form nodules by themselves has been reported (Peix et al. 2015; Aeron et al. 2015; Martínez-Hidalgo and Hirsch 2017; Velázquez et al. 2017). There are several denominations for bacteria that co-exist with symbiotic rhizobia; they have been called non-rhizobial bacteria (Tariq et al. 2012; Dhole et al. 2016), non-rhizobial endophytes (De Meyer et al. 2015), nodule endophytes (Velázquez et al. 2013), bacteria associated with the nodule (Rajendran et al. 2012), and non-symbiotic bacteria with disinfected nodules (Da Costa et al. 2013). In this review, we will refer to them as Non-Nodular Endophytic Bacteria (NNEB).

The first report published on the presence of bacteria different from the classic rhizobia recovered from the interior of healthy nodules is more than 100 years old (Beijerinck and Van Delden 1902) and was associated to *Agrobacterium radiobacter* isolated from root nodules of *Trifolium* plants. These results did not go unnoticed by Philipson and Blair (1957) who, in addition to *Rhizobium*, reported the presence of *Aerobacter, Bacillus* and *Flavobacterium* from the interior of *Trifolium* plant nodules. For a long time, based on the morphocolonial characteristics and biochemical tests as well as on the absence of nodulation traits after reinfection tests with the original host, many authors underestimated the presence of endophytic bacteria different to those of classical rhizobia (Fred et al. 1932; Vincent 1970; Somasegaran and Hoben 1994). Even, they were catalogued as 'contaminants' by Somasegaran and Hoben (1994). For many years, endophytic bacteria were discarded based mainly on two tests: the absorption capacity of Congo Red when they grow in Petri plates containing yeast extract mannitol medium, and the colour change of bromocresol purple indicator on Petri plates containing peptone glucose agar (Martínez-Hidalgo and Hirsch 2017). Nowadays it is well stablished that nodule disinfection is required for the correct isolation of NNEB. There are several methodologies to confirm the efficiency of surface-sterilized nodules: (a) incubation of disinfected nodules on plates containing YEM medium (Tariq et al. 2012); (b) plating aliquots of the rinsing water after the disinfection process in growth media (Bai et al. 2002; Leite et al. 2017); (c) rolling the surface-sterilized nodules on plates containing growth media (De Meyer et al. 2015).

In any case, the appearance of bacterial growth invalidates the experiment and bacteria grown in the planes can be associated to external contamination coming from the outer surface of the nodule (Bai et al. 2002). About a hundred of different genera considered NNEB have been isolated from root legumes worldwide, of which 29.7% belong to the Alfaproteobacteria, 20.8% to the Actinobacteria, 18.8% to the

Gammaproteobacteria, 14.9% to the Betaproteobacteria, 9.9% to the Bacilli and 5.9% to other classes Table 6.1. The greatest presence of some groups inside the nodule is unknown, though believed to be associated with the genotype of the host plant (De Meyer et al. 2015; Leite et al. 2017; Trabelsi et al. 2017). Soil physicochemical characteristics also influence structure and composition of NNEB; De Meyer et al. (2015) showed that genera *Paenibacillus*, *Kocuria* and *Leifsonia* were dominant in legumes grown in extensive moors, tracts of sandy heath and wetlands, *Bosea*, *Moraxella* and *Microbacterium* were in dune ecoregions and *Brevibacterium*, *Mycobacterium* and *Micromonospora* were more abundant in sandy soils. Moreover, the diversity of NNEB within nodules from *Vigna unguiculata* growing in different types of soil was dependent on the soil pH (Leite et al. 2017).

6.4 Role of Non-nodulating Endophytic Bacteria in the Promotion of Legumes Growth

More than 30% of the NNEB and symbiotic bacteria present in leguminous nodules are potential Plant Growth Promoting Bacteria (PGPB) (Trabelsi et al. 2017 and references therein). It is also known that the endophytic microbiome promotes plant growth and health and that this beneficial effect is mediated by secondary metabolites (Brader et al. 2014). NNEB aids in plant-growth promotion by several mechanisms such as N_2-fixation, inorganic phosphate solubilization, release of siderophores, production of phytohormones, biocontrol activity, etc. A list of NNEB with PGPB properties is presented in Table 6.2.

Many NNEB are N_2-fixers capable of converting atmospheric dinitrogen (N_2) into ammonium (NH_4^+) without causing morphological changes in roots but increasing the growth and yield of the plants (Hayat et al. 2010) Table 6.2. NNEB also improves the availability of P, with the concomitant increase in the yield and nutritional efficiency of the plants, even under effective biological N_2-fixation conditions (Santoyo et al. 2016) Table 6.2. Siderophores act as iron scavenging molecules and create nutrient-limiting conditions for phytopathogenic microorganisms (Berg et al. 2005) Table 6.2. Iron is considered an important cofactor for enzymes involved in many biochemical pathways such as photosynthesis in plants and nitrogen fixation in bacteria. In this sense, Khandelwal et al. (2002) reported that the production of siderophores by NNEB improve the symbiosis through the increase of ferric ion necessary for an efficient nitrogenase activity. Microbial production of phytohormones such as auxins, and auxin-like compounds, is known to trigger cell elongation, division and differentiation in plants (Santoyo et al. 2016). They also play an important role in the ontogeny of the nodules in the legume-rhizobium symbiosis and many studies indicate that changes in the content of indole acetic acid (IAA), or in its balance with cytokinins, are a prerequisite for the organogenesis of the nodule (Downie 2014).

The hormone ethylene produced within the plant endosphere has a major consequence on the bacterial microbiota residing within. NNEB with

Table 6.1 Non-nodular endophytic bacteria (NNEB) isolated from the interior of worldwide grown legume nodules

Host legume		NNEB—Origin (Reference)
Caesalpinioideae	*Acacia salicina, A. stenophylla, A. ehrenbergiana, A. nilotica, A. seyal, A. tortilis, A. laeta.*	*Acinetobacter haemolyticus, Ancylobacter* sp., *Arthrobacter* sp., *Bacillus fusiformis, B. senegalensis, Brevibacillus brevis, Caulobacter vibriodes, Chitinophaga sancti, Comamonas testosteroni, Enterobacter* sp., *Herbaspirillum frisingense, Kaistia* sp., *Microbacterium flavescens, Mycobacterium frederiksbergensis, Paenibacillus agarexedens, P. amylolyticus, P. glycanilyticus, P. pabuli, Pseudomonas fluorescens, P. putida, Rhizobium radiobacter, Roseateles depolymerans, Stenotrophomonas maltophilia, Variovorax paradoxus*—Australia (Hoque et al. 2011); *Advenella kashmirensis, Agrobacterium tumefasciens, Brevibacillus nitrificans, Microbacterium oxydans, Ochrobactrum anthropi, O. intermedium, Paenibacillus glycanilyticus, P. humicus, Pseudomonas aeruginosa, Rhizobium* sp., *Stenotrophomonas maltophila*—Algeria (Boukhatem et al. 2016)
	Enterolobium saman	*Shinella* sp.—Philippines (Bautista et al. 2017)
	Leucaena leucocephala	*Ensifer morelense*-México (Wang et al. 2002)
	Mimosa pudica	*Pantoea agglomerans*—Phillipines (Bautista et al. 2017)
	Neptunia natans	*Labrys neptuniae*—Taiwan (Chou et al. 2007)
	Prosopis farcta	*Paenibacillus prosopidis*—Tunisia (Valverde et al. 2010)
Papilionioideae	*Adesmia emarginata*	*Labrys methylaminiphilus* sp.—Chile (Gerding et al. 2017)
	Aeschynomene sensitiva	*Pseudomonas moselli, Rhizobium rhizogenes*—Phillipines (Bautista et al. 2017)
	Alysicarpus spp.	*Stenotrophomonas* sp.—Phillipines (Bautista et al. 2017)
	Arachis duranensis	*Agrobacterium* sp., *Burkholderia cepacea, Herbaspirillum frisingense, Rhizobium mesosinicum, R. multihospitium*—China (Chen et al. 2014)
	Arachis hypogaea	*Enterobacter* spp., *Klebsiella* spp., *Pseudomonas* spp.—Argentina (Ibáñez et al. 2009); *Chryseobacterium indologenes, Enterobacter cloacae, E. ludwiigii, Klebsiella pneumoniae, Pseudomonas aeruginosa*—India (Dhole et al. 2016)
	Argyrolobium uniflorum	*Agromyces* sp., *Phyllobacterium* sp.—Tunisia (Zakhia et al. 2006)
	Aspalathus abietina	*Burkholderia aspalathi*—South Africa (Mavengere et al. 2014)
	Astragalus gombiformis, As. algerianus, As. armatus, As. chrysopterus.	*Bosea* sp., *Inquilinus* sp., *Mycobacterium frederiksbergense, Phyllobacterium* sp., *Sphyngomonas* sp.—Tunisia (Zakhia et al. 2006); *Nocardioides astragali*—China (Xu et al. 2018)
	Calicotome villosa	*Bacillus circulans, Inquilinus* sp., *Phyllobacterium* sp., *Sphingomonas* sp.—Tunisia (Zakhia et al. 2006)
	Calopogonium mucunoides	*Enterobacter cloacae, Herbaspirillum putei*—Phillipine (Bautista et al. 2017)
	Cicer arietinum	*Ochrobactrum ciceri*—Pakistan (Imran et al. 2010); *Agrobacterium tumefasciens*—Tunisia (Säidï et al. 2011); *Paenibacillus endophyticus*—Spain (Carro et al. 2013); *Bacillus subtilis*—India (Saini et al. 2015); *Enterobacter* sp.—India (Koli and Swarnalakshmi 2017)

(continued)

Table 6.1 (continued)

Host legume		NNEB—Origin (Reference)
	Colutea arborescens L.	*Buttiauxella* sp.—Belgium (De Meyer et al. 2015)
	Crotalaria incana, *C. spinosa*	*Agrobacterium radiobacter*, *Bacillus simplex*, *Burkholderia* sp., *Cronobacter dublinensi Enterobacter cancerogenus*, *E. cloacae*, *Pantoea agglomerans*, *Rhizobium lusitanum*—Ethiopia (Aserse et al. 2013)
	Cytisus scoparius (L.) Link	*Bacillus* sp., *Caulobacter* sp., *Kocuria* sp., *Leifsonia* sp., *Microbacterium* sp., *Paenibacillus* sp., *Paracoccus* sp., *Phyllobacterium* sp., *Rhodococcus* sp., *Sphingomonas* sp., *Streptomyces* sp.—Belgium (De Meyer et al. 2015)
Papilionioideae	*Dalbergia* spp.	*Burkholderia caribensis*, *Phyllobacterium* sp., *Ralstonia picketti*—Madagascar (Rasolomampianina et al. 2005)
	Dipogon lignosus	*Burkholderia dipogonis*—New Zealand and Western Australia (Sheu et al. 2015)
	Erythrina brucei	*Agrobacterium radiobacter*, *Rahnella aquatilis*, *Rhizobium* sp., *Variovorax paradoxus*—Ethiopia (Aserse et al. 2013)
	Glycine max	*Bacillus subtilis*—Canada (Bai et al. 2002); *Acinetobacter* sp., *Agrobacterium* sp., *Bacillus* sp., *Burkholderia* sp., *Pantoea* sp., *Serratia* sp.—China (Li et al. 2008); *Diaphorobacter ruginosibacter*—China (Wei et al. 2015); *Acinetobacter* sp., *Bacillus* sp., *Enterobacter* sp., *Ochrobactrum* sp., *Pseudomonas* sp.—China (Zhao et al. 2017); *Acinetobacter calcoaceticus*, *Burkholderia cepacia*, *Enterobacter cloacae*, *Ochrobactrum anthropi*, *Pantoea agglomerans*, *Proteus mirabilis*, *Pseudomonas* spp.—United States of America (Tokgöz 2018)
	Hedysarum carnosum	*Pseudomonas* sp.—Tunisia (Zakhia et al. 2006)
	Indigofera amorphoide, *I. arrecta*	*Enterobacter* sp., *Paenibacillus sonchi*, *Planomicrobium glaciei*, *Rhizobium leguminosarum*—Ethiopia (Aserse et al. 2013)
	Lathyrus latifolius L., *L. numidicus*, *L. pratensis* L., *L. silvestris* L.	*Phyllobacterium* sp.—Tunisia (Zakhia et al. 2006); *Bosea lathyri*—Belgium (De Meyer et al. 2012); *Bacillus* sp., *Enterobacter* sp., *Erwinia* sp., *Microbacterium* sp., *Paenibacillus* sp., *Pantoea* sp., *Promicromonospora* sp., *Pseudomonas* sp., *Staphylococcus* sp., *Streptomyces* sp.—Belgium (De Meyer et al. 2015)
	Lespedeza sp.	*Arthrobacter nitroguaiacolicus*, *Bacillus megaterium*, *Burkholderia phenazinium*, *B. phytofirmans*, *B. caledonica*, *B. semidinicola*, *B. glathei*, *B. sordidicola*, *Dyella koreensis*, *D. marensis*, *D. japonica*, *Methylobacterium fujisawaense*, *Microbacterium ginsengisoli*, *Staphylococcus warneri*—Korea (Palaniappan et al. 2010; Subramanian et al. 2015)
	Lotus argenteus	*Paenibacillus* sp., *Sphingomonas* sp.—Tunisia (Zakhia et al. 2006)
	Lotus corniculatus	*Phyllobacterium loti*—Uruguay (Sánchez et al. 2014); *Actinoplanes* sp., *Ancylobacter* sp., *Arthrobacter* sp., *Bacillus* sp., *Corynebacterium* sp., *Inquilinus* sp., *Mycobacterium* sp., *Oerskovia* sp., *Phyllobacterium* sp., *Staphylococcus* sp., *Xanthomonas* sp.—Belgium (De Meyer et al. 2015)
	Lotus pedunculatus	*Acinetobacter* sp., *Aeromicrobium* sp., *Bacillus* sp., *Chryseobacterium* sp., *Curtobacterium* sp., *Paenibacillus* sp., *Pantoea* sp., *Plantibacter* sp., *Pseudomonas* sp., *Sphingomonas* sp., *Staphylococcus* sp.—Belgium (De Meyer et al. 2015)
	Lupinus albus	*Cohnella lupini*—Spain (Flores-Félix et al. 2014); *Paenibacillus lupini*—Spain (Carro et al. 2014)
	Lupinus angustifolius	*Micromonospora lupini*, *M. saelisecensis*—Spain (Trujillo et al. 2007)

(continued)

Table 6.1 (continued)

Host legume		NNEB—Origin (Reference)
	Lupinus polyphyllus	*Bosea lupini*—Belgium (De Meyer et al. 2012a); *Arthrobacter* sp., *Bacillus* sp., *Brevibacillus* sp., *Cohnella* sp., *Enhydrobacter* sp., *Kocuria* sp., *Lysinibacillus* sp., *Microbacterium* sp., *Microbispora* sp., *Roseateles* sp., *Staphylococcus* sp.—Belgium (De Meyer et al. 2015).
	Macroptilium lathyroides	*Pleomorphomonas oryzae*, *Rhizobium rizogenes*—Phillipines (Bautista et al. 2017)
Papilionioideae	*Medicago falcata* L., *M. lupulina* L., *M. polymorpha.*	*Bacillus* sp., *Corynebacterium* sp., *Massilia* sp., *Microbacterium* sp., *Moraxella* sp., *Pantoea* sp., *Paracoccus* sp., *Pseudomonas* sp., *Staphylococcus* sp.—Belgium (De Meyer et al. 2015); *Pseudomonas brassicacearum*—China (Kong et al. 2017); *Bacillus megaterium*—Spain (Chinnaswamy et al. 2018)
	Medicago hispida	*Kaistia* sp., *Pseudomonas* sp, *Achromobacter* sp., *Stenotrophomonas* sp., *Xanthomonas* sp., *Duganella* sp, *Rheinheimera* sp.—Peru (Arone et al. 2014)
	Medicago sativa	*Endobacter medicaginis*- Spain (Ramírez-Bahena et al. 2013); *Micromonospora* spp. (Martínez-Hidalgo et al. 2014); *Paenibacillus medicaginis*—Taiwan (Lai et al. 2015); *Streptomyces* sp., *Variovorax* sp.—Belgium (De Meyer et al. 2015); *Pseudomonas* sp., *Variovorax* sp.—New Zealand (Wigley et al. 2015); *Bacillus* sp., *Novosphingobium* sp., *Methylibium* sp., *Mycobacterium* sp., *Shinella* sp.—New Zealand (Wigley et al. 2017).
	Medicago truncatula	*Ornithinicoccus* sp., *Pseudomonas* sp.—Tunisia (Zakhia et al. 2006)
	Melilotus albus Medicus., *M. altissimus* Thuillier., *M. indicus* (L.) Allioni; *Melilotus officinalis* (L.) Pallas	*Agrobacterium* sp.—China (Kan et al. 2007); *Bacillus* sp., *Brevibacterium.*, *Lysinibacillus* sp., *Microbacterium* sp., *Moraxella* sp., *Paenibacillus* sp., *Pantoea* sp., *Phyllobacterium* sp., *Pseudomonas* sp., *Sphaerisporangium* sp., *Streptomyces* sp.—Belgium (De Meyer et al. 2015).
	Ononis vaginalis, O. natrix, O. repens L.	*Bosea* sp., *Microbacterium* sp., *Rhodopseudomonas* sp.—Tunisia (Zakhia et al. 2006); *Bacillus* sp., *Promicromonospora* sp., *Streptomyces* sp.—Belgium (De Meyer et al. 2015)
	Ornithopus perpusillus L.	*Bacilllus* sp., *Herbaspirillum* sp., *Leifsonia* sp., *Microbacterium* sp., *Micromonospora* sp., *Paenibacillus* sp., *Sphingomonas* sp.—Belgium (De Meyer et al. 2015)
	Oxytropis ochrocephala	*Acinetobacter bouvetti.*, *A. junii.*, *Bacillus amyloliquefasciens.*, *B. aryabhattai.*, *B. litoralis.*, *B. infantis.*, *B. psychrosaccharolyticus.*, *B.safensis.*, *B. simplex.*, *B. subtilis.*, *Brevibacterium frigoritolerans.*, *Cohnella ferri.*, *Erwinia piriflorinigrans.*, *Leclercia adecarboxilata.*, *Microvirga aerophila.*, *Mycobacterium monacense.*, *Paenibacillus sepulcri.*, *P. costaneae.*, *Paracoccus chinensis.*, *Phyllobacterium trifolii.*, *P. bourgognense.*, *Pseudomonas baetica.*, *P. salomonii.*, *P. frederiksbergensis.*, *Sphingomonas astaxanthinifaciens.*, *Sporosarcina termotolerans.*, *Staphylococcus epidermidis*—China (Xu et al. 2014); *Bacillus radicibacter*—China (Wei et al. 2015)
	Oxytropis triphylla	*Bosea vestrisii*—Russian federation (Safronova et al. 2017); *Phyllobacterium zundukense* (Safronova et al. 2018b)
	Oxytropis popoviana	*Bradyrhizobium* sp.—Russian federation (Safronova et al. 2018a)
	Phaseolus coccineus	*Cohnella phaseoli*—Spain (García-Fraile et al. 2008)

(continued)

Table 6.1 (continued)

Host legume		NNEB—Origin (Reference)
	Phaseolus vulgaris	*Agrobacterium* sp.—Tunisia (Mhamdi et al. 2005); *Agrobacterium tumefasciens*—Tunisia (Säidï et al. 2011); *Achromobacter* sp., *Enterobacter* sp., *Serratia proteamaculans*, *S. liquefaciens*, *Pseudomonas* sp., *Pantoea* sp.—Ethiopia (Aserse et al. 2013); *Herbaspirillum lusitanum*—Spain (Valverde et al. 2013); *Phyllobacterium endophyticum*—Spain (Flores-Félix et al. 2013); *Fontibacillus phaseoli*—Spain (Flores-Félix et al. 2014); *Acinetobacter* sp., *Enterobacter* sp., *Delftia* sp., *Klebsiella* sp., *Pseudomonas* sp., *Providencia* sp., *Rhizobium* sp.—Kenya (Wekesa et al. 2017); *Rhizobium hidalgonense*—México (Yan et al. 2017b); *Bacillus megaterium*, *Paenibacillus polimixa*—Kenya (Korir et al. 2017)
Papilionioideae	*Periandra mediterranea*	*Paenibacillus periandrae*—Brazil (Menéndez et al. 2016)
	Pisum sativum	*Micromonospora pisi*—Spain (García et al. 2010); *Ochrobactrum* sp., *Enterobacter* sp.—Pakistan (Tariq et al. 2014); *Micromonospora aurantiaca.*, *M. carbonacea.*, *M. chalcea.*, *M. chokoriensis.*, *M. coxiensis.*, *M. halophytica.*, *M. humi.*, *M. krabiensis.*, *M. lupini.*, *M. marina.*, *M. matsumotoense.*, *M. mirobrigensis.*, *M. purpureochromogenes.*, *M. rifamycinica.*, *M. saelicesensis*, *M. siamensis*—Spain (Carro et al. 2012); *Micromonospora noduli*, *M. ureilytica*, *M. vinacea*—Spain (Carro et al. 2016); *Micromonospora luteifusca*—Spain (Carro et al. 2016); *Micromonospora phytophila*, *M. luteiviridis*—Spain (Carro et al. 2018).
	Pterocarpus indicus	*Labrys neptuniae*, *Rhizobium rhizogenes*—Phillipines (Bautista et al. 2017)
	Pueraria candollei	*Rhizobium puerariae*—Taiwan (Boonsnongcheep et al. 2016)
	Pueraria lobata	*Devosia yakushimensis*—Japan (Bautista et al. 2010)
	Pueraria thunbergiana	*Bacillus thuringiensis*, *Enterobacter asburiae*, *Serratia marcesens*—India (Selvakumar et al. 2008)
	Retama raetam	*Bosea* sp., *Microbacterium flavescens*, *M. barkeri*, *Ochrobactrum* sp., *Paracraurococcus* sp., *Starkeya novella*—Tunisia (Zakhia et al. 2006)
	Robinia pseudoacacia	*Bosea robiniae*- Belgium (De Meyer et al. 2012a), *Tardiphaga robiniae*—Belgium (De Meyer et al. 2012b); *Paenibacillus enshidis*—China (Yin et al. 2015); *Arthrobacter* sp., *Chryseobacterium* sp., *Leifsonia* sp., *Pantoea* sp., *Rahnella* sp., *Stenotrophomonas* sp., *Xanthomonas* sp.—Belgium (De Meyer et al. 2015); *Mitsuaria noduli*—China (Fan et al. 2017); *Herbaspirillum robiniae*—China (Fan et al. 2018).
	Scorpiurus muricatus	*Phyllobacterium endophyticum*, *P. ifriqiyense*; *Pseudomona* sp., *Rhizobium vignae*, *R. radiobacter*, *R. leguminosarum*, *Starkeya* sp.—Algeria (Bouchiba et al. 2017)
	Sesbania cannabina	*Agrobacterium deltaense*—China (Yan et al. 2017a)
	Sphaerophysa salsula	*Paracoccus sphaerophysae*—China (Deng et al. 2011)
	Sophora alopecuroides	*Bacillus cereus*—China (Zhao et al. 2011)
	Sulla capitata, S.pallida	*Arthrobacter* sp., *Neorhizobium galegae*, *Phyllobacterium* sp., *Pseudomona fluorescens*, *Rhizobium nepotum*, *Sinorhizobium* sp., *Variovorax* sp.—Tunisia-Algeria (Beghalem et al. 2017)
	Trifolium pratense	*Agrobacterium rhizogenes*, *Bacillus brevis*, *B. insolitus*, *B. megaterium*, *B. subtilis*, *Bordetella avium*, *Curtobacterium luteum*, *C. citreum*, *C, flaccumfaciens*, *Mesorhizobium loti*, *Phyllobacterium myrsinacearum*, *Pseudomonas corrugata*, *P. fragi*, *Rhizobium leguminosarum*—Canada (Sturz et al. 1997); *Xanthomonas* sp.—Belgium (De Meyer et al. 2015)

(continued)

Table 6.1 (continued)

Host legume		NNEB—Origin (Reference)
	Trifolium arvense L., *T. dubium* Sihthorp., T. *hybridum* L., *T. repens* L.	*Arthrobacter* sp., *Bacillus* sp., *Corynebacterium* sp., *Dyadobacter* sp., *Microbacterium* sp., *Novosphingobium* sp., *Paenibacillus* sp., *Pantoea* sp., *Pseudomonas* sp., *Streptomyces* sp.—Belgium (De Meyer et al. 2015)
	Trigonella foenum-graecum	*Exiguobacterium* sp.—India (Rajendran et al. 2012).
	Vavilovia formosa	*Bosea vaviloviae*—Russian Federation (Safronova et al. 2015a); *Tardiphaga* sp.—Armenia (Safronova et al. 2015b)
	Vicia faba	*Agrobacterium* sp.—China (Kan et al. 2007); *Agrobacterium tumefasciens, Shinella* sp.—Tunisia (Säidï et al. 2011); *Bacillus* sp., *Clostridium* sp., *Desulfovibrio* sp., *Desulfatimicrobium* sp., *Methylobacter* sp., *Phyllobacterium* sp., *Rhizobium* sp., *Sphingomonas* sp., *Streptomyces* sp.—Tunisisa (Trabelsi et al. 2017); *Pseudomonas brenneri, Ps. fluorescens, Ps. frederikbergensis, Ps. fragi., Ps. putida, Ps. rhodesiae, Ps. yamanorum*—Tunisia (Bahroun et al. 2018)
	Vicia alpestris	*Microvirga ossetica*—Russian federation (Safronova et al. 2017)
	Vicia cracca (L.)., *V. hirsuta* (L.) S.F. Gray., *V. Lathyroides., V. sativa* (L.)., *V. sepium* (L.)., *V. tetrasperma* (L.) Schereber	*Agrobacterium sp.*—China (Kan et al. 2007); *Arthrobacter* sp., *Bacillus* sp., *Curtobacterium* sp., *Exigobacterium* sp., *Kocuria* sp., *Leifsonia* sp., *Lysinibacillus* sp., *Paenibacillus* sp., *Pantoea* sp., *Pseudomonas* sp., *Staphylococcus* sp., *Stenotrophomonas* sp.—Belgium (De Meyer et al. 2015)
	Vigna radiata	*Agrobacterium tumefasciens, Bacillus subtilis, B. simplex* —Pakistan (Tariq et al. 2012); *Paenibacillus* spp., *Klebsiella* spp., *Ensifer* spp., *Agrobacterium* spp., *Blastobacter* spp., *Dyadobacter* spp., *Chitinophaga* spp. —India (Pandya et al. 2015); *Pseudomonas* sp.—Pakistan (Noreen et al. 2015); *Chryseobacterium indologenes*—India (Dhole et al. 2017)
	Vigna unguiculata	*Bacillus* sp., *Brevibacillus* sp., *Paenibacillus* sp., *Enterobacter* sp.—Brazil (Da Costa et al. 2013); *Acinetobacter* sp., *Chitinophaga* sp., *Dyella* sp., *Herbaspirillum* sp., *Novosphingobium* sp., *Pseudomonas* sp., *Stenotrophomonas* sp.—Brazil (Castro et al. 2017); *Enterobacter* sp., *Chryseobacterium* sp., *Sphingobacterium* sp.—Brazil (Leite et al. 2017)
	Wisteria sinensis (Sims) Sweet	*Bacillus* sp.—Belgium (De Meyer et al. 2015)

1-aminocyclopropane-1-carboxylate deaminase (ACC deaminase) Table 6.2 are involved in reduction of ethylene levels by cleaving ACC into alpha-ketobutyrate and ammonium (Shah et al. 1998). This activity is also related with the tolerance to various types of stress such as flooding, drought, salinity, wilting and organic contamination (Glick 2014). It is likely that NNEB with ACC deaminase activity plays an important role in the microaerophilic metabolism of the nodule by reducing the level of ethylene and thus leading to a better ontogeny and organogenesis (Murset et al. 2012). Biocontrol is the process of suppressing pathogenic living organisms by using other living organisms. Antagonism and antibiosis, and production of hydrogen cyanide, exopolysaccharides and lytic enzymes (cellulase, pectinase and chitinase) are main mechanisms involved in biocontrol activity (Hayat et al. 2010; Bhattacharyya and Jha 2012; Bulgarelli et al. 2013; Santoyo et al. 2016). Methodologies to evaluate PGPB traits have been reported by Arora et al. (2001) and Castellano-Hinojosa and Bedmar (2017).

Table 6.2 Direct and indirect mechanisms of plant growth promotion in non-nodular endophytic bacteria (NNEB)

PGPB trait	NNEB strain	References
Direct mechanism		
N_2 fixation	*Burkholderia* sp. CCBAU 15508, *Pantoea* sp. CCBAU 15488, *Serratia* sp. CCBAU 15465	Li et al. (2008)
	Bacillus cereus MQ23	Zhao et al. (2011)
	Enterobacter sp. MSP10	Tariq et al. (2014)
	Bacillus megaterium LNL6	Subramanian et al. (2015)
	Enterobacter sp. A3CK	Ghosh et al. (2015)
	Enterobacter ludwigii ABG6	Dhole et al. (2016)
	Acinetobacter calcoaceticus DD161, *Bacillus cereus* DD176, *Bacillus amyloliquefasciens* DD222, *Enterobacter cloacae* DD198, *Pseudomonas putida* DD201	Zhao et al. (2017)
Phosphate solubilization	*Serratia* sp. CCBAU 15465	Li et al. (2008)
	Serratia marcesens KR-4	Selvakumar et al. 2008
	Microbacterium ginsengisoli BLN6	Palaniappan et al. (2010)
	Exiguobacterium sp. M2N2c	Rajendran et al. (2012)
	Bacillus subtilis M6	Tariq et al. (2012)
	Bacillus UFPI CB1-8	Da Costa et al. (2013)
	Burkholderia sp. CSR2, *Enterobacter cancerogenus* CIR17C, *Paenibacillus sonchi* IAR22, *Pantoea* sp. HBR8, *Rahnella aquatilis* ERR5, *Rhizobium leguminosarum* IAR30, *R. phaseoli* ERR17, *R. lusitanum* CPSR4B,	Aserse et al. (2013)

(continued)

Table 6.2 (continued)

PGPB trait	NNEB strain	References
	Serratia proteamaculans HBR15	
	Pseudomonas sp. NCHA35	Ibáñez et al. (2014)
	Ochrobactrum sp. MSP9	Tariq et al. (2014)
	Enterobacter sp. A3CK	Ghosh et al. (2015)
	Bacillus megaterium LNL6	Subramanian et al. (2015)
	Paenibacillus sp. M15	Pandya et al. (2015)
	Bacillus subtilis CNE215	Saini et al. (2015)
	Klebsiella pneumoniae AG4	Dhole et al. (2016)
	Enterobacter sp. NAB69	Koli and Swarnalakshmi (2017)
	Bacillus megaterium NMp082	Chinnaswamy et al. (2018)
ACC deaminase activity	*Burkholderia* sp. CCBAU 15508, *Pantoea* sp. CCBAU 15488;	Li et al. (2008)
	Bacillus thuringiensis KR-1, *Serratia marcesens* KR-4	Selvakumar et al. (2008)
	Microbacterium ginsengisoli BLN6	Palaniappan et al. (2010)
	Bacillus cereus MQ23	Zhao et al. (2011)
	Exiguobacterium sp. M2N2c	Rajendran et al. (2012)
	Bacillus subtilis M2	Tariq et al. (2012)
	Enterobacter sp. CIR19	Aserse et al. (2013)
	Paenibacillus UFPI CB7-8	Da Costa et al. (2013)
	Micromonospora saelicesensis AL16	Martínez-Hidalgo et al. (2014)
	Ochrobactrum sp. MSP9	Tariq et al. (2014)
	Enterobacter sp. A3CK	Ghosh et al. (2015)
	Paenibacillus sp. M10	Pandya et al. (2015)
	Bacillus megaterium LNL6	Subramanian et al. (2015)

(continued)

Table 6.2 (continued)

PGPB trait	NNEB strain	References
	Enterobacter ludwigii ABG6	Dhole et al. (2016)
	*Bacillus subtilis*EB10	Egamberdieva et al. (2017)
	Pseudomonas brassicacearum Zy-2-1	Kong et al. (2017)
	Pseudomonas putida DD201.	Zhao et al. (2017)
Ammonia Production	*Enterobacter asburiae* KR-3	Selvakumar et al. (2008)
	Bacillus subtilis CNE215	Saini et al. (2015)
Siderophore production	*Microbacterium ginsengisoli* BLN6	Palaniappan et al. (2010)
	Bacillus megaterium LNL6	Subramanian et al. (2015)
	Enterobacter sp. A3CK	Ghosh et al. (2015)
	Chryseobacterium indologenes AM2, *Enterobacter cloacae* ACP3, *Pseudomonas aeruginosa* ABG5	Dhole et al. (2016)
	Pseudomonas brassicacearum Zy-2-1	Kong et al. (2017)
	Bacillus megaterium NMp082	Chinnaswamy et al. (2018)
	Microbacterium ginsengisoli BLN6	Palaniappan et al. (2010)
	Bacillus cereus MQ23	Zhao et al. (2011)
	Exiguobacterium sp. M2N2c	Rajendran et al. (2012)
	Burkholderia sp. CIR1, *Cronobacter dublinensis* CIR9B, *Enterobacter cancerogenus* CIR17C, *Panteoa agglomerans* CSR8A, *Pseudomonas* sp. HBR44, *Serratia proteamaculans* HBR15	Aserse et al. (2013)
	Pseudomonas sp. NCHA35, NVAM24, *Klebsiella* sp. TT001	Ibáñez et al. (2014)
	Enterobacter sp. A3CK	Ghosh et al. (2015)
	Bacillus anthracis M1	Pandya et al. (2015)

(continued)

Table 6.2 (continued)

PGPB trait		NNEB strain	References
		Pseudomonas aeruginosa ABG5	Dhole et al. (2016)
		Pseudomonas brassicacearum Zy-2-1	Kong et al. (2017)
		Acinetobacter calcoaceticus DD161	Zhao et al. (2017)
		Pseudomonas yamanorum B12, *Ps. fluorescens* B8P, *Rahnella aquatilis* B16C,	Bahroun et al. (2018)
		Bacillus megaterium NMp082	Chinnaswamy et al. (2018)
Root colonization and biofilm production		*Agrobacterium tumefasciens* M5, *Bacillus simplex*, *B. subtilis* M2, M6	Tariq et al. (2012)
		Ochrobactrum sp. MSP9	Tariq et al. (2014)
Indirect mecanisms			
Extracellular enzymatic activity	Protease, lipase y celulase	*Paenibacillus sonchi* IAR22; *Rahnella aquatilis* ERR5; *Pseudomonas* sp. HBR44; *Serratia* sp. HBR25; *Burkholderia* sp. CSR2; *Serratia liquefaciens* HBR16B; *Variovorax paradoxus* CIR11Bs.	Aserse et al. (2013)
	Pectinase, celulase, xilanase, caseinase, gelatinase, amilase y lipase, alkaline phosphatase	*Micromonospora saelicesensis* AL16, ALFb1, ALF7, *M. chokoriensis*AL20, *M. echinospora* ALFb4, *M. aurantica* ALFb5, *M. lupini* ALFpr18C	Martínez-Hidalgo et al. (2014)
	Pectinase, quitinase	*Bacillus* sp. M11, M17, *Paenibacillus* sp. M15	Pandya et al. (2015)
	Lipase, protease, celulase y quitinase	*Bacillus cereus* EB2, *B. subtilis* EB10	Egamberdieva et al. (2017)
Antifungic activity	HCN production	*Bacillus thuringiensis* KR-1	Selvakumar et al. (2008)
		Rahnella aquatilis B16C	Bahroun et al. (2018)

(continued)

Table 6.2 (continued)

PGPB trait		NNEB strain	References
	Fusarium oxysporum, *Magnaporthe grisea*, *Botrytis cinerea*, *Alternaria alternata*	*Bacillus cereus* MQ23	Zhao et al. (2011)
	Alternaria burnsii	*Exiguobacterium* sp. M2N2c	Rajendran et al. (2012)
	Macrophomina phaseolina	*Bacillus* sp. M11, M17 y *Paenibacillus* sp. M15	Pandya et al. (2015)
	Macrophomina phaseolina, *Rhizoctonia solanii*, *Fusarium solanii*, *F. oxysporum*	*Pseudomonas* sp. NAFP-4, NAFP-7, NAFP-27	Noreen et al. (2015)
	Phytophthora sojae	*Acinetobacter calcoaceticus* DD161	Zhao et al. (2017)
	Fusarium solanii	*Bacillus subtilis* EB10	Egamberdieva et al. (2017)
		Rahnella aquatilis B16C	Bahroun et al. (2018)
	Sclerotinia sp., *Botrytis* sp.	*Bacillus megaterium* NMp082	Chinnaswamy et al. (2018)
Nematicidal activity	*Meloydogine javanica*	*Pseudomonas* sp. NAFP-12 to NAFP-32	Noreen et al. (2015)

6.5 Utilization of Non-nodulating Endophytic Bacteria for Plant Inoculation

Although agricultural production heavily depends on the use of synthetic fertilizers to provide essential plant nutrients (e.g. as nitrogen, phosphorus and potassium), the overuse of fertilizers is causing unexpected environmental problems (Erisman et al. 2015). In this sense, the use of efficient inoculants based on PGPB, by decreasing the use of chemical fertilizers, is a vital strategy for sustainable agricultural management and mitigation of environmental impacts (Bulgarelli et al. 2013; Santoyo et al. 2016; Erisman et al. 2015). The success of NNEB as inoculants for agricultural crops is dependent of various factors such as their ability to: colonize plant roots where co-exist microbial competition and survival in the soil (Alquéres et al. 2013; Beauregard et al. 2013); increase the production of plant exudates which interfere with the plant-bacteria interaction (Carvalhais et al. 2013); improve soil health which is highly dependent of management practices, etc. (Hayat et al. 2010).

Examples of NNEB used as inoculants for different plant species, including cereal and legumes, under greenhouse and field experiments have been reported (Lugtenberg and Kamilova 2009; Compant et al. 2010; Hayat et al. 2010; Bhattacharyya and Jha 2012; Gamalero and Glick 2011; Souza et al. 2015; Santoyo et al. 2016 and references therein). The effect of NNEB on growth promotion has been widely demonstrated in oilseed legumes, such as soybean (Li et al. 2008; Bisht and Mishra 2013; Zhao et al. 2017) and peanut (Ibáñez et al. 2009; Dhole et al. 2016), as well as grain legumes such as lentil (Bisht and Mishra 2013), chickpea (Egamberdieva et al. 2017; Koli and Swarnalakshmi 2017), cowpea (Da Costa et al. 2013) and mung bean (Noreen et al. 2015) Table 6.3. Forage legumes such as clover (Sturz et al. 1997) and alfalfa (Martínez-Hidalgo et al. 2014; Chinnaswamy et al. 2018) showed a grow-positive effect after inoculation with NNEB Table 6.3. A list of NNEB and their effects of plant growth promotion after single inoculation is shown in Table 6.3.

Nevertheless, utilization of NNEB in plant inoculation experiments has been mostly confined to greenhouse assays and in vitro screenings. Assessment of the feasibility of beneficial bacteria under field conditions is difficult since the response of microorganisms differs from laboratory to greenhouse and field conditions (Glick 2012). Inoculation methods are essential for the success of NNEB as biofertilizers and several studies show the differences found in relation with the method used (Herrmann and Lesueur 2013; Schoebitz et al. 2013; Bashan et al. 2014). A study aimed to test the NNEB revealed that genera *Azospirillum*, *Bacillus*, *Pseudomonas*, *Azotobacter*, *Serratia*, *Rhanella* and *Herbaspirillum* are associated with yield promotion in cereal and other crops (Szilagyi-Zecchin et al. 2014).

Because the ancestral relationship between legumes and rhizobia, it is assumed that the bacterial partners are not pathogenic for their cognate plants, and legumes are used without any prejudice for human and animal feeding. In this sense, co-inoculation of NNEB with rhizobial species is a common practice aimed to increase crops productivity due to their positive effect related with the increase in root development and nodule biomass (Korir et al. 2017). *Bacillus* and *Paenibacillus* promoted plant growth when used together with *Rhizobium* in chickpea plants and co-inoculation of P-solubilizing NNEB and *Rhizobium* stimulated plant growth more than their separate inoculation (Souza et al. 2015). Bai et al. (2002) also reported that co-inoculation of soybeans with *Bacillus* and *Bradyrhizobium* increased nodulation, plant growth, nitrogen content and the yield of the grain harvest. Results by Tariq et al. (2012) showed that co-inoculation of crop specific rhizobia together with NNEB improve nodulation and grain yield of *Vigna radiata* plants using strains of *Bacillus subtilis*, *Bacillus simplex* and *Agrobacterium tumefaciens*.

Nevertheless, NNEB strains do not necessarily increase the efficiency of the rhizobia among different leguminous plants (Camacho et al. 2001). Thus, the different responses to co-inoculation underscore the need to describe appropriate rhizobia-NNEB combinations to enhance growth of legumes under particular environmental conditions and soil types, but co-inoculation, frequently, increases growth and yield, compared to single plant inoculation (Araújo et al. 2009). In some cases, however, inoculation with NNEB had a negative effect on growth and yield parameters. Whereas inoculation with many *Agrobacterium* strains reduced the nodulation of *Rhizobium gallicum* in common beans, nodulation of alfalfa plants by *Ensifer meliloti* was not affected (Mrabet et al. 2006). Table 6.4 shows the effects of co-inoculation with NNEB and rhizobia under greenhouse conditions.

Table 6.3 Growth promotion effects of single inoculation of non-nodular endophytic bacteria (NNEB) under greenhouse conditions

Host	Type of effect	NNEB	References
Trifolium pratense, Red trebol	Increase plant height (1.6%), Shoot length (45.8%), Plant dry weight (57.6%), Shoot dry weight (35.3%), and Root dry weight (112%).	*Curtobacterium luteum*	Sturz et al. (1997)
Glycine max, Soybean	Increase shoot lenght (11.2%)	*Serratia* sp. CCBAU 15460	Li et al. (2008)
	Shoot fresh weight (4.3%)	*Bacillus* sp. CCBAU 15518	
Arachys hypogaea, peanut	Increase shoot fresh weight (52.9%)	*Klebsiella* sp. (NTI31, TT001)	Ibáñez et al. (2009)
	Increase shoot dry weight (58.6%)	*Pseudomonas* sp. (NCHA33, NCHA35, NVAM24)	
	Increase shoot dry weight (48.6%)	*Enterobacter* sp. (*NMAN11, NONC13*)	
Sophora alopecuroides	Shoot lenght (34.8%), Root lenght (8.9%), shoot Fresh Weight (53.6%), Shoot Dry Weight (116.7%) and Root Dry Weight (73.4%)	*Bacillus cereus* MQ23	Zhao et al. (2011)
Trigonella foenum-graecum	Increase in shoot and root lenght, chlorophyll content, Nodule number per plant and Nodule Dry Weight.	*Exiguobacterium* sp. M2N2c	Rajendran et al. (2012)
Vigna unguiculata, cowpea	Shoot dry weight (166.7%), Root dry weight (109.2%), plant dry weight (149.4%) and accumulation nitrogen in the shoot (350%)	*Paenibacillus* UFPI B3-9	Da Costa et al. (2013)
Glycine max, Soybean	Increase root dry weight (7.5%), Shoot dry weight (32.5%); Increase root lenght (11.6%), shoot lenght (13.9%) and root fresh weight (7.8%)	*B. thuringiensis* VL572.1, VL4C	Bisht and Mishra (2013)
Lens culinaris, Lentil	Increase root lenght (8.1%), shoot lenght (25.4%), root dry weight (46.5%) and shoot dry weight (42.1%)	*B. thuringiensis* VLG15	
Macrotyloma uniflorum	Increase root lenght (20.0%), shoot lenght (40.6%), root dry weight (53.0%) and shoot dry weight (85.0%)	*B. thuringiensis* VL4C	
Vigna umbellata, Ricebean	Increase root lenght (9.7%), shoot lenght (13.9%) and shoot dry weight (35.8%)	*B. thuringiensis* VL4C	

(continued)

Table 6.3 (continued)

Host	Type of effect	NNEB	References
Medicago sativa, Lucerne	Shoot to root ratio (75.7%)	*Micromonospora saelicesensis*ALF7	Martínez-Hidalgo et al. (2014)
	Shoot Dry Weight (30.0%), Shoot Carbon (26.2%), Shoot Nitrogen (101%), shoot P (37.7%) and shoot K (41.9%)	*Micromonospora aurantiaca* ALFr5	
	Root Dry Weight (22.5%)	*Micromonospora aurantiaca* ALFb5	
Vigna radiata, mung bean	Increase in seddling Vigour Index and Plant Lenght	*Bacillus safensis* M11	Pandya et al. (2015)
Vigna radiata, mung bean	Increase Shoot length (14.0%), Shoot weight (51.3%)	*Pseudomonas* sp. NAFP-19	Noreen et al. (2015)
	Increase root lenght (24.2%), Root weight (65.4%)	*Pseudomonas* sp. NAFP-32	
Arachis hypogaea, peanut	Increase germination (38.5%), Shoot fresh weight (45.8%), dry biomass (68.54%)	*Chryseobacterium indologenes* AM2	Dhole et al. (2016)
	Increase germination (38.5%), Root lenght (116.6%), nodules per plant (87.4%)	*Enterobacter cloacae* ACP3	
	Shoot lenght (48.1%)	*Klebsiella pneumoniae* AG4	
	Increase chlorophyll content (16.1%)	*Pseudomonas aeruginosa* ABG5	
Adesmia emarginata	Increase plant height (28.1%)	*Labrys* sp. AG-45	Gerding et al. (2017)
	Increase root dry weight (9.8%), shoot dry weight (38.1%)	*Labrys* sp. AG-49	
Adesmia tenella	Increase root dry weight (171.4%), shoot dry weight (62.5%)		
Cicer arietinum, Chickpea	Increase in shoot lenght (3.6%), root dry weight (40.0%), shoot dry weight (20.0%)	*B. subtilis* NUU4	Egamberdieva et al. (2017)
Cicer arietinum, Chickpea	Increase in shoot dry weight (40.63%), root dry weight (45.09%)	*Enterobacter* sp. NAB69	Koli and Swarnalakshmi (2017)
Glycine max, Soybean	Increase Shoot lenght (19.2%), root lenght (38.3%), Fresh weight per plant (36.5%) and chlorophyll content (36.7%)	*Bacillus cereus* DD176	Zhao et al. (2017)
Vicia faba, fababean	Increase shoot/root lenght and weight	*Rahnella aquatilis* B16C	Bahroun et al. (2018)
Medicago sativa, Lucerne	Increase shoot weight (68.3%), root weight (62.2%)	*Bacillus megaterium* NMp082	Chinnaswamy et al. (2018)
M. polymorpha	Increase shoot weight (100.0%), root weight (89.2%)		
M. lupulina	Increase shoot weight (89.2%), root weight (60.0%)		
M. truncatula	Increase shoot weight (125.0%), root weight (66.7%)		

Table 6.4 Growth promotion effect of co-inoculation with non-nodular endophytic bacteria (NNEB) and rhizobia under greenhouse conditions

NNEB	rhizobia	Host legume	Growth promotion effect	References
Curtobacterium luteum	*Rhizobium leguminosarum* bv. trifolii	*Trifolium pratense*, red trebol	Increase shoot height (cm) (60.7%) and root lenght (cm) (97.8%),	Sturz et al. (1997)
Bordetella avium			Root dry weight (g/plant) (442.9%)	
Phyllobacterium myrsinacearum			Increase shoot dry weight (cm) (47.9%), total plant weight (g/plant) (118.7%), shoot dry weight (g/plant) (107.7%).	
Bacillus insolitus			Increase Number nodule (442.9%)	
Streptomyces lydicus	*Rhizobium* sp.	*Pisum sativum*, pea	Increase shoot lenght (83.85), root lenght (78.4%), shoot dry weight (61.0%) and number nodules (297.5%)	Tokala et al. (2002)
Bacillus thuringiensis NEB17	*Bradyrhizobium japonicum* 532C	*Glycine max*, soybean	Increase nodule numbers and nodule dry weight	Bai et al. (2002)
Bacillus sp. CCBAU 15518	*Bradyrhizobium japonicum* B15	*Glycine max*, soybean	Increase shoot lenght (cm) (11.0%), Fresh weight (g/plant) (18.3%) and nodule numbers (15.2%)	Li et al. (2008)
Klebsiella sp. (NTI31, TT001)	*Bradyrhizobium* sp. SEMIA6144	*Arachys hypogaea*, peanut	Increase number nodules (43.9%) Shoot dry weight (17.1%)	Ibáñez et al. (2009)
Pseudomonas sp. (NCHA33, NCHA35, NVAM24)			Increase number nodules (10.3%), Shoot dry weight (14.3%)	
Enterobacter (NMAN11, NONC13)			Increase number nodules (79.9%), shoot dry weight (12.4%)	
Bacillus thuringiensis KR1	*Bradyrhizobium japonicum* SB1	*Glycine max, soybean*	Increase in root fresh weight (22.2%), root dry weight (40%), root lenght (35.9%), nodule number (73.3%), shoot fresh weight (12.5%) and shoot dry weight (12%)	Mishra et al. (2009)

(continued)

Table 6.4 (continued)

NNEB	rhizobia	Host legume	Growth promotion effect	References
Bacillus cereus MQ23	*Mesorhizobium* sp. MQ23II	*Sophora alopecuroides*	Increase in shoot lenght (21.3%), root lenght (19.5%), shoot fresh weight (84.9%), shoot dry weight (33.7%), root dry weight (21.4%), number nodules (85.8%)	Zhao et al. (2011)
Exiguobacterium sp. M2N2c	*Sinorhizobium meliloti*	*Trigonella foenum-graecum*	Increase in shoot and root lenght, chlorophyll content, nodule number per plant and nodule dry weight	Rajendran et al. (2012)
Bacillus subtilis M6	*Bradyrhizobium* sp. MN-S	*Vigna radiata*, mung bean	Increase nodule number per plant (78.5%), Nodule dry weight per plant (127.4%) and total plant dry weight (35.6%)	Tariq et al. (2012)
Micromonospora aurantiaca (ALFb5)	*Ensifer meliloti* 1021	*Medicago sativa*, Lucerne	Shoot to root ratio (54.2%), shoot dry weight (26.0%), shoot carbon (23.7%), shoot nitrogen (24.9%), shoot K (35.0%). increase nodulation (107.1%)	Martínez-Hidalgo et al. (2014)
Klebsiella sp. TT001	*Bradyrhizobium* sp. SEMIA6144	*Arachys hypogaea*, peanut	Increase Root dry weight (38.2%), Nodule number per plant (23.2%)	Ibáñez et al. (2014)
Enterobacter sp. NMAN11			Increase dry weight (4.7%)	
Bacillus megaterium LNL6	*Bradyrhizobium japonicum* MN110	*Glycine max*, soybean	Shoot lenght (43.9%), Root lenght (33.8%), Dry Weight (51.0%)	Subramanian et al. (2015)
Bacillus safensis M11	*Ensifer adhaerens* M8	*Vigna radiata*, mung bean	Increase in seddling Vigor Index and Plant Lenght	Pandya et al. (2015)
Pseudomonas sp. NAFP-32	*Rhizobium* sp. NFB109		Increase shoot lenght (5.3%), shoot weight (24.4%), root weight (22.5%), number nodules per plant (8.3%)	Noreen et al. (2015)

(continued)

Table 6.4 (continued)

NNEB	rhizobia	Host legume	Growth promotion effect	References
B. subtilis NUU4	*Mesorhizobium ciceri* IC53	*Cicer arietinum*	Increase in shoot Lenght (35.0%), nodule number (141.7%), Root Dry Weight (20.8%), Shoot Dry Weight (24.3%)	Egamberdieva et al. (2017)
B. megaterium HK6	*Rhizobium* sp. IITA-PAU 987	*Phaseolus vulgaris*	Increase in nodule fresh weight (25.0%), Fixation Biological nitrogen (42.1%), N derived from atmosphere (31.1%)	Korir et al. (2017)
Paenibacillus polimixa HK1	*Rhizobium* sp. IITA-PAU 983		Increase in Shoot dry weight (39.1%), root dry weight (52.3%)	
Bacillus megaterium NMp082	*Ensifer medicae*	*Medicago sativa*, Lucerne	Increase shoot weight (28.2%), root weight (66.1%), number nodules (53.4%)	Chinnaswamy et al. (2018)
		M. polymorpha	Increase shoot weight (71.1%), root weight (196.7%), number nodules (71.1%)	
		M. lupulina	Increase shoot weight (45.1%), root weight (76.5%), number nodule (81.3%)	
		M. truncatula	Increase shoot weight (43.9%), root weight (37.9%), number nodule (81.5%)	

Despite the potential to improve plant growth, the use of NNEB as biofertilizers is not widespread, with the exception of *Azospirillum* which is used for a variety of crops in Europe, South America and Africa (Dobbelaere et al. 1999). The most challenging aspect that limits commercialization of NNEB is the inconsistency in the response to inoculation depending of the plant variety and field location (Bulgarelli et al. 2013; Souza et al. 2015; Santoyo et al. 2016).

6.6 Utilization of Endophytic Bacteria in the Recuperation of Degraded Soils

The increase of the global human population demands greater extensions of arable land to cover the production of food, fodder, fibre and biomass which, in turn, is causing biotic and abiotic stresses to the soil (Dubey et al. 2017). About 30% of the global soil area is degraded or contaminated due to anthropogenic activities (Abhilash et al. 2013). Soil degradation implies reduction and even loss of its physicochemical and biological characteristics, poor agricultural practices and the use of xenobiotic contaminants being the most common cause of the loss of soil quality (Schloter et al. 2018). Among the cultural practices that generate soil degradation are the burning of the vegetation cover, deforestation and monoculture farming, leading to a loss of organic matter by reducing the soil microbial activity and biomass (Kaschuk et al. 2010). Many soils are contaminated with heavy metals and chemical organic compounds that prevent soil utilization for agricultural practices (Dubey et al. 2017 and references therein). NNEB is being used as a biotechnological tool for the recovery of degraded soils and of those contaminated with xenobiotic compounds (Barac et al. 2004; Brader et al. 2014; Bao et al. 2015; Fernández-González et al. 2017; Kong et al. 2017; Pawlik et al. 2017; Dubey et al. 2017; Imran et al. 2017).

In association with NNEB, legumes manage to tolerate high concentrations of heavy metals, improving the phytoremediation process, preventing the entry of heavy metal into the food chain (Kong et al. 2017) and restoring soil fertility by increasing the nitrogen content of the soil (Dary et al. 2010). In legumes inoculated with NNEB, a hyperaccumulation of heavy metals has been observed due to the production of siderophores, biosurfactants and acid phosphatases produced by NNEB (Ray et al. 2017). Excessive use of xenobiotics in agriculture, such as pesticides and chemical fertilizers, results in severe soil contamination (Dubey et al. 2017). Phytoremediation using NNEB is a low-cost technology that can be applied to the restoration of contaminated soils (Ahemad and Khan 2011; Tétard-Jones and Edwards 2016). The recovery of degraded soils can be facilitated by using microorganisms that increase the vegetation cover. Several studies have shown that NNEB has a critical role in the restoration of marginal, degraded and contaminated soils in the prevention of soil erosion and in the afterwards reforestation events (Bashan and de-Bashan 2010). In this sense, soils contaminated with hydrocarbons were successfully recovered using *Lotus corniculatus* and *Oenothera biennis* plants inoculated with *Rhizobium*-, *Pseudomonas*-, *Stenotrophomonas*- and *Rhodococcus*-NNEB (Pawlik et al. 2017). Germaine et al. (2006) showed that the inoculation of *Pisum sativum* with *Pseudomonas putida* VM1450 diminished the content of the herbicide 2, 4-D in the soil and prevented its accumulation in the aerial part of the plant. Similarly, Wani and Khan (2010) showed that treatment with *Bacillus* sp. PSB10 of *Cicer arietinum* seeds mitigated the toxic effects of the hexavalent chromium present in contaminated soils. Moreover, inoculation of *Cytisus striatus* with *Rhodococcus erythropolis* ET54b and *Sphingomonas* sp. D4 resulted in a better dissipation of hexachlorocyclohexane contaminated soils as compared to the control treatment (Becerra-Castro et al. 2013). Inoculation of legumes with their specific microsymbiont together with an NNEB is also a common practice. Kong et al. (2017) showed that the co-inoculation of *Sinorhizobium meliloti* and *Pseudomonas brassicacearum* Zy-2-1 improved the growth of *Medicago lupulina* in the presence of toxic Cu^{2+} concentrations, and Chinnaswamy et al. (2018) reported that the co-inoculation of *Medicago polymorpha*, *M. lupulina*, *M. truncatula* and *M. sativa* with *Ensifer medicae* and *Bacillus megaterium* NMp082 increased nodulation and plant growth in salty soils. Based on the influence on the physical, chemical and microbiological properties of a soil contaminated with heavy metals, inoculation with NNEB resulted in the improvement of the ecosystem services such as nutrient cycling, microbial biomass and basal soil respiration (Burges 2017). There are still few greenhouse studies, and even less field experiments, directed to demonstrate the ability of NNEB isolated from root nodules of legumes in the recovery of degraded soils; the aforementioned background, however, allows us to infer they have an excellent potential to improve soil health.

6.7 Conclusions and Perspectives

The presence of symbiosis specific N_2-fixing nodule bacteria in legume roots is known since the seventeenth century, albeit it was at the end of the 19th century when Hellriegel and Wilfarth showed that the legume nodules were responsible for nitrogen fixation. In 1888, Beijerinck isolated bacterial cells from nodules of *Vicia faba* that were called *Bacillus redicicola* and later renamed *Rhizobium leguminosarum* by Frank in 1889 (revised in Leigh 2004). Since then, the bacteria nodulating legumes, best known as rhizobia, were studied, and all other morphologically different cells were discarded as contaminants. In the last 20 years, however, evidence accumulated to demonstrate the existence in the nodule interior of other non-nodulating bacteria which are recognized as nodule endophytes. Using confocal microscopy, NNEB have been shown to reach the inner of nodules together with the rhizobia (Pandya et al. 2015) and that accession and accommodation within the nodule is regulated genetically by the host plant (Zgadzaj et al. 2016). Thus, the legume nodule is a complex ecological niche whose microbiome is made up of rhizobia, responsible for nodule formation, and by other bacterial endophytes, named here as NNEB, whose function is poorly known.

The NNEB inside the nodules are analyzed using culture-dependent (culturomic) methods aimed to the obtaining of bacterial species whose plant growth promotion abilities and biocontrol activities could be studied to allow selection of rhizobia and NNEB for the formulation of biofertilizers. The metagenomic techniques, on the other hand, are of valuable interest not only in the sequencing of the 16S rRNA core genes, but also in that of the nodulation genes which will make it possible to define which bacterium was involved in nodule formation and what are the NNEB. Like rhizospheric bacteria, NNEB have direct and indirect mechanisms of plant growth promotion, and their utilization, whether alone or in co-inoculation with rhizobia, could constitute an ecological and healthy alternative to synthetic fertilizers for legume and non-legume plants.

Most of the research on the mechanisms of plant growth promotion has been done in rhizosphere bacteria, and it is generally assumed that similar mechanisms occur in NNEB. However, the environment in the rhizosphere is quite different to that in the nodule interior, so that differences in the biotic and abiotic stresses outside and outside the nodule (soil type, temperature, pH, oxygen concentration, nutrient availability, etc.) could affect the survival and lifestyle of NNEB. Thus, it is likely that new mechanism of plant growth promotion could be found in NNEB that are unknown in rhizobacteria. Despite several genes have been sequenced that provide some clues on the endophytic lifestyle, the question about what turns a rhizobacteria into a bacterial endophyte is not answered yet, and identification of undiscovered genes involved in endophytism has not been pursued systematically. All this designed to engineer NNEB with improved plant-probiotic capabilities and their use as biofertilizers to increase plant growth and development and to recover degraded soils.

Acknowledgments This study was supported by the ERDF-cofinanced grants PEAGR2012-1968 from Consejería de Economía, Innovación y Ciencia (Junta de Andalucía, Spain). ACH is recipient of a grant of MECD (FPU 2014/01633). The authors thank CONCYTEC (Consejo Nacional de Ciencia y Tecnología) and MINEDU (Ministerio de Educación, Perú) for the financing of projects 8000-2015 and 9500-2015, in recovery of degraded soils, as well as the National University of San Martin.

References

Abhilash PC, Dubey RK, Tripathi V, Srivastava P, Verma JP, Singh HB (2013) Remediation and management of POPs-contaminated soils in a warming climate: challenges and perspectives. Environ Sci Pollut Res 20:5879–5885. https://doi.org/10.1007/s11356-013-1808-5

Aeron A, Chauhan PS, Dubey RC, Maheshwari DK, Bajpai VK (2015) Root nodule bacteria from *Clitoria ternatea* L. are putative invasive nonrhizobial endophytes. Can J Microbiol 142:131–142. https://doi.org/10.1139/cjm-2014-0483

Ahemad M, Khan MS (2011) Insecticide-tolerant and plant growth promoting *Bradyrhizobium* sp. (*Vigna*) improves the growth and yield of greengram [*Vigna radiata* (L.) Wilczek] in insecticide-stressed soils. Symbiosis 54:17–27. https://doi.org/10.1007/s13199-011-0122-6

Alquéres S, Meneses C, Rouws L, Rothballer M, Baldani I, Schmid M, Hartmann A (2013) The bacterial superoxide dismutase and glutathionereductase are crucial for endophytic colonization of rice roots by *Gluconacetobacter diazotrophicus* PAL5. Mol Plant-Microbe Interact 26:937–945. https://doi.org/10.1094/MPMI-12-12-0286-R

Araújo F, Sérgio A, Figueiredo MVB (2009) Role of plant growth-promoting bacteria in sustainable agriculture. In: Salazar A, Rios I (eds) Sustainable agriculture: technology, planning. Nova Science Publishers, Hauppauge

Arone G, Calderón C, Moreno S, Bedmar EJ (2014) Identification of *Ensifer* strains isolated from root nodules of *Medicago hispida* grown in association with *Zea mays* in the Quechua region of the Peruvian Andes. Biol Fertil Soils 50:185–190. https://doi.org/10.1007/s00374-013-0825-3

Arora NK, Kang SC, Maheshwari DK (2001) Isolation of siderophore producing strains of Rhizobium meliloti and their biocontrol potential against *Macrophomina phaseolina* that causes charcoal rot of groundnut. Curr Sci 81:673–677

Aserse AA, Räsänen LA, Aseffa F, Hailemariam A, Lindström K (2013) Diversity of sporadic symbionts and nonsymbiotic endophytic bacteria isolated from nodules of woody, shrub, and food legumes in Ethiopia. Appl Microbiol Biotechnol 97:10117–10134. https://doi.org/10.1007/s00253-013-5248-4

Bahroun A, Jousset A, Mhamdi R, Mrabet M, Mhadhbi H (2018) Anti-fungal activity of bacterial endophytes associated with legumes against *Fusarium solani*: assessment of fungi soil suppressiveness and plant protection induction. Appl Soil Ecol 124:131–148. https://doi.org/10.1016/j.apsoil.2017.10.025

Bai Y, D'Aoust F, Smith DL, Driscoll BT (2002) Isolation of plant-growth-promoting *Bacillus* strains from soybean root nodules. Can J Microbiol 48:230–238. https://doi.org/10.1139/w02-014

Bao G, Saikkonen K, Wang H, Zhou L, Chen S, Li C, Nan Z (2015) Does endophyte symbiosis resist allelopathic effects of an invasive plant in degraded grassland? Fungal Ecol 17:114–125. https://doi.org/10.1016/j.funeco.2015.05.006

Barac T, Taghavi S, Borremans B (2004) Engineered endophytic bacteria improve phytoremediation of water-soluble, volatile, organic pollutants. Nat Biotechnol 22:583. https://doi.org/10.1038/nbt960

Bashan Y, de-Bashan LE (2010) Microbial Populations of arid lands and their potential for restoration of deserts. In: Dion P. (ed) Soil biology and agriculture in the tropics. Soil biology, vol 21. Springer, Berlin

Bashan Y, de-Bashan LE, Prabhu SR, Hernandez JP (2014) Advances in plant growth-promoting bacterial inoculant technology: formulations and practical perspectives (1998–2013). Plant Soil 378:1–33. https://doi.org/10.1007/s11104-013-1956-x

Bautista VV, Monsalud R G, Yokota A (2010) *Devosia yakushimensis* sp. nov., isolated from root nodules of *Pueraria lobata* (Willd.) Ohwi. Int J Syst Evol Microbiol 60:627–632. https://doi.org/10.1099/ijs.0.011254-0

Bautista VV, Monsalud RG, Yokota A (2017) Isolation and Identification of Bacteria from Root Nodules of Philippine Legumes Using 16S rRNA Gene Sequencing. Philipp Agric Sci 100:103–117

Beauregard PB, Chai Y, Vlamakis H, Losick R, Kolter R (2013) *Bacillus subtilis* biofilm induction by plant polysaccharides. Proc Natl Acad Sci USA 110:E1621–E1630. https://doi.org/10.1073/pnas.1218984110

Becerra-Castro C, Kidd PS, Rodríguez-Garrido B, Monterroso C, Santos-Ucha P, Prieto-Fernández Á (2013) Phytoremediation of hexachlorocyclohexane (HCH)-contaminated soils using *Cytisus striatus* and bacterial inoculants in soils with distinct organic matter content. Environ Pollut 178:202–210. https://doi.org/10.1016/j.envpol.2013.03.027

Beghalem H, Aliliche K, Chriki A, Landoulsi A (2017) Molecular and phenotypic characterization of endophytic bacteria isolated from sulla nodules. Microb Pathog 111:225–231. https://doi.org/10.1016/j.micpath.2017.08.049

Beijerinck MW, Van Delden A (1902) Ueber die Assimilation des freien Stickstoffs durch Bakterien. Zentralbl Bakteriol B 9:3–43

Berg G, Eberl L, Hartmann A (2005) The rhizosphere as a reservoir for opportunistic human pathogenic bacteria. Environ Microbiol 71:4203–4213. https://doi.org/10.1111/j.1462-2920.2005.00891.x

Bhattacharjee RB, Singh A, Mukhopadhyay SN (2008) Use of nitrogen-fixing bacteria as biofertiliser for non-legumes: prospects and challenges. Appl Microbiol Biotechnol 80:199–209. https://doi.org/10.1007/s00253-008-1567-2

Bhattacharyya PN, Jha DK (2012) Plant growth-promoting rhizobacteria (PGPR): emergence in agriculture. World J Microbiol Biotechnol 28:1327–1350. https://doi.org/10.1007/s11274-011-0979-9

Bisht SC, Mishra PK (2013) Ascending migration of endophytic *Bacillus thuringiensis* and assessment of benefits to different legumes of NW Himalayas. Eur J Soil Biol 56:56–64. https://doi.org/10.1016/j.ejsobi.2013.02.004

Boonsnongcheep P, Prathanturarug S, Takahashi Y, Matsumoto A (2016) *Rhizobium puerariae* sp. nov., an endophytic bacterium from the root nodules of the medicinal plant *Pueraria candollei* var. candollei. Int J Syst Evol Microbiol 66:1236–1241. https://doi.org/10.1099/ijsem.0.000863

Bouchiba Z, Boukhatem ZF, Ighilhariz Z (2017) Diversity of nodular bacteria of *Scorpiurus muricatus* in western Algeria and their impact on plant growth. Can J Microbiol 63:450–463. https://doi.org/10.1139/cjm-2016-0493

Boukhatem ZF, Merabet C, Bekki A et al (2016) Nodular bacterial endophyte diversity associated with native *Acacia* spp. in desert region of Algeria. Afr J Microbiol Res 10:634–645. https://doi.org/10.5897/AJMR2015.7678

Brader G, Compant S, Mitter B, Trognitz F, Sessitsch A (2014) Metabolic potential of endophytic bacteria. Curr Opin Biotechnol 27:30–37. https://doi.org/10.1016/j.copbio.2013.09.012

Bulgarelli D, Schlaeppi K, Spaepen S, Loren Ver, van Themaat E, Schulze-Lefert P (2013) Structure and functions of the bacterial microbiota of plants. Annu Rev Plant Biol 64:807–838. https://doi.org/10.1146/annurev-arplant-050312-120106

Burges A, Epelde L, Blanco F, Becerril JM, Garbisu C (2017) Ecosystem services and plant physiological status during endophyte-assisted phytoremediation of metal contaminated soil. Sci Total Environ 584:329–338. https://doi.org/10.1016/j.scitotenv.2016.12.146

Camacho M, Santamaría C, Temprano F, Rodriguez-Navarro DN, Daz A (2001) Co-inoculation with *Bacillus* sp CECT 450 improves nodulation in *Phaseolus vulgaris*. Can J Microbiol 47:1058–1062. https://doi.org/10.1139/w01-107

Carvalhais LC, Dennis PG, Fan B et al (2013) Linking plant nutritional status to plant-microbe interactions. PLoS ONE 8:e68555. https://doi.org/10.1371/journal.pone.0068555

Castellano-Hinojosa A, Bedmar EJ (2017) Methods for Evaluating Plant Growth-Promoting Rhizobacteria Traits. In: Singh HB, Sarma BK, Keswani C (eds) Advances in PGPR Research. CABI international, Oxford, pp 255–274

Carro L, Spröer C, Alonso P, Trujillo ME (2012) Diversity of *Micromonospora* strains isolated from nitrogen fixing nodules and rhizosphere of *Pisum sativum* analyzed by multilocus sequence analysis. Syst Appl Microbiol 35:73–80. https://doi.org/10.1016/j.syapm.2011.11.003

Carro L, Flores-Félix JD, Cerda-Castillo E (2013) *Paenibacillus endophyticus* sp. nov., isolated from nodules of *Cicer arietinum*. Int J Syst Evol Microbiol 63:4433–4438. https://doi.org/10.1099/ijs.0.050310-0

Carro L, Flores-Félix JD, Ramírez-Bahena MH (2014) *Paenibacillus lupini* sp. nov., isolated from nodules of *Lupinus albus*. Int J Syst Evol Microbiol 64:3028–3033. https://doi.org/10.1099/ijs.0.060830-0

Carro L, Riesco R, Spröer C, Trujillo ME (2016) *Micromonospora ureilytica* sp. nov., *Micromonospora noduli* sp. nov. and *Micromonospora vinacea* sp. nov., isolated from *Pisum sativum* nodules. Int J Syst Evol Microbiol 66:3509–3514. https://doi.org/10.1099/ijsem.0.001231

Carro L, Veyisoglu A, Riesco R, Spröer C, Klenk HP, Sahin N, Trujillo ME (2018) *Micromonospora phytophila* sp. nov. and *Micromonospora luteiviridis* sp. nov., isolated as natural inhabitants of plant nodules. Int J Syst Evol Microbiol 68:248–253. https://doi.org/10.1099/ijsem.0.002490

Castro JLD, Souza MG, Rufini M, Guimarães AA, Rodrigues TL, Moreira FMDS (2017) Diversity and efficiency of rhizobia communities from iron mining areas using cowpea as a trap plant. R Bras Ci Solo 41. http://dx.doi.org/10.1590/18069657rbcs20160525

Chen JY, Gu J, Wang ET et al (2014) Wild peanut *Arachis duranensis* are nodulated by diverse and novel *Bradyrhizobium* species in acid soils. Syst Appl Microbiol 37:525–532. https://doi.org/10.1016/j.syapm.2014.05.004

Chinnaswamy A, Coba de la Peña T, Stoll A et al (2018) A nodule endophytic *Bacillus megaterium* strain isolated from *Medicago polymorpha* enhances growth, promotes nodulation by *Ensifer medicae* and alleviates salt stress in alfalfa plants. Ann Appl Biol 172:295–308. https://doi.org/10.1111/aab.12420

Chou YJ, Elliott GN, James EK et al (2007) *Labrys neptuniae* sp. nov., isolated from root nodules of the aquatic legume *Neptunia oleracea*. Int J Syst Evol Microbiol 57:577–581. https://doi.org/10.1099/ijs.0.64553-0

Compant S, Clément C, Sessitsch A (2010) Plant growth-promoting bacteria in the rhizo and endosphere of plants: their role, colonization, mechanisms involved and prospects for utilization. Soil Biol Biochem 42:669–678. https://doi.org/10.1016/j.soilbio.2009.11.024

Da Costa EM, Nóbrega RSA, de Carvalho F, Trochmann A, Ferreira LDVM, de Souza Moreira FM (2013) Plant growth promotion and genetic diversity of bacteria isolated from cowpea nodules. Pesq Agropec Bras 48:1275–1284. https://doi.org/10.1590/S0100-204X2013000900012

Dary M, Chamber-Pérez MA, Palomares AJ, Pajuelo E (2010) "In situ" phytostabilisation of heavy metal polluted soils using *Lupinus luteus* inoculated with metal resistant plant-growth promoting rhizobacteria. J Hazard Mater 177:323–330. https://doi.org/10.1016/j.jhazmat.2009.12.035

De Meyer SE, Willems A (2012) Multilocus sequence analysis of *Bosea* species and description of *Bosea lupini* sp. nov., *Bosea lathyri* sp. nov. and *Bosea robiniae* sp. nov., isolated from legumes. Int J Syst Evol Microbiol 62:2505–2510. https://doi.org/10.1099/ijs.0.035477-0

De Meyer SE, Coorevits A, Willems A (2012) *Tardiphaga robiniae* gen. nov., sp. nov., a new genus in the family Bradyrhizobiaceae isolated from *Robinia pseudoacacia* in Flanders (Belgium). Syst Appl Microbiol 35:205–214. https://doi.org/10.1016/j.syapm.2012.02.002

De Meyer SE, De Beuf K, Vekeman B, Willems A (2015) A large diversity of non-rhizobial endophytes found in legume root nodules in Flanders (Belgium). Soil Biol Biochem 83:1–11. https://doi.org/10.1016/j.soilbio.2015.01.002

Deng ZS, Zhao LF, Xu L (2011) *Paracoccus sphaerophysae* sp. nov., a siderophore-producing, endophytic bacterium isolated from root nodules of *Sphaerophysa salsula*. Int J Syst Evol Microbiol 61:665–669. https://doi.org/10.1099/ijs.0.021071-0

Dhole A, Shelat H, Vyas R, Jhala Y, Bhange M (2016) Endophytic occupation of legume root nodules by nifH-positive non-rhizobial bacteria, and their efficacy in the groundnut (*Arachis hypogaea*). Ann Microbiol 66:1397–1407. https://doi.org/10.1007/s13213-016-1227-1

Dhole A, Shelat H, Panpatte D (2017) *Chryseobacterium indologenes* A Novel Root Nodule Endophyte in *Vigna radiata*. Int J Curr Microbiol App Sci 6:836–844. https://doi.org/10.20546/ijcmas.2017.604.104

Dobbelaere S, Croonenborghs A, Thys A, Vande Broek A, Vanderleyden J (1999) Phytostimulatory effect of *Azospirillum brasilense* wild type and mutant strains altered in IAA production on wheat. Plant Soil 212:155–164. https://doi.org/10.1023/A:1004658000815

Downie JA (2014) Legume nodulation. Curr Biol 24:R184–R190. https://doi.org/10.1016/j.cub.2014.01.028

Dubey RK, Tripathi V, Edrisi et al (2017) Role of plant growth-promoting microorganisms in sustainable agriculture and environmental remediation. In: Singh HB, Sarma BK and Keswani C (eds) Advances in PGPR research. CABI international, Oxford, pp 75–126

Egamberdieva D, Wirth SJ, Shurigin VV, Hashem A, Abd-Allah EF (2017) Endophytic bacteria improve plant growth, symbiotic performance of chickpea (*Cicer arietinum* L.) and induce suppression of root rot caused by *Fusarium solani* under salt stress. Front Microbiol 8:1887. https://doi.org/10.3389/fmicb.2017.01887

Erisman JW, Galloway; JN, Dice; NB (2015) Nitrogen: too much of a vital resource. Science Brief. WWF Netherlands, Zeist, The Netherlands

Estrela S, Kerr B, Morris JJ (2016) Transitions in individuality through symbiosis. Curr Opin Microbiol 31:191–198. https://doi.org/10.1016/j.mib.2016.04.007

Fan MC, Nan LJ, Zhu YM, Chen WM, Wei GH, Lin YB (2017) *Mitsuaria noduli* sp. nov., isolated from the root nodules of *Robinia pseudoacacia* in a lead–zinc mine. Int J Syst Evol Microbiol 68:87–92. https://doi.org/10.1099/ijsem.0.002459
Fan MC, Guo YQ, Zhang LP, Zhu YM, Chen WM, Lin YB, Wei GH (2018) *Herbaspirillum robiniae* sp. nov., isolated from root nodules of *Robinia pseudoacacia* in a lead–zinc mine. Int J Syst Evol Microbiol 68:1300–1306. https://doi.org/10.1099/ijsem.0.002666
Fernández-González AJ, Martínez-Hidalgo P, Cobo-Díaz J et al (2017) The rhizosphere microbiome of burned holm-oak: potential role of the genus *Arthrobacter* in the recovery of burned soils. Sci Rep 7:6008. https://doi.org/10.1038/s41598-017-06112-3
Flores-Félix JD, Carro L, Velázquez E, Valverde Á, Cerda-Castillo E, García-Fraile P, Rivas R (2013) *Phyllobacterium endophyticum* sp. nov., isolated from nodules of *Phaseolus vulgaris*. Int J Syst Evol Microbiol 63:821–826. https://doi.org/10.1099/ijs.0.038497-0
Flores-Félix JD, Carro L, Ramírez-Bahena MH (2014) *Cohnella lupini* sp. nov., an endophytic bacterium isolated from root nodules of *Lupinus albus*. Int J Syst Evol Microbiol 64:83–87. https://doi.org/10.1099/ijs.0.050849-0
Fred EB, Baldwin IL, McCoy E (1932) Root-nodule bacteria and leguminous plants. University of Wisconsin, Madison, Wisconsin
Gamalero E, Glick BR (2011) Mechanisms used by plant growth-promoting bacteria. In: Maheshwari DK (ed) Bacteria in agrobiology. Plant nutrient management, Springer, Berlin
García-Fraile P, Velazquez E, Mateos| PF, Martínez-Molina E, Rivas R (2008) *Cohnella phaseoli* sp. nov., isolated from root nodules of *Phaseolus coccineus* in Spain, and emended description of the genus *Cohnella*. Int J Syst Evol Microbiol 58:1855–1859. https://doi.org/10.1099/ijs.0.65468-0
García LC, Martínez-Molina E, Trujillo ME (2010) *Micromonospora pisi* sp. nov., isolated from root nodules of *Pisum sativum*. Int J Syst Evol Microbiol 60:331–337. https://doi.org/10.1099/ijs.0.012708-0
Gerding M, Oyarzúa P, García L (2017) Diversity and symbiotic effectiveness of *Adesmia* spp. root nodule bacteria in central and southern Chile. Symbiosis 72:61–72. https://doi.org/10.1007/s13199-016-0440-9
Germaine KJ, Liu X, Cabellos GG, Hogan JP, Ryan D, Dowling DN (2006) Bacterial endophyte-enhanced phytoremediation of the organochlorine herbicide 2, 4-dichlorophenoxyacetic acid. FEMS Microbiol Ecol 57:302–310. https://doi.org/10.1111/j.1574-6941.2006.00121.x
Ghosh PK, De TK, Maiti TK (2015) Ascorbic acid production in root, nodule and *Enterobacter* spp. (Gammaproteobacteria) isolated from root nodule of the legume *Abrus precatorius* L. Biocatal Agric Biotechnol 4:127–134. https://doi.org/10.1016/j.bcab.2017.11.017
Glick BR (2012) Plant growth-promoting bacteria: mechanism and applications. Scientifica 2012:1–15. https://doi.org/10.6064/2012/963401
Glick BR (2014) Bacteria with ACC deaminase can promote plant growth and help to feed the world. Microbiol Res 169:30–39. https://doi.org/10.1016/j.micres.2013.09.009
Gresshoff PM, Ferguson BJ. (2017) Molecular signals in nodulation control. Int J Mol Sci 18:125. https://doi.org/10.3390/ijms18010125
Hayashi T, Shimoda Y, Sato S, Tabata S, Imaizumi-Anraku H, Hayashi M (2014) Rhizobial infection does not require cortical expression of upstream common symbiosis genes responsible for the induction of calcium spiking. Plant J 77:146–159. https://doi.org/10.1111/tpj.12374
Hayat R, Ali S, Amara U, Khalid R, Ahmed I (2010) Soil beneficial bacteria and their role in plant growth promotion: a review. Ann Microbiol 60:579–598. https://doi.org/10.1007/s13213-010-0117-1
Herrmann L, Lesueur D (2013) Challenges in formulation and quality of biofertilisers for successful inoculation. Appl Microbiol Biotechnol 97:8859–8873. https://doi.org/10.1007/s00253-013-5228-8

Hoque MS, Broadhurst LM, Thrall PH (2011) Genetic characterization of root-nodule bacteria associated with *Acacia salicina* and *A. stenophylla* (Mimosaceae) across south-eastern Australia. Int J Syst Evol Microbiol 61:299–309. https://doi.org/10.1099/ijs.0.021014-0

Ibáñez F, Angelini J, Taurian T, Tonelli ML, Fabra A (2009) Endophytic occupation of peanut root nodules by opportunistic Gammaproteobacteria. Syst Appl Microbiol 32:49–55. https://doi.org/10.1016/j.syapm.2008.10.001

Ibáñez F, Arroyo ME, Angelini J et al (2014) Non-rhizobial peanut nodule bacteria promote maize (*Zea mays* L.) and peanut (*Arachis hypogaea* L.) growth in a simulated crop rotation system. Appl Soil Ecol 84:208–212. https://doi.org/10.1016/j.apsoil.2014.08.002

Imran A, Hafeez FY, Frühling A, Schumann P, Malik KA, Stackebrandt E (2010) *Ochrobactrum ciceri* sp. nov., isolated from nodules of *Cicer arietinum*. Int J Syst Evol Microbiol 60:1548–1553. https://doi.org/10.1099/ijs.0.013987-0

Imran AFK, Naveed M, Afzal M (2017) Plant-bacteria synergism: an innovative approach for the remediation of crude oil-contaminated soils. Soil Environ 36:93–113. https://doi.org/10.25252/SE/17/51346

Kan FL, Chen ZY, Wang ET, Tian CF, Sui XH, Chen WX (2007) Characterization of symbiotic and endophytic bacteria isolated from root nodules of herbaceous legumes grown in Qinghai-Tibet plateau and in other zones of China. Arch Microbiol 188:103–115. https://doi.org/10.1007/s00203-007-0211-3

Kaschuk G, Alberton O, Hungria M (2010) Three decades of soil microbial biomass studies in Brazilian ecosystems: lessons learned about soil quality and indications for improving sustainability. Soil Biol Biochem 42:1–13. https://doi.org/10.1016/j.soilbio.2009.08.020

Khandelwal SR, Manwar AV, Chaudhari BL, Chincholkar SB (2002) Siderophoregenic bradyrhizobia boost yield of soybean. Appl Biochem Biotechnol 102:155–168. https://doi.org/10.1385/ABAB:102-103:1-6:155

Koli DK, Swarnalakshmi K (2017) Isolation and characterization of nodule associated bacteria from chickpea and their potential for plant growth promotion. Int J Curr Microbiol App Sci 6:1992–2004. https://doi.org/10.20546/ijcmas.2017.605.223

Korir H, Mungai NW, Thuita M, Hamba Y, Masso C (2017) Co-inoculation effect of rhizobia and plant growth promoting rhizobacteria on common bean growth in a low phosphorus soil. Front Plant Sci 8:141. https://doi.org/10.3389/fpls.2017.00141

Kong Z, Deng Z, Glick BR, Wei G, Chou M (2017) A nodule endophytic plant growth-promoting *Pseudomonas* and its effects on growth, nodulation and metal uptake in *Medicago lupulina* under copper stress. Ann Microbiol 67:49–58. https://doi.org/10.1007/s13213-016-1235-1

Lace B, Ott T (2018) Commonalities and differences in controlling multipartite intracellular infections of legume roots by symbiotic microbes. Plant Cell Physiol 59:661–672. https://doi.org/10.1093/pcp/pcy043

Lai WA, Hameed A, Lin SY (2015) *Paenibacillus medicaginis* sp. nov. a chitinolytic endophyte isolated from a root nodule of alfalfa (*Medicago sativa* L.). Int J Syst Evol Microbiol 65:3853–3860. https://doi.org/10.1099/ijsem.0.000505

Leigh GJ (2004) The world's greatest fix: a history of nitrogen and agriculture. Oxford University Press, NewYork

Leite J, Fischer D, Rouws LFM et al (2017) Cowpea nodules harbor non-rhizobial bacterial communities that are shaped by soil type rather than plant genotype. Front Plant Sci 7:2064. https://doi.org/10.3389/fpls.2016.02064

Li JH, Wang ET, Chen WF, Chen WX (2008) Genetic diversity and potential for promotion of plant growth detected in nodule endophytic bacteria of soybean grown in Heilongjiang province of China. Soil Biol Biochem 40:238–246. https://doi.org/10.1016/j.soilbio.2007.08.014

Lugtenberg B, Kamilova F (2009) Plant-growth-promoting rhizobacteria. Annu Rev Microbiol 63:541–556. https://doi.org/10.1146/annurev.micro.62.081307.162918

Martínez-Hidalgo P, Galindo-Villardón P, Trujillo ME, Igual JM, Martínez-Molina E (2014) *Micromonospora* from nitrogen fixing nodules of alfalfa (*Medicago sativa* L.). A new promising Plant Probiotic Bacteria. Sci Rep 4:6389. https://doi.org/10.1038/srep06389

Martínez-Hidalgo P, Hirsch AM (2017) The nodule microbiome: N_2-fixing rhizobia do not live alone. Phytobiomes 1:70–82. https://doi.org/10.1094/PBIOMES-12-16-0019-RVW

Mavengere NR, Ellis AG, Le Roux JJ (2014) *Burkholderia aspalathi* sp. nov., isolated from root nodules of the South African legume *Aspalathus abietina* Thunb. Int J Syst Evol Microbiol 64:1906–1912. https://doi.org/10.1099/ijs.0.057067-0

Menéndez E, Ramírez-Bahena MH, Carro L (2016) *Paenibacillus periandrae* sp. nov., isolated from nodules of *Periandra mediterranea*. Int J Syst Evol Microbiol 66:1838–1843. https://doi.org/10.1099/ijsem.0.000953

Mhamdi R, Mrabet M, Laguerre G, Tiwari R, Aouani ME (2005) Colonization of *Phaseolus vulgaris* nodules by *Agrobacterium*-like strains. Can J Microbiol 51:105–111. https://doi.org/10.1139/w04-120

Mishra PK, Mishra S, Selvakumar G, Kundu S, Shankar-Gupta H (2009) Enhanced soybean (*Glycine max* L.) plant growth and nodulation by *Bradyrhizobium japonicum*-SB1 in presence of *Bacillus thuringiensis*-KR1. Acta Agric Scand BSP Sci 59:189–196. https://doi.org/10.1080/09064710802040558

Mousavi SA, Willems A, Nesme X, De Lajudie P, Lindström K (2015) Revised phylogeny of Rhizobiaceae: proposal of the delineation of *Pararhizobium* gen. nov., and 13 new species combinations. Syst Appl Microbiol 38:84–90. https://doi.org/10.1016/j.syapm.2014.12.003

Mrabet M, Mnasri B, Laguerre G, Aouani ME, Mhamdi R (2006) *Agrobacterium* strains isolated from root nodules of common bean specifically reduce nodulation by Rhizobium gallicum. FEMS Microbiol Ecol 56:304–309. https://doi.org/10.1111/j.1574-6941.2006.00069.x

Murset V, Hennecke H, Pessi G (2012) Disparate role of rhizobial ACC deaminase in root-nodule symbioses. Symbiosis 57:43–50. https://doi.org/10.1007/s13199-012-0177-z

Noreen R, Ali SA, Hasan KA, Sultana V, Ara J, Ehteshamul-Haque S (2015) Evaluation of biocontrol potential of fluorescent *Pseudomonas* associated with root nodules of mungbean. Crop Prot 75:18–24. https://doi.org/10.1016/j.cropro.2015.04.018

Oldroyd GE, Downie JA (2008) Coordinating nodule morphogenesis with rhizobial infection in legumes. Annu Rev Plant Biol 59:519–546. https://doi.org/10.1146/annurev.arplant.59.032607.09283

Palaniappan P, Chauhan PS, Saravanan VS, Anandham R, Sa T (2010) Isolation and characterization of plant growth promoting endophytic bacterial isolates from root nodule of *Lespedeza* sp. Biol Fertil Soils 46:807–816. https://doi.org/10.1007/s00374-010-0485-5

Pandya M, Rajput M, Rajkumar S (2015) Exploring plant growth promoting potential of non rhizobial root nodules endophytes of *Vigna radiata*. Microbiology 84:80–89. https://doi.org/10.1134/S0026261715010105

Pawlik M, Cania B, Thijs S, Vangronsveld J, Piotrowska-Seget Z (2017) Hydrocarbon degradation potential and plant growth-promoting activity of culturable endophytic bacteria of *Lotus corniculatus* and *Oenothera biennis* from a long-term polluted site. Environ Sci Pollut Res 24:19640–19652. https://doi.org/10.1007/s11356-017-9496-1

Peix A, Ramírez-Bahena MH, Velázquez E, Bedmar EJ (2015) Bacterial associations with legumes. Crit Rev Plant Sci 34:17–42. https://doi.org/10.1080/07352689.2014.897899

Philipson MN, Blair ID (1957) Bacteria in clover root tissue. Can J Microbiol 3:125–129. https://doi.org/10.1139/m57-016

Poole P, Ramachandran V, Terpolilli J (2018) Rhizobia: from saprophytes to endosymbionts. Nat Rev Microbiol. https://doi.org/10.1038/nrmicro.2017.171

Rajendran G, Patel MH, Joshi SJ (2012) Isolation and characterization of nodule-associated *Exiguobacterium* sp. from the root nodules of fenugreek (*Trigonella foenum-graecum*) and their possible role in plant growth promotion. Int J Microbiol 693982. https://doi.org/10.1155/2012/693982

Ramírez-Bahena MH, Tejedor C, Martín I, Velázquez E, Peix A (2013) *Endobacter medicaginis* gen. nov., sp. nov., isolated from alfalfa nodules in an acidic soil. Int J Syst Evol Microbiol 63:1760–1765. https://doi.org/10.1099/ijs.0.041368-0

Rasolomampianina R, Bailly X, Fetiarison R (2005) Nitrogen-fixing nodules from rose wood legume trees (*Dalbergia* spp.) endemic to Madagascar host seven different genera belonging to α- and β-proteobacteria. Mol Ecol 14:4135–4146. https://doi.org/10.1111/j.1365-294X.2005.02730.x

Ray SS, Ali MN, Mukherjee S, Chatterjee G, Banerjee M (2017) Elimination and molecular identification of endophytic bacterial contaminants during in vitro propagation of *Bambusa balcooa*. World J Microbiol Biotechnol 33:31. https://doi.org/10.1007/s11274-016-2196-z

Safronova VI, Kuznetsova IG, Sazanova AL (2015a) *Bosea vaviloviae* sp. nov., a new species of slow-growing rhizobia isolated from nodules of the relict species *Vavilovia formosa* (Stev.) Fed. Antonie Van Leeuwenhoek 107:911–920. https://doi.org/10.1007/s10482-015-0383-9

Safronova VI, Kuznetsova IG, Sazanova AL (2015b) Extra-slow-growing *Tardiphaga* strains isolated from nodules of *Vavilovia formosa* (Stev.) Fed. Arch Microbiol 197:889–898. https://doi.org/10.1007/s00203-015-1122-3

Safronova V, Belimov A, Sazanova A (2017) Does the Miocene-Pliocene relict legume *Oxytropis triphylla* form nitrogen-fixing nodules with a combination of bacterial strains? Int J Environ Stud 74:706–714. https://doi.org/10.1080/00207233.2018.1479564

Safronova V, Belimov A, Sazanova A (2018a) Taxonomically different co-microsymbionts of a relict legume *Oxytropis popoviana* have complementary sets of symbiotic genes and together increase the efficiency of plant nodulation. Mol Plant Microbe Interact MPMI-01-18-0011-R. https://doi.org/10.1094/MPMI-01-18-0011-R

Safronova VI, Sazanova AL, Kuznetsova IG (2018b) *Phyllobacterium zundukense* sp. nov., a novel species of rhizobia isolated from root nodules of the legume species *Oxytropis triphylla* (Pall.) Pers. Int J Syst Evol Microbiol https://doi.org/10.1099/ijsem.0.002722

Säidï S, Mnasri B, Mhamdi R (2011) Diversity of nodule-endophytic agrobacteria-like strains associated with different grain legumes in Tunisia. Syst Appl Microbiol 34:524–530. https://doi.org/10.1016/j.syapm.2011.01.009

Saini R, Dudeja SS, Giri R, Kumar V (2015) Isolation, characterization, and evaluation of bacterial root and nodule endophytes from chickpea cultivated in Northern India. J Basic Microbiol 55:74–81. https://doi.org/10.1002/jobm.201300173

Sánchez M, Ramírez-Bahena MH, Peix A, Lorite MJ, Sanjuán J, Velázquez E, Monza J (2014) *Phyllobacterium loti* sp. nov. isolated from nodules of *Lotus corniculatus*. Int J Syst Evol Microbiol 64:781–786. https://doi.org/10.1099/ijs.0.052993-0

Santoyo G, Moreno-Hagelsieb G, del Carmen Orozco-Mosqueda M, Glick BR (2016) Plant growth-promoting bacterial endophytes. Microbiol Res 183:92–99. https://doi.org/10.1016/j.micres.2015.11.008

Schoebitz M, López MD, Roldan A (2013) Bioencapsulation of microbial inoculants for better soil–plant fertilization. A review. Agron Sustain Dev 33:751–765. https://doi.org/10.1007/s13593-013-0142-0

Schloter M, Nannipieri P, Sørensen SJ, van Elsas JD (2018) Microbial indicators for soil quality. Biol Fertil Soils 54:1–10. https://doi.org/10.1007/s00374-017-1248-3

Schneijderberg M, Schmitz L, Cheng X, Polman S, Franken C, Geurts R, Bisseling T (2018) A genetically and functionally diverse group of non-diazotrophic *Bradyrhizobium* spp. colonizes the root endophytic compartment of *Arabidopsis thaliana*. BMC Plant Biol 18:61. https://doi.org/10.1186/s12870-018-1272-y

Selvakumar G, Kundu S, Gupta AD, Shouche YS, Gupta HS (2008) Isolation and characterization of nonrhizobial plant growth promoting bacteria from nodules of Kudzu (*Pueraria thunbergiana*) and their effect on wheat seedling growth. Curr Microbiol 56:134–139. https://doi.org/10.1007/s00284-007-9062-z

Shah S, Li J, Moffatt BA, Glick BR (1998) Isolation and characterization of ACC deaminase genes from two different plant growth-promoting rhizobacteria. Can J Microbiol 44:833–843

Shamsheldin A, Abdelkhalek A, Sadowsky MJ (2017) Recent changes to the classification of symbiotic, nitrogen-fixing, legume-associating bacteria: a review. Symbiosis 71:91–109. https://doi.org/10.1007/s13199-016-0462-3

Sheu SY, Chen MH, Liu WY (2015) *Burkholderia dipogonis* sp. nov., isolated from root nodules of *Dipogon lignosus* in New Zealand and Western Australia. Int J Syst Evol Microbiol 65:4716–4723. https://doi.org/10.1099/ijsem.0.000639

Somasegaran P, Hoben HJ (1994) Handbook for rhizobia: methods in legume-Rhizobium technology. Springer Science and Business Media

Souza RD, Ambrosini A, Passaglia LMP (2015) Plant growth-promoting bacteria as inoculants in agricultural soils. Genet Mol Biol 38:401–419. https://doi.org/10.1590/S1415-475738420150053

Sturz AV, Christie BR, Matheson BG, Nowak J (1997) Biodiversity of endophytic bacteria which colonize red clover nodules, roots, stems and foliage and their influence on host growth. Biol Fertil Soils 25:13–19. https://doi.org/10.1007/s003740050273

Subramanian P, Kim K, Krishnamoorthy R, Sundaram S, Sa T (2015) Endophytic bacteria improve nodule function and plant nitrogen in soybean on co-inoculation with *Bradyrhizobium japonicum* MN110. Plant Growth Regul 76:327–332. https://doi.org/10.1007/s10725-014-9993-x

Szilagyi-Zecchin VJ, Ikeda AC, Hungria M, Adamoski D, Kava-Cordeiro V, Glienke C, Galli-Terasawa LV (2014) Identification and characterization of endophytic bacteria from corn (Zea mays L.) roots with biotechnological potential in agriculture. AMB Express 4:2–9. https://doi.org/10.1186/s13568-014-0026-y

Tariq M, Hameed S, Yasmeen T, Ali A (2012) Non-rhizobial bacteria for improved nodulation and grain yield of mung bean [*Vigna radiata* (L.) Wilczek]. Afr J Biotechnol 11:15012. https://doi.org/10.5897/AJB11.3438

Tariq M, Hameed S, Yasmeen T, Zahid M, Zafar M (2014) Molecular characterization and identification of plant growth promoting endophytic bacteria isolated from the root nodules of pea (*Pisum sativum* L.). World J Microbiol Biotechnol 30:719–725. https://doi.org/10.1007/s11274-013-1488-9

Tétard-Jones C, Edwards R (2016) Potential roles for microbial endophytes in herbicide tolerance in plants. Pest Manag Sci 72:203–209. https://doi.org/10.1002/ps.4147

Tokala RK, Strap JL, Jung CM et al (2002) Novel plant-microbe rhizosphere interaction involving *Streptomyces lydicus* WYEC108 and the pea plant (*Pisum sativum*). Appl Environ Microbiol 68:2161–2171. https://doi.org/10.1128/AEM.68.5.2161-2171.2002

Tokgöz S (2018) Exploration of soybean nodule microbiome for plant health management. University of Nebraska

Trabelsi D, Chihaoui SA, Mhamdi R (2017) Nodules and roots of *Vicia faba* are inhabited by quite different populations of associated bacteria. Appl Soil Ecol 119:72–79. https://doi.org/10.1016/j.apsoil.2017.06.002

Trujillo ME, Kroppenstedt RM, Fernandez-Molinero C, Schumann P, Martinez-Molina E (2007) *Micromonospora lupini* sp. nov. and *Micromonospora saelicesensis* sp. nov., isolated from root nodules of *Lupinus angustifolius*. Int J Syst Evol Microbiol 57:2799–2804. https://doi.org/10.1099/ijs.0.65192-0

Valverde A, Velazquez E, Gutierrez C, Cervantes E, Ventosa A, Igual JM (2003) *Herbaspirillum lusitanum* sp. nov., a novel nitrogen-fixing bacterium associated with root nodules of *Phaseolus vulgaris*. Int J Syst Evol Microbiol 53:1979–1983. https://doi.org/10.1099/ijs.0.02677-0

Valverde A, Fterich A, Mahdhi M (2010) *Paenibacillus prosopidis* sp. nov., isolated from the nodules of *Prosopis farcta*. Int J Syst Evol Microbiol 60:2182–2186. https://doi.org/10.1099/ijs.0.014241-0

Velázquez E, Martínez-Hidalgo P, Carro L, Alonso P, Peix A, Trujillo ME, Martínez-Molina E (2013) Nodular endophytes: an untapped diversity. In: Gonzalez R, Gonzalez-Lopez J (eds). Beneficial plant–microbial interactions: ecology and applications. Taylor and Francis Group, pp 215–235

Velázquez E, Carro L, Flores-Félix JD (2017) The legume nodule microbiome: a source of plant growth-promoting bacteria. In: Kumar V, Kumar M, Sharma S, Prasad R (eds) Probiotics and Plant Health. Springer, Singapore

Vincent JM (1970) A manual for the practical study of the root-nodule bacteria. Blackwell Scientific Publications, Oxford

Wang ET, Tan ZY, Willems A, Fernández-López M, Reinhold-Hurek B, Martínez-Romero E (2002) *Sinorhizobium morelense* sp. nov., a *Leucaena leucocephala*-associated bacterium that is highly resistant to multiple antibiotics. Int J Syst Evol Microbiol 52:1687–1693. https://doi.org/10.1099/00207713-52-5-1687

Wani PA, Khan MS (2010) *Bacillus* species enhance growth parameters of chickpea (*Cicer arietinum* L.) in chromium stressed soils. Food Chem Toxicol 48:3262–3267. https://doi.org/10.1016/j.fct.2010.08.035

Wekesa C, Muoma J, Ombori O et al (2017) Genetic characterization of rhizosphere bacteria that inhabit common bean nodules in Western Kenya soils. Appl Microbiol 3:1. https://doi.org/10.4172/2471-9315.1000128

Wei XL, Han MS, Xia CC (2015) *Diaphorobacter ruginosibacter* sp. nov., isolated from soybean root nodule, and emended description of the genus *Diaphorobacter*. Arch Microbiol 197:683–692. https://doi.org/10.1007/s00203-015-1102-7

Wigley K, Liu WYY, Khumalo Q, Moot DJ, Brown DS, Ridgway HJ (2015) Effectiveness of three inoculation methods for lucerne (*Medicago sativa* L.) in two Canterbury soils. N Z J Agric Res 1:10. https://doi.org/10.1080/00288233.2015.1028652

Wigley K, Moot D, Wakelin SA, Laugraud A, Blond C, Seth K, Ridgway H (2017) Diverse bacterial taxa inhabit root nodules of lucerne (*Medicago sativa* L.) in New Zealand pastoral soils. Plant Soil 420:253–262. https://doi.org/10.1007/s11104-017-3395-6

Xu L, Zhang Y, Wang L, Chen W, Wei G (2014) Diversity of endophytic bacteria associated with nodules of two indigenous legumes at different altitudes of the Qilian Mountains in China. Syst Appl Microbiol 37:457–465. https://doi.org/10.1016/j.syapm.2014.05.009

Xu L, Zhang Y, Li C, Wang X, Liu J, Friman VP (2018) *Nocardioides astragali* sp. nov., isolated from a nodule of wild *Astragalus chrysopterus* in northwestern China. Antonie van Leeuwenhoek 1–7. https://doi.org/10.1007/s10482-018-1020-1

Yan J, Li Y, Han XZ, Chen WF, Zou WX, Xie Z, Li M (2017a) *Agrobacterium deltaense* sp. nov., an endophytic bacteria isolated from nodule of *Sesbania cannabina*. Arch Microbiol 199:1003–1009. https://doi.org/10.1007/s00203-017-1367-0

Yan J, Yan H, Liu LX et al (2017b) *Rhizobium hidalgonense* sp. nov., a nodule endophytic bacterium of *Phaseolus vulgaris* in acid soil. Arch Microbiol 199:97–104. https://doi.org/10.1007/s00203-016-1281-x

Yin J, He D, Li X, Zeng X, Tian M, Cheng G (2015) *Paenibacillus enshidis* sp. nov., isolated from the nodules of *Robinia pseudoacacia* L. Curr Microbiol 71:321–325. https://doi.org/10.1007/s00284-015-0854-2

Zakhia F, Jeder H, Willems A, Gillis M, Dreyfus B, De Lajudie P (2006) Diverse bacteria associated with root nodules of spontaneous legumes in Tunisia and first report for nifH-like gene within the genera *Microbacterium* and *Starkeya*. Microb Ecol 51:375–393. https://doi.org/10.1007/s00248-006-9025-0

Zgadzaj R, Garrido-Oter R, Jensen DB, Koprivova A, Schulze-Lefert P, Radutoiu S (2016) Root nodule symbiosis in *Lotus japonicus* drives the establishment of distinctive rhizosphere, root, and nodule bacterial communities. Proc Natl Acad Sci USA 113:E7996–E8005. https://doi.org/10.1073/pnas.1616564113

Zhao L, Xu Y, Sun R, Deng Z, Yang W, Wei G (2011) Identification and characterization of the endophytic plant growth prompter *Bacillus cereus* strain MQ23 isolated from *Sophora alopecuroides* root nodules. Braz J Microbiol 42:567–575. https://doi.org/10.1590/S1517-83822011000200022

Zhao L, Xu Y, Lai X (2017) Antagonistic endophytic bacteria associated with nodules of soybean (*Glycine max* L.) and plant growth-promoting properties. Braz J Microbiol 49:269–278. https://doi.org/10.1016/j.bjm.2017.06.007

Chapter 7
Plant–Microbes Relationships in Soil Ecological System and Benefits Accruable to Food Health

Lateef Bamidele Taiwo, Adedayo Omowumi Oyedele, Bukola Victoria Ailenokhuoria and Oladapo Titus Okareh

Abstract The excessive use of chemicals in agricultural production gives rise to various issues such as unanticipated environmental impacts, soil biological degradation as well as water contamination. This in many instances has led to eutrophication as well as human health hazards. The concern has raised the question on the safety of food products obtained from this conventional method. It has, therefore, become imperative to adopt biological fertilization strategy that may minimize the use of these inputs. Exploiting the relationships among plants and rhizospheric soil microbes is a rational option. Such interactions are the major factors that determine the health of a plant, plants' yield, and fertility of the soil. The Plant Growth Promoting Rhizobacteria (PGPR) are being used as bio-inoculants for the provision of nutrients, plant growth promotion and to combat plant diseases. The PGPR includes 72 bacterial genera including *Agrobacterium*, *Azotobacter*, *Azospirillum*, *Bacillus*, *Burkholderia*, *Erwinia*, *Flavobacterium*, *Micrococcus*, *Pseudomonas*, *Serratia*, etc. With the use of these genera, the chemical inputs, and agrochemicals are bound to be reduced in order to sustain benefits to human health. The application of effective PGPR in inoculant technology, therefore, is considered as a vital approach for sustainable soil management and solving environmental issues.

Keywords Plant growth promoting rhizobacteria · Plant-microbe interrelationship · Soil ecology · Food health

---

L. B. Taiwo (✉) · A. O. Oyedele · B. V. Ailenokhuoria
Institute of Agricultural Research and Training, Obafemi Awolowo University, Moor Plantation, P. M. B. 5029 Ibadan, Nigeria
e-mail: lateefbtaiwo@gmail.com

A. O. Oyedele
e-mail: adeedii@yahoo.com

B. V. Ailenokhuoria
e-mail: bukvic2008@yahoo.com

O. T. Okareh
Faculty of Public Health, Department of Environmental Health Sciences, College of Medicine, University of Ibadan, Ibadan, Nigeria
e-mail: dapsy2001@yahoo.co.uk

D. K. Maheshwari and S. Dheeman (eds.), *Field Crops: Sustainable Management by PGPR*, Sustainable Development and Biodiversity 23,
https://doi.org/10.1007/978-3-030-30926-8_7

7.1 Introduction

Globally, the significance of sustainable agricultural production cannot be overemphasized, in order to achieve food security for a growing populace. Chemical fertilizers which have become an integral part of present-day agriculture are being used extensively to supply essential plants nutrients in agricultural production. Indiscriminate application of these fertilizers used in the bid to improve crop production has led to soil ecological disturbances such as a reduction in soil biota, nutrient imbalances in plants, increased plants' vulnerability to pests and diseases, reduced nodulation in leguminous plants, plant–mycorrhizal relationships as well as constituting environmental hazards. In addition, the gradual decrease in the population of soil microbes that are beneficial has been linked to the indiscriminate use of chemical fertilizers. The overdependence on chemical fertilizers and its excessive use have also brought about other environmental related issues including degradation of soil and its components, eutrophication caused by excessive runoff of the nutrients into water bodies and also several health issues in humans arising from the consumption of plants having residual chemicals.

7.2 Biological Degradation of Soil

This has been defined as the impairment or elimination of soil microbial population which brings about alterations in biogeochemical processing within the associated ecosystem (Sims 2018). One of the sources of pollutants is the agrochemicals which have shown an imbalance in some of the ecological processes disruption due to the introduction of these chemicals. Aside from polluting the soil environment, it also leads to a reduction in soil fertility. Annual bush fire especially within the savanna ecology for game hunting in the zone also destroyed the ecological biodiversity, depletes the accumulating soil organic matter that has seriously affected the soil macro and microflora and fauna. Soil degradation also manifests by a decrease in the organic matter, as well as the total and available nitrogen and phosphorus forms.

More sustainable alternatives are being exploited in the form of use of plant growth promoting bacteria (PGPR) to boost crop production, maintaining the fertility of soils, as well as decreasing the overdependence on chemical fertilizers. Thus, PGPRs are seen as promising alternatives because of their unique characteristics. Due to the distinctive characteristics of microbes to be able to display several spontaneous biosynthetic activities in response to particular environmental and growth conditions, they have been considered suitable for providing solutions to challenging issues related to life sciences and other aspects. Some soil microbes especially in the rhizosphere, have been shown to secrete plant beneficial substances that enhance crop production such as growth promoters, siderophores, antibiotics, and assisting in the uptake of essential plant nutrients such as phosphorus, potassium and zinc among several others. Plant-microbe interrelationship within the rhizosphere is the basis for plant well-being, production, as well as the fertility of soils.

In cropping systems, the continuous application of chemical fertilizers decreases the fertility status of soils and also has health implications in humans. Therefore, the integration of PGPR inoculants with chemical fertilizers is a necessity in cropping systems so as to reduce the use of chemicals but at the same time increase productivity. Several authors have proposed an integrated approach in fertilizer technology, which will involve the use of PGPR-based inoculation alongside the right and required levels of chemical fertilizer, this will not only reduce the high usage and overdependence on chemical fertilizers but will also contribute to improved plant growth. Kumar et al. (2009) reported a research carried out using effective plant growth promoting bacteria with a reduction in the amount of chemical fertilizer applied, as an approach for reducing the environmental issues that arise as a result of excessive input of chemical fertilizers while at the same time improving crop production.

7.3 Plant Growth Promoting Rhizobacteria (PGPR)

Different authors have provided various but similar definitions of PGPR. Maheshwari et al. (2010) defined PGPR as the bacteria which through diverse mechanisms can enhance plant growth, reduce plants' susceptibility to disease as well as confer on the plants, a form of defense from abiotic stresses. Grover et al. (2011) noted that PGPRs belong to a beneficial and diverse group of microbes domiciled within the root environment including the rhizosphere and rhizoplane. Ahemad and Kibret (2014) described PGPR as the soil bacteria present within or associated with root surfaces, which synthesize a wide range of regulatory chemicals within the rhizosphere environment thereby directly or indirectly enhancing the growth and development of plants.

The PGPR includes the extracellular PGPR and the intracellular PGPR. The extracellular PGPR could inhabit the rhizosphere surrounding including the rhizosphere itself, root surfaces and even within cellular spaces of root tissues, examples of extracellular PGPR includes *Arthrobacter*, *Azospirillum*, *Bacillus*, *Burkholderia* and, *Pseudomonas* (Gray and Smith 2005). *Frankia* species and the endophytes which include rhizospheric bacteria belonging to the Rhizobiaceae family are examples of intracellular PGPR (Aeron et al. 2014).

7.4 Mechanisms of Action of PGPR

A key attribute of PGPR is the possession of some traits and the mechanisms for influencing plant growth. These traits include nitrogen fixation, nutrient solubilization, phytohormone, and antibiotics production among several others. These bacterial species enhance plant growth via various systematic ways which include the production of plant hormones (e.g., Indole acetic acid (IAA) and Gibberellin), solubilization of major nutrients (e.g., phosphorus and potassium) and micronutrients such as zinc and also siderophore synthesis. Rocheli de Souza et al. (2015) also stated that PGPR has several mechanisms through which they incite plant growth, these involve genomic processes to make nutrients available for plant use, such as biological nitrogen fixation and solubilization of complex forms of phosphorus, regulation of ACC deaminase to reduce plant stress, as well as production of beneficial compounds that contribute to plant growth, and these include siderophores and phytohormones and several other plant stimulating mechanisms. In addition, Kumar et al. (2012), also listed some ways through which this set of microbes stimulates the growth of plants; these include nutrient availability through biological nitrogen fixation and phosphate solubilization, release of plant beneficial metabolites such as phytohormone, siderophore, 1-Aminocyclopropane-1- carboxylate deaminase and volatile organic compounds (VOCs), bio-control activity such as antifungal activity, defense mechanisms such as disrupting the production of toxic substances disease-causing organisms, stimulation of systemic resistance, rhizosphere engineering, intrusion of sensing signals and prevention of the establishment of biofilm as well as enhancing beneficial plant-microbe symbioses.

Although various researchers have proposed some of the mechanisms of action of PGPR for plant growth promotion, the exact mechanisms are complex and not entirely unraveled. Different species and different strains of bacteria enhance the growth of plants through any of the mechanisms and some strains have also been seen to show multiple mechanisms of plant growth. Earlier researchers have also reported some of the mechanisms to include cytokinins and ethylene production (Glick 1995), non-symbiotic biological nitrogen fixation by free-living bacteria (Boddey and Dobereiner 1995), as well as cyanide production (Flaishman et al. 1996).

Biological nitrogen fixation, solubilization of minerals, resource acquisition, regulation of plant hormone levels, phytohormone productions, synthesis of vitamins, siderophores, enzymes and stimulation of systemic resistance are categorized as direct mechanisms of plant growth promoting activities. These plant growth promoting bacteria reduce plant development impeding impacts caused by various disease-causing organisms through bio-control activities such as antibiotic secretion, chelation of Fe, production of extracellular enzymes which break down the cell walls of fungi and competition for niches within the rhizosphere, these activities are however categorized as indirect methods.

7.4.1 Biological Nitrogen Fixation

A major plant beneficial mechanism that can occur either through symbiotic or asymbiotic association with rhizospheric soil bacteria is through biological nitrogen fixation (BNF). The process of BNF usually occurs through the conversion of unreactive atmospheric nitrogen through the action of nitrogenase enzyme by bacteria to ammonium (NH_3) which is a form that can be utilized by plants (Bhattacharjee et al. 2008).

7.4.2 Nutrient Solubilization/Availability

The soil is a sink of nutrients from which plants acquire nutrients needed for their growth. A wide range of soil microbes including the major groups plays major roles in the solubilization process depending on the soil environment. Therefore, PGPR is a major source of these microorganisms that accelerate the solubilization process. Solubilization of calcium phosphate is an example, which leads to the release of phosphorus, a vital nutrient required for plant growth but exists in insoluble and inaccessible forms in the soil; as a result, it is the second major chemical fertilizer being applied for crop production. Available P exists in the soil majorly as insoluble organic and inorganic compounds in complex forms.

Plant Growth Promoting Rhizobacteria can solubilize complex soil nutrients by secreting organic acids, sugar acids and carbon dioxide, resulting in a reduced soil pH making it a suitable condition for the solubilization of inorganic compounds. Several phosphate solubilizing soil bacteria including both aerobic and anaerobic strains have been isolated and identified around the rhizosphere and its vicinity. Earlier reports have shown the ability of some isolated and characterized rhizospheric bacteria such as *Agrobacterium*, *Bacillus*, *Burkholderia*, *Erwinia Pseudomonas* and *Rhizobium*, to solubilize complex forms of phosphate such as tricalcium phosphate, dicalcium phosphate, hydroxyapatite and rock phosphate (Hayat et al. 2010). PGPR/plant–root relationship is fundamental in the availability and uptake of phosphorus, especially in low phosphorus soils.

A large percentage of soil potassium which is the third major essential macronutrient is present as complex insoluble forms such as silicate minerals, hence the amount of potassium accessible to plant are usually inadequate. Therefore, the discovery of other easily accessible and/or readily available sources of soil potassium is crucial for sustainable cropping. Potassium solubilizing bacteria can solubilize complex potassium compounds by the synthesis of organic acids. Several rhizospheric bacteria such as *Acidothiobacillus* sp., *Bacillus* sp., *Pseudomonas* sp., *Burkholderia* sp. and *Paenibacillus* sp. are effective soil Potassium solubilizers (Liu et al. 2012).

Other essential soil minerals are the micronutrients, required for plant nutrition although, in very small quantities. While the requirement of these micronutrients is usually in minute quantities, the absence or shortage of one or more of these in the soil may lead to critical issues that may hamper the totality of plant growth in addition to making the macronutrients available as well as accessible for plant use, PGPRs also enhance the availability of soil micronutrients, thereby making them available for plant use. Abaid-Ullah et al. (2011) reported the effectiveness of PGPR in enhancing the availability of soil macronutrients (phosphorus and potassium), as well as some micronutrients Fe, Zn, and silicate) for plant uptake. Biological plant promoting technology provides a cheaper and eco-friendly alternative, coupled with its ability to carry out soil nutrient bioavailability can, therefore, be a valuable tool in solving micronutrient deficiency-related issues. Aside from the macro- and micronutrients, Plant growth promoting regulators and hormones play crucial roles in plant development. Plant hormones also referred to as phytohormones have been defined as a class of organic substances that are produced during plant metabolism (Shi et al. 2017). Quite a large amount of plant metabolic processes such as nutrient uptake, spatial differences in cellular processes, cell elongation, embryo formation and development and many more biological cellular processes necessary for plant growth are modulated by these phytohormones (Sauer et al. 2013). Phytohormones are classified into five main groups which are: auxins (e.g. indole acetic (IAA) acid), cytokinins, ethylene, gibberellic acid and abscisic acid which reportedly have evident physiological impacts on plant growth when present even at very little concentrations (Shi et al. 2017). Generally, plant hormones are produced by microorganisms as secondary metabolites, which though are not the basis for plant survival but are an integral part of a number of plant metabolic processes which includes competition and defense mechanisms vital for proper plant growth and development. Several rhizosphere bacteria are able to synthesize the five classes of phytohormones, and this has been validated through various laboratory studies, while majority of soil microorganisms can synthesize auxin as well as ethylene, the production of gibberellic acid has however been noted to be limited to a minute category of bacterial and fungal strains (Hedden and Thomas 2012).

Several beneficial bacteria secrete the auxin phytohormone, i.e., IAA which is of great physiological importance in the interrelationship between plant and bacteria. Rhizospheric bacteria have been reported to secrete more auxin-related compounds compared to bulk soil-associated bacteria. In a study carried out by Taiwo et al. (2017), different Plant growth promoting microorganisms isolated from rhizosphere of *Zea mays* were reported to be positive for the production of some plant growth promoting hormones such as IAA, gibberellin and cytokinin and the most promising among the isolates were identified as *Pseudomonas putida, Klebsiella varricola,* and *Bacillus thuringiensis* which produced 2.693, 19.697, and 15.091 mg/l of gibberellic acid; 0.152, 0.348, and 0.132 mg/l of IAA and 5.066, 1.804, and 2.410 mg/l of cytokinin, respectively.

Other studies have shown the production of these phytohormones by different rhizosphere associated bacterial strains of different crops in different countries dating back to several decades and up till now. Among other plant enhancement traits possessed by PGPR, their capability to secrete plant growth hormones in addition to their nutrient solubilizing ability as well as other plant enhancement beneficial substances synthesis such as siderophores and, cyanogens makes them the most suitable group of microorganisms to be considered in crop production.

Another important metabolite produced by PGPR is siderophore, which are extracellular iron transport agents that solubilize and sequester iron which availability to plant is inhibited by its chemical oxidation to highly insoluble ferric salts (Buyer and Sikora 1990). These siderophores play critical roles in the composition and structure of soil microbiota, in bio-control of soil-borne plant disease-causing organisms and in enhancing growth and grain yield (Sahu and Sindhu 2011).

Ethylene which is a stress hormone is an important growth hormone that can be produced aggressively by plants under stress conditions, however, its uncontrolled production can affect plant root and hence the plant aging process. Plant Growth Promoting Rhizobacteria modulates ethylene production in plants by secreting a deaminase enzyme known as 1-aminocyclopropane-1-carboxylate (ACC).

7.4.3 Induced Systemic Resistance (ISR)

This is a mechanism by which some PGPR enhance the protection of plants against pests and diseases. Prathab and Ranjitha (2015) defined ISR as a physiological state by which the defensive ability induced as a reaction to a specific environmental stimulus is enhanced, The authors further stated that PGPRs through this mechanism bring about resistance against several environmental stressors. Induced System Resistance is not targeted against specific pathogens and a diverse range of soil microbes are able to help plants to be defensive against a broad range of pathogens (Kamal and Yogendra 2014).

As stated earlier, the plant growth promoting bacteria protects plants against pathogens through direct or indirect mechanism. Induced systemic resistance is an indirect method of plant protection mechanisms as against secretion of several antagonistic metabolites like siderophores, bacteriocins, and antibiotics which are direct methods of plant protection (Amar et al. 2013).

7.4.4 Production of Disease Resistance Antibiotics

Bio-control activities through the production of various antibiotic substances are an indirect way by which PGPR promotes plant growth. Many of these rhizobacteria, like *Bacillus* spp. and *Pseudomonas* sp. inhibit pathogens by producing antibiotics which are extracellular metabolites capable of inhibiting plant pathogens even at low concentrations. A large percentage of *Pseudomonas* species and *Bacillus* sp. secretes a wide range of both antifungal and antibacterial metabolites (Chauhan et al. 2016). Antimicrobial metabolites produced by *Pseudomonas* include phenazines, sulfonamide, pyocyanin and pseudomonic acid and azomycin (Ramadan et al. 2016) while *Bacillus* sp. produce metabolites such as polymyxin, circulin and colistin (Maksimov et al. 2011). In addition to siderophore and antibiotics secretion, the majority of bacteria also produce at least one bactericidal activity with some having a broad inhibition spectrum.

7.4.5 Production of Protective Enzymes

As indicated earlier, one of the mechanisms by which Plant Growth Promoting Rhizobacteria promote plant growth is by the secretion of bio-control compounds that acts against plant pathogens. The ability of the majority of PGPR to release cell wall lysing enzymes is one of the devices employed by this set of microbes to control soil-borne pathogens. Chitinase, cellulase, and protease are examples of the cell wall degrading enzymes secreted by PGPR, which through their actions suppress the growth of pathogenic fungi by degrading their cell wall. For instance, chitinase degrades chitin, the major component of the fungal cell wall (Goswami et al. 2016). Their cell wall degrading ability are highly effective such that they can inhibit a wide spectrum of fungi pathogens even *Phytophthora capsici* and *Rhizoctonia solani*, which are considered to be the utmost disastrous plant pathogens worldwide e (Islam et al. 2016). A range of microorganisms including both Gram-positive and Gram-negative bacteria such as *Bacillus* species (e.g., *B. licheniformis, B. cereus, B. subtilis and B. thuringiensis), Serratia marcescens, Enterobacter agglomerans, Pseudomonas aeruginosa,* and *P. fluorescens* have chitinolytic activities which makes them to be considered as potential bio-control agents (Sadfi et al. 2001).

7.4.6 Exopolysaccharide Production

Exopolysaccharides (EPSs) are produced by a diverse group of bacteria as well as algae and plants. They have been defined as high molecular weight, biodegradable polymers formed of monosaccharide residues and their derivatives (Sanlibaba and Çakmak 2016). Exopolysaccharides play several major roles that are directly involved in plant development, these include the maintenance of water potential, soil particles aggregation, facilitating plant root-rhizobacteria interrelationship, sustenance of plants when exposed to pathogens or under stress conditions (salinity, drought, or waterlogging) (Pawar et al. 2013).

7.5 Factors Affecting the Efficiency of PGPR

The success and efficiency of PGPR as effective inoculants is determined by a number of factors and these range from the soil environment, including the soil status, interactions with plant roots and effects of root secretions as well as the capability of the PGPR itself to colonize and establish itself in the rhizosphere. In determining the suitability of bacterial strains to be classified as PGPR, the bacterial strains should have the ability to exhibit at least two of three key traits which are the ability to establish and colonize the soil ecosystem in aggressive manner, enhance plant growth and exhibition of biological control activities (Vessey 2003). It is, however, noteworthy to state that though inoculating plants with PGPR may temporarily promote the rhizosphere microbial population, sustaining a dense population of the microbes in the rhizosphere for a long time is fundamental in achieving the effectiveness of PGPR in the rhizosphere. Various biotic and abiotic factors including the soil type, pH and microflora have a large impact on PGPR efficiency. Soil health is also an important factor affecting the inoculation efficiency and is determined by several characteristics such as the type of soil, nutrient reservoir, concentrations of toxic metals, soil moisture, microbial composition, and soil perturbations resulting from management activities (Rocheli de Souza et al. 2015). Zahir et al. (2004) noted that the modification of the rhizosphere increases the production and health of the plant by supplementing/replacing the resident microflora with beneficial microorganisms. Roots exudates have also been stated to play a central role in the success and effectiveness of PGPR as inoculants. The plant roots produce quite a limitless variety of compounds in reaction to the prevailing factors in their habitat which includes both the biotic and abiotic factors; this action consequently affects the plant-bacteria interaction.

7.6 Effect of PGPR on Major Crops

Maheshwari et al. (2012) highlighted some ways PGPR technology can contribute to the achievement of agricultural sustainability and these include reducing the dose of chemical fertilizers usage, hence a reduction in production cost and environmental hazards, improved soil fertility through growth enhancement activities that promote sustainable crop production and soil ecology management. Several reports have documented that most of the PGPR strains isolated from rhizospheric soils of several crops in different locations significantly increased plant growth parameters and biomass of various agricultural crops like maize (Gholami et al. 2009), wheat (Cakmakci et al. 2014) and legumes such as common beans, runner beans and beans and soybean (Stefan et al. 2013; Korir et al. 2017). Ashrafuzzaman et al. (2009) reported that most of the PGPR isolates have a significant effect on the emergence, growth parameters and biomass production of rice seedlings. Çakmakçia et al. (2014) showed the effect of PGPR on yield of two cereal crops (spring wheat and barley) under both greenhouse conditions and field studies and noted that they significantly affected their yield, yield components and quality parameters and concluded that this satisfied the nitrogen requirements of the crops under greenhouse and field conditions even irrespective of land elevation above sea levels.

Under field conditions, Zahir et al. (1998) carried out biological growth promoting studies on maize seeds with a combination of four isolates (two *Azotobacter* spp. and two *Pseudomonas* spp.) after NPK fertilizer application and observed that collective inoculation of the isolates considerably increased grain yield by 19.8%, cob weight, cob length, 1000 grain weight, plant height, nitrogen content in the straw and grain by 21.3%, 20.6%, 9.6%, 8.5%, 18% and 19.8% respectively compared to the un-inoculated control plants. Other reports on maize inoculation were documented by Vedder-Weiss et al. (1999) who showed that maize seed inoculated with *Azospirillum* spp. at a concentration of 10^6 cfu/ml enhanced fresh root and shoot weight of seedlings while Stancheva et al. (1992) showed that maize seeds inoculated with *Azospirillum brasilense* strain 1774 in a blend with 100 kg N/ha fertilizer gave the same result as 200 kg N/ha of non-inoculated plants.

7.7 Influence of PGPR on Soil Respiration

Plant Growth Promoting Rhizobacteria (PGPR) influence some soil microbial activities which include soil respiration. Findings indicate that soil inoculation using some PGPR increased soil respiration by (18%) when compared to the control (Taiwo et al. 2018) as shown in Fig. 7.1. Increasing microbial numbers through inoculation with appropriate and N_2 fixing microbes that are infective and effective in nodulation might lead to an increase in nodulation activities on the rhizoplane. The production of exudates, infection threads arising from plant-microbe interaction and presence of molybdenum (Mo) nitrogenase enzymes could be responsible for the increase in activities of nodulating bacteria and hence improvement in nodulation of cowpea by about 70% as shown in Table 7.1.

7.8 Safety Regulation Considerations

The usage of a wide range of synthetic products including chemical fertilizers, pesticides and growth regulators which are currently being employed for agricultural production calls for public health concerns as they have been implicated to

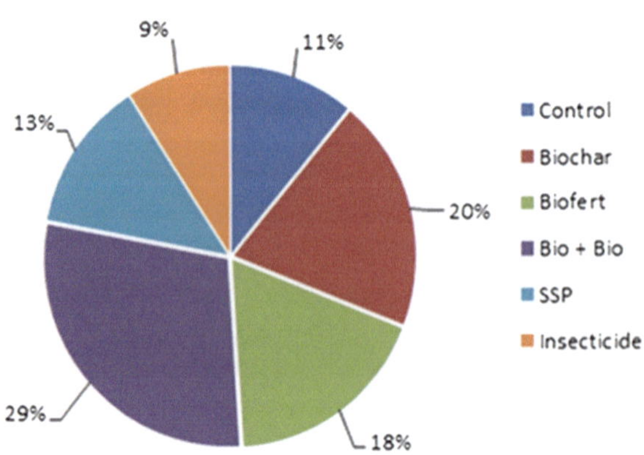

Fig. 7.1 CO_2 efflux evaluation (million/g) in the screenhouse soil (Post-Harvest Analysis) (Taiwo et al. 2018)

Table 7.1 Effect of treatments on the number of nodules of cowpea on the field (Taiwo et al. 2018)

Treatments	Number of nodules
Control	6.0^{e}
Biofertilizer	8.0^{d}
Biofertilizer + Biochar	10.0^{a}
Biochar	9.3^{b}
SSP	8.7^{c}
Insecticide	9.0^{b}

Values followed by different letters are significantly different from each other ($p \leq 0.05$)
SSP = Single Super Phosphate applied at a rate of 60 kg P_2O_5/ha

pose threats to human health and the environment. Owing to the volatile nature of the composition of these products, they cause contamination of groundwater, are taken up by plants and subsequently transferred to the food chain thereby causing public health hazards.

In the search for a more viable biotechnological approach to overcoming the challenges mentioned above, the use of biological growth promoting technique has been explored. These are known to be exogenous bacteria employed in agricultural production to generate a positive impact on the agrobiology of plants. Compared to chemical products, Plant Growth Promoting Rhizobacteria are also effective, but environmentally safe and non-toxic to naturally occurring microorganisms. Up till recently, the focus of inoculant technology was more on the use of nitrogen-fixing bacteria such as *Rhizobium*, *Azospirillum*, *Azotobacter*, however with the recent advances in inoculant technology, there has been a remarkable expansion in the range of microorganisms that are used and this includes a wider range of rhizosphere associated bacteria and fungi. The use of a wide variety of bacteria in inoculant technology has brought about safety concerns thereby necessitating the need for biosafety considerations in the formulation of microbial inoculants. Quite a number of common microbes associated with the environment are opportunistic pathogens, therefore having a knowledge that most PGPR used for inoculant technology originates from the environment, detailed screening and analysis should be carried out in the selection of novel bacterial for their eventual usage as microbial inoculants (Selvakumar et al. 2014).

Plant and human pathogenic rhizobacteria and the likes cannot be considered for application in the field; therefore, safe PGPR should be given utmost priority. In earlier research carried out by Taiwo et al. (2017), some PGPR such as *Chryseobacterium* spp., *K. pneumonia, P. vulgaris, B. cereus, P. monteilli, E. asburiae,* and *Myroides odoratimimus* isolated were eliminated from further study due to their known human pathogenic characteristics. Safety consideration in the formulation and use of PGPR is of paramount importance, not only for human safety and not to impart toxicity to mammals but also to protect the environment. As earlier stated, some of the microbial strains may be opportunistic pathogens with the possibility of causing health hazards to humans, animals or plants, therefore, urgent attention should be given to the establishment of holistic procedures that take into consideration the predictability, effectiveness, consistency, and especially the biosafety of PGPR for human and animal health and the environment (Vílchez et al. 2015). Important and imperative efforts are required in the field of microbial technology. The scientists are to be aware of the advances in biosafety, to enable them to make up-to-date decisions in their work routine.

7.9 Conclusion

All the beneficial traits of PGPR such as nitrogen fixation, phosphorus and potassium solubilization, micronutrient availability and bio-control activities through various defense mechanisms are key important traits in the formulation of effective bio-fertilizers. Plant Growth Promoting Rhizobacteria when exploited as bio-fertilizers can enhance plant growth, increase grain yield, lower malnutrition rates, ameliorate degraded land, improve soil fertility and reduce over-dependence on chemical fertilizers. The characteristics of PGPR besides economical and eco-friendly attributes can also represent a natural and inexpensive alternative that can mitigate the hazards related to the continuous use of chemical fertilizers. Current practices in agriculture are tilting toward focusing on the reduction of chemical fertilizers and pesticides input; therefore, the exploitation of PGPR inoculants as bio-inoculants offers a promising alternative for more sustainable agriculture.

References

Abaid-Ullah M, Hassan MN, Nawaz MK, Hafeez FY (2011) Biofortification of wheat (*Triticum aestivum* L.) through Zn mobilizing PGPR. In: Proceedings of international science conference on prospects and challenges to sustainable agriculture. Azad Jammu and Kashmir University, Pakistan, p 298

Aeron A, Chauhan PS, Dubey RC, Maheshwari DK, Bajpai VK (2014) Root nodule bacteria from *Clitoria ternatea* L. are putative invasive non-rhizobial endophytes. Can J Microbiol 61(2): 131–142

Ahemad M, Kibret M (2014) Mechanisms and applications of plant growth promoting rhizobacteria: current perspective. J King Saud Univ Sci 26(1):1–20

Amar JD, Manoj K, Rajesh K (2013) Plant growth promoting rhizobacteria (PGPR) an alternative of chemical fertilizer for sustainable environment-friendly agriculture. Res. J. Agric For Sci 1:21–23

Ashrafuzzaman M, Hossen FA, Ismail MR, Hoque A, Islam MZ, Shahidullah SM, Meon S (2009) Efficiency of plant growth-promoting rhizobacteria (PGPR) for the enhancement of rice growth. Afr J Biotechnol 8(7):1247–1252

Bhattacharjee RB, Singh A, Mukhopadhyay SN (2008) Use of nitrogen-fixing bacteria as biofertilizer for non-legumes: prospects and challenges. Appl Microbiol Biot 80:199–209

Boddey RM, Dobereiner J (1995) Nitrogen fixation associated with grasses and cereals: recent progress and perspectives for the future. Fert Res 42:241–250

Buyer JS, Sikora LJ (1990) Rhizosphere interactions and siderophore. Plant Soil 129:101–107

Cakmakci R, Turan M, Gulluce M, Sahin F (2014) Rhizobacteria for reduced fertilizer inputs in wheat (*Triticum aestivum* spp. vulgare) and barley (*Hordeum vulgare*) on Aridisols in Turkey. Int J Plant Prod 8(2):163–182

Chauhan AK, Maheshwari DK, Kim K, Bajpai VK (2016) Termitarium-inhabiting *Bacillus endophyticus* TSH42 and *Bacillus cereus* TSH77 colonizing *Curcuma longa* L.: isolation, characterization, and evaluation of their biocontrol and plant-growth-promoting activities. Can J Microbiol 62(10):880–892

Flaishman MA, Eyal ZA, Zilberstein A, Voisard C, Hass D (1996) Suppression of *Septoria tritci* blotch and leaf rust of wheat by recombinant cyanide producing strains of *Pseudomonas putida*. Mol Plant Microbe Interact 9:642–645

Gholami A, Shahsavani S, Nezarat S (2009) The effect of plant growth promoting rhizobacteria (PGPR) on germination, seedling growth and yield of maize. Proc World Acad Sci Eng Technol 49:19–24

Glick BR (1995) The enhancement of plant growth by free-living bacteria. Can J Microbiol 41:109–114

Goswami Dweipayan, Thakker Janki N, Dhandhukia Pinakin C, Moral Manuel Tejada (2016) Portraying mechanics of plant growth promoting rhizobacteria (PGPR): A review. Cogent Food Agric 2:1. https://doi.org/10.1080/23311932.2015.1127500

Gray EJ, Smith DL (2005) Intracellular and extracellular PGPR: commonalities and distinctions in the plant-bacterium signaling processes. Soil Biol Biochem 37:395–412

Grover M, Ali SKZ, Sandhya V, Rasul A, Venkateswarlu B (2011) Role of microorganisms in adaptation of agriculture crops to abiotic stresses. World J Microbiol Biotechnol 27:1231–1240

Hayat R, Safdar A, Ummay A, Rabia K, Iftikhar A (2010) Soil beneficial bacteria and their role in plant growth promotion: A review. Ann Microbiol 60:579–598

Hedden P, Thomas GS (2012) Gibberellin biosynthesis and its regulation. Biochem J 444:11–25

Islam MT, Croll D, Gladieux P, Soanes DM, Persoons A, Bhattacharjee P, Hossain MS, Gupta DR, Rahman MM, Mahboob MG, Cook N, Salam MU, Surovy MZ, Sancho VB, Maciel JL, NhaniJúnior A, Castroagudín VL, Reges JT, Ceresini PC, Ravel S, Kellner R, Fournier E, Tharreau D, Lebrun MH, McDonald BA, Stitt T, Swan D, Talbot NJ, Saunders DG, Win J, Kamoun S (2016) Emergence of wheat blast in Bangladesh was caused by a South American lineage of *Magnaporthe oryzae*. BMC Biol 14:84

Kamal RG, Yogendra Kumar V (2014) Interaction and symbiosis of AM fungi, actinomycetes and plant growth promoting rhizobacteria with plants: strategies for the improvement of plants health and defense system. Int J Curr Microbiol App Sci 3:564–585

Korir H, Mungai NW, Thuita M, Hamba Y, Masso C (2017) Co-inoculation effect of rhizobia and plant growth promoting rhizobacteria on common bean growth in a low phosphorus soil. Front Plant Sci 8:141

Kumar P, Dubey RC, Maheshwari DK (2012) *Bacillus* strains isolated from rhizosphere showed plant growth promoting and antagonistic activity against phytopathogens. Microbiol Res 167 (8):493–499

Kumar S, Pandey P, Maheshwari DK (2009) Reduction in dose of chemical fertilizers and growth enhancement of sesame (*Sesamum indicum* L.) with application of rhizospheric competent *Pseudomonas aeruginosa* LES4. Euro J Soil Biol 45(4):334–340

Liu H, Li X, Xiao J, Wang S (2012) A convenient method for simultaneous quantification of multiple phytohormones and metabolites: application in the study of rice-bacterium interaction. Plant Methods 8(2). https://doi.org/10.1186/1746-4811-8-2

Maheshwari DK, Dubey RC, Aeron A, Kumar B, Kumar S, Tewari S, Arora NK (2012) Integrated approach for disease management and growth enhancement of *Sesamum indicum* L. utilizing *Azotobacter chroococcum* TRA2 and chemical fertilizer. World J Microbiol Biotechnol 28 (10):3015–3024

Maheshwari DK, Kumar S, Kumar B, Pandey P (2010) Co-inoculation of urea and DAP tolerant *Sinorhizobium meliloti* and *Pseudomonas aeruginosa* as integrated approach for growth enhancement of *Brassica juncea*. Indian J Microbiol 50(4):425–431

Maksimov IV, Abizgil'dina RR, Pusenkova LI (2011) Plant growth promoting rhizobacteria as an alternative to chemical crop protectors from pathogens. Appl Biochem Microbiol 47:373–385

Pawar ST, Bhosale AA, Gawade TB, Nale TR (2013) Isolation, screening and optimization of exopolysaccharide producing bacterium from saline soil. J Microbiol Biotechnol Res 3(3): 24–31

Prathap M, Ranjitha Kumari BD (2015) A critical review on plant growth promoting rhizobacteria. J Plant Pathol Microb 6. https://doi.org/10.4172/2157-7471.1000266

Ramadan Elshahat M, AbdelHafez Ahmed A, Hassan Enas A, Saber Fekria M (2016) Review: plant growth promoting rhizobacteria and their potential for biocontrol of phytopathogens. Afr J. Microbiol Res 10(15):486–504

de Souza R, Mbrosini AA, Passaglia LMP (2015) Plant growth promoting bacteria as inoculants in agricultural soils. Genet Mol Biol 38(4):401–419

Sadfi N, Cherif M, Fliss I, Boudabbous A, Antoun H (2001) Evaluation of bacterial isolates from salty soils and *Bacillus thuringiensis* strains for the biocontrol of *Fusarium* dry rot of potato tubers. J Plant Pathol 83:101–117

Sahu GK, Sindhu SS (2011) Disease control and plant growth promotion of green gram by siderophore producing *Pseudomonas* sp. Res J Microbiol 6:735–749

Sanlibaba P, Çakmak GA (2016) Exopolysaccharides production by lactic acid bacteria. Appl Microbiol 2(2):1–5

Sauer M, Robert S, Kleine-Vehn J (2013) Auxin: simply complicated. J Exp Bot 64:2565–2577

Selvakumar G, Panneerselvam P, Ganeshamurthy AN (2014) Biosafety of novel bioinoculants. J Biofertil Biopestic 5:145. https://doi.org/10.4172/2155-6202.1000145

Shi TQ, Peng H, Zeng SY, JiRY Shi K, Huang H, JiX J (2017) Microbial production of plant hormones: opportunities and challenges. Bioengineered 8(2):124–128

Sims GK (2018) Biological degradation of soil. Advances in Soil Science (book series (SOIL, volume II). Springer Nature, Switzerland AG, pp 289–330

Stancheva I, DimitrovI KaloyanovaN, DimitrovaA AngelovM (1992) Effect of inoculation with *Azospirillum brasilense* on photosynthetic enzyme activities and grain yield in maize. Agron 12:319–324

Stefan M, Munteanu N, Stoleru V, Mihasan M (2013) Effects of inoculation with plant growth promoting rhizobacteria on photosynthesis, antioxidant status and yield of runner bean. Rom Biotechnol Lett 18(2):8132–8143

Taiwo LB, Adesokun KT, Olatoberu FT, Oyedele AO, Ojo AO, Olayinka AA (2018) Effect of Plant Growth Promoting Rhizobacteria (PGPR) and biochar on soil properties and performance of cowpea (*Vigna unguiculata (L.) Walp*). Ife J Agric 30(3) (in press)

Taiwo LB, Ailenokhuoria BV, Oyedele AO (2017) Profiling rhizosphere microbes on the root of maize (*Zea mays*) planted in an Alfisol for selection as Plant Growth Promoting Rhizobacteria (PGPR). Microbiol Res J Int 21(5):1–10

Vedder-Weiss D, Jurkevitch E, Burdman S, Weiss D, Okon Y (1999) Root growth, respiration and β-glucosidase activity in maize (*Zea mays)* and common bean (*Phaseolus vulgaris*) inoculated with *Azospirillum brasilense*. Symbiosis 26:363–377

Vessey JK (2003) Plant growth promoting rhizobacteria as biofertilizers. Plant Soil 255:571–586. https://doi.org/10.1023/a:1026037216893

Vílchez JI, Navas A, González-López J, Arcos SC, Manzanera M (2015) Biosafety test for plant growth-promoting bacteria: proposed Environmental and Human Safety Index (EHSI) protocol. Front Microbiol 6:1514

Zahir A, Muhammad A, Frankenberger WT Jr (2004) Plant growth promoting rhizobacteria: applications and perspectives in agriculture. Adv Agron 81:97–168

Zahir ZA, Akram M, Arshad M, Khalid A (1998) Improving maize yield by inoculation with plant growth promoting rhizobacteria. Pak J Soil Sci 15:7–11

Chapter 8
The Role of Rhizobacterial Volatile Organic Compounds in a Second Green Revolution—The Story so Far

Darren Heenan-Daly, Siva L. S. Velivelli and Barbara Doyle Prestwich

Abstract The role of microbial-emitted volatiles (mVOCs) also termed 'info-chemicals' in agriculture is an emerging area of research with many perceived attributes including but not limited to the alleviation of abiotic and biotic stress factors. Several reports in the literature to date have demonstrated the potential of these mVOCs in plant growth-promotion and disease-suppression, albeit mainly under artificial conditions. The mVOCs are low molecular mass compounds with a high vapour pressure and low boiling point and through diffusion can affect a response over a long distance both above and below ground. They belong to many different classes of chemicals that include terpenes, alcohols, alkenes and ketones amongst others. This review examines recent literature in this area and cites examples of mVOCs, or more particularly; bacterial-derived volatile compounds hereby referred to as 'BVCs', that have plant growth promoting and biocontrol effects. The multifaceted role of BVCs can be viewed as an integral part of a second green revolution in agriculture where alternative environmentally-friendly solutions are being sought for crop protection and bio-stimulation. Their ability to modulate plant photosynthetic and ISR pathways may provide the agricultural sector with more sustainable solutions for increased crop protection and production in the face of increasing climate and population changes.

Keywords PGPR · Volatile organic compounds · BVCs · Biocontrol

D. Heenan-Daly · B. D. Prestwich (✉)
School of Biological Earth and Environmental Science, University College Cork,
Butler Building, Distillery Fields, North Mall Campus, Cork, Ireland
e-mail: b.doyle@ucc.ie

D. Heenan-Daly
e-mail: 108590052@umail.ucc.ie

S. L. S. Velivelli
Donald Danforth Plant Science Center, 975 North Warson Road, 63132 St Louis, MO, USA
e-mail: svelivelli@danforthcenter.org

D. K. Maheshwari and S. Dheeman (eds.), *Field Crops: Sustainable Management by PGPR*, Sustainable Development and Biodiversity 23,
https://doi.org/10.1007/978-3-030-30926-8_8

8.1 Introduction

The plant rhizosphere is a highly competitive environment with plants releasing as much as 40% of their photosynthetic carbon through their roots with these secreted nutrients enabling the creation of a 'hot-spot' of microbial activity (including rhizobacterial activity) (Kai et al. 2016). Rhizobacteria enhance plant growth through a number of different mechanisms, some known, others as yet unknown (Velivelli et al. 2015). Over the past four decades these microbes have been commonly referred to as 'Plant Growth-Promoting Rhizobacteria' (PGPR) (Kloepper and Schroth 1978; Ryu et al. 2005a). Different mechanisms are involved in the enhancement of plant growth by rhizobacteria. Some of these mechanisms are termed direct, others indirect. Examples of direct mechanisms can include the biosynthesis of chemicals analogous to plant hormones involved in the plant growth process such as indole-3-acetic acid (Shao et al. 2015). The optimisation of plant nutrient-uptake is also facilitated through phosphorus solubilisation (Oteino et al. 2015), nitrogen fixation (Singh 2014) and/or by modulating the levels of ethylene in the plant through the activity of enzymes such as aminocyclopropane-1-carboxylic acid (ACC) deaminase (Glick 2014). Indirect mechanisms include the synthesis of non-volatile antibiotics such as pyoluterin, surfactin and fengycin (Dimkić et al. 2017); competition for nutrients mediated by siderophore production for enhanced iron-uptake from soil (Ahmed and Holmström 2014); the secretion of lytic enzymes (e.g. chitinase, β-1,3-glucanase) and the modulation of plant immunity via activation of the induced systemic resistance (ISR) pathway (Compant et al. 2005; Lugtenberg and Kamilova 2009; Velivelli et al. 2015; Tahir et al. 2017a) Fig. 8.1. In recent years, there has been increased interest in the effects of rhizobacterial volatiles on plants (Weisskopf et al. 2016). The metabolic activity of the soil microbiota involves the synthesis of a broad variety of infochemicals, of which volatile organic compounds (VOCs) comprise a large proportion (Kanchiswamy et al. 2015; Velivelli et al. 2015). VOCs are characterised as having a relatively low molecular weight (<300 Da), a low boiling point and high vapour pressure (Vespermann et al. 2007; Velivelli et al. 2014). Microbial-emitted volatiles (mVOCs) belong to a number of different chemical classes including, but not limited to; alcohols, ketones, alkenes and terpenes (Schulz-Bohm 2017) and to date bacteria have been found to produce over 1000 VOCs (Sharifi and Ryu 2018a). The profile of VOCs emitted depends to a large extent on the external environment, be that soil properties or media components (Fincheira and Quiroz 2018). Infochemicals are of great importance, because volatiles can facilitate both the intra and inter-kingdom interaction between many organisms including plants and microbes (Farag et al. 2017). Due to their capacity to disperse in the atmosphere and to circulate through permeable soil structures, volatiles can exert their effects on plants above and below ground (Sharifi and Ryu 2018a). There are two types of VOCs—organic (e.g. 2,3-butanediol) and inorganic (e.g. HCN, CO_2). The complex blends of VOCs that rhizobacteria are capable of generating have been the focus of numerous studies over the past decade (Yuan et al. 2017; Song and Ryu 2018; Tahir et al. 2017b; Blom et al. 2011a) These VOC blends can have beneficial or detrimental effects on the growth of plants, fungi and other associated organisms within the environment of the respective VOC-emitter (Effmert et al. 2012) Table 8.1. Exposure to BVCs can enhance plant growth under certain conditions, but can induce phytotoxic effects in others (Rath et al. 2018). Different blends of volatiles have been implicated in seed germination, flowering time and number, and in fruit and seed production (Sharifi and Ryu 2018a). A variety of chemical signalling molecules are produced by both rhizobacteria and plants when grown together, demonstrating that active communication exists between these kingdoms during plant development (Leach et al. 2017; Farag et al. 2017). For example, microbe-derived compounds are detected by plants, which can then adapt their defence and growth responses to specific types of microorganism. Furthermore, for the exploitation of BVCs in agronomical contexts, we need to determine both their activity and validity-for-use in the field (Rosier et al. 2018). Therefore, it is essential to have a comprehensive understanding of the biocontrol mechanisms of these agents so as to facilitate efficient and effective agronomic application.

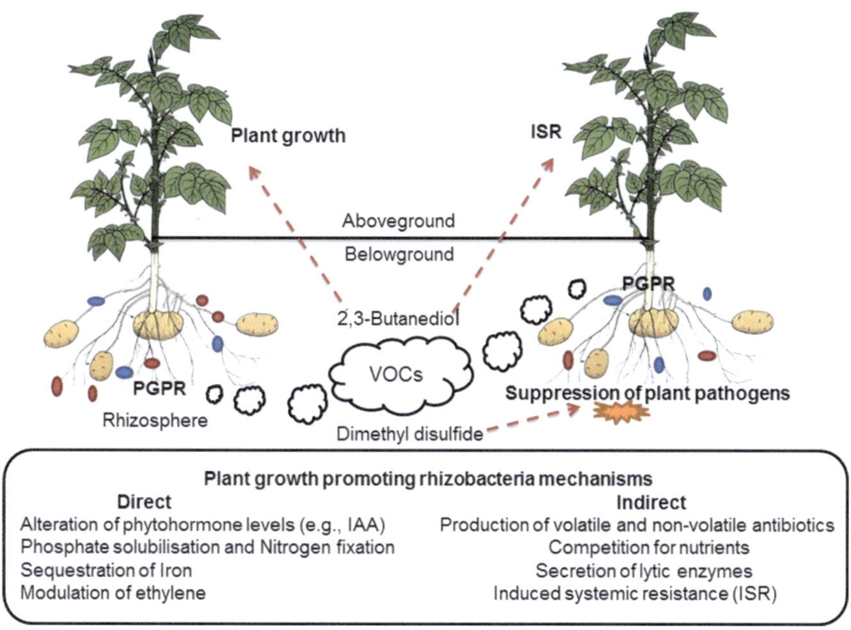

Fig. 8.1 Representation of interactions of bacterial volatile organic compounds (VOCs) on plants (reproduced with permission from Velivelli et al. 2014)

Table 8.1 List of some of volatile-mediated effects of bacteria on plants, fungi and nematodes

Bacteria	Volatile metabolites	Functions	References
B. subtilis GB03,	2,3-butanediol	Growth promotion and ISR	Ryu et al. (2003, 2004); Song et al. (2019)
B. amyloliquefaciens IN937a *P. chlororaphis* O6	2,3-butanediol	Protection against drought stress	Cho et al. (2008)
B. subtilis FB17	Acetoin	Induced systemic resistance (ISR)	Rudrappa et al. (2010)
B. megaterium XTBG34	2-pentylfuran	Growth promotion	Zou et al. (2010)
A. agilis UMCV2	Dimethylhexadecylamine	Growth Promotion	Velázquez-Becerra et al. (2011)
P. polymyxa E681	Tridecane	Induced systemic resistance (ISR)	Lee et al. (2012)
B. ambifaria	Dimethyl disulphide, Acetophenone, 3-hexanone and 2,5-dimethyl pyrazine	Growth promotion	Groenhagen et al. (2013)
P. vulgaris JBLS202	Indole	Growth promotion	Yu and Lee (2013)
S. plymuthica HRO-C48	Dimethyl disulphide	Fungal growth inhibition	Muller et al. (2009)
Bacillus sp.	Acetoin	Fungal pathogen inhibition	Arrebola et al. (2010)
Flavobacterium sp. GSE09 *Lysobacter enzymogenes* ISE13	2,4-di-tert-butylphenol	Fungal growth inhibition	Sang et al. (2011)
P. polymyxa BMP-11	1-octen-3-ol,benzothiazole and citronellol	Fungal growth inhibition	Zhao et al. (2011)
X. campestris pv. *vesicatoria* 85-10	Decan-2-one	Fungal growth inhibition	Weise et al. (2012)
B. ambifaria	Dimethyl disulphide, 2-nonanone, 1-phenyl-1,2-propanedione, 2-undecanone, dimethyl trisulfide, 4-octanone, S-methylmethanethiosulphonate and acetophenone	Fungal growth inhibition	Groenhagen et al. (2013)

(continued)

Table 8.1 (continued)

Bacteria	Volatile metabolites	Functions	References
A. agilis UMCV2	Dimethylhexadecylamine	Fungal growth inhibition	Velázquez-Becerra et al. (2013)
B. atrophaeus CAB-1	O-anisaldehyde	Fungal growth inhibition	Zhang et al. (2013)
B. megaterium YMF3.25	Benzeneacetaldehyde, 2-nonanone, decanal, 2-undecanone and dimethyl disulphide	Nematicidal activity	Huang et al. (2010)
Streptomyces spp.,	Caryolan-1-ol	Fungal growth inhibition	Cho et al. (2017)
P. flourescens SS101	13-Tetradecadien-1-ol	Plant growth promotion	Park et al. (2015)
B. subtilis SYST2	Albuterol 1,3-propanediol	Plant growth promotion and ISR	Tahir et al. (2017a, b)
B. amyloliquifaciens FZB42, *B. artrophaeus* LSSC22	1,2-Benzisothiazol-3(2 H)-one, Benzaldehyde, 1,3-butadiene	Bacterial growth/ motility inhibition and ISR	Tahir et al. (2017c)
S. odorifera 4Rx13	Carbon dioxide	Growth promotion	Kai and Piechulla (2009)

8.2 Rhizobacterial Volatiles and Plant Growth

The direct physical interaction between rhizobacteria and their respective plant host underpins most PGPR-plant interactions. However, an emerging field over the past number of years has examined the long-distance relationships which plants and rhizobacteria can achieve via the medium of VOCs (Velivelli et al. 2014). The first observation of this phenomenon was by Choong-Min Ryu and co-workers and this has led to the opening of many new avenues in the field of plant–microbe interactions. Where physical contact with the plant is not possible, certain rhizobacteria rely on the production of BVCs, a classic example being the volatile alcohol, 2,3-butanediol or more specifically its stereoisomer '2R, 3R- butanediol' to stimulate plant development or activate ISR (Ryu et al. 2003, 2004; Lee et al. 2012; Fincheira and Quiroz 2018) Fig. 8.1. The identification of various plant growth promoting infochemicals and the determination of their structures and their associated functions have been ground-breaking moments in the study of plant-microbe interactions and pinning down the roles of VOCs in the intricate signalling systems between plants and rhizobacteria has been the key area of interest among a number of research groups (Bailly and Weisskopf 2017; Sharifi and Ryu 2018a).

Nevertheless, the contribution of rhizobacterial VOCs to plant development and the importance of these compounds in agricultural systems are still topics of significant debate and speculation. In addition, only a handful of VOCs that are secreted by rhizobacteria have been identified to date. Therefore, a comprehensive understanding of the biological and ecological functions of BVCs—let alone a full understanding of their potential uses—has yet to be achieved.

However, according to Fincheira and Quiroz (2018), mVOCs can influence plant growth in at least four ways: 1. Modulation of nutrients; 2. Alteration of hormone levels; 3. Influencing plant metabolism and 4. Changing sugar concentrations.

Most importantly, future studies will need to address the types of responses and signalling cascades that are induced in plants by BVCs. Research focusing on the effects of plant exposure to different BVCs has uncovered a wide range of effects, including significant plant growth, the induction of ISR and even plant phytotoxicity. In particular, 2,3-butanediol (Ryu et al. 2003, 2004) dimethylhexadecylamine (Velázquez-Becerra et al. 2011), 2-pentylfuran (Zou et al. 2010), indole (Blom et al. 2011a; Yu and Lee 2013; Fincheira and Quiroz 2018) and dimethyl disulphide (DMDS) (Groenhagen et al. 2013), are amongst some of the infochemicals that have been shown to increase plant growth, whereas negative effects are at least in part due to the presence of high levels of hydrogen cyanide (HCN) (Blom et al. 2011b), DMDS and ammonia (Kai et al. 2010; Weise et al. 2013). To determine the extent to which BVCs stimulate plant growth, Choong-Min Ryu and co-workers used two-compartment Petri dishes, hereby by referred to as 'I-plates', to physically separate *Arabidopsis thaliana* from rhizobacteria under laboratory conditions. In this way, the dispersal of non-volatile metabolites through the medium was prevented, allowing for only the exchange of volatile organic compounds.

The researchers observed that plant growth was most strongly stimulated by the bacterial strains *Bacillus subtilis* GB03 and *Bacillus amyloliquefaciens* IN937a. When the volatiles produced by these two strains were examined, it was found that these bacteria produced the compounds 3-hydroxy-2-butanone (acetoin) and 2, 3-butanediol, which were not detected in bacterial strains that were unable to induce volatile-mediated plant growth. The exogenous application of these two compounds in pure solutions induced similar effects in a dose-dependent manner, and *Bacillus* spp. mutants defective in 2,3-butanediol and acetoin synthesis showed no plant growth-promotion, which confirmed the role of these compounds in mediating plant growth. A set of hormonal mutant *A. thaliana* lines impaired in specific regulatory pathways was then tested to identify the signalling networks necessary for these growth-promoting activities. It was found that exposure to volatiles from the *B. subtilis* GB03 strain did not promote growth in cytokinin receptor-deficient (*cre1*) or cytokinin/ethylene-insensitive (*ein2*) mutants. On the other hand, the *B. subtilis* GB03 volatiles did promote growth in ethylene-insensitive (*etr1*), auxin-transporter-deficient/ethylene-insensitive (*eir1*), gibberellic acid-insensitive (*gai2*), and brassinosteroid-insensitive (*cbb1*) mutants, suggesting that the promotion of growth elicited by GB03 VOCs is mediated by the cytokinin-signalling pathway (Ryu et al. 2003). Further experiments demonstrated that disease severity caused by the necrotrophic bacterial pathogen *Erwinia carotovora* subsp. *carotovora* was significantly decreased when *A. thaliana* seedlings were exposed to VOCs produced by *B. subtilis* GB03 and *B. amyloliquefaciens* IN937a. This phenomenon, called 'induced systemic resistance' (ISR), occurred in as little as 4 days. The exogenous application of pure 2,3-butanediol induced similar effects in a dose-dependent manner. Seedlings exposed to *Bacillus* mutants defective in 2,3-butanediol synthesis showed no disease protection, which confirmed the priming activity of this compound in ISR-induction. A set of mutant *A. thaliana* lines impaired in specific regulatory pathways, including a jasmonic acid (JA)-insensitive (*coi1*), an ethylene-insensitive (*ein2*), a salicylic acid (SA)-degrading line (NahG), and a line that is SA-insensitive or non-expressor of pathogenesis-related (PR) genes (*npr*1), was then tested to identify the signalling networks necessary for ISR. Pre-exposure to VOCs from *B. subtilis* GB03 did not trigger ISR in the ethylene-insensitive line (*ein2*) and did not show pathogen resistance. In addition, to further test whether these VOCs induced known signalling pathways in *A. thaliana*, transgenic plants with β-glucuronidase (GUS) fusions to *Pr-1a* (a gene activated by SA), *Pdf1.2* (a gene activated by JA and ethylene), and *Jin14* (a gene activated by JA) were exposed to VOCs released by *B. subtilis* GB03 (Ryu et al. 2004). Of these lines, the JA/ethylene-activated *Pdf1.2*-GUS line showed increased GUS activity compared with untreated control plants.

The plants carrying an ectopic copy of the JA-activated *Jin14* gene were unaffected by *B. subtilis* GB03 VOCs; thus, ethylene signalling may be required for the activation of ISR in *A. thaliana,* independently of the SA and JA signalling pathways. Surprisingly, the VOCs of *B. amyloliquefaciens* IN937a functioned independently of all of the signaling pathways, indicating that some VOCs utilised alternative pathways to trigger ISR (Ryu et al. 2004), and are still not known. Meanwhile, a proteomics study revealed that ethylene biosynthetic enzymes were significantly up-regulated in *A. thaliana* plants exposed to *B. subtilis* GB03 VOCs, and a transcriptomic analysis showed the up-regulation of genes related to ethylene biosynthesis (*SAM-2*, *ACS4*, *ACS12* and *ACO2*) as well as ethylene response genes (*ERF1*, *CHIB* and *GST1*) following exposure to *B. subtilis* GB03 VOCs (Kwon et al. 2010; Velivelli et al. 2015). It was demonstrated that the long-chain volatiles produced by *Paenibacillus polymyxa* E681 stimulated plant growth in *A. thaliana* and induced systemic resistance against *Pseudomonas syringae* pv. *maculicola* ES4326 (Lee et al. 2012; Velivelli et al. 2014).

A set of hormonal mutant *A. thaliana* lines impaired in specific regulatory pathways was tested to identify the signalling networks necessary for plant growth promoting activities. Lee et al. (2012) found that exposure to volatiles from *P. polymyxa* E681 did not promote growth in cytokinin/ethylene-insensitive (*ein2*) mutants. On the other hand, the *P. polymyxa* E681 volatiles did promote growth in jasmonic acid-insensitive line (*coi1*), a salicyclic acid-degrading line (NahG), and gibberellic acid-insensitive line (*gai2*) mutants, indicating that cytokinin/ethylene signalling is essential for the promotion of plant growth in response to *P. polymyxa* E681 volatiles. Further experiments demonstrated that the severity of the disease caused by the hemibiotrophic bacterial pathogen *Pseudomonas syringae* pv. *maculicola* ES4326 was significantly decreased where *A. thaliana* seedlings were pre-exposed to VOCs produced by *P. polymyxa* E681. In addition, to further test whether *P. polymyxa* E681 VOCs induced known signalling pathways in *A. thaliana*, transgenic plants with β-glucuronidase (GUS) fusions to *Pr-1a* (a gene activated by SA), and *Pdf1.2* (a gene activated by JA and ethylene), were exposed to VOCs released by *P. polymyxa* E681. Of these lines, the SA-activated *Pr-1a*-GUS line showed increased GUS activity compared with untreated control plants; thus indicating SA signalling may be required for the activation of ISR. A further transcriptomic study of *A. thaliana* exposed to VOCs from *P. polymyxa* E681 followed by pathogen challenge revealed the induction of salicylic acid (SA), jasmonic acid (JA) and ethylene (ET) signalling marker genes, *PR1*, *ChiB*, and *VSP2*, respectively. When the volatiles produced by *P. polymyxa* E681 were examined, it was found that this rhizobacteria produced the long-chain volatile compound tridecane. The exogenous application of pure tridecane induced *PR1* and *VSP2* in a dose-dependent manner after pathogen challenge; thus indicating that SA/ET signalling is essential for the activation of ISR in response to tridecane. The researchers performed additional tests to demonstrate whether the observed growth promotion of *A. thaliana* when exposed to *P. polymyxa* E681 was correlated to carbon dioxide (CO_2). Plant growth was still enhanced when exposed to barium hydroxide ($Ba(OH)_2$) which traps CO_2, indicating that some other unknown VOCs are involved in the promotion of growth (Lee et al. 2012; Jeong et al. 2019).

In addition to the impact of BVCs, it has been speculated that increased plant growth could be due to the increase in CO_2 concentrations that is seen to rise when using the sealed petri dish method. Based on co-cultivation studies involving *Arabidopsis* and *Serratia odorifera* in a closed system, it was observed that growth promotion was closely linked to carbon dioxide enrichment. In particular, the growth of *A. thaliana* was stimulated in a closed system, where carbon dioxide was the dominant component of the volatile mixture (390–3000 ppm). By contrast, in an open system under ambient carbon dioxide concentrations, volatiles with negative effects on plant growth became dominant. It is possible that activation of the tricarboxylic acid cycle (TCA) triggers the emission of carbon dioxide, although this molecule does not accumulate to higher-than-ambient concentrations in an open system (Kai and Piechulla 2009). Similarly, the growth of *Physcomitrella patens* (moss) was stimulated in a closed system in which carbon dioxide was the dominant component of an *S. odorifera*-derived volatile mixture, growth of this moss was inhibited in an open system due to the negative influence of volatiles (Kai and Piechulla 2010). At most, high levels of carbon dioxide have been observed to increase plant biomass by as much as 25%, although this was primarily the result of increased starch accumulation rather than biomass expansion. However, it is likely that volatiles are present at much higher concentrations in closed systems than in any found under natural circumstances (Van der Kooij et al. 1999; Ward and Strain 1999; Blom et al. 2011a).

In another study, C16 hexadecane, a long-chain hydrocarbon emitted by *P. polymyxa* E681, also protected *Arabidopsis* plants from infection by the necrotrophic pathogen *Pectobacterium carotovorum* and the hemibiotrophic pathogen *Pseudomonas syringae* pv. *maculicola* ES4326 (Park et al. 2013). Certain volatiles produced by rhizobacteria regulate plant auxin homeostasis, and can promote growth in *Arabidopsis*. Genes for auxin biosynthesis were up-regulated when aerial parts of the plant were exposed to *B. subtilis* GB03. As observed in a transgenic (DR5:auxin-responsive reporter) *Arabidopsis* line expressing a DR5:: GUS fusion, exposure to *B. subtilis* GB03 volatiles induced a decrease in auxin accumulation in leaves and an increase in roots, indicative of basipetal auxin transport activation. The decrease in auxin accumulation in leaves resulted in enhanced leaf cell elongation, whereas the increase in auxin accumulation in roots led to the development of lateral roots. Thus, despite the fact that auxin is not produced by *B. subtilis* GB03, auxin signalling must be present in the root architecture response elicited by one or more BVC produced by *B. subtilis* GB03. Auxin accumulation to the sites of synthesis was impeded by the application of the auxin transport inhibitor 1-napthylphthalamic acid (NPA), which prevented the *B. subtilis* GB03-induced reduction in shoot auxin levels and the associated growth promotion. Moreover, modifications in cell wall-loosening were observed during transcriptional analysis, which might explain the accelerated cell expansion and leaf growth associated with exposure to *B. subtilis* GB03 volatiles (Zhang et al. 2007). Further experiments revealed that exposure to *B. subtilis* GB03 volatiles caused *Arabidopsis* to increase both its photosynthetic activity and chlorophyll content.

Exposure to BVCs also enhanced endogenous sugar accumulation and also led to the partial suppression of sugar sensing in plants. In contrast to wild-type plants, enhanced photosynthetic capacity (that was not additionally increased by exposure to *B. subtilis* GB03) was observed in the two glucose-insensitive *Arabidopsis* mutants, *gin1* and *gin2*, which lack hexokinase-dependent sugar signalling. Photosynthesis is promoted by BVCs through repression of the hexokinase-dependent sugar signalling pathway. Exposure to *B. subtilis* GB03 causes an overlap in sugar/ABA sensing in plants, as ABA-synthetic transcripts, ABA-responsive genes and ABA levels in leaves become reduced. Furthermore, the increase in photosynthetic efficiency and chlorophyll content induced by *B. subtilis* GB03 can be abolished by exogenous ABA treatment. Therefore, to enhance photosynthesis, some modulate endogenous sugar/ABA signalling and use soil symbionts as regulators of energy procurement by plants (Zhang et al. 2008b). Common photosynthetic markers, which are enhanced by high carbon dioxide levels, include increase of photosynthetic efficiency, chlorophyll content, and sugar accumulation (Kai and Piechulla 2009). Indeed, *Arabidopsis* exhibited enhanced photosynthetic capacity (e.g. chlorophyll content) and iron accumulation following protracted exposure to *B. subtilis* GB03 (Xie et al. 2009). Under normal growth conditions, *B. subtilis* GB03 volatiles triggered a rise in the mRNA levels of Fe-deficiency-induced transcription factor 1 (FIT1), as well as two of its downstream targets, ferric reductase *FRO2* and the iron transporter gene *IRT1*. However, in *Arabidopsis fit1-2* knockout mutants, volatile-induced increases in iron assimilation and photosynthetic efficiency are impaired, suggesting that volatile-induced iron assimilation is mediated by FIT1 (Zhang et al. 2009). To stimulate iron assimilation, *Sinorhizobium meliloti* VOCs induced acidification of the *Medicago truncatula* rhizosphere, which triggers enhanced photosynthetic activity (e.g.

chlorophyll content), an indicator of nutritional Fe status in plants (Carmen Orozco-Mosqueda et al. 2013). *B. subtilis* GB03 volatiles enhance salt tolerance through the sodium transporter HKT1, which is down-regulated in roots and up-regulated in shoots, resulting in a plant-wide reduction in Na+ accumulation versus control plants not exposed to GB03 volatiles (Zhang et al. 2008a). It was demonstrated that tolerance to abiotic stress could be induced in *Arabidopsis* by *Pseudomonas chlororaphis* O6 and the results suggested that this phenomenon was largely due to production of the volatile compound 2,3-butanediol. A *P. chlororaphis* O6 mutant defective in 2,3-butanediol production showed no drought resistance upon bacterial colonisation, which confirmed the role of this compound in induction of drought tolerance.

Interestingly, it was shown that 2,3-butanediol produced by this rhizobacterium facilitated both stomatal closure and drought tolerance through an ABA-1- and OST-1 kinase-dependent manner. A set of mutant *A. thaliana* lines impaired in specific regulatory pathways, including a mutant with reduced ABA synthesis (*aba1*) and a mutant deficient in the protein kinase mediating stomatal regulation in response to drought (*ost-1*) showed no drought tolerance upon *P. chlororaphis* O6 root colonisation. When drought-stressed plants were exposed to 2,3-butanediol, the plants accumulated greater levels of SA than unexposed plants, indicating that the SA signalling pathways are involved in *P. chlororaphis* O6-induced drought tolerance (Cho et al. 2008). *Arbidopsis* plants treated with *Bacillus subtilis* FB17 significantly reduced the severity of the disease caused by the hemibiotrophic bacterial pathogen *Pseudomonas syringae* pv. tomato DC3000. This phenomenon, called 'induced systemic resistance' (ISR), also occurred when *Arabidopsis* plants were exposed to acetoin. *B. subtilis* FB17 mutants defective in acetoin biosynthesis showed reduced disease protection, and this result confirmed the priming activity of this compound in ISR. A set of mutant *A. thaliana* lines impaired in specific regulatory pathways, including a jasmonic acid (JA) mutant (*jar1-1*), an ethylene mutant (*etr1-3*), a salicylic acid (SA) deficient mutant (*NahG*), and a line that is SA-insensitive or non-expressor of pathogenesis-related (PR) genes (*npr1-1*), was then tested to identify the signalling networks necessary for ISR. Of these lines, treatment with *B. subtilis* FB17 and acetoin did not trigger ISR in the *etr1-3*, *NahG* and *npr1-1* lines and did not show the pathogen resistance against *Pseudomonas syringae* pv. tomato DC3000; thus indicating that ISR elicitation is mediated via NPR1 and SA/ET signalling pathways to activate ISR in *Arabidopsis* independently of JA signalling pathway (Rudrappa et al. 2010).

In species such as *B. subtilis,* 2,3-butanediol synthesis is mediated by the transformation of pyruvate into acetocholate by the enzyme 'acetocholate synthase'. Following this, acetocholate decarboxylase converts alpha-acetocholate into acetoin which is subsequently converted to 2,3-butanediol via catalysis mediated by the acetoin reductase/2,3-butanediol dehydrogenase (AR/BDH) (Nicholson 2008).

The conversion of glucose into 2,3-butanediol and acetoin occurs under hypoxic conditions and serves as an electron sink for the generation of NAD+ when aerobic respiration is restricted. Furthermore, low partial oxygen pressure (as generally exists in soil surrounding roots) induces the bacterial acetoin pathway that controls the production of 2,3-butanediol. Consequently, it is likely that 2,3-butanediol and/ or other biologically active molecules are produced by certain root-colonising rhizobacteria at concentrations appropriate to elicit plant reactions (Ryu et al. 2003, 2004). The Methyl Red Voges Proskauer (MR-VP) medium is generally used to determine the ability of bacteria to ferment 2,3-butanediol (Nicholson 2008). As a by-product of the fermentation pathway employed by some rhizobacteria to avoid acidification, the biosynthesis of 2,3-butanediol is often induced on low pH Murashige and Skoog (MS) medium containing sucrose (Ryu et al. 2003). The growth of *Penicillium* spp. was suppressed both in vitro and in vivo by volatiles released by *Bacillus* spp., and citrus fruit inoculated with *Penicillium crustosum* showed reduced disease incidence and severity due to the presence of acetoin. Furthermore, it was observed that longer exposure times led to stronger volatile-mediated antifungal effects; this was attributed to the extended incubation period within the closed system, leading to the restriction of oxygen over time (Arrebola et al. 2010).

On the other hand, it was also demonstrated that the acetoin pathway is optimal in the lifecycle of the necrotrophic bacterial pathogen *Pectobacterium carotovorum* subsp. *carotovorum* WPP14. Mutants defective in the 2,3-butanediol pathway were unable to alkalinise growth media and also showed reduced virulence on potato tubers (Marquez-Villavicencio et al. 2011). Furthermore, the capacity of *Bacillus megaterium* XTBG34 to promote growth in *A. thaliana* was validated by Zou and Co-workers. In particular, a number of compounds produced by this organism, including 2-pentylfuran, were identified by GC/MS analysis, and they showed that plant growth was significantly enhanced by 2-pentylfuran in a dose-dependent manner. The lowest dose at which this compound could enhance plant growth was 0.1 μg, and maximum growth was achieved with 10 μg; by contrast, doses greater than 10 μg inhibited growth (Zou et al. 2010). Santoro and Co-workers examined the effects of BVCs on growth promotion and the enhanced biosynthesis of essential oils (EO), such as pulegone and menthone in *Mentha piperita* (peppermint). The results of this study indicated that BVCs exhibit species-specific effects on plants. BVCs not only trigger secondary metabolite production but also impact pathway flux during certain stages of monoterpene metabolism (Santoro et al. 2011). It was demonstrated that volatiles released by *Proteus vulgaris* JBLS202 stimulate growth in Chinese cabbage and GC/MS analysis showed that indole was the primary headspace volatile compound produced by this bacteria. They showed that plant growth was significantly enhanced by indole in a dose-dependent manner. When plants were exposed to 0.63 μg of synthetic indole, growth was significantly enhanced. Indole and its derivatives are known to be involved in the auxin signalling pathway (Yu and Lee 2013).

Blom and colleagues analysed the effects of different bacterial strains cultured on four distinct media on the growth of *A. thaliana*. Of the bacterial strains tested, one strain promoted growth on all four media tested. GC/MS analysis revealed the presence of a range of compounds, including indole, 1-hexanol and pentadecane, and they showed that plant growth was affected in a concentration-dependent manner by these compounds. Indole promoted growth at low concentrations but showed lethal effects when used at high concentrations. Furthermore, when 1-hexanol was applied in moderate amounts, it showed weak growth promotion, and pentadecane promoted growth when applied at high concentrations (Blom et al. 2011a).

In another study, rhizobacterial strains were isolated from the rhizosphere of lemon plants (*Citrus aurantifolia*) and then analysed to determine whether their VOCs had an effect on the development of *A. thaliana* roots. Using a simple experimental system involving I-plates, the authors observed several morphological changes in root architecture due to VOCs. It is interesting to note that some rhizobacterial strains stimulated primary root growth and lateral root development. Several compounds were detected by GC/MS, including aldehydes, ketones and alcohols. However, short-chain alcohols, such as 2,3-butanediol and acetoin were not identified in this study, indicating that other VOCs can also trigger plant growth (Gutiérrez-Luna et al. 2010). Exposure to *A. thaliana* plants with *Bacillus megaterium* UMCV1 modified the architecture of the root system in this plant. In particular, the authors observed an inhibition of primary root growth as well as increases in lateral root number, lateral root growth, and root hair length, and they found that reduced cell elongation and cell proliferation in the root meristem was the cause of the inhibition of primary root growth. The analysis of *Arabidopsis* mutant lines defective in either ethylene (*etr1 and ein2*) or auxin (*aux1-7*, *axr4*, *eir1*) signalling revealed that, modifications in root architecture caused by *B. megaterium* UMCV1 may involve either auxin- or ethylene-independent mechanisms. Furthermore, transgenic *Arabidopsis* line expressing a DR5:*uidA* (a reporter line for auxin and ethylene-inducible gene expression) GUS fusion showed reduced expression in root tips (López-Bucio et al. 2007).

Velázquez-Becerra and Co-workers tested the effects of *Arthrobacter agilis* UMCV2 volatiles on alfalfa (*Medicago sativa*), and they found that *A. agilis* volatiles decreased taproot growth and increased lateral root formation, indicating that the BVCs emitted by this bacterium play an important role in root development. Analysis of BVCs produced by this organism revealed a range of compounds, at least one of which, N-N-dimethyl-hexadecanamine, may act as a growth-promoting trigger, affecting root development in *Medicago sativa* in a dose-dependent manner (Velázquez-Becerra et al. 2011).

A study by Tahir and colleagues determined the effect of BVCs from *Bacillus subtilis* SYST2 on tomato was examined. Two compounds, albuterol and 1,3-propanediol were identified as having a positive effect on plant growth with observed increases in auxin and cytokinin in the plant tissues and noticeable increases in expansin gene transcripts. Variations in VOC concentrations and/or plant exposure times can dictate whether an inoculation has positive or negative effects on primary root growth and/or lateral root formation (Tahir et al. 2017b). *Arabidopsis* plants exposed to *Burkholderia ambifaria* volatiles show enhanced biomass, greater numbers of secondary roots and shorter main roots. Analysis of the VOCs produced by this bacterium revealed a range of compounds, including dimethyl disulphide (DMDS), acetophenone and 3-hexanone, 4-methyl-2-pentanone, 4-octanone, and 2,5-dimethyl pyrazine. These compounds were shown to affect plant growth in a concentration-dependent manner, and indeed, treatment with very high amounts could inhibit plant growth. Plants showed greater biomass when exposed to some concentrations of dimethyl disulphide (DMDS), acetophenone and 3-hexanone. By contrast, high concentrations of 4-methyl-2-pentanone and 4-octanone were lethal to the plants. Finally, 2,5-dimethyl pyrazine promoted growth at lower concentrations, whereas higher concentrations were deleterious to *Arabidopsis* plants (Groenhagen et al. 2013).

Many studies have shown that BVC interact with the host root system and in addition to its role as a structural support for the plant, it is also crucial for the acquisition of water and nutrients from the soil. From an ecological perspective, BVC-mediated alterations in the root system proteome and root architecture may have beneficial effects through increasing bacterial root colonisation and optimising symbiotic interactions (Yaoyao et al. 2017). These symbiotic relationships mediate biochemical interactions which stimulate root growth and development which leads to enhanced levels of bioavailable nutrients for the plant (Velivelli et al. 2015; Hérnandez-Calderón et al. 2018). In reciprocation, PGPR gain access to richer sources of nutrition and carbon through root exudates produced by a healthy plant host (Gutiérrez-Luna et al. 2010; Velivelli et al. 2015).

8.3 Inhibitory Effects of Rhizobacterial Volatiles on Plants

In addition to promoting plant growth, it has been demonstrated that certain BVC have negative effects on plants such as *A. thaliana* (Vespermann et al. 2007). One of the most important sources of nitrogen is ammonia, although this compound has recently been shown to play a number of other biological roles. When *S. odorifera* 4Rx13 is grown on a peptone-rich medium, it produces high levels of ammonia, and when this plant was exposed to *A. thaliana* plants in an I-plate, the bacterium caused the neighbouring plant medium to become alkalinised, leading to reduced plant growth (Weise et al. 2013). Under conditions of low oxygen, such as in a closed system, the production of HCN by some *Pseudomonas* sp. is enhanced (Athukorala et al. 2010). Deleterious effects were observed when *A. thaliana* was exposed to HCN, with a four-fold reduction in growth following exposure to 1 μmol HCN (Blom et al. 2011b). In has also been shown that rhizobacterial volatile compounds such as ammonia and DMDS have negative effects in higher concentrations on *A. thaliana* growth (Kai et al. 2010).

8.4 Effect of Rhizobacterial Volatiles on Fungi and Other Organisms

For truly sustainable agriculture, the strategies we employ to combat plant diseases must become more environmentally-friendly with lower inputs of synthetic chemicals. The use of beneficial microbes as a biological input to sustainable agricultural systems offers an alternative, and potentially more environmentally stable approach, to conventional agri-chemical-based solutions for the suppression of plant pathogens and the treatment of plant diseases in an integrated pest management system (Velivelli et al. 2015). Despite their powerful antifungal activities, non-volatile antibiotics are unable to spread over long distances, making them only effective at preventing infection by pathogenic microbes/fungi when applied directly to plant roots. The ability of BVC to suppress the growth and proliferation of plant pathogens has attracted ample attention with regard to biological applications and rhizobacterial VOCs have displayed antagonistic activity against pathogenic fungi, which may classify them as novel antibiotic compounds (Velivelli et al. 2015). The best known example of one such inorganic volatile metabolite is hydrogen cyanide (HCN).

Hydrogen cyanide (HCN) is a secondary metabolite produced by some gram-negative *Pseudomonas* spp. upon the hydrolysis of glycine by HCN synthase. *Pseudomonas fluorescens* CHA0 was shown to inhibit the development of *Thielaviopsis basicola*, which causes black root rot in tobacco plants, through the production of HCN (Bailly and Weisskopf 2017). The biocontrol potential of these antibiotics has been experimentally validated through the use of mutant rhizobacterial strains with altered antibiotic production. A hydrogen cyanide negative mutant (*hcn*), *P. fluorescens* CHA0 strain was no longer able to protect tobacco against black root rot (Voisard et al. 1989; Blumer and Haas 2000). A more recent study (Rijavec and Lapanje 2016) proposed that the main contribution of HCN to biocontrol is more indirect and is related to the sequestration of metals and the associated beneficial increase of nutrients to the plant and rhizobacteria.

The volatile inorganic compound ammonia, which is released by the rhizobacteria *Enterobacter cloacae*, suppressed the growth of *Pythium ultimum* in dual-culture assays, thus describing its possible role in the biological control of *Pythium* pre-emergence damping-off (Howell et al. 1988). In addition to HCN and ammonia, the antifungal nature of the organic volatiles has been demonstrated in several experiments. *Pseudomonas* spp. isolated from canola and soybean plants were reported to produce volatile antibiotics including; n-decanal, nonanal, 2-ethyl-1-hexanol, benzothiazole and dimethyl trisulfide, that inhibit the fungal pathogen *Sclerotinia sclerotiorum* in I-plate assays (Fernando et al. 2005). Furthermore, the growth of *S. sclerotiorum* was also suppressed in antifungal bioassays performed in sealed plates containing pure synthetic volatiles such as furfural, benzaldehyde, 1-octanol, 1-octen-3-ol, 3,7-dimethyl-1-ol, 6-octadien-3-ol, 2-ethyl-1-hexanol (Liu et al. 2009).

The growth of *Fusarium oxysporum* f. sp. *cubense* was suppressed in a divided plate assay by BVC produced by *Bacillus amyloliquefaciens* NJN-6. The volatile organic compounds emitted by this organism were diverse and included; benzothiazole, phenol, 2,3,6-trimethyl-phenol, 2-ethyl-1-hexanol, 2-undecanol, 2-nonanone, 2-decanone, nonanal, naphthalene, naphthalene, 2-methyl and naphthalene 1-methyl. The application of pure, synthetic volatiles in the same bioassay revealed strong antifungal activities against *F. oxysporum* f. sp. *cubense* (Yuan et al. 2012).

Jasmonic acid BVC produced by *B. subtilis* significantly inhibited the spore germination of *B. cinerea* in an I-plate assay. An analysis of VOCs revealed a range of compounds, such 4-Hydroxybenzaldehyde, 2-nonanone, Ammonium acetate, 1,2,4,5-Tetramethyl-pyrazine, 9-Methyl-nonadecane, 2,6,11,15-Tetramethyl-hexadecane, 2,6,10,15-Trimethyl-tetradecane and 8-Hexyl-pentadecane; however, the authors did not evaluate the antagonistic potential of these compounds against *Botrytis cinerea* (Chen et al. 2008). The BVC 2-nonanone showed an inhibitory effect towards *B. cinerea* fungal decay of strawberries in closed containers, thus suggesting its potential role in reducing post-harvest diseases of agricultural products, an area in which there is a significant increase in research activity worldwide (Almenar et al. 2007; Sharifi and Ryu 2018b).

The growth of the soil-borne pathogenic fungi *Rhizoctonia solani* was strongly inhibited in an I plate assay by BVC emitted by a number of common soil bacterial genera such as *Bacillus* spp., *Pseudomonas* spp., *Serratia* spp., and *Stenotrophomonas* spp. Further molecular analysis revealed a wide array of compounds including: β-phenylethanol, trans-9-hexadecene-1-ol, undecene, undecadiene, dodecanal, benzylnitrile, benzyloxybenzonitrile, and dimethyl trisulfide (Kai et al. 2007). Dimethyl disulphide was shown to inhibit the growth of *Fusarium culmorum* in a dual-culture assay and this inhibitory effect on mycelial growth was observed to occur in a dose-dependent manner; less obvious effects were also observed with the use of pure 1-undecene (Kai et al. 2009). Antifungal volatile metabolites produced by *A. agilis* UMCV2 inhibited the growth of *B. cinerea* on sealed plates. The volatile organic compound, dimethylhexadecylamine (DMHDA), which is released by the rhizobacteria *A. agilis*, inhibited the growth of both *B. cinerea* and *P. cinnamomi* in dual-culture assays when provided to the culture medium at low concentrations (Velázquez-Becerra et al. 2013). It has been demonstrated that BVC produced by *Streptomyces plantesis* F-1 could inhibit the growth of *R. solani*, *B. cinerea* and *S. sclerotiorum*. Exposure to *S. plantesis* F-1 BVC significantly reduced the incidence and severity of leaf blight/seedling blight caused by *R. solani*, leaf blight of oilseed rape caused by *S. sclerotiorum* and fruit rot of strawberry caused by *B. cinerea*; thus indicating its possible role as a bio-fumigant in the biological control of fungal diseases. The analysis of volatile organic compounds from this organism revealed diverse compounds, including but not limited to phenylethyl alcohol, phenol, 2,5-bis(1,1-dimethylethyl)-, (+)-epi-bicyclesesquiphellandrene and cyclohexane carboxylic acid; however, the potential role of these compounds remains to be investigated. As suggested by the complex nature of BVC, significant growth inhibition may require the synergistic activity of multiple compounds or the activity of extremely potent infochemicals normally present at low concentrations (Wan et al. 2008). It has been demonstrated that the volatile metabolites produced by fungistatic soils suppress the growth of *Paecilomyces lilacinus*, *Pochonia chlamydosporia* and *Clonostachys rosea*. In particular, VOCs produced by fungistatic soils, including trimethylamine, benzaldehyde, and N,N-dimethyloctylamine elicit strong antifungal activity in a sealed petri plate assay containing known amounts of fungal spore suspension, autoclaved soil and/or pure synthetic compounds (Chuankun et al. 2004).

In an I plate assay, the volatile organic compound, O-anisaldehyde, which is released by *Bacillus atrophaeus* CAB-1 inhibited the growth of *B. cinerea*; thus suggesting its possible role in soil fungistatis and the subject of further study (Zhang et al. 2013). The pathogenic activity of various fungal pathogens was suppressed by volatiles emitted by *P. polymyxa* strain BMP-11. The mycelial growth of various fungal pathogens, including *R. solani*, was suppressed in sealed Petri dish antifungal bioassays involving pure, synthetic volatiles such as 1-octen-3-ol, benzothiazole and citronellol; mycelial morphological deformities were also observed (Zhao et al. 2011). Antifungal metabolites produced by *B. subtilis* inhibited the growth of two strains of *R. solani* in sealed I plate assays. Interestingly, these two fungal strains reacted differently to exposure to the *B. subtilis* BVC mixtures, indicating that BVC-mediated interactions between bacteria and fungi can be species/genus specific. Therefore, it is possible that complex volatile mixtures may trigger significantly different responses in different fungi, which may for example be due to differentially conserved molecular activities of fungi for detoxifying metabolites.

Fungal growth modifications due to BVC exposure are not uncommon and indeed are closely linked to growth medium and inoculum dose, and in soil, volatile emissions have been linked to nutrient availability, pH, temperature and oxygen availability (Schulz-Bohm 2017). Rhizobacterial strains emit volatiles that can have distinct effects on fungi or inhibit different fungal types to varying degrees. This could be due to the fact that different BVC blends are released by various rhizobacteria, and inhibitory effects may be induced due to the synergistic effects of several compounds within those respective blends (Velivelli et al. 2015). For example, the fungistatic BVC emitted by *Bacillus cereus* were shown to more strongly repress a *Trichoderma viride* strain than a *Gelasinospora cerealis* strain. Similarly, the BVC released by a strain of *Aerobacter aerogenes* were not as effective against *F. oxysporum* f. sp. *conglutinans* as they were against *T. viride* and *Penicillium* sp. (Fiddaman and Rossall 1994).

Using an I plate assay, it was shown that exposure to *Burkholderia ambifaria* VOCs reduced the growth of fungi, and this inhibitory effect was observed to be stronger against *Alterneria alternata* than it was for *R. solani*. In particular, the growth of *A. alternata* and *R. solani* was decreased by higher doses of dimethyl trisulphide, 2-nonanone, 1-phenyl-1,2-propanedione, and 2-undecanone. Moreover, *R. solani* was also suppressed by acetophenone, phenylpropan-1-one, DMDS, 4-octanone and S-methyl methanthiosulphonate. The growth of *Fusarium solani* was not reduced by any of the volatiles tested, indicating that fungi react to BVCs differently (Groenhagen et al. 2013; Velivelli et al. 2015).

Bacterial volatile patterns can also be affected by growth media. For example, BVC emission profiles differed based on whether the media (nutrient broth) they were grown on contained glucose. The growth of *R. solani* on nutrient broth (NB) was more strongly suppressed by the volatiles produced from *Xanthomonas campestris* pv. *vesicatoria* 85-10 when the latter was grown on NB than when it was grown on nutrient broth with glucose (NBG). *A. nidulans* and *F. solani* exhibited similar growth patterns. The inhibition of *A. nidulans* was stronger when *X. campestris* pv. *vesicatoria* 85-10 was grown on NB than when it was grown on NBG. When *X. campestris* pv. *vesicatoria* 85-10 was grown on NB, the growth of

A. nidulans and *F. solani* was inhibited by 85% and 14%, respectively, whereas growth inhibition on NBG was only 11% and 3.5%, respectively. The BVCs of *X. campestris* pv. *vesicatoria* 85-10 showed only weak effects on *F. solani*, indicating species-specific activity. These results demonstrate that when grown on glucose-containing media, *X. campestris* pv. *vesicatoria* 85-10 BVCs have only a weak effect on fungi, indicating that nutrient levels influence growth inhibition. There are several potential explanations for the observed reduction in growth suppression by *X. campestris* pv. *vesicatoria* when grown on NBG: (1) growth on NB media stimulates the production of a larger amount of suppressive volatiles; (2) the production of suppressive volatiles relies on peptone-rich NB media; (3) the production of suppressive volatiles is restricted by glucose through a mechanism such as catabolite repression; or (4) there is a delay in the production of suppressive volatiles on NBG media (Weise et al. 2012). Fiddaman and Rossall (1994) observed that *B. subtilis* only produced suppressive BVCs in the presence of D-glucose, and not L-glucose (Fiddaman and Rossall 1994). Previously, Fiddaman and Rossall (1993) had assumed that agar containing high levels of glucose, i.e. PDA (potato dextrose agar) and SGA (Sabouraud's glucose agar) with *B. subtilis* would stimulate significant antifungal activity in vitro. By contrast, limited or no in vitro antifungal activity was produced by agar containing little or no glucose (VJA (V8 juice agar), NA (nutrient agar) or 10% TSA (tryptic soy agar) (Fiddaman and Rossall 1993).

Serratia plymuthica strains emitting DMDS as the primary headspace BVC, were seen to inhibit the growth of *Agrobacterium tumefaciens* and *Agrobacterium vitis* strains in dual-culture assays. When *Solanum lycopersicum* plants were inoculated with *S. plymuthica*, it inhibited the growth of *Agrobacterium*, and resulted in the emission of DMDS by the *S. lycopersicum* plants (Dandurishvili et al. 2011). It was demonstrated that *S. plymuthica* HRO-C48 inhibited the growth of *R. solani* using I plates (Kai et al. 2007). When oilseed rape cv. Talent was treated with *S. plymuthica* HRO-C48, disease severity of *Verticillium dahlia* was significantly reduced. This strain also produces DMDS and is registered and distributed by RhizoStar®, E-nema GmbH Raisdorf, Germany (Müller et al. 2009). This strain was shown to suppress *V. dahliae* in strawberry and *R. solani* in lettuce (Kurze et al. 2001; Grosch et al. 2005). The development of DMDS has been targeted as a possible alternative to the fumigant methyl bromide. DMDS was observed to supress phytopathogenic nematodes and fungi in the soil, offering strong evidence for the effect of DMDS against plant-pathogenic fungi (Fritsch 2005) and nematodes (Coosemans 2005). Based on these findings, rhizobacteria that produce DMDS can be classified as natural fumigants and this biocontrol capacity is closely associated with VOC production. Therefore, antifungal VOCs should be considered as significant tools for the control of plant pathogens, in addition to more conventional agrichemicals. At present, Paladin®, a new and effective soil fumigant based on DMDS, had already been registered in the USA, (http://www.arkema.com/en/media/news/news-details/Arkema-receives-U.S.-registration-for-Paladin-soil-fumigant-00001/) and the development of new soil fumigants based on dimethyl disulphide (DMDS) is progressing in Europe.

The growth of *F. oxysporum* was suppressed in an I-plate assay by the BVCs produced by endophytic bacteria. A range of compounds was observed during the analysis of these volatiles, such as 2-pentanone 3-methyl, methanethiol and 3-undecene. However, the authors did not evaluate the antifungal properties of the individual compounds on *F. oxysporum* (Ting et al. 2011). It has been demonstrated that the volatile metabolites produced by endophytic *Burkholderia tropics* strains inhibited the growth of four phytopathogenic fungi: *Colletotrichum gloeosporioides*, *Fusarium culmorum*, *F. oxysporum* and *Sclerotium rolfsii*. Further analysis revealed the existence of numerous compounds, including α-pinene, ocimene, limonene and fencona, indicating that these compounds may also play important antifungal roles (Tenorio-Salgado et al. 2013). Considering the effects of BVC on the development of fungi and plants, an obvious question is whether these effects extend to animals living within the soil as well.

In an I-plate assay, volatiles released by *Bacillus megaterium* YMF3.25 were shown to inhibit the hatching of nematode eggs and reduce nematode infections. GC-MS analysis revealed a variety of compounds, such as benzeneacetaldehyde, 2-nonanone, decanal, 2-undecanone and DMDS that exhibited strong nematicidal activities against both juveniles and eggs. Additional research will be required to create integrated management systems for lowering root-knot nematode inocula and enhancing crop yields in the field, perhaps by improving bacterial formulation. (Huang et al. 2010).

An analysis of *Lysinibacillus mangiferahumi* BVCs showed nematicidal activity against the root-knot nematode *Meloidogyne incognita*. Volatile analysis revealed the presence of numerous compounds, including 2-octanol, cyclohexene, 3-chloro-4-fluorobenzaldehyde, dibutyl phthalate, 2-nitro-2-chloropropane, dimethachlore, and DMDS. When nematodes were exposed to these pure volatiles in a three-compartment Petri dish assay, the compounds showed growth-inhibitory effects towards nematodes (Yang et al. 2012). Similarly, BVCs produced by soil bacteria showed nematicidal activity against *Panagrellus redivivus* and *Bursaphelenchus xylophilus* in three-compartment Petri dish bioassays. Further analysis revealed significant difference in the nematicidal activity of the VOCs, indicating that VOC-mediated interactions between bacteria and nematodes can be species-specific/isolate-specific. Numerous distinct volatiles were identified, including phenol, octanol, benzaldehyde, benzene, acetaldehyde, decanal, 2-nonanone, 2-decanone, cyclohexene, and dimethyl disulphide. Finally, exposure to pure synthetic VOCs using the same bioassay revealed growth-inhibitory activities against both *B. xylophilus* and *P. redivivus* (Gu et al. 2007).

In a recent study by Cho et al. (2017) it was observed that a novel antifungal volatile, caryolan-1-ol, produced by *Streptomyces* spp., was effective at inhibiting the growth of the fungal pathogen *B. cinerea*. By using a homozygous profiling (HOP) assay in *Saccharomyces cerevisiae,* in which both copies of non-essential genes are deleted, it was determined that caryolan-1-ol most likely inhibits fungal growth by targeting membrane lipid processes and intracellular transport systems in fungi (Cho et al. 2017)

For many bacterial phytopathogens an integral component of their ability to infect susceptible plant hosts is the capacity to be motile, in order to actively seek out sites of colonisation and subsequent infection. To this end, Tahir and colleagues investigated the potential of BVCs from six *Bacillus* spp., to inhibit the motility of the bacterial phytopathogen *Ralstonia solanacaerum (Rsc)* TBBS1, the causative agent of bacterial wilt disease in tobacco. They observed a particularly strong inhibitive effect of BVCs from two strains; *Bacillus amyloliquefaciens* FZB42 and *Bacillus artrophaeus* LSSC22. Three BVCs; 1,2-benzisothiazol-3(2 H)-one, Benzaldehyde and 1,3-butadiene all negatively influenced cell viability, colony size, motility and chemotaxis. Transcriptomic investigation of this activity identified alterations in the expression of pathogenesis-related genes such as *PhcA* which is a global regulator of virulence factors in *Rsc,* as well as genes involved in type III and IV secretion systems, production of extracellular polysaccharides and chemotaxis. Defence genes in tobacco were also up-regulated with over-expression of the proteins *EDS1* and *NPR1* suggesting the activation of the SA pathway in the ISR response to *Rsc*-challenge (Tahir et al. 2017c). Interestingly, in a separate study by Tahir and colleagues it was observed that the BVCs produced by *Rsc* could not elicit a significant plant growth promoting effect. However they did inhibit the growth-promoting potential of *B. subtilis* SYST2 BVCs when plants were exposed to BVCs emitted by both SYST2 and *Rsc.* Co-culture of both bacteria together revealed that they inhibited the growth of one another, but the effect of inhibition by *Rsc* of SYST2 was not as great as SYST2 versus *Rsc* (Tahir et al. 2017c)

8.5 Species-Specific Rhizobacterial Volatiles

The qualitative and quantitative complexity of the VOC profiles of various rhizobacteria can vary significantly. For example, common rhizobacterial strains such as *B. subtilis* GB03 and *B. amyloliquefaciens* IN937a, release a mixture of volatile compounds (e.g. 2,3-butanediol and acetoin) that promote growth in *Arabidopsis thaliana*, whereas other rhizobacterial strains that do not promote growth via VOCs produce different mixtures of compounds, indicating that synthesis of VOCs is a strain-specific phenomenon. Studies have shown that whereas some compounds are isolate-specific, others are produced by more than one type of bacteria (Ryu et al. 2003; Kai et al. 2007). Furthermore, BVC profiles may be affected by growth phase and environmental conditions. For example, different qualitative and quantitative VOC profiles are generated by growth on nutrient broths with and without glucose in *Stenotrophomonas rhizophila* P69. In particular, the production of dimethyl pyrazine and beta-phenylethanol occurs under both growth conditions, whereas trimethyl pyrazine, tetramethyl-pyrazine and beta-phenylethyl acetate are produced when glucose is absent from the growth medium (Kai et al. 2009).

The effect of different bacterial strains cultured on four distinct types of media on the growth of *A. thaliana* were analysed by Blom and Co-workers. They found that more nutrient-rich media Luria-Bertani (LB) caused plant death, whereas all other media tested—including two less nutrient-rich media, Murashige and Skoog (MS) and the soil mimicking Angle medium—promoted growth (Blom et al. 2011a). MS and LB have significantly different compositions; MS is a mineral medium that contains sugar as a carbon (C) source and has low pH and agar concentrations, whereas LB is a nutrient-rich medium containing hydrolysed proteins with relatively high pH and agar concentrations. Therefore, it is perhaps unsurprising that the same bacteria cultured on these two media types can emit different types of BVCs that elicit different plant responses. Furthermore, these two media types also affect the growth kinetics (maximal volatile production, which is assumed to take place during the stationary phase of bacterial growth) of bacterial strain on these two media, with LB supporting faster growth than MS (Bailly and Weisskopf 2012). The primary factors that determine the qualitative and quantitative distribution of compounds in highly complex mixtures are the metabolic abilities of the bacterial species, as well as the nutrients available in the specific growth conditions. Indeed, differences in the types of BVCs produced by pathogenic and non-pathogenic mycobacteria grown on different types of media have been observed; furthermore, variations were sometimes observed, not only between different media but also within individual analyses (Nawrath et al. 2012). However, it is important to note that results obtained under artificial conditions may not accurately reflect natural conditions and an in-depth understanding of this area of plant-microbe research cannot be acquired solely from in vitro experiments of VOC-mediated interactions between all organisms involved in these interactions. Results obtained in controlled, artificial environments may not, and indeed most likely do not, represent interactions which would be observed in the field (Velivelli et al. 2015).

This issue of the interpretation of results from field versus lab is relevant for metabolic experiments utilising artificial growth media and nutrient sources and also for interactions where BVCs are outside of a detectable range. Such concentrations of BVCs observed under in vitro conditions are unlikely to be naturally present in the soil. Nevertheless, in vitro experiments contribute greatly to our understanding of the molecular interplay between bacterial volatiles and other organisms. This can give an insight as to what may be observed within these complex relationships in the field (Ryu et al. 2003; Blom et al. 2011a; Effmert et al. 2012; Park et al. 2015; Besset-Manzoni et al. 2017; Brilli et al. 2019; Song et al. 2019). Furthermore, not all plant species respond similarly to the same group of volatiles produced by a given bacterial strain, which could be due to several reasons: (1) fundamental differences in the pathways used by plants to respond to BVCs (2) differences in reactive sites and (3) differences in the capacity to metabolise volatiles (Santoro et al. 2011). The volatiles produced by a given bacterial strain elicited different responses in different fungi; in other words, differences exist in the interactions between different fungi and bacteria. For example, although *Arabidopsis* growth was significantly promoted by DMDS; the responses of different fungal species to this compound varied considerably: the growth of *R. solani* was suppressed, whereas *A. alternata* and *F. solani* were unaffected (Groenhagen et al. 2013). Therefore, it is necessary to employ a number of different approaches and to test bacteria under distinct growth conditions to determine comprehensive bacterial emission patterns (Farag et al. 2017).

8.6 Analytical Approaches to Identify VOCs

The capture of VOCs which are emitted as a result of microbial metabolic activity is the first and most crucial step in the analysis of biological VOCs, and research in this area of 'separation science' has advanced significantly in the last half a century (Velivelli et al. 2015). SPME-GC/MS and PTR-MS are just two of the numerous approaches designed for the capture and subsequent identification of volatiles. Although each approach has its own advantages, no single method is currently capable of completely surveying bacterial-produced volatile profiles, either in terms of quantity or quality. To successfully identify volatiles, a number of sampling techniques are generally employed, such as purge and trap, solid phase microextraction (SPME) followed by Gas chromatography/Mass spectrometry (GC/MS) identification. The purge-and-trap method involves passing a given volume of purified air over the sample, collecting it onto an absorbent filter and then either directly releasing it using an organic solvent to rinse the filter or transferring it straight to the GC/MS (Ryu et al. 2003, 2004; Yuan et al. 2012). Solid phase microextraction (SPME) has become the method of choice for the extraction of bacterial volatiles in recent years because it minimises preparation time and has greater sensitivity than other extraction techniques.

The process of SPME for the analysis of bacterial volatiles is relatively fast and can be conducted under low oxygen conditions. Although SPME has many advantages over other methods, it is necessary to carefully consider the particular fibre coatings used in each experiment, as they can either absorb or exclude specific analytes based on polarity or size, leading to reduced sensitivity. For instance, non-polar metabolites are absorbed by Polydimethylsiloxane (PDMS) fibre, whereas short-chain polar compounds are absorbed by divinylbenzene/carboxen/PDMS (DCP) fibre. With respect to rhizobacterial volatiles, the best recovery is offered by DCP fibre, as it absorbs polar low molecular weight volatiles (Farag et al. 2006). Yuan and Co-workers tested three different fibres from Supelco—PDMS, 7 μm, stable flex DCP, 50/30 μm, and polydimethylsiloxane/divinylbenzene (PDMS-DVB, 65 μm)—and they found that DCP fibre performed best (Yuan et al. 2012). The volatile extraction methods discussed above are often used in combination with other analytical techniques, such as GC/MS. Considering its effectiveness for both separation and identification, GC/MS is the foremost method for the detection of bacterial volatiles. GC-MS can be used to separate, identify and quantify the volatiles present in a given sample. However, this technique has one significant drawback; namely, it does not allow for the identification of new compounds. Furthermore it can be difficult to obtain quantitative results using SPME, as volatile compounds compete for binding sites within the SPME fibre. Small bacterial molecules are characterised by high polarity and a strong tendency to co-elute. As a result, overlapping MS spectra can be produced, limiting the precision of volatile detection and adversely affecting the process of matching peak quality against database entries.

The identification of existing volatiles is carried out using software programs and libraries of different mass spectra such as NIST/or reference standards. Nevertheless, as similar mass spectra can be produced by related compounds, particular care must be taken in performing these analyses (Kai et al. 2009). A recently developed technique with the advantage of real-time analysis without requiring sample preparation is PTR-MS. In proton transfer reaction-mass spectrometry (PTR-MS), the headspace air is drawn directly into the instrument, where interactions occur between protonated water (H_3O^+) and any molecules with a proton affinity greater than that of water. Next, a quadruple mass spectrometer in mass-to-charge (m/z) ratio is used to mass analyse and identify the resulting protonated organic molecules. Although PTR-MS cannot accurately detect the individual volatiles produced, it is advantageous because it ensures real-time emission observation.

In addition, PTR-MS cannot be used to distinguish between analytes of the same mass because it employs single reagent ions (Spinelli et al. 2012). PTR-MS was used to determine the volatile profiles of various bacteria and infected plants, and it allowed for the observation of pathogenic emissions in real time (Spinelli et al. 2011). Several researchers have analysed the volatiles emitted by bacteria, and a comparison of these signature compounds indicates that GC/MS analysis can be used to distinguish between different bacterial species. Indeed, the VOC-profile data obtained from GC/MS could be harnessed to develop so-called 'electronic-nose' (e-nose) sensor technology for the detection of different bacterial species, or they may be useful for real-time disease monitoring in agricultural settings and post-harvest monitoring where e-nose technology could detect ring and brown rot in commercial potato storage with a detection efficiency of 90% of samples (Biondi et al. 2014). Currently, the identification of bacterial species, including phytopathogenic bacterial isolates, is achieved through 'classical' molecular techniques such as polymerase chain reaction followed by dideoxy sequencing of 16S rRNA and these methods, although reliable, are expensive (Velivelli et al. 2015). The efficacy of e-nose technology has been demonstrated for the detection of a single phytopathogen within a wider microbial community; using a metal-oxide-based semiconductor sensor the pathogen that causes fire blight (*E. amylovora*)—could be differentiated based on volatile mixes alone from other bacteria (Spinelli et al. 2012). However, this technique has one significant drawback; it does not identify and quantify each compound; rather, it uses metal-oxide semiconductor sensors to characterise an aroma based on the signature profile that is generated.

The e-nose distinguishes between *Botrytis*- and *Sclerotinia*-induced rots on both green and yellow kiwifruits. In addition, it was successfully used to detect asymptomatic apple and pear plants that had been infected with *Erwinia amylovora*. More research is needed before e-nose applications in open fields becomes standard practice (Spinelli et al. 2010). In 2015, Cernava et al. described a novel assay for the detection of bacterial volatiles from lichen-associated bacteria using headspace analysis and indicator assays (for HCN production) followed by the determination of the effect of the VOCs on the growth of *E. coli* and *B. cinerea*. The initial test was combined with a qPCR-based quantification assay (Cernava et al. 2015).

8.7 Discussion and Conclusive Remarks

Research into volatile-mediated plant–bacteria interactions is relatively new, and this field has the potential to yield significant agricultural applications, particularly in the context of growth promotion and induced systemic resistance (ISR) (Liu and Brettell 2019; Romera et al. 2019). This niche of plant-bacteria research is still emerging and at present, the biological effects of BVCs have only been assessed in a relatively small cohort of plant species with *Arabidopsis thaliana* being one of the main candidates, along with more economically relevant plants such as *Capsicum annum* and *Nicotiana tabacum* (Choi et al. 2014; Kim et al. 2015). How these bacterial volatiles will affect the growth and disease-suppression in other major crops remains to be seen. In general, I plates are used to perform experiments on the effects of bacterial volatiles on plants and fungi. This setup allows for the exchange of volatile compounds while preventing non-volatile metabolites from dispersing throughout the respective growth media. However, following identification of the bio-active compounds, further in vitro laboratory tests should be conducted under conditions that are either in, or similar to, field conditions (e.g. nutrient media that closely resembles soil environment), as opposed to the highly artificial Petri dishes used in the majority of existing studies (Choi et al. 2014; Brilli et al. 2019; Song et al. 2019).

To gain a comprehensive understanding of the role of BVCs, it is necessary to study mutant strains that do not produce such compounds. Adverse effects from chemical treatments and difficulties in establishing optimal treatment concentrations restrict the use of chemicals for growth promotion and resistance induction in plants. Therefore, future research programs should focus on measuring VOC levels produced by rhizobacteria in soil ecosystems (Ryu et al. 2005b; Sharifi and Ryu 2018a). An understanding of the relevant signalling pathways and the extent to which they resemble plant responses to pathogens, and whether the physiological and molecular plant responses vary between bacterial species is crucial to aid our understanding and will determine future technology transfer in this area (Velivelli et al. 2015). In addition, genetically altering rhizobacterial strains to produce greater amounts of volatiles that promote plant growth or that induce systemic resistance may represent promising avenues of future research. However, the use of GM organisms is becoming increasingly controversial in terms of human health and the environment. Phytopathogenic fungi—which are responsible for the majority of economically important crop diseases—are inhibited by BVCs, indicating that BVCs have the potential for use in agriculture as biological control agents. However, research on the effects of bacterial volatiles is still in the early stages (Sharifi and Ryu 2016).

In particular, it has yet to be determined whether the effects of volatiles are limited to specific plant tissues or if they affect plant development in general. Also, an understanding (positive or negative) of the impact of bacterial volatiles on plant metabolism is essential. Furthermore, it will be necessary to address whether effects observed in the laboratory transfer to the field (Velivelli et al. 2015). An alternative strategy that may prove beneficial for agriculture is the application of bio-active but affordable compounds, such as 2,3-butanediol, to aerial plant components to stimulate growth and ISR. Additional limitations on the use of rhizobacteria volatiles such as 2,3-butanediol and acetoin include the fact that they evaporate very quickly following application in the open field, however solutions are being developed to tackle this problem such as the use of microcapsules to slowly release BVCs to the environment (Song and Ryu 2013; Sharifi and Ryu 2018b). Nevertheless, significant positive effects for BVCs have been comprehensively analysed under growth-chamber, greenhouse and open-field conditions (Velivelli et al. 2015; Choi et al. 2014). For example, pathogen growth in *A. thaliana* by *Pseudomonas syringae* was found to be considerably suppressed by the direct application of acetoin to roots under growth-chamber conditions (Rudrappa et al. 2010). Under greenhouse conditions, the soil-drench application of volatile metabolites, such as DMDS (released by *B. cereus* C1L), inhibited the activities of *Botrytis cinerea* and *Cochliobolus heterostrophus* against tobacco and corn plants, respectively (Huang et al. 2012). Pre-treatment of cucumber plants with 3-pentanol and 2-butanone showed protective benefits against the biotrophic bacterial pathogen *P. syringae* in open-field trials. (Song and Ryu 2013; Velivelli et al. 2015). Notwithstanding the challenges of transitioning from laboratory to field with respect to the application of mVOCs (Bailly and Weisskopf 2017) the studies conducted to date should lead to new discoveries regarding the use of VOCs to control microbial pathogens under open-field conditions and can be expected to act as big players in any second green revolution.

Conflict of Interest Author(s) have no conflict of interest

References

Ahmed E, Holmström SJM (2014) Siderophores in environmental research: roles and applications. Microb. Biotech 7:196–208

Almenar E, Del Valle V, Catala R, Gavara R (2007) Active package for wild strawberry fruit (*Fragaria vesca* L.). J Agric Food Chem 55:2240–2245

Arrebola E, Sivakumar D, Korsten L (2010) Effect of volatile compounds produced by *Bacillus* strains on postharvest decay in citrus. Biol Cont 53:122–128

Athukorala SNP, Fernando WGD, Rashid KY, De Kievit T (2010) The role of volatile and non-volatile antibiotics produced by *Pseudomonas chlororaphis* strain PA23 in its root colonization and control of *Sclerotinia sclerotiorum*. Biocont Sci Technol 20:875–890

Bailly A, Weisskopf L (2012) The modulating effect of bacterial volatiles on plant growth: current knowledge and future challenges. Plant Signal Behav 7:79–85

Bailly A, Weisskopf L (2017) Mining the volatilomes of plant-associated microbiota for new biocontrol solutions. Front Microbiol 8:1638

Besset-Manzoni Y, Rieusset L, Joly P, Comte G, Prigent-Combaret C (2018) Exploiting rhizosphere microbial cooperation for developing sustainable agriculture strategies. Environ Sci Pollut Res 25:29953

Biondi E, Blasioli S, Galeone A, Spinelli F, Cellini A, Luchese C, Braschi I (2014) Detection of potato brown rot and ring rot by electronic nose: From laboratory to real scale. Talanta 129:422–430

Blom D, Fabbri C, Connor EC, Schiestl FP, Klauser DR, Boller T, Eberl L, Weisskopf L (2011a) Production of plant growth modulating volatiles is widespread among rhizosphere bacteria and strongly depends on culture conditions. Environ Microbiol 13:3047–3058

Blom D, Fabbri C, Eberl L, Weisskopf L (2011b) Volatile-mediated killing of *Arabidopsis thaliana* by bacteria is mainly due to hydrogen cyanide. Appl Environ Microbiol 77:1000–1008

Blumer C, Haas D (2000) Iron regulation of the hcnABC genes encoding hydrogen cyanide synthase depends on the anaerobic regulator ANR rather than on the global activator GacA in *Pseudomonas fluorescens* CHA0. Microbiol 146:2417–2424

Brilli F, Loreto F, Baccelli I (2019) Exploiting Plant Volatile Organic Compounds (VOCs) in Agriculture to Improve Sustainable Defense Strategies and Productivity of Crops. Front Plant Sci 10:264

Carmen Orozco-Mosqueda M, Macías-Rodríguez L, Santoyo G, Farías-Rodríguez R, Valencia-Cantero E (2013) *Medicago truncatula* increases its iron-uptake mechanisms in response to volatile organic compounds produced by *Sinorhizobium meliloti*. Folia Microbiol 58:579–585

Cernava T, Müller H, Aschenbrenner IA, Grube M, Berg G (2015) Analysing the antagonistic potential of the lichen microbiome against pathogens by bridging metagenomic with culture studies. Front Microbiol 6:620

Chen H, Xiao X, Wang J, Wu L, Zheng Z, Yu Z (2008) Antagonistic effects of volatiles generated by *Bacillus subtilis* on spore germination and hyphal growth of the plant pathogen, *Botrytis cinerea*. Biotechnol Lett 30:919–923

Cho G, Kim J, Park CG, Nislow C, Weller DM, Kwak Y-S (2017) Caryolan-1-ol, an antifungal volatile produced by *Streptomyces* spp., inhibits the endomembrane system of fungi. Open Biol 7:170075

Cho SM, Kang BR, Han SH, Anderson AJ, Park J-Y, Lee Y-H, Cho BH, Yang K-Y, Ryu C-M, Kim YC (2008) 2R,3R-Butanediol, a bacterial volatile produced by *Pseudomonas chlororaphis* O6, is involved in induction of systemic tolerance to drought in *Arabidopsis thaliana*. Mol Plant Microbe Interact 21:1067–1075

Choi HK, Song GC, Yi HS, Ryu C-M (2014) Field evaluation of the bacterial volatile derivative 3-pentanol in priming for induced resistance in pepper. J Chem Ecol 40:882–892

Chuankun X, Minghe M, Leming Z, Keqin Z (2004) Soil volatile fungistasis and volatile fungistatic compounds. Soil Biol Biochem 36:1997–2004

Compant S, Duffy B, Nowak J, Clément C, Barka EA (2005) Use of plant growth-promoting bacteria for biocontrol of plant diseases: principles, mechanisms of action, and future prospects. Appl Environ Microbiol 71:4951–4959

Coosemans J (2005) Dimethyl disulphide (DMDS): a potential novel nematicide and soil disinfectant. Acta Hortic (ISHS) 698:57–64

Dandurishvili N, Toklikishvili N, Ovadis M, Eliashvili P, Giorgobiani N, Keshelava R, Tediashvili M, Vainstein A, Khmel I, Szegedi E, Chernin L (2011) Broad-range antagonistic rhizobacteria *Pseudomonas fluorescens* and *Serratia plymuthica* suppress *Agrobacterium* crown gall tumours on tomato plants. J Appl Microbiol 110:341–352

Dimkić I, Stanković S, Nišavić M, Petković M, Ristivojević P, Fira D, Berić T (2017) The profile and antimicrobial activity of *bacillus* lipopeptide extracts of five potential biocontrol strains. Front Microbiol 8:925

Effmert U, Kalderás J, Warnke R, Piechulla B (2012) Volatile mediated interactions between bacteria and fungi in the soil. J Chem Ecol 38:665–703

Farag MA, Ryu C-M, Sumner LW, Paré PW (2006) GC–MS SPME profiling of rhizobacterial volatiles reveals prospective inducers of growth promotion and induced systemic resistance in plants. Phytochemistry 67:2262–2268

Farag MA, Song GC, Park Y-S, Audrain B, Lee S, Ghigo JM, Kloepper JW, Ryu C-M (2017) Biological and chemical strategies for exploring inter- and intra-kingdom communication mediated via bacterial volatile signals. Nat Protoc 12:1359–1377

Fernando WGD, Ramarathnam R, Krishnamoorthy AS, Savchuk SC (2005) Identification and use of potential bacterial organic antifungal volatiles in biocontrol. Soil Biol Biochem 37:955–964

Fiddaman PJ, Rossall S (1993) The production of antifungal volatiles by *Bacillus subtilis*. J Appl Bacteriol 74:119–126

Fiddaman PJ, Rossall S (1994) Effect of substrate on the production of antifungal volatiles from *Bacillus subtilis*. J Appl Bacteriol 76:395–405

Fincheira P, Quiroz A (2018) Microbial volatiles as plant growth inducers. Microbiol Res 208:63–75

Fritsch J (2005) Dimethyl disulfide as a new chemical potential alternative to methyl bromide in soil disinfestation in France. Acta Hortic (ISHS) 698:71–76

Glick BR (2014) Bacteria with ACC deaminase can promote plant growth and help to feed the world. Microbiol Res 69:30–39

Groenhagen U, Baumgartner R, Bailly A, Gardiner A, Eberl L, Schulz S, Weisskopf L (2013) Production of bioactive volatiles by different *Burkholderia ambifaria* Strains. J Chem Ecol 39:892–906

Grosch R, Faltin F, Lottmann J, Kofoet A, Berg G (2005) Effectiveness of 3 antagonistic bacterial isolates to control *Rhizoctonia solani* Kühn on lettuce and potato. Can J Microbiol 51:345–353

Gu Y-Q, Mo M-H, Zhou J-P, Zou C-S, Zhang K-Q (2007) Evaluation and identification of potential organic nematicidal volatiles from soil bacteria. Soil Biol Biochem 39:2567–2575

Gutiérrez-Luna F, López-Bucio J, Altamirano-Hernández J, Valencia-Cantero E, Cruz H, Macías-Rodríguez L (2010) Plant growth-promoting rhizobacteria modulate root-system architecture in *Arabidopsis thaliana* through volatile organic compound emission. Symbiosis 51:75–83

Hérnandez-Calderón E, Aviles-Garcia ME, Castulo-Rubio DY, Macías-Rodríguez L, Ramírez VM, Santoyo G, López-Bucio J, Valencia-Cantero E (2018) Volatile compounds from beneficial or pathogenic bacteria differentially regulate root exudation, transcription of iron transporters, and defense signaling pathways in *Sorghum bicolor*. Plant Mol Biol 96:291–304

Howell CR, Beier RC, Stipanovic RD (1988) Production of ammonia by *Enterobacter cloacae* and its possible role in the biological control of *Pythium preemergence* damping-off by the bacterium. Phytopathology 78:1075–1078

Huang C-J, Tsay J-F, Chang S-Y, Yang H-P, Wu W-S, Chen C-Y (2012) Dimethyl disulfide is an induced systemic resistance elicitor produced by *Bacillus cereus* C1L. Pest Manag Sci 68:1306–1310

Huang Y, Xu C, Ma L, Zhang K, Duan C, Mo M (2010) Characterisation of volatiles produced from *Bacillus megaterium* YFM3.25 and their nematicidal activity against *Meloidogyne incognita*. Euro J Plant Pathol 126:417–422

Jeong H, Choi S-K, Ryu C-M, Park S-H (2019) Chronicle of a soil bacterium: *paenibacillus polymyxa* E681 as a tiny guardian of plant and human health. Front Microbiol 10:467

Kai M, Crespo E, Cristescu SM, Harren FJ, Francke W, Piechulla B (2010) *Serratia odorifera*: analysis of volatile emission and biological impact of volatile compounds on *Arabidopsis thaliana*. Appl Microbiol Biotechnol 88:965–976

Kai M, Effmert U, Berg G, Piechulla B (2007) Volatiles of bacterial antagonists inhibit mycelial growth of the plant pathogen *Rhizoctonia solani*. Arch Microbiol 187:351–360

Kai M, Haustein M, Molina F, Petri A, Scholz B, Piechulla B (2009) Bacterial volatiles and their action potential. Appl Microbiol Biotechnol 81:1001–1012

Kai M, Piechulla B (2009) Plant growth promotion due to rhizobacterial volatiles – An effect of CO_2? FEBS Lett 583:3473–3477

Kai M, Piechulla B (2010) Impact of volatiles of the rhizobacteria *Serratia odorifera* on the moss *Physcomitrella patens*. Plant Signal Behav 5:444–446

Kai M, Effmert U, Piechulla B (2016) Bacterial-plant-interactions: approaches to unravel the biological function of bacterial volatiles in the rhizosphere. Front. Microbiol. 7

Kanchiswamy CN, Malnoy M, Maffei M (2015) Chemical diversity of microbial volatiles and their potential for plant growth and productivity. Front Plant Sci 6:151

Kim J-S, Lee J, Seo S-J, Lee C, Woo SY, Kim S-H (2015) Gene expression profile affected by volatiles of new plant growth promoting rhizobacteria, *Bacillus subtilis* strain JS, in tobacco. Genes Genom 37:387–397

Kloepper JW, Rodríguez-Kábana R, Zehnder AW, Murphy JF, Sikora E, Fernández C (1999) Plant root-bacterial interactions in biological control of soilborne diseases and potential extension to systemic and foliar diseases. Australas Plant Pathol 28:21–26

Kloepper JW and Schroth MN (1978) Plant growth promoting rhizobacteria on radishes. In: Proceedings of the 4th international conference on plant pathogenic bacteria, angers, France, vol 2, pp 879–882

Kurze S, Bahl H, Dahl R, Berg G (2001) Biological control of fungal strawberry diseases by *Serratia plymuthica* HRO-C48. Plant Dis 85:529–534

Kwon Y, Ryu C-M, Lee S, Park H, Han K, Lee J, Lee K, Chung W, Jeong M-J, Kim H, Bae D-W (2010) Proteome analysis of *Arabidopsis* seedlings exposed to bacterial volatiles. Planta 232:1355–1370

Leach J, Triplett LR, Argueso CT, Trivedi P (2017) Communication in the phytobiome. Cell 169 (4):587–596

Lee B, Farag MA, Park HB, Kloepper JW, Lee SH, Ryu C-M (2012) Induced resistance by a long-chain bacterial volatile: elicitation of plant systemic defense by a C13 volatile produced by *Paenibacillus polymyxa*. PLoS One 7:e48744

Liu H, Brettell LE (2019) Plant defense by VOC-induced microbial priming. Trends Plant Sci 24:187–189

Liu W, Zhao L, Wang C, Mu W, Liu F (2009) Bioactive evaluation and application of antifungal volatiles generated by five soil bacteria. Acta Phytophyl Sin 36:97–105

López-Bucio J, Campos-Cuevas JC, Hernández-Calderón E, Velásquez-Becerra C, Farías-Rodríguez R, Macías-Rodríguez LI, Valencia-Cantero E (2007) *Bacillus megaterium* rhizobacteria promote growth and alter root-system architecture through an auxin- and ethylene-independent signaling mechanism in *Arabidopsis thaliana*. Mol Plant Microbe Interact 20:207–217

Lugtenberg B, Kamilova F (2009) Plant-growth-promoting rhizobacteria. Ann Rev Microbiol 63:541–556

Marquez-Villavicencio MDP, Weber B, Witherell RA, Willis DK, Charkowski AO (2011) The 3-Hydroxy-2-Butanone pathway is required for *Pectobacterium carotovorum* pathogenesis. PLoS One 6:e22974

Müller H, Westendorf C, Leitner E, Chernin L, Riedel K, Schmidt S, Eberl L, Berg G (2009) Quorum-sensing effects in the antagonistic rhizosphere bacterium *Serratia plymuthica* HRO-C48. FEMS Microbiol Ecol 67:468–478

Nawrath T, Mgode GF, Weetjens B, Kaufmann SH, Schulz S (2012) The volatiles of pathogenic and nonpathogenic mycobacteria and related bacteria. Beilstein J Org Chem 8:290–299

Nicholson WL (2008) The *Bacillus subtilis* ydjL (bdhA) gene encodes acetoin reductase/ 2,3-Butanediol dehydrogenase. Appl Environ Microbiol 74:6832–6838

Oteino N, Lally RD, Kiwanuka S, Lloyd A, Ryan D, Germaine KJ, Dowling DN (2015) Plant growth promotion induced by phosphate solubilizing endophytic *Pseudomonas* isolates. Front Microbiol 6:745

Park Y-S, Dutta S, Ann M, Raajmakers JM, Park K (2015) Promotion of plant growth by *Pseudomonas fluorescens* strain SS101 *via* novel volatile organic compounds. Biochem Biophys Res Commun 461:361–365

Park HB, Lee B, Kloepper JW, Ryu CM (2013) One shot-two pathogens blocked: exposure of *Arabidopsis* to hexadecane, a long chain volatile organic compound, confers induced resistance against both *Pectobacterium carotovorum* and *Pseudomonas syringae*. Plant Signal Behav 8: e24619

Rath M, Mitchell TR, Gold SE (2018) Volatiles produced by *Bacillus mojavensis* RRC101 act as plant growth modulators and are strongly culture-dependent. Microbiol Res 208:76–84

Rijavec T, Lapanje A (2016) Hydrogen cyanide in the rhizosphere: not suppressing plant pathogens, but rather regulating availability of phosphate. Front Microbiol 7:1785

Romera FJ, García MJ, Lucena C, Martínez-Medina A, Aparicio MA, Ramos J, Alcántara E, Angulo M, Pérez-Vicente R (2019) Induced systemic resistance (ISR) and fe deficiency responses in dicot plants. Front Plant Sci 10:287

Rosier A, Medeiros FHV, Bais HP (2018) Defining plant growth promoting rhizobacteria molecular and biochemical networks in beneficial plant-microbe interactions. Plant Soil 428:35

Rudrappa T, Biedrzycki ML, Kunjeti SG, Donofrio NM, Czymmek KJ, Paul WP, Bais HP (2010) The rhizobacterial elicitor acetoin induces systemic resistance in *Arabidopsis thaliana*. Commun Integr Biol 3:130–138

Ryu C-M, Farag MA, Hu C-H, Reddy MS, Kloepper JW, Paré PW (2004) Bacterial volatiles induce systemic resistance in *Arabidopsis*. Plant Physiol 134:1017–1026

Ryu CM, Farag MA, Hu CH, Reddy MS, Kloepper JW, Paré PW (2004) Bacterial volatiles induce systemic resistance in Arabidopsis. Plant Physiol 134(3):1017–1026 (Published correction appears in Plant Physiol. 2005 Apr;137(4):1486)

Ryu C-M, Farag MA, Hu C-H, Reddy MS, Wei H-X, Paré PW, Kloepper JW (2003) Bacterial volatiles promote growth in *Arabidopsis*. Proc Natl Acad Sci 100:4927–4932

Ryu C-M, Hu C-H, Locy R, Kloepper J (2005a) Study of mechanisms for plant growth promotion elicited by rhizobacteria in *Arabidopsis thaliana*. Plant Soil 268:285–292

Ryu C-M, Farag MA, Paré PW, Kloepper JW (2005b) Invisible signals from the underground: bacterial volatiles elicit plant growth promotion and induce systemic resistance. Plant Pathol J 21:7–12

Sang MK, Kim JD, Kim BS, Kim KD (2011) Root treatment with rhizobacteria antagonistic to Phytophthora blight affects anthracnose occurrence, ripening, and yield of pepper fruit in the plastic house and field. Phytopathology 101(6):666–678

Santoro MV, Zygadlo J, Giordano W, Banchio E (2011) Volatile organic compounds from rhizobacteria increase biosynthesis of essential oils and growth parameters in peppermint (*Mentha piperita*). Plant Physiol Biochem 49:1177–1182

Schulz-Bohm K (2017) Microbial volatiles: small molecules with an important role in intra- and inter-kingdom interactions. Front Microbiol 8:2484

Shao J, Xu Z, Zhang N, Shen Q, Zhang R (2015) Contribution of indole-3-acetic acid in the plant growth promotion by the rhizospheric strain *Bacillus amyloliquefaciens* SQR9. Biol Fertil Soils 51:321

Sharifi R, Ryu CM (2016) Making healthier or killing enemies? Bacterial volatile-elicited plant immunity plays major role upon protection of *Arabidopsis* than the direct pathogen inhibition. Commun Integr Biol 9:e1197445

Sharifi R, Ryu C-M (2018a) Revisiting bacterial volatile-mediated plant growth promotion: lessons from the past and objectives for the future. Ann Bot 122:349–358

Sharifi R, Ryu C-M (2018b) Biogenic volatile compounds for plant disease diagnosis and health improvement. Plant Pathol J 34:459–469

Singh S (2014) A review on possible elicitor molecules of cyanobacteria: their role in improving plant growth and providing tolerance against biotic or abiotic stress. J Appl Microbiol 117:1221–1244

Song G, Ryu C-M (2013) Two volatile organic compounds trigger plant self-defense against a bacterial pathogen and a sucking insect in cucumber under open field conditions. Int J Mol Sci 14:9803–9819

Song GC, Ryu CM (2018) Evidence for Volatile Memory in Plants: Boosting Defence Priming through the Recurrent Application of Plant Volatiles. Mol. cells 41(8):724

Song GC, Riu M, Ryu C-M (2019) Beyond the two compartments Petri-dish: optimising growth promotion and induced resistance in cucumber exposed to gaseous bacterial volatiles in a miniature greenhouse system. Plant Methods 15:9

Spinelli F, Cellini A, Vanneste J, Rodriguez-Estrada M, Costa G, Savioli S, Harren FM, Cristescu S (2012) Emission of volatile compounds by *Erwinia amylovora*: biological activity in vitro and possible exploitation for bacterial identification. Trees 26:141–152

Spinelli F, Costa G, Rondelli E, Busi S, Vanneste JL, Rodriguez EMT, Savioli S, Harren FJM, Crespo E, Cristescu SM (2011) Emission of volatiles during the pathogenic interaction between *Erwinia amylovora* and *Malus domestica*. Acta Hortic (ISHS) 896:55–63

Spinelli F, Noferini M, Vanneste JL, Costa G (2010) Potential of the electronic-nose for the diagnosis of bacterial and fungal diseases in fruit trees. EPPO Bull 40:59–67

Tahir JAS, Gu Q, Wu J, Niu Y, Huo R, Gao X (2017a) Bacillus volatiles adversely affect the physiology and ultra-structure of *Ralstonia solanacearum* and induce systemic resistance in tobacco against bacterial wilt. Sci Rep 7:40481

Tahir JAS, Gu Q, Wu J, Raza W, Hanif A, Wu L, Colman MV, Gao X (2017b) Plant growth promotion by volatile organic compounds produced by *Bacillus subtilis* SYST2. Front Microbiol 8:171

Tahir JAS, Gu Q, Wu H, Raza W, Safdar A, Huang Z, Rajer FU, Gao X (2017c) Effect of volatile compounds produced by *Ralstonia solanacearum* on plant growth promoting and systemic resistance inducing potential of *Bacillus* volatiles. BMC Plant Biol 17:133

Tenorio-Salgado S, Tinoco R, Vazquez-Duhalt R, Caballero-Mellado J, Perez-Rueda E (2013) Identification of volatile compounds produced by the bacterium *Burkholderia tropica* that inhibit the growth of fungal pathogens. Bioengineered 4:236–243

Ting A, Mah S, Tee C (2011) Detection of potential volatile inhibitory compounds produced by endobacteria with biocontrol properties towards *Fusarium oxysporum* f. sp. cubense race 4. World J Microbiol Biotechnol 27:229–235

Van Der Kooij LAW, Kok LJD, Stulen I (1999) Biomass production and carbohydrate content of *Arabidopsis thaliana* at atmospheric CO_2 concentrations from 390 to 1680 μl l-1. Plant Biol 1:482–486

Velázquez-Becerra C, Macías-Rodríguez L, López-Buci J, Altamirano-Hernández J, Flores-Cortez I, Valencia-Cantero E (2011) A volatile organic compound analysis from *Arthrobacter agilis* identifies dimethylhexadecylamine, an amino-containing lipid modulating bacterial growth and *Medicago sativa* morphogenesis in vitro. Plant Soil 339:329–340

Velázquez-Becerra C, Macías-Rodríguez L, López-Bucio J, Flores-Cortez I, Santoyo G, Hernández-Soberano C, Valencia-Cantero E (2013) The rhizobacterium *Arthrobacter agilis* produces dimethylhexadecylamine, a compound that inhibits growth of phytopathogenic fungi in vitro. Protoplasma 250:1251–1262

Velivelli SL, Sessitch A, Doyle Prestwich B (2014) The role of microbial inoculants in integrated crop management systems. Potato Res 57:291–309

Velivelli SL, Kromann P, Lojan P, Rojas M, Franco J, Suarez JP, Doyle Prestwich B (2015) Identification of mVOCs from Andean rhizobacteria and field evaluation of bacterial and mycorrhizal inoculants on growth of potato in its center of origin. Microb Ecol 69:652–667

Vespermann A, Kai M, Piechulla B (2007) Rhizobacterial volatiles affect the growth of fungi and *Arabidopsis thaliana*. Appl Environ Microbiol 73:5639–5641

Voisard C, Keel C, Haas D, Defago G (1989) Cyanide production by *Pseudomonas fluorescens* helps suppress black root rot of tobacco under gnotobiotic conditions. EMBO J 8:351–358

Wan M, Li G, Zhang J, Jiang D, Huang H-C (2008) Effect of volatile substances of *Streptomyces platensis* F-1 on control of plant fungal diseases. Biol Control 46:552–559

Ward JK, Strain BR (1999) Elevated CO_2 studies: past, present and future. Tree Physiol 19:211–220

Weise T, Kai M, Gummesson A, Troeger A, Von Reuss S, Piepenborn S, Kosterka F, Sklorz M, Zimmermann R, Francke W, Piechulla B (2012) Volatile organic compounds produced by the phytopathogenic bacterium *Xanthomonas campestris* pv. vesicatoria 85-10. Beilstein J Org Chem 8:579–596

Weise T, Kai M, Piechulla B (2013) Bacterial ammonia causes significant plant growth inhibition. PLoS One 8:e63538

Weisskopf L, Ryu CM, Raaijmakers JM, Garbeva P (2016) Smelly fumes-volatile-mediated communication between bacteria and other organisms. Front. Microbiol. 7

Xie X, Zhang H, Pare PW (2009) Sustained growth promotion in *Arabidopsis* with long-term exposure to the beneficial soil bacterium *Bacillus subtilis* (GB03). Plant Signal Behav 4:948–953

Yang L-L, Huang Y, Liu J, Ma L, Mo M-H, Li W-J, Yang F-X (2012) *Lysinibacillus mangiferahumi* sp. nov., a new bacterium producing nematicidal volatiles. Antonie Van Leeuwenhoek 102:53–59

Yaoyao E, Yuan J, Yang F, Wang L, Ma J, Li J, Pu X, Raza W, Hwang Q, Shen Q (2017) PGPR strain *Paenibacillus polymyxa* SQR-21 potentially benefts watermelon growth by re-shaping root protein expression. AMB Expr 7:104

Yu SM, Lee Y (2013) Plant growth promoting rhizobacterium *Proteus vulgaris* JBLS202 stimulates the seedling growth of Chinese cabbage through indole emission. Plant Soil 370:485–495

Yuan J, Raza W, Shen Q, Huang Q (2012) Antifungal activity of *Bacillus amyloliquefaciens* NJN-6 volatile compounds against *Fusarium oxysporum* f. sp. *cubense*. Appl Environ Microbiol 78:5942–5944

Yuan J, Zhao M, Li R, Huang Q, Raza W, Rensing C, Shen Q (2017) Microbial volatile compounds alter the soil microbial community. Environ Sci Pollut Res 24:22485–22493

Zhang H, Kim MS, Sun Y, Dowd SE, Shi H, Paré PW (2008a) Soil bacteria confer plant salt tolerance by tissue-specific regulation of the sodium transporter HKT1. Mol Plant Microbe Interact 21:737–744

Zhang H, Kim MS, Krishnamachari V, Payton P, Sun Y, Grimson M, Farag MA, Ryu CM, Allen R, Melo IS, Pare PW (2007) Rhizobacterial volatile emissions regulate auxin homeostasis and cell expansion in *Arabidopsis*. Planta 226:839–851

Zhang H, Sun Y, Xie X, Kim MS, Dowd SE, Paré PW (2009) A soil bacterium regulates plant acquisition of iron via deficiency-inducible mechanisms. Plant J 58:568–577

Zhang H, Xie X, Kim MS, Kornyeyev DA, Holaday S, Pare PW (2008b) Soil bacteria augment *Arabidopsis* photosynthesis by decreasing glucose sensing and abscisic acid levels in planta. Plant J 56:264–273

Zhang X, Li B, Wang Y, Guo Q, Lu X, Li S, Ma P (2013) Lipopeptides, a novel protein, and volatile compounds contribute to the antifungal activity of the biocontrol agent *Bacillus atrophaeus* CAB-1. Appl Microbiol Biotechnol 97:9525–9534

Zhao L-J, Yang X-N, Li X-Y, Mu W, Liu F (2011) Antifungal, insecticidal and herbicidal properties of volatile components from *Paenibacillus polymyxa* strain BMP-11. Agric Sci China 10:728–736

Zou C, Li Z, Yu D (2010) *Bacillus megaterium* strain XTBG34 promotes plant growth by producing 2-pentylfuran. J Microbiol 48:460–466

Chapter 9
Potential of PGPR in Improvement of Environmental-Friendly Vegetable Production

Haluk Caglar Kaymak

Abstract Plant growth promoting rhizobacteria (PGPR) can directly cause enhanced plant growth, early seed emergence or improvement in crop yields by supplying biologically fixed nitrogen and increasing minerals uptake, producing and secreting plant growth regulators and other beneficial substances. Also, PGPR indirectly influences the plant growth promotion by suppressing of pest and diseases in vegetables. Hence, replacement of chemical fertilizers and pesticides is required because of the adverse effect of these chemicals and their residues seriously pollute the environment, impart and also threaten the health of human and animals. Thus, people are focused on healthy products not only for growing vegetables but also for all agricultural products. In recent years because of increasing food-borne illnesses, the importance and potential of PGPR is hereby discussed for improvement of sustainable environment-friendly vegetable production for healthy human nutrition with special reference selected vegetable species, such as tomato, pepper, melon, radish and lettuce.

Keywords Plant growth promoting rhizobacteria · Vegetables · Sustainable production

9.1 Introduction

Although at least 10,000 plant species are used as vegetables worldwide, only 50 or so are considered of great economic and commercial importance (Decoteau 2000). These 50 species are vital for healthy diets and essential for human health, hence are not a luxury (Palada et al. 2006). Also, the widely used definition of the term vegetable is: herbaceous edible plants, or edible, usually succulent part of a plant that is eaten whole or in part such as roots, leaves or fruits, raw or cooked. In fact,

---

H. C. Kaymak (✉)
Department of Horticulture, Faculty of Agriculture, Atatürk University,
25240 Erzurum, Turkey
e-mail: hckaymak@atauni.edu.tr

D. K. Maheshwari and S. Dheeman (eds.), *Field Crops: Sustainable Management by PGPR*, Sustainable Development and Biodiversity 23,
https://doi.org/10.1007/978-3-030-30926-8_9

all over the world over 1150 million tons of vegetables are produced annually. China produces over 50% of the world vegetable production. The United States of America, India and Turkey are the other important vegetable producer countries in the world. The production of tomato, potato, garlic, onion and watermelon accounted for over 60% of the total world vegetable production. In addition, tomato is one of the major vegetable produced and consumed throughout the globe. With worldwide production reaching near 170 million tons, tomato is the seventh most important crop species after maize, rice, wheat, potatoes, soybeans and cassava (Bergougnoux 2014; FAOSTAT 2016).

When farmers began to produce or cultivate vegetables, soil fertility be taken into consideration for selected vegetable crops, disease prevention, precisely, "damping off" control and pest management. Although the global nitrogen cycle pollutes groundwater and increases the risk of chemicals, the excessive use of chemical fertilizers affects the soil health, soil physicochemical characteristics and also pollutes the ecosystem in addition to the degradation in natural resources and high cost. It is now an established fact that high-input farming practices achieving high yields have created environmental problems (Şahin et al. 2004; Nosheen and Bano 2014). Phosphorous (P), next to nitrogen, is one of the major and key nutrients limiting plant growth (Kumar and Narula 1999; Sundara et al. 2002; Podile and Kishore 2006). The sustainable, environmental-friendly and cost-effective measures, such as PGPR, can replace chemical fertilizers (Nosheen and Bano 2014) and pesticides that cause environmental pollution and degradation (Robacer, et al. 2016), resulting in contamination of the vegetables with toxic pesticides, fungicides, herbicides and insecticides residues. Eventually, the growers had to turn to new methods of alternative fertilization, especially biological new sources, so as to meet out the requirement of minerals besides, assisting in diseases suppression so as not to harm the environment and do not leave any toxin behind. Recently, consumers have focused on healthy products because of increasing food-borne illnesses. This pressure has led to these new quests mentioned by producers. Also, the excessive usage of chemical fertilizers and pesticides not only affect the soil properties and residues in vegetables such as nitrate accumulation but also pollute the environment in addition to the depleting resources and high cost. Therefore, efforts are being made to replace chemical fertilizers and pesticides with organically, environmental-friendly and cost-effective resources such as plant growth promoting rhizobacteria (PGPR). It is known that biological control of plant diseases has gained importance for healthy production of vegetables and other crops in recent years.

Rhizosphere, a hotspot of microbial diversity which houses a wide range of soil bacteria, belongs to the different group and with diverse properties influence the growth and health of plants directly or indirectly (Gowtham et al. 2017). Long back, pioneer work of Kloepper and Schroth (1978) stated that microbial communities exert beneficial for plant growth, yield, crop quality and plant health which have been called "plant growth promoting rhizobacteria (PGPR)" including numerous species of the genera *Acinetobacter*, *Aeromonas*, *Alcaligenes*, *Arthrobacter*, *Azospirillium*, *Azotobacter*, *Azoarcus*, *Bacillus*, *Beijerinckia*, *Burkholderia*,

Clostridium, *Enterobacter*, *Erwinia*, *Flavobacterium*, *Gluconacetobacter*, *Klebsiella*, *Pseudomonas*, *Serratia*, *Rhizobium* and so on (Kaymak 2010). PGPR plays an important role in plant growth by one or more mechanisms in direct plant growth promotion such as nitrogen fixation, phosphate solubilization, early seedling emergence, secreting of plant growth regulators and indirect plant growth promotion such as suppression of pest and disease (Tailor and Joshi 2014). In other words, Tenuta (2004) reported that PGPR improves plant health that can occur through three ways, namely: (a) decrease development of pest/disease (bioprotectant), (b) by secreting phytohormones (biostimulant) and (c) due to increasing availability of nutrients for plants (biofertilizer). For example, phosphate solubilizing bacteria (PSB) secretes organic acids and enzyme phosphatases, and converts the insoluble phosphates into soluble inorganic phosphates through the process of acidification, chelation, exchange reactions and production of organic acids (Whitelaw 2000; Gyaneshwar et al. 2002; Rodriguez et al. 2004; Chung et al. 2005; Hameeda et al. 2008). Generally, rhizobacteria has positive effects in occupying the ecological niche in order to provide compounds beneficial to the plants.

The use of plant–PGPR interactions for various crops is important and is useful not only for the enhancement of crop production but also for restoration of environment (Vessey 2003). Most of the free-living rhizobacteria confer protection against plant diseases via induced systemic resistance and promote growth and yield of numerous agronomically and ecologically important plant species, including vegetables. Therefore, the importance and potential of PGPR is discussed for improvement of environment-friendly vegetable production for human nutrition with special references to utilization of vegetable species such as tomato, pepper, melon, radish and lettuce.

9.2 Potential of PGPR in Some Selected Vegetable Crops

The potential of PGPR for environment-friendly vegetable production and healthy human nutrition in some selected vegetable species is discussed below.

*9.2.1 Tomato (*Lycopersicon esculentum *Mill.)*

Tomato, *Lycopersicon esculentum* Mill., is an annual shrubby member of Solanaceae family but it is a herbaceous perennial in a protected environment. The tropical coastal areas of Ecuador, Peru or Bolivia and portions of Northern Chile are considered originating areas and also native to tropical America. It was introduced to Central America and Mexico and also used by the Aztec and Toltec people, and seeds of tomato were taken back to Italy by the returning explorers as early as in the year 1554, and the plant quickly found favor in the warm Mediterranean climate. From there, tomatoes moved to Northern Europe and throughout the world. It was considered poisonous in early times because of the toxic alkaloid tomatine. After finding out its non-toxic nature, tomato cultivation was spread worldwide (Decoteau 2000; Peralta and Spooner 2007; Welbaum 2015). Today, tomato is not only produced for fresh market but also used widely in the food processing industry. Besides, a source of nutrients such as lycopene, β-carotene and vitamin C, its production and consumption are increasing in various countries (Bergougnoux 2014). It is an important crop for both the greenhouse and field vegetable growers throughout the world with a production area nearly 5 million ha and over 170 million metric tons of total production produced by China, India, USA and Turkey (FAOSTAT 2018).

Tomato is a warm season vegetable and injured by light frosts. The nitrogen is important for vegetative growth with a form of NH_3^+. Phosphorus is also important for plant development and flowering (Decoteau 2000; Welbaum 2015). On the other hand, numerous diseases adversely affect the crop production, and many bacterial, fungal and virus diseases attack tomato. Bacterial speck, bacterial soft rot, bacterial wilt, syringae leaf spot and so on are the most important bacterial diseases. In addition, some important fungal diseases of tomato are early blight, black leaf mold, powdery mildew, Fusarium wilt and Verticillium wilt. A large number of viruses such as cucumber mosaic, tobacco mosaic, tobacco leaf curl virus and tobacco ring spot also cause deleterious disease-causing entities.

The direct and indirect effects of PGPR have a great potential for improving growth and yield of tomato under different environmental conditions. Protection of vegetables against different plant diseases is quite important for environmental-friendly production. As mentioned before, Fusarium wilt caused by *Fusarium oxysporum* f. sp. *lycopersici* (Sacc.) W.C. Snyder & H.N. Hans, is common and important disease for tomato. Recently, Boukerma et al. (2017) reported that *Pseudomonas fluorescens* PF15 and *Pseudomonas putida* PP27 showed a significant protection against Fusarium wilt in tomato. Similarly, Fatima and Anjum (2017) observed that *Pseudomonas aeruginosa* PM12 has a great potential against Fusarium wilt in tomato, and act as a biocontrol agent that provides a complementary tactic for sustainable integrated pest management. Pastor et al. (2016) reported that inoculation of tomato seeds with *P. putida* strain PCI2 increases the resistance of plants to root rot caused by *F. oxysporum* and that *P. putida* PCI2 produces compounds such as 1-aminocyclopropane-1-carboxylic acid (ACC) deaminase, involved at different levels in increasing resistance. More recently, *Bacillus amyloliquefaciens* SN16-1 promotes plant growth and has great potential as a biocontrol of *F. oxysporum* f. sp. *lycopersici* in tomato (Wan et al. 2018). Another important disease of tomato is bacterial wilt caused by *Ralstonia solanacearum* that infects roots and multiplies in xylem vessels. Bacterial wilt results in huge economic losses each year in many tropical and subtropical areas because an effective control measures yet not been developed (Peng et al. 2017). According to Peng et al. (2017) report, the combination of saisentong (N, N-methylene-bis-(2-amino-5-sulfhydryl-1,3,4-thiadiazole) copper) and *Bacillus subtilis* B-001 effectively controls tomato bacterial wilt. On the other hand, *B. amyloliquefaciens* SQY 162 efficiently colonized the tomato rhizosphere, which directly suppressed *R. solanacearum* by secreting antibiotic surfactin in the tomato rhizosphere soil. Li et al. (2017) reported that *B. amyloliquefaciens* strain SQRT3 reduced tomato bacterial wilt by 68% and its biocontrol efficiency suppressed *R. solanacearum* populations in the rhizosphere soil significantly.

PGPR is being explored to harness their traits against *Meloidogyne incognita* causing root-knot in tomato. Sharma and Sharma (2017) reported that *Pseudomonas jessenii* strain R62 and *Pseudomonas synxantha* strain R81 reduced nematode infection in tomato under glasshouse conditions. Also, *P. putida*, *P. fluorescens*, *Serratia marcescens*, *B. amyloliquefaciens*, *B. subtilis* and *Bacillus cereus* reduced galling and egg masses on the roots by root-knot nematodes in tomato crops and increased yield (Almaghrabi et al. 2013).

Most of the vegetables are exposed to many abiotic stresses such as salinity and drought. The high level of salinity limited the crop productivity by influencing seed germination, growth, flowering and fruit setting not only in tomatoes but also in most of the vegetables that change according to the developing stages. PGPR can mitigate the adverse effect of salinity in tomato. For example, *Bacillus megaterium* strain A12 and *P. putida* strain A20 increased shoot length of tomato plants and dry biomass grown under varying salinity stress (100 and 200 mM NaCl). Further, *B. megaterium* strain A12 reduced endogenous ethylene production and increased water use efficacy better than *P. putida* strain A20 under field conditions (Aslam et al. 2018). In addition, Shen et al. (2012) suggested that *Erwinia persicinus* RA2 and *Bacillus pumilus* WP8 proved promising PGPR strains which are suited for application in salt marsh planting. Calvo-Polanco et al. (2016) reported that inoculation of *Variovorax paradoxus* 5C-2 and arbuscular mycorrhizal fungus *Rhizophagus irregularis* has a great potential in order to improve plant yield under conditions of drought stress.

It is well known that PGPR improves plant growth and yield via producing phytohormones and increases the availability of nutrients. There are many researches about the direct effect of PGPR in vegetable crops. Esquivel-Cote and Ramírez-Gama (2015) observed that *Azospirillum lipoferum* AZm5 proved useful for increasing N uptake in N-deficient soil by production of cytokinins, increase in ACC deaminase activity in tomato. Similarly, *Azospirillum brasilense* Sp7-S and *A. brasilense* Sp245 strongly enhanced root and shoot growth, seed germination index and vigor of tomato (Mangmang et al. 2015) Also, *B. fortis* IAGS162 and *B. subtilis* IAGS174 increased shoot length, root length, total biomass, total chlorophyll, carotenoid and sugar concentrations in tested tomato varieties under greenhouse conditions (Akram et al 2015). *B. circulans* CB7 increased seed germination, shoot length, root length, shoot dry weight and root dry weight, nitrogen, potassium and phosphorus content of shoot biomass in tomato (Mehta et al. 2015). IAA producing *P. putida* FA-8, FA-56 and FA-60 also increased the plant height, stem diameter, dry biomass and fruit yield of tomato (Hernández-Montiel et al. 2017). Similarly, phosphate solubilizing *P. putida* PCI2 exhibited a clear growth-promoting effect on shoot growth of tomato in the presence of L-tryptophan (Pastor et al. 2014). On the other hand, a commercial product, Bioroot®, containing *B. subtilis*, *P. fluorescens*, *Trichoderma harzianum*, yeast, algae and *Nocardia* improved the leaf area, shoot dry weight, root dry weight, volume of roots and root forks compared with the control, and also recommended as an alternative to tomato seedling growers' dependence on synthetic agrochemicals (Brutti et al. 2015). Earlier, Lee et al. (2008) have reported that *Rhodopseudomonas* sp. KL9 acts as an eco-friendly biofertilizer for cultivation of tomato and other lycopene-containing vegetable crops. *Burkholderia* sp. 7016 also enhanced the tomato yield and significantly promoted activities of soil urease, phosphatase, sucrase and catalase and can be used as a biofertilizer (Gao et al. 2015). Tripti et al. (2017) reported *Burkholderia* sp. L2 as inoculum which can tremendously enhance the productivity of tomato, soil fertility, and can also act as a sustainable substitute for chemical fertilizers. Various examples about the improving plant growth and yield of tomato by PGPR are given in Table 9.1.

Table 9.1 Examples of promoting effect of PGPR in tomato (*Lycopersicon esculentum* Mill.)

PGPR	Promoting effect	References
Rhodopseudomonas KL9	Seed germination, total length and dry mass of germinated seedling	Koh and Song (2007)
Mixture of Pseudomonas *putida* and *Trichoderma atroviride* (fungi)	Fresh weight, shoot and roots of seedlings	Gravel et al. (2007)
Bacillus subtilis BS13	Yield per plant, marketable yield, fruit weight, length and quality	Mena-Violante and Olalde-Portugal (2007)
Pseudomonas putida, Azotobacter chroococcum, Azosprillum lipoferum	Lycopene and antioxidant activity, shoot and fruit potassium content	Ordookhani et al. (2010)
Pseudomonas sp. RFNB3, *Serratia* sp. RFNB14	Early growth, root and shoot length, seedling vigor and dry biomass	Islam et al. (2013)
Pseudomonas putida UW4	Shoot length, shoot fresh and dry mass, and the chlorophyll concentration of tomato seedlings under salinity stress	Yan et al. (2014)
Bacillus amyloliquefaciens	Fruit yield suppression of *Fusarium oxysporum* f. sp *lycopersici*	Patakioutas et al. (2015)
Bacillus pumilus	Seedling growth	Anith et al. (2015)
Herbaspirillum seropedicae	Fruit biomass, nitrate uptake and nitrate reductase activity	Olivares et al. (2015)
Pseudomonas spp., *Bacillus* sp.	Seed germination, time of germination seedling growth	Widnyana and Javandira (2016)
Bacillus amyloliquefaciens IN937a *Bacillus pumilus* T4	N and P uptake	Fan et al. 2017 (2017)

*9.2.2 Pepper (*Capsicum annuum *L.)*

Pepper (*Capsicum annuum* L.) is an economically important genus of family Solanaceae It is a warm season, frost-sensitive crop that requires somewhat similar conditions as tomato and eggplant, that is, higher temperature requirement and commercially cultivated around the world due to its better adaptation to different agro-climatic regions. This crop is native to the Americas, and likely domesticated from central to southern Mexico. The other major species of pepper are *Capsicum frutescens*, *Capsicum chinense*, *Capsicum baccatum* and *Capsicum pubescens*. Although there are many commercial names of peppers, cultivars can be classified into two main types: mild or sweet fleshed fruit and hot or pungent fleshed fruit. It is a good source of vitamin A and C, in addition to potassium, phosphorus, calcium and iron, and can also be processed by canning or drying. It is an important crop for both the greenhouse and field vegetable growers worldwide. The total world production of fresh peppers including chillies was 34 million metric tons in 2 million ha production area (Decoteau 2000; Welbaum 2015; FAOSTAT 2018).

Fertilizer requirement of peppers is not as much as tomatoes. Fertilization is performed according to the soil and foliar testing results, and nitrogen is often applied prior to transplanting and again at first bloom. Phosphorus is often applied as fertigation because of limited solubility of phosphorus at transplanting to improve early growth (Welbaum 2015).

Peppers are also affected by serious phytopathogenic diseases. During pepper cultivation in greenhouses and fields, bacterial spot, bacterial soft rot, gray leaf spot, *Phytophthora* blight, wet rot and virus diseases such as tobacco mosaic virus are few examples among these diseases. Also, aphids make serious damages on pepper plants and reduced plant growth and yield (Herman et al. 2008; Mardani-Talaee et al. 2017).

Although chemical control method is common for disease control approaches to protect pepper plants from various diseases, but chemical control method accumulates residual pesticides in the ecosystem, pollutes the natural environment and causes human toxicity. Therefore, eco-friendly farming methods, such as crop rotation, and use of biological control agent are essential for sustainable agricultural development (Jung et al. 2015). Biological control agents in the form of PGPR can also be used to inhibit pathogens-causing diseases. Hahm et al. (2012) reported that *Ochrobactrum lupini* KUDC1013 and *Novosphingobium pentaromativorans* KUDC1065 are capable of eliciting induced systemic resistance (ISR) in pepper against bacterial spot disease. PGPR mixtures have a potential as biological control agent on pepper under greenhouse and field conditions. In addition, three bio-products, *Bacillus vallismortis* EXTN-1, *Bacillus* sp. and *Paenibacillus* sp. (ESSC) and *B. subtilis* (MFMF) are examined in greenhouse to reduce bacterial wilt, Fusarium wilt and foot rot. Thus, ESSC and MFMF reduced disease severity under greenhouse conditions and EXTN-1 provided a mean level of disease reduction under field conditions (Thanh et al. 2009). Infection of pathogenic fungi *Rhizoctonia solani* causes disease and decree in the yield of pepper. *B. subtilis*

SL-44 elicited strong inhibitory effect on *R. solani* and enhanced the biomass and length of pepper seedlings (Huang et al. 2017). On the other hand, *Kluyvera cryocrescens* KUDC1771 and *Brevibacterium iodinum* KUDC1716 have the ability of phosphate solubilization and production of phytohormones; *B. iodinum* KUDC1716 significantly decreased gray leaf spot disease in pepper hence, used as a potential agent for biological control (Son et al. 2014). In addition, Sang and Kim (2012) reported that Phytophthora blight of pepper caused by *Phytophthora capsici* is the most destructive disease in the commercial production of pepper. The biocontrol agent *Pseudomonas corrugata* CCR80 and *Chryseobacterium indologenes* ISE14 played a significant role in the suppression of *Phytophthora* blight of pepper. Similarly, Xu and Kim (2016) reported that *Paenibacillus polymyxa* SC0921 induced a defensive response against Phytophthora blight and promotes growth in pepper plants (Xu and Kim 2016).

The green peach aphid, *Myzus persicae*, is a polyphagous aphid that caused chlorosis in several vegetable plants. In one of the studies, *B. subtilis* and *B. amyloliquefaciens* proven useful in *Myzus persicae* management for pepper plants grown in locations having consistently high aphid pressure (Herman et al. 2008). Mardani-Talaee et al. (2017) investigated that the effect of zinc sulfate and vermicompost (30%) along with *B. subtilis*, *P. fluorescens*, *Glomus intraradices*, and combined effect of *G. intraradices* with *B. subtilis*, and *P. fluorescens* against biological parameters of *Myzus persicae*, and concluded that the best plant growth was attained due to application of zinc sulfate and *B. subtilis*.

The PGPR can directly cause modifications on plant metabolism such as N fixation, phosphate (P) solubilization, Fe sequestration, and cytokinin, gibberellin, indoleacetic acid, ethylene production and so on (Lucas et al. 2014). For example, inoculation with *Bacillus* spp. M9 and *B. cereus* K46 positively affected the performance of the photosynthetic mechanism in pepper plants. However, plants inoculated with *Bacillus* spp. M9 showed better performance in all vegetative plant growth characteristics in sustainable agriculture programs (Samaniego-Gámez et al. (2016). Similarly, *Bacillus licheniformis* increased the height of plants, leaf area and had less disease that of non-treated plants, and hence, it could be used as a biofertilizer or biocontrol agent in greenhouses (Garcia et al. 2004). Reduction of inorganic fertilizer application up to 25% with *P. putida* increased the growth and productivity of King Pakal hybrid habanero pepper (Chiquito-Contreras et al. 2017; Pastor-Bueis et al. 2017). During a field experiment, it is observed that the *Bacillus siamensis* that fertilized with decreased mineral N (80%) produced significantly better pepper yields. In addition, Supanjani et al. (2006) reported that phosphate and potassium solubilizing bacteria *B. megaterium* var. *phosphaticum* provides a sustainable alternative to the use of synthetic fertilizers for hot pepper production in soil with low fertility.

PGPR has potential to alleviate some stress factors such as salt and drought stress in different crops including pepper. The inoculation of pepper plants with *Microbacterium oleivorans* KNUC7074, *B. iodinum* KNUC7183 and *Rhizobium massiliae* KNUC7586 can alleviate the destructive effects of salt stress on plant growth (Hahm et al. 2017). Drought stress is also one of the important affecting

Table 9.2 Examples of promoting effect of PGPR in pepper (*Capsicum annuum* L.)

PGPR	Promoting effect	References
Formulation LS255 (*Bacillus subtilis* GBO3 + *Bacillus subtilis* IN937b), Formulation LS256 (*Bacillus subtilis* GBO3 + *Bacillus pumilis* INR7)	Yield of extra-large tomato fruit and total yield	Kokalis–Burelle et al. (2002)
Bacillus subtilis	Suppressing of *Phytophthora* blight	Lee et al. (2008)
Bacillus aryabhattai RS341 *Brevibacterium epidermidis* RS15 *Micrococcus yunnanensis*	Growth promotion under salt stress	Siddikee et al. (2012)
Pseudomonas corrugata CCR80 *Chryseobacterium indologenes* ISE14	Suppressing of *Phytophthora* blight	Sang and Kim (2012)
Pseudomonas sp. RFNB3 *Serratia* sp. RFNB14 *Novosphingobium* sp. RFNB21	Solubilizing tri-calcium phosphate and zinc oxide, nitrogen fixation, chlorophyll content, macro- and micro-nutrient nutrient uptake	Islam et al. (2013)
Bacillus sp.	Root and fruit weights and total yield	Hernandez-Castillo et al. (2014)
Bacillus amyloliquefaciens, *Bacillus subtilis*	Suppressing of *Fusarium* wilt	Wu et al. (2015)
Chryseobacterium sp. ISE14	Phosphate-solubilizing suppressing of *Phytophthora* blight	Sang et al. (2018)

factors for pepper plant, and auxin and ACC deaminase producing *B. licheniformis* K11 reduces drought stress in pepper in drought-affected regions without the need for overusing agrochemicals and chemical fertilizer (Lim and Kim 2013). Similarly, *Burkholderia seminalis* and arbuscular mycorrhizal fungi, *Rhizophagus intraradices* and *Rhizophagus fasciculatum* increase biomass, root length, shoot length and chlorophyll content of pepper plants under drought stress conditions (Tallapragada et al. 2016). Examples about direct and indirect effect of PGPR on pepper are given in Table 9.2.

9.2.3 *Eggplant (*Solanum melongena* L.)*

There are more than 1000 species of eggplant (*Solanum melongena* L.); however, a few of them are produced commercially as vegetable for human nutrition. Total world production of eggplant is over 45 million metric tons with an average yield of 10 tons/acre (Welbaum 2015). Production of eggplant is limited in a few region of world because it is a traditional vegetable crop in tropical, subtropical and

Mediterranean countries (Abd El-Azeem et al. 2012) but is an important vegetable in China, India, Turkey and Egypt. Fertilization is important in production of eggplant because of its fairly high nutrient requirement. Eggplants have better drought tolerance than other crops of the family and generally quite sensitive to salinity (Welbaum 2015).

The effect of PGPR in raising high yielding eggplant is insufficient when compared to that of pepper and tomato. PGPRs are focused as suppressive agent to pathogens and tolerate salt stress. For example, *Xanthobacter autotrophicus* BM13, *Enterobacter aerogenes* BM10 and *Bacillus brevis* FK2 from the salt-affected maize kidney bean reduced the negative effects of salinity while *E. aerogenes* was capable of promoting eggplant growth and yield when compared to an uninoculated control (Abd El-Azeem et al. 2012). Similarly, inoculating with *Pseudomonas* sp. DW1 increased shoot Ca^{2+} of eggplant compared to the non-inoculating eggplant plants under salinity. It is interesting to note that the salinity decreased superoxide dismutase activities, and increased peroxidase activities. Numerous soil-borne organisms are harmful to the crop.

R. solanacearum (Smith) Yabuuchi is an important soil-borne bacterial plant pathogen causing bacterial wilt. The talc formulation of two species of *Pseudomonas* (RBh41 and RBh42) completely suppressed the disease incidence of eggplant wilt up to 36 days of inoculation under greenhouse conditions. Also, *Pseudomonas mallei* (RBG4, ET17) and one *Bacillus* spp. RCh6 reduced eggplant wilt incidence by 83% compared to control (Ramesh and Phadke 2012). Ramesh et al. (2009) reported that *Pseudomonas* EB9 and EB67; *Enterobacter* EB44 and EB89 and *Bacillus* EC4 and EC13 reduced the wilt incidence by more than 70%. Similarly, *F. oxysporum* f. sp. *melongenae (Fomg)* caused the most destructive disease in eggplant (*Solanum melongena* L.). Various strains of *Pseudomonas* and *B. cereus* significantly reduced disease and induced resistance to that of Fusarium wilt of eggplant (Altinok et al. 2013). Yıldız et al. (2012) reported that *Bacillus* and *Pseudomonas* species showing good performance were tested against *F. oxysporum* f. sp. *melongenae* for the antagonistic activities in eggplant.

*9.2.4 Melon (*Cucumis melo *L.)*

Melon (*Cucumis melo* L.), belongs to the family Cucurbitaceae and is an important vegetable, widely grown in the world with a production of about 27 million mt. The crop is mainly cultivated in China, Iran, Turkey, Egypt and United States (Welbaum 2015; Sharma et al. 2017). It is originated from Iran (Decoteau 2000) with rich genetic diversity (Balkaya and Karaağaç 2006). The economically important groups of melons are Cantaloupensis and Indorous. A number of disease such as Fusarium wilt, powdery mildew, downy mildew, bacterial wilt and bacterial blotch seriously affect melons in conventional production.

Fertilizer requirements of melons are moderate compared with other vegetable species due to development of extensive root systems to uptake need-based water and available nutrients (Welbaum 2015). Nitrogen and phosphorus are essential key nutrients for adequate growth and yield for melon. Welbaum (2015) suggested that nitrogen applied in two side-dressing or through fertigation must be low during flowering to promote fruit set. The phosphorus must be banded lightly but producers should be careful for alkaline mineral soils because of chemically tie up phosphorus, unavailable to the plants.

Available literature revealed that PGPRs are inadequate for melons and more work is necessary to carry out eco-friendly melon production. The effects of PGPR on melons reveal their suppressive effect on some plant diseases and plant growth promotion. For example, *B. subtilis* UMAF6639 was able to induce systemic resistance (ISR) in melon and provide additional protection against powdery mildew because of producing lipopeptides by activation of jasmonate- and salicylic acid-dependent defense responses (García-Gutiérrez et al. 2013). Further, García-Gutiérrez et al. (2012) observed that *B. subtilis* UMAF6614, UMAF6639 and *cereus* UMAF8564, and *P. fluorescens* strains, UMAF6031 and UMAF6033 provide protection for melon against powdery mildew and angular leaf spot. *Bacillus* sp. RAB9 and *B. lentimorbus* MEN2 can be used as biocontrol agent for both melon seedlings and for preventing against bacterial blotch (Medeiros et al. 2009). Suppression of various soil-borne disease caused by *F. oxysporum* and *R. solani* also occurred due to appreciation of *B. subtilis* (Singh et al. 2017). Preventing plants against fungal and bacterial diseases yet to explore require new, safer and effective formula as an alternative to chemical pesticides for growing melon and other vegetable production.

The early vegetative and reproductive plant, parameters such as plant height, root dry weight, nitrogen and sodium concentrations in foliage and fruit weight of melons increased when *B. cereus* (N198) and *P. fluorescens* applied under greenhouse conditions (Rodriguez Mendoza et al. 2013). Strains of *B. subtilis* UMAF6614 and UMAF6639, *B. cereus* UMAF8564, and *P. fluorescens* strains, UMAF6031 and UMAF6033 promote plant growth and increase fresh weight up to 30% of melon seedlings as stated by García-Gutiérrez et al. (2012). Lee et al. (2015) suggested that *Enterococcus faecium* LKE12 promotes plant growth by producing GAs and IAA. Application of its cell-free culture extract proved to enhance plant growth. The root fresh/dry weight, shoot dry weight, plant height, foliar area, fruit yield and quality increased when *B. subtilis* LAL-36, BEB-23, BEB22 and BEB-13 applied and finally concluded that *B. subtilis* has potential to obtain high-quality melon fruits with increased profits (Abraham-Juárez et al. 2018). Researches clearly showed that PGPR has a great potential not only to improve yield and quality of melons but also suppressive to different diseases.

9.2.5 *Cucumber (*Cucumis sativus *L.)*

Cucumber (*Cucumis sativus* L.) belongs to the family Cucurbitaceae and is an important crop for both greenhouse and field growers throughout the world. It is cultivated in a production area near 2 million ha with average 37–40 million mt. total production. The centre of origin of cucumber is believed to be India and also cultivated in the ancient Egypt. Cultivation of cucumber spread to China and Greece from India about 2000 years ago before Italy, France and other European countries and USA (Welbaum 2015).

Cucumber is a good source of vitamin A, vitamin C, potassium and poor source of protein, carbohydrates and fats; however, it is highly sensitive to environmental conditions (Radhakrishnan and Lee 2013; Welbaum 2015). Cucumbers are warm season, cold-sensitive crop and require warm soil temperatures both for germination, early harvest and high yield (Decoteau 2000). Cucumbers respond to high levels of fertilizer, and nitrogen is the key nutrient for successful cucumber production. In fact, from seedling to the maturity, the crop experiences several influences of bacterial and fungal community able to cause diseases like downy mildew, powdery mildew, damping-off, angular leaf spot, bacterial wilt, Fusarium wilt, and so on (Folman et al. 2003), and can be given as examples for important fungal and bacterial diseases of cucumber.

Pesticides are generally the first choice for the control of the diseases for growers in conventional cucumber production, and gained importance at suppressing of cucumber pathogens because of ISR by triggered preinoculating with PGPR (Van Loon et al. 1998). Jeun et al. (2004, 2007) inoculation with *S. marcescens* and *P. fluorescens* induced systemic protection of cucumber plants against the anthracnose pathogen, *Colletotrichum orbiculare*. The application of *B. subtilis* strain B4 combined with acibenzolar-S-methyl resulted in promotion of plant growth and systemic resistance against anthracnose infection (Park et al. 2013). Recently, Akköprü and Özaktan (2018) reported that *P. putida* AA11/1 significantly reduced average severity of angular leaf spot in disease-sensitive cultivar. The plant defense enzymes stimulated in cucumber roots colonized due to *P. corrugata* and *P. aureofaciens*, thus reduced *Pythium aphanidermatum* (Chen et al. 2000). Utkhede et al. (1999) reported that *B. subtilis* BACT-0 increased shoot growth and weight of cucumber plants in terms of fruit yield, and fruit number under greenhouse conditions. According to Ongena et al. (2000) positive effects of *P. putida* BTP1 and its sid$^-$ mutant M3 inhibited *P. aphanidermatum* by eliciting phytoalexins systemically. Du et al. (2016) showed that *P. polymyxa* NSY50 influenced amino acid metabolism to increase biomass of the plant and regulate jasmonic acid pathway that trigger resistance against *F. oxysporum* f. sp. *cucumerinum*. The plant defense-related enzymes such as peroxidase and polyphenol oxidase were significantly increased in seed treated with *B. subtilis* B579 against *F. oxysporum* f. sp. *cucumerinum* in cucumber plants (Chen et al. 2010).

Organic farming is a technique designed to produce healthy food for human nutrition, therefore biofertilization with PGPR has attained importance. Kang et al. (2010) reported an environment-friendly alternative of chemical fertilizers in the agriculture industry as beneficial bacteria such as *Burkholderia* sp. KCTC 11096BP if applied. The growth attributes of cucumber such as shoot fresh weight and biomass, chlorophyll contents, soluble sugar contents and crude protein content were promoted. Similarly, other genera, namely *Rhodobacter sphaeroides*,

Lactobacillus plantarum and *Saccharomyces cerevisiae*, induced the shoot length, root length, shoot fresh weight, shoot dry weight, chlorophyll contents and total amino acids via secretion of IAA and/or organic acids. In addition, *R. sphaeroides* increased the calcium, potassium, magnesium and phosphate contents involved to regulate the mineral content in cucumber (Kang et al. 2015). The elemental nutrients were increased due to application of PGPR (Orhan et al. 2006; Kaymak et al. 2013). Pii et al. (2015) observed the presence of *A. brasilense* in soil, which facilitate a rapid and faster recovery of cucumber plants from iron deficiency symptoms by increasing chlorophyll content, biomass and iron content of cucumber leaves.

Almost majority of vegetables including cucumber are exposed to many stresses including both biotic and abiotic stresses such as attach of pathogens, low temperature and soil salinity. Diverse bacterial genera such as *B. cepacia* SE4, *Promicromonospora* sp. SE188 and *A. calcoaceticus* SE370 not only reduced sodium ion concentration but also increased potassium and phosphorus contents. In such cases, the increased biomass and chlorophyll contents of cucumber under salinity and drought stress have also been observed (Kang et al. 2014). It was reported that auxin-producing *K. planticola* strain TSKhA 91 exhibited protective effect on development of cucumber increased seed germination and root formation under conditions of low temperature (Blinkov et al. 2014).

*9.2.6 Watermelon [*Citrullus lanatus *(Thunb.) Matssumura and Nakai]*

Similar to melon, watermelon (*Citrullus lanatus* (Thunb.) Matsum & Nakai) belongs to the family Cucurbitaceae. The origin of watermelon is unclear but it was known that its cultivation started in Nile valley before 2000 BC. It is the most common cucurbit cultivated in many regions of the world. It has over 3,500,000 ha production area with approximately 117 million mt production. It is generally grown for the sweet juicy fruit pericarp tissue and used fresh as a dessert fruit and fruit salads. Watermelon consists mostly of water and contain rich source of vitamins, minerals lycopene and soluble sugars. The crop is of warm season, frost-sensitive that requires up to 4 months of frost-free weather. Fertilizer requirements are moderate because watermelons tend to develop extensive root systems in the upper profile of the soil for available nutrients. Cultivators prefer direct-seeded production in warm regions but transplanting is more useful in temperate regions (Decoteau 2000; Zohary et al. 2012; Welbaum 2015; FAOSTAT 2018).

The most common diseases of watermelon are bacterial wilt, Fusarium wilt, leaf spot, anthracnose, alternaria leaf blight, downy and powdery mildew. Bacterial fruit blotch is known as a serious disease to watermelon growers in the world. Both *P. polymyxa* (SN-22) and *Sinomonas atrocyanea* (NSB27) showed inhibitory effect on bacterial fruit blotch resulting in increase in growth parameters of watermelon under greenhouse conditions (Adhikari et al. 2017). Strains of *B. subtilis* GBO3, *B. amyloliquefaciens* IN937, *B. pumilus* INR7 and SE34 reduced angular leaf spot lesions and gummy stem blight and exhibit an increased shoot weight seedlings growth and health (Kokalis-Burelle et al. 2003). Nga et al. (2010) reported that gummy stem blight, caused by *Didymella bryoniae* (Auersw.) Rehm., can be checked by *P. aeruginosa* 23_{1-1} by antibiosis and induced resistance under greenhouse and field conditions. Recently, Yaoyao et al. (2017) recommended that *P. polymyxa* (SQR-21) is not only PGPR but also an effective biocontrol agent against Fusarium wilt. Kokalis-Burelle (2004) examined few strains of PGPR reduced angular leaf spot lesions caused by *P. syringae* pv. *lachrymans*, and gummy stem blight, caused by *D. bryoniae* resulted in formulation of a product (Bio yield (TM)).

*9.2.7 Lettuce (***Lactuca sativa** *L.)*

Lettuce (*Lactuca sativa* L.) has four morphological forms: crisphead, butterhead, romaine or cos, and loose leaf is a member of family Asteraceae. It is a major world salad crop (Jeffrey 2007; Welbaum 2015), and is a source of vitamins and nutrients required for human health (Chamangasht et al. 2012). The crop is best in cool growing environments and can be grown in temperate regions and requires a well fertilized soil for high yield because of a weak, shallow root system (Decoteau 2000). Lettuce's nutritional requirements are fulfilled during 3–4 weeks before harvest. Nitrogen is the most important fertilizer because of the plant's limited root system for rapid and continuous growth (Swaider et al. 1992). On the other hand, lettuce mosaic virus, turnip mosaic virus, bottom root, lettuce drop, downy mildew, bacterial leaf spot, varnish spot verticillium wilt and so on are some of the most important diseases that adversely affected the lettuce growth and quality.

There are a lot of agricultural practices applied for increasing yield components of lettuce. One of them is application of PGPR for promoting growth and environment-friendly and sustainable production of lettuce. In one of the first reports, phosphate-solubilizing *Rhizobium leguminosarum* bv. *phaseoli* strains P31 and R1, *Serratia* sp. strain 22b, *Pseudomonas* sp. strain 24 and *Rhizopus* sp. strain 68 showed their PGP potential on lettuce and *R. leguminosarum* bv. *phaseoli* induced the growth of lettuce maximally under field conditions (Chabot et al. 1996). According to Flores-Félix et al. (2013), microbial bioinoculants are an effective way for sustainable and environment-friendly agriculture due to their reduction of the use of chemical fertilizers and declared that *R. leguminosarum* PEPV16 actively colonizes the rhizosphere of lettuce resulting in increase in plant growth as well as N and P contents in lettuce. Similarly, Chamangasht et al. (2012) reported that species of *Azotobacter, Azospirillum* and *Pseudomonas* strain 187 proved effective on the enhancement of vegetative characters of lettuce. Among these genera, *Azospirillum* spp. was most effective in increasing plant height and yield due to its salt stress management (Fasciglione et al. 2012). In fact, *Azospirillum*-inoculated lettuce seeds yield a higher number of superior quality lettuce than non-inoculated controls grown under NaCl stress. A field-trial was undertaken to compare the effect of inorganic fertilizer. A PGPR, *Pseudomonas mendocina*, alone or in combination with inorganic fertilizer, on plant growth and nutrient uptake in lettuce exhibited a significant effect of PGPR bacteria on the dehydrogenase and phosphatase activities (Kohler et al. 2006). Hoffmann-Hergarten et al. (1998) demonstrated that *Pseudomonas* sp. W34 and *B. cereus* S18 inhibited *M. incognita* on lettuce and seed treatment by significant reductions in root galling resulting in enhanced seedling biomass, besides having a suppressive effect on pathogen *Fusarium* sp. for biological control in lettuce (Sottero et al. 2006).

*9.2.8 Spinach (*Spinacia oleracea *L.)*

Similar to other foliage vegetable crops, spinach (*Spinacia oleracea* L.) belongs to the Chenopodiaceae family and the group of leafy vegetables contains important minerals and vitamins, has good cooking adaptability, and hence consumer demand increases annually (Nishihara et al. 2001). This vegetable is consumed steamed, fried or even raw at the early growth stages (Krężel and Kołota 2014). Spinach is produced in all regions of the world, and it has over 900,000 ha production area with approximately 26 million mt agricultural field. China alone produced over 90% of total world production (FAOSTAT 2018). It is produced by direct-seeded method. The optimum seed germination occurs at about 15–21 °C. Spinach plants are sensitive to acidity because of this reason: rainy regions are not suitable for spinach production because of low soil pH (Vural et al. 2000; Welbaum 2015).

Nitrogen is the key nutrient for spinach, and easily accumulates in the form of nitrate which is harmful for human health. N_2-fixing plant growth promoting rhizobacteria has a great potential for eco-friendly and healthy production of spinach. N_2-fixing, phytohormone producing and P-solubilizing bacterial species *B. cereus* RC18, *B. licheniformis* RC08, *B. megaterium* RC07, *B. subtilis* RC11, *Bacillus* OSU-142, *Bacillus* M-13, *P. putida* RC06, *P. polymyxa* RC05 and RC14 increased spinach shoot fresh weight, leaf area and plant height (Çakmakçı et al. 2007). Hou and Oluranti (2013) reported that *P. putida* NWU12, *P. fluorescence* NWU65, *P. fluvialis* NWU37 and *Ewingella americana* NWU59 are all positive for ammonia, catalase, hydrogen cyanide and phosphate solubilization, thus helpful in increasing plant heights. Although growth of spinach was not promoted when only the organic materials were applied, multiple applications with fluorescent *Pseudomonas* strains and organic materials proved better for promoting the growth because fluorescent *Pseudomonas* strains aggressively colonized the roots when the organic materials were applied (Urashima et al. 2005). Interestingly, Jiménez-Gómez et al. (2018) reported *Rhizobium laguerreae* as an excellent plant probiotic, in increasing yield and quality of spinach.

9.2.9 *Radish (***Raphanus sativus** *L.)*

Radish, *Raphanus sativus* L., is cultivated in three distinct types: spring or summer, daikon and winter. It is a member of Brassicaceae family, known as a cool season vegetable because it does not grow well in hot and dry weather, and considered hardy in cold temperatures. An optimal fertilization with N (100 kg h^{-1}) and P (80 kg h^{-1}) is sufficient to promote growth and yield of this salt-sensitive crop. The plant suffers from various diseases, such as rhizoctonia root rot, Fusarial wilt, black spot, powdery mildew and downy mildew. However, the most important physiological factor limited radish growth and yields are day length and vernalization. Especially in long days, most of the radish cultivars bear flowers without forming a root. In other words, radish cultivars have a facultative vernalization response because cold exposure is not required for flowering, but flowering will occur more rapidly after cold treatment. In contrast, radish cultivars have an obligate requirement for cold treatment and do not flower without prior cold exposure (Kaymak and Güvenç 2010).

Nitrate accumulation and heavy metal toxicity are major problems in production of most of the vegetables including radish. Heavy metals such as cadmium, nickel and lead contamination are ubiquitous and cause toxicity to radish. According to Ahmad et al. (2018), application of compost and PGPR ameliorates Pb toxicity in radish. *Bacillus* sp. CIK-512 and compost synergy improved growth and physiology of radish in contaminated soil, possibly by antioxidant activities and reduction in lead accumulation in shoot. Similarly, use of both PGPR *Bacillus* sp. CIK-516 and *Stenotrophomonas* sp. CIK-517Y improved the growth of radish under four different levels of Ni contamination. The plant growth, dry biomass, chlorophyll and nitrogen contents were significantly reduced in the plant due to exposure of Ni, but *Bacillus* sp. CIK-516 and *Stenotrophomonas* sp. CIK-517Y overcame the negative effects of Ni stress on radish by improving the overall growth parameters (Akhtar et al. 2018).

Bio-priming with PGPR genera (*Agrobacterium rubi* strain A 16, *Burkholderia gladii* strain BA 7, *P. putida* strain BA 8, *B. subtilis* strain BA 142 and *B. megaterium* strain M 3) significantly improved the percentage of seed germination of radish under saline conditions (Kaymak et al. 2009). Hong and Lee (2017) reported that *Arthrobacter scleromae* SYE-3, an isolate from local plants grown in saline soil, evidenced in enhancing yield in salinized environments. On the other hand, the evidence of gibberellin (GA) production by PGPR is rare (Vessey 2003). The new isolate *Leifsonia soli* sp. SE134 significantly promoted the growth of cucumber, tomato and radish due to significant GA production capacity (Kang et al. 2014).

It is known that PGPR commonly use to suppress the plant diseases. In this crop, seed and pre-plant applications of cell suspension of *P. fluorescens* Pf 9A-14, *Pseudomonas* sp. Psp. 8D-45 and *B. subtilis* Bs 8B-1 suppressed rhizoctonia damping-off of radish (Khabbaz et al. 2015). Long back, De Boer et al. (2003)

observed that combination of *P. putida* strains WCS358 and RE8 leads to more effective and more reliable biocontrol of Fusarium wilt in radish.

9.2.10 *Cabbage (***Brassicae oleraceae** *var.* **capitata** *L.,) and Chinese Cabbage (***Brassica campestris** *L. subsp.* **pekinensis***)*

Cruciferous crops, such as broccoli, cabbage, cauliflower, are important vegetable crops and widely grown in different regions of world with a production of cabbages and other brassicas including cauliflowers and broccoli 96 million mt. in 3,816,000 ha production area (FAOSTAT 2018). Different fungi, harmful bacteria, pest and pathogens cause various seed and soil-borne disease. *Xanthomonas campestris* pv. *campestris* (Xcc) causes black rot disease in cabbage. The most significant bacteria *Bacillus velezensis* AP218 significantly reduced disease incidence and disease severity and exhibited biocontrol of black rot of cabbage under field and greenhouse conditions (Liu et al 2016a). Application of the *P. fluorescens* as seed treatment significantly reduced black rot of cabbage under greenhouse conditions (Umesha and Roohie 2017). Different species of *Bacillus,* namely *B. cereus*, *B. subtilis* and *B. amyloliquefaciens* treatments efficiently suppressed the cabbage aphid (*Brevicoryne brassicae*) field populations in varying magnitudes (Gadhave et al. 2016). Samancioglu et al. (2016a) reported that *B. pumilus* TV-67C strains increase drought stress tolerance in cabbage seedlings by accelerating the accumulation of inherent levels of superoxide dismutase, catalase and peroxidase, amino acid and hormone production. In addition, combined effect of *B. megaterium* TV6D and *Pantoea agglomerans* RK-92 + *Brevibacillus choshiensis* TV-53D mitigates drought stress tolerance in cabbage plants by accumulating antioxidant enzymes, osmolytes, hormone production and decreased electrolyte leakage in treated plants under water-deficit conditions (Samancioglu et al. 2016b). Yildirim et al. (2015, 2016) observed that seed and seedling inoculation with *P. polymyxa* RC14 increased shoot fresh and dry weight, root fresh and dry weight, macro- and micro-nutrient content of leaf and root of cabbage seedlings, yield and yield parameters as well as chlorophyll reading value and stomatal conductance versus controls. Turan et al. (2014) showed that treatments with *B. megaterium* TV-91C, *Pantoea agglomerans* RK-92 and *B. subtilis* TV-17C improved seedling growth and quality in cabbages.

Different strains of *Bacillus velezensis* AP136, AP188, AP213, AP218, AP295 and AP305 reduced the number of foliar lesions of black rot; other bacilli such as *Bacillus safensis* AP7 and *Bacillus altitudinis AP18* increased shoot fresh and dry weight, root fresh and dry weight of Chinese cabbage (Liu et al 2016b). Shrestha et al. (2009) reported that *Lactobacillus*, *Lactococcus* and *Paenibacillus* spp. can be further developed as a stable biological control agent to manage soft rot caused by *Pectobacterium carotovorum* subsp. *carotovorum*, which greatly affect the Chinese cabbage production worldwide.

9.2.11 *Broccoli and Cauliflower (***Brassica oleracea** *L. var.* **italica** *Plenck and* **Brassica oleracea** *var botrytis L.)*

Broccoli (*Brassica oleracea* L. var. *italica* Plenck) belongs to the Brassicaceae family member, and is spread in the Mediterranean region (Decoteau 2000). This crop is an important vegetable because of its high selling price and its vitamin-rich, high fibers and low calorie properties (El-Nemr et al. 2011). In other words, broccoli is a low sodium food and is fat and cholesterol free, high in vitamin C and a good source of folate (Decoteau 2000). USA is the world's largest producer of broccoli in 60,000 ha (Welbaum 2015). Heads of broccoli are sold fresh and boiled or steamed for eating, and also uncooked broccoli is eaten in salads without cooking. Broccoli is also processed by freezing and dried (Welbaum 2015). There are many diseases such as black rot, blackleg, downey mildew, bacterial leaf spot, Verticillium wilt, Fusarium wilt, which adversely affect broccoli (Decoteau 2000; Rimmer et al. 2007; Welbaum 2015).

Available literature revealed that different groups of PGPR have a great potential for improving growth and yield of broccoli under different environment conditions. More recently, Altuntaş (2018) reported that *B. subtilis* strain QST 713 increased the yield, plant growth parameters and nutrient uptake of broccoli and ascorbic acid contents. Both Gram-negative and Gram-positive bacteria such as *B. cereus*, *Brevibacillus reuszeri* and *Rhizobium rubi* inoculations with manure increased yield, plant weight, head diameter, chlorophyll content, nitrogen, potassium, calcium, phosphorus, magnesium and iron content of broccoli (Yildirim et al. 2011). Further, *P. fluorescens* have great potential in enhancing the growth, phosphatase activity, chlorophyll content, nutrient uptake and yield of broccoli when combined with 80 kg ha^{-1} superphosphate (Tanwar et al. 2014).

Among vegetable crops, cauliflower (*Brassica oleracea* var. *botrytis* L.) is one of the most important winter vegetable crops (Devi et al. 2017). It is a cool season, frost-tolerant crop widely adapted and grown in both temperate and subtropical regions. It is a low calorie crop but a good source of vitamin C (Decoteau 2000; Welbaum 2015). Sold fresh and boiled, baked or steamed and is also processed by freezing and drying. In Asian countries like China and India, it is produced on large scale and about 70% of its total world production occurs in Asia. Cauliflower shares most of the same insect pest and disease similar to that of broccoli. The crop suffered from black rot, black leg, alternaria leaf spot, bacterial leaf spot, soft rot verticillium wilt and Fusarium wilt at various stages of development (Decoteau 2000; Rimmer et al. 2007; Welbaum 2015).

Cauliflower requires high nitrogen as nutrient, and thus deficiency of nitrogen causes low yield, delayed maturation and low productivity. Its requirements are also high for boron and molybdenum (Swaider et al. 1992; Welbaum 2015). The N_2-fixing PGPR has a great importance for this reason in healthy crop production.

Khan et al. (2018) reported that nitrogen-fixing by rhizobacteria has a great potential to improve the yield of economic plants like cauliflower. Bhardwaj et al. (2017) observed that *Bacillus* sp. SB_{11} has tremendous potential to be used as a biofertilizer/bioprotectant for growth promotion and natural protection of cauliflower under low hill conditions of Himachal Pradesh in India. Further, *B. pumilus* with inorganic P-source have been used to enhance growth, yield and quality of cauliflower under mid hill conditions (Dipta et al. 2017). The use of PGPR isolates as biofertilizers proved beneficial for cauliflower cultivation because of the enhanced growth occurred mainly due to bacteria-mediated IAA production and phosphorus solubilization (Kushwaha et al. 2013). Similarly, the role of *Bacillus* spp. MK_5 with N and P as biofertilizer established their nature as beneficial bacteria for cauliflower cultivation as evidenced by induced IAA production and phosphorus solubilization (Kaushal et al. 2011). The application of *Bacillus* spp. MK_5 combined with recommended dose of 75% NP fertilizers significantly and increased the cauliflower yield (Kaushal et al. 2013). In addition, Ekinci et al. (2014) reported that *B. megaterium* TV-3D, TV-91C, TV-87A and KBA-10, *P. agglomerans* RK-92 and *B. subtilis* TV-17C increased fresh shoot weight, dry shoot weight, root diameter, root length, fresh root weight, dry root weight, plant height, stem diameter, leaf area and chlorophyll contents of cauliflower transplants. In addition, combined application of fertilizers, manures and PGPR is not only a cost-effective nutrient module but also helps in getting higher yield and quality with 25% net saving of fertilizers, as reported by Thakur et al. (2018).

*9.2.12 Mint (*Mentha sps*)*

Mentha sps (Mint) belong to the family Lamiaceae aromatic. It is one of the most important essential oils crops with a specific flavor, and the crop is perennial. Its leaves are picked, dried and used until needed (Abbas 2009). It is also cultivated worldwide for production of essential oils (Lawrence 2007). Besides its dry usage, fresh leaves are also utilized as a leafy vegetable in Turkey and in some Middle-East countries (Kaymak et al. 2013). Total world production of mint or peppermint (*Mentha* × *piperita*) was 106,252 mt. in 3390 ha production area (FAOSTAT 2018).

Various groups of PGPR induced growth and reproduction in medicinal and aromatic plants Kaymak et al. (2013) reported that *P. putida* biotype B C3/101, *P. polymyxa* RC105 and urea treatments induced growth, which resulted in significant increase in total fresh and dry yields of mint compared with the control. Although yield obtained due to treated plant by bacterial inoculation was lower than urea treatment, but it was more than that of control. Both *P. putida* biotype B C3/101 and *P. polymyxa* RC105 have the potential to increase the yield, growth and mineral composition of leaves. In addition, *B. subtilis* GB03, *P. fluorescens* WCS417r and *P. putida* SJ04 increased shoot and root biomass, leaf area, node number, trichome and stomatal density. There were marked qualitative and quantitative changes in monoterpene content of *M. piperita* (Cappellari et al. 2015). Similarly, *P. fluorescens* WCS417r, *B. subtilis* GB03 and *P. putida* SJ04 enhanced synthesis of phenolic compounds in leaves of *M. piperita* and have the potential to improve the productivity (Cappellari et al. 2017). Santoro et al. (2011) reported that volatile organic compounds of *P. fluorescens*, *B. subtilis* and *A. brasilense*, besides inducing biosynthesis of secondary metabolites, affect pathway flux or specific steps of monoterpene metabolism in *M. piperita.* The bacterial volatile organic compounds are rich source for new natural compounds that may increase crop productivity and essential oils yield of this economically important plant species. Inoculated with the native *P. putida* strains showed increased shoot fresh weight and root dry weight of *M. piperita* (peppermint) because indigenous bacteria which are generally more effective in terms of their acclimitization and adaptation to the environment have a competitive advantage over non-native strains (Santoro et al. 2015; Aeron et al. 2010).

Wide range of PGPR group of bacteria *pumilus* STR, *Halomonas desiderata* STR8 and *Exiguobacterium oxidotolerans* STR36 reduced the negative effects of salinity while cultivating *M. arvensis*, and application of *Halomonas desiderata* STR8 in sodic soil increased herb yield and oil content of *M. arvensis* under salt stress (Bharti et al. 2014). Ebstam and Nadia (2013) reported that the maximum values of growth parameters of *M. viridis* L. (plant height, number of branches, plant fresh and dry weight) as well as volatile oil percentage occurred in plants treated with *B. subtilis* and *P. fluorescence* as individual and consortia as well, which also showed biocontrol phenomenon against *Rhizoctonia solani.*

9.3 Conclusion and Future Prospects

Numerous plant growth promoting rhizobacteria (PGPR) have been examined under laboratory, greenhouse and field conditions for vegetable production. PGPR has a great potential for increasing yield, reducing of fertilizer requirement and synthetic pesticides. In the last few years, PGPR studies focused on the biocontrol of destructive diseases of vegetables such as Fusarium wilt and bacterial wilt in tomatoes, *Phytophthora* blight in peppers, Fusarium wilt in melons and angular leaf spot in cucumber. Similar to oil seed, pulses and maize, researches about the effects of PGPR on growth and yield of vegetables are also continuing.

PGPR replaces the chemical fertilizers and pesticides by reducing pollution, save the resources and cut down the high production cost. Evolvement of environment-friendly and cost-effective strategies with PGPR is still to be continued for the commercialization of PGPR for healthy production of vegetables. In other words, the chemical inputs are being used to increase yields, suppress pathogens and pests but their excessive, uncontrolled and unconscious use over the years resulted in an accumulation of toxic chemical resides in the crop production areas. To overcome this adverse affects, an effective and sustainable alternative method is the application of PGPR in all stages of crop production.

References

Abbas JA (2009) The effect of nitrogenous and phosphate fertilizers of the properties on the vegetative growth and aromatical oil yield of local mint (*Mentha spicata* L.). Am-Eurasian J Sustain Agric 3:262–265

Abd El-Azeem SA, Elwan MW, Sung JK, Ok YS (2012) Alleviation of salt stress in eggplant (*Solanum melongena* L.) by plant-growth-promoting rhizobacteria. Comm Soil Sci Plant Anal 43:1303–1315. https://doi.org/10.1080/00103624.2012.666305

Abraham-Juárez MDR, Vázquez IE, Mendoza RG et al (2018) Development, yield, and quality of melon fruit (*Cucumis melo* L.) inoculated with mexican native strains of *Bacillus subtilis* (ehrenberg). Agrociencia 52:91–102

Aeron A, Pandey P, Maheshwari DK (2010) Differential response of sesame under influence of indigenous and non-indigenous rhizosphere competent fluorescent pseudomonads. Curr Sci 99 (2):166–168

Adhikari M, Yadav DR, Kim SW et al (2017) Biological control of bacterial fruit blotch of watermelon pathogen (*Acidovorax citrulli*) with rhizosphere associated bacteria. Plant Pathol 33:170. https://doi.org/10.5423/PPJ.OA.09.2016.0187

Ahmad I, Akhtar MJ, Mehmood S et al (2018) Combined application of compost and *Bacillus* sp. CIK-512 ameliorated the lead toxicity in radish by regulating the homeostasis of antioxidants and lead. Ecotoxicol Environ Safe 148:805–812. https://doi.org/10.1016/j.ecoenv.2017.11.054

Akhtar MJ, Ullah S, Ahmad I et al (2018) Nickel phytoextraction through bacterial inoculation in *Raphanus sativus*. Chemosphere 190:234–242. https://doi.org/10.1016/j.chemosphere.2017.09.136

Akram W, Anjum T, Ali B (2015) Co-cultivation of tomato with two *Bacillus* strains: effects on growth and yield. J Anim Plant Sci 25:1644–1651

Altinok HH, Dikilitas M, Yildiz HN (2013) Potential of *Pseudomonas* and *Bacillus* isolates as biocontrol agents against fusarium wilt of eggplant. Biotechnol Biotechnol Equip 27:3952–3958. https://doi.org/10.5504/BBEQ.2013.0047

Altuntaş O (2018) A comparative study on the effects of different conventional, organic and bio-fertilizers on broccoli yield and quality. Appl Ecol Env Res 16:1595–1608. https://doi.org/10.15666/aeer/1602_15951608

Anith KN, Sreekumar A, Sreekumar J (2015) The growth of tomato seedlings inoculated with co-cultivated *Piriformospora indica* and *Bacillus pumilus*. Symbiosis 65:9–16. https://doi.org/10.1007/s13199-015-0313-7

Almaghrabi OA, Massoud SI, Abdelmoneim TS (2013) Influence of inoculation with plant growth promoting rhizobacteria (PGPR) on tomato plant growth and nematode reproduction under greenhouse conditions. Saudi J Biol Sci 20:57–61. https://doi.org/10.1016/j.sjbs.2012.10.004

Aslam H, Ahmad SR, Anjum T, Akram W (2018) Native halotolerant plant growth promoting bacterial strains can ameliorate salinity stress on tomato plants under field conditions. Int J Agric Biol 20:315–322. https://doi.org/10.17957/IJAB/15.0491

Balkaya A, Karaağaç O (2006) Vegetable genetic resources of Turkey. J Veg Sci 11:81–102. https://doi.org/10.1300/J484v11n04_08

Bergougnoux V (2014) The history of tomato: from domestication to biopharming. Biotechnol Adv 32:170–189. https://doi.org/10.1016/j.biotechadv.2013.11.003

Bharti N, Barnawal D, Awasthi A, Yadav A, Kalra A (2014) Plant growth promoting rhizobacteria alleviate salinity induced negative effects on growth, oil content and physiological status in *Mentha arvensis*. Acta Physiol Plant 36:45–60. https://doi.org/10.1007/s11738-013-1385-8

Bhardwaj S, Dipta B, Kirti S, Kaushal R (2017) Screening of efficient rhizobacteria associated with cauliflower (*Brassica oleracea* var. *botrytis* L.) for plant growth promoting traits J Appl Nat Sci 9:167–172

Blinkov EA, Tsavkelova EA, Selitskaya OV (2014) Auxin production by the *Klebsiella planticola* Strain TSKhA-91 and its effect on development of cucumber (*Cucumis sativus* L.) seeds. Microbiology 83:531–538. https://doi.org/10.1134/S0026261714050063

Boukerma L, Benchabane M, Charif A, Khélifi L (2017) Activity of Plant Growth Promoting Rhizobacteria (PGPRs) in the Biocontrol of Tomato Fusarium Wilt. Plant Protect Sci 53:78–84. https://doi.org/10.17221/178/2015-PPS

Brutti L, Alvarado P, Rojas T, Martensson A (2015) Tomato seedling development is improved by a substrate inoculated with a combination of rhizobacteria and fungi. Acta Agric Scand B Soil Plant Sci 65:170–176. https://doi.org/10.1080/09064710.2014.977338

Calvo-Polanco M, Sánchez-Romera B, Aroca R, Asins MJ, Declerck S, Dodd IC, Martínez-Andújare C, Albacete A, Ruiz-Lozano JM (2016) Exploring the use of recombinant inbred lines in combination with beneficial microbial inoculants (AM fungus and PGPR) to improve drought stress tolerance in tomato. Environ Exp Bot 131:47–57. https://doi.org/10.1016/j.envexpbot.2016.06.015

Chabot R, Antoun H, Cescas MP (1996) Growth promotion of maize and lettuce by phosphate-solubilizing *Rhizobium leguminosarum* biovar. *phaseoli*. Plant Soil 184:311–321. https://doi.org/10.1007/BF00010460

Chen C, Belanger RR, Benhamou N, Paulitz TC (2000) Defense enzymes induced in cucumber roots by treatment with plant growth-promoting rhizobacteria (PGPR) and *Pythium aphanidermatum*. Physiol Mol Plant Pathol 56:13–23. https://doi.org/10.1006/pmpp.1999.0243

Chamangasht S, Ardakani MR, Khavazi K, Abbaszadeh B, Mafakheri S (2012) Improving lettuce (*Lactuca sativa* L.) growth and yield by the application of biofertilizers. Ann Biol Res 3:1876–1879

Chen F, Wang M, Zheng Y, Luo J, Yang X, Wang X (2010) Quantitative changes of plant defense enzymes and phytohormone in biocontrol of cucumber Fusarium wilt by *Bacillus subtilis* B579. World J Microbiol Biotechnol 26:675–684. https://doi.org/10.1007/s11274-009-0222-0

Chiquito-Contreras RG, Murillo-Amador B, Chiquito-Contreras CJ, Marquez-Martinez JC, Cordoba-Matson MV, Hernandez-Montiel LG (2017) Effect of *Pseudomonas putida* and inorganic fertilizer on growth and productivity of habanero pepper (*Capsicum Chinense* Jacq.) in greenhouse. J Plant Nutr 40:2595–2601. https://doi.org/10.1080/01904167.2017.1381119

Chung H, Park M, Madhaiyan M, Seshadri S, Song J, Cho H, Sa T (2005) Isolation and characterization of phosphate solubilizing bacteria from the rhizosphere of crop plants of Korea. Soil Biol Biochem 37:1970–1974. https://doi.org/10.1016/j.soilbio.2005.02.025

Cappellari LD, Santoro MV, Reinoso H, Travaglia C, Giordano W, Banchio E (2015) Anatomical, morphological, and phytochemical effects of ınoculation with plant growth- promoting rhizobacteria on peppermint (*Mentha piperita*). J Chem Ecol 41:149–158. https://doi.org/10.1007/s10886-015-0549-y

Cappellari LDR, Chiappero J, Santoro MV, Giordano W, Banchio E (2017) Inducing phenolic production and volatile organic compounds emission by inoculating *Mentha piperita* with plant growth-promoting rhizobacteria. Sci Hortic 220:193–198. https://doi.org/10.1016/j.scienta.2017.04.002

Çakmakçı R, Erat M, Erdoğan Ü, Dönmez MF (2007) The influence of plant growth–promoting rhizobacteria on growth and enzyme activities in wheat and spinach plants. J Plant Nutr Soil Sci 170:288–295. https://doi.org/10.1002/jpln.200625105

de Boer M, Bom P, Kindt F, Keurentjes JJ, van der Sluis I, Van Loon LC, Bakker PAHM (2003) Control of Fusarium wilt of radish by combining *Pseudomonas putida* strains that have different disease-suppressive mechanisms. Phytopathol 93:626–632. https://doi.org/10.1094/phyto.2003.93.5.626

Decoteau RD (2000) Vegetable crops. Prentice-Hall Inc., Upper Saddle River, New Jersey
Devi M, Upadhyay GP, Garima SR (2017) Biological properties of soil and nutrient uptake in cauliflower (*Brassica oleracea* var *botrytis* L.) as influenced by integrated nutrient management. J Pharmacog Phytochem 6:325–328
Dipta B, Kirti S, Bhardwaj S, Gupta S, Kaushal R (2017) Phosphate solubilizing potential of *Bacillus pumilus* for the enhancement of Cauliflower (*Brassica oleracea* var. *botrytis* L.). Ecol Environ Cons 23:1541–1548
Du N, Shi L, Yuan Y, Li B, Shu S, Sun J, Guo S (2016) Proteomic analysis reveals the positive roles of the plant-growth-promoting rhizobacterium NSY50 in the response of cucumber roots to Fusarium *oxysporum* f. sp. *cucumerinum* inoculation. Front Plant Sci 7:1859. https://doi.org/10.3389/fpls.2016.01859
Ebstam MM, Nadia AM (2013) Efficincy of *Bacillus subtilis* and *Pseudomonas fluorescence* as biocontrol agents against rhizoctonia solanı on spearmint plants (*Mentha viridis* L.). Sci J Flower Ornamen Plant 2:213–225. https://doi.org/10.21608/SJFOP.2015.5114
Ekinci M, Turan M, Yildirim E, Güneş A, Kotan R, Dursun A (2014) Effect of plant growth promoting rhizobacteria on growth, nutrient, organic acid, amino acid and hormone content of cauliflower (*Brassica oleracea* L. var. *botrytis*) transplants. Acta Sci Pol-Hortorum Cultus 13:71–85
El-Nemr MA, Zaki MF, Tantawy AS, Abdel-Mawgoud AMR (2011) Enhancement of growth and production of broccoli crop using bio-nutritional foliar compound. Australian J Basic Appl Sci 5:2578–2583
Esquivel-Cote R, Ramírez-Gama RM, Tsuzuki-Reyes G, Orozco-Segovia A, Huante P (2010) *Azospirillum lipoferum* strain AZm5 containing 1-aminocyclopropane-1-carboxylic acid deaminase improves early growth of tomato seedlings under nitrogen deficiency. Plant Soil 337:65–75. https://doi.org/10.1007/s11104-010-0499-7
Fan X, Zhang S, Xiaodan MO, Yuncong LI, Yuqing FU, Zhiguang LIU (2017) Effects of plant growth-promoting rhizobacteria and N source on plant growth and N and P uptake by tomato grown on calcareous soils. Pedosphere 27:1027–1036. https://doi.org/10.1016/S1002-0160(17)60379-5
FAOSTAT (2018) Online database of crop production statistics. Available via DIALOG. http://www.fao.org/faostat/en/#data/QC. Accessed 25 June 2018
Fasciglione G, Casanovas EM, Yommi A, Sueldo RJ, Barassi CA (2012) *Azospirillum* improves lettuce growth and transplant under saline conditions. J Sci Food Agric 92:2518–2523. https://doi.org/10.1002/jsfa.5661
Fatima S, Anjum T (2017) Identification of a potential ISR determinant from *Pseudomonas aeruginosa* PM12 against *Fusarium* wilt in tomato. Front Plant Sci 8:848. https://doi.org/10.3389/fpls.2017.00848
Flores-Félix JD, Menéndez E, Rivera LP et al (2013) Use of *Rhizobium leguminosarum* as a potential biofertilizer for *Lactuca sativa* and *Daucus carota* crops. J Plant Nutr Soil Sci 176:876–882. https://doi.org/10.1002/jpln.201300116
Folman LB, Postma J, van Veen JA (2003) Characterisation of *Lysobacter enzymogenes* (Christensen and Cook 1978) strain 3.1 T8, a powerful antagonist of fungal diseases of cucumber. Microbiol Res 158(2):107
Gadhave KR, Finch P, Gibson TM, Gange AC (2016) Plant growth-promoting *Bacillus* suppress *Brevicoryne brassicae* field infestation and trigger density-dependent and density-independent natural enemy responses. J Pest Sci 89:985–992. https://doi.org/10.1007/s10340-015-0721-8
García-Gutiérrez L, Romero D, Zeriouh H, Cazorla FM, Torés JA, de Vicente A, Pérez-García A (2012) Isolation and selection of plant growth-promoting rhizobacteria as inducers of systemic resistance in melon. Plant Soil 358:201–212. https://doi.org/10.1007/s11104-012-1173-z
Gao M, Zhou JJ, Wang ET, Chen Q, Xu J, Sun JG (2015) Multiphasic characterization of a plant growth promoting bacterial strain, *Burkholderia* sp. 7016 and its effect on tomato growth in the field. J Integr Agric 14:1855–1863. https://doi.org/10.1016/S2095-3119(14)60932-1
García-Gutiérrez L, Zeriouh H, Romero D, Cubero J, de Vicente A, Pérez-García A (2013) The antagonistic strain Bacillus subtilis UMAF6639 also confers protection to melon plants against

cucurbit powdery mildew by activation of jasmonate-and salicylic acid-dependent defence responses. Microb Biotechnol 6:264–274. https://doi.org/10.1111/1751-7915.12028
Garcia JAL, Probanza A, Ramos B, Palomino MR, Manero FJG (2004) Effect of inoculation of *Bacillus licheniformis* on tomato and pepper. Agronomie 24(169–176):169. https://doi.org/10.1051/agro:2004020
Gravel V, Antoun H, Tweddell RJ (2007) Growth stimulation and fruit yield improvement of greenhouse tomato plants by inoculation with *Pseudomonas putida* or *Trichoderma atroviride*: possible role of indole acetic acid (IAA). Soil Biol Biochem 39:1968–1977. https://doi.org/10.1016/j.soilbio.2007.02.015
Gyaneshwar P, Naresh Kumar G, Parekh LJ, Poole PS (2002) Role soil microorganisms in improving P nutrition of plant. Plant Soil 245:83–93
Gowtham HG, Duraivadivel P, Hariprasad P, Niranjana SR (2017) A novel split-pot bioassay to screen indole acetic acid producing rhizobacteria for the improvement of plant growth in tomato (Solarium lycopersicum L.). Sci Hortic 224:351–357. https://doi.org/10.1016/j.scienta.2017.06.017
Hahm MS, Son JS, Hwang YJ, Kwon DK, Ghim SY (2017) Alleviation of salt stress in pepper (*Capsicum annum* L.) plants by plant growth-promoting rhizobacteria. J Microbiol Biotechnol 27:1790–1797. https://doi.org/10.4014/jmb.1609.09042
Hahm MS, Sumayo M, Hwang YJ, Jeon SA, Park SJ, Lee JY, Ahn JH, Kim BS, Ryu CM, Ghim SY (2012) Biological control and plant growth promoting capacity of rhizobacteria on pepper under greenhouse and field conditions. J Microbiol 50:380–385. https://doi.org/10.1007/s12275-012-1477-y
Hameeda B, Harini G, Rupela OP, Wani SP, Reddy G (2008) Growth promotion of maize by phosphate solubilizing bacteria isolated from composts and macrofauna. Microbiol Res 163:234–242. https://doi.org/10.1016/j.micres.2006.05.009
Hernández-Montiel LG, Chiquito Contreras CJ, Murillo Amador B, Vidal Hernández L, Aguilar Q, Evanjelina E, Chiquito Contreras RG (2017) Efficiency of two inoculation methods of *Pseudomonas putida* on growth and yield of tomato plants. J Soil Sci Plant Nutr 17:1003–1012. https://doi.org/10.4067/S0718-95162017000400012
Herman MAB, Nault BA, Smart CD (2008) Effects of plant growth-promoting rhizobacteria on bell pepper production and green peach aphid infestations in New York. Crop Protec 27:996–1002. https://doi.org/10.1016/j.cropro.2007.12.004
Hernandez-Castillo FD, Lira-Saldivar RH, Gallegos-Morales G, Hernandez-Suarez M, Solis-Gaona S (2014) Biocontrol of pepper wilt with three *Bacillus* species and its effect on growth and yield. Phyton-Int J Exp Bot 83:49–55
Hoffmann-Hergarten S, Gulati MK, Sikora RA (1998) Yield response and biological control of Meloidogyne incognita on lettuce and tomato with rhizobacteria. J Plant Dis Protec 105:349–358
Hong SH, Lee EY (2017) Phytostabilization of salt accumulated soil using plant and biofertilizers: field application. Int Biodeterior Biodegr 124:188–195. https://doi.org/10.1016/j.ibiod.2017.05.001
Hou MP, Oluranti BO (2013) Evaluation of plant growth promoting potential of four rhizobacterial species for indigenous system. J Cent South Univ 20:164–171. https://doi.org/10.1007/s11771-013-1472-4
Huang YY, Wu ZS, He YH, Ye BC, Li C (2017) Rhizospheric bacillus subtilis exhibits biocontrol effect against rhizoctonia solani in pepper (*Capsicum annuum*). Hindawi BioMed Research International Article ID: 9397619. https://doi.org/10.1155/2017/9397619
Islam MR, Sultana T, Joe MM, Yim W, Cho JC, Sa T (2013) Nitrogen-fixing bacteria with multiple plant growth-promoting activities enhance growth of tomato and red pepper. J Basic Microbiol 53:1004–1015. https://doi.org/10.1002/jobm.201200141
Islam MR, Sultana T, Joe MM, Yim W, Cho JC, Sa T (2013b) Nitrogen-fixing bacteria with multiple plant growth-promoting activities enhance growth of tomato and red pepper. J Basic Microbiol 53:1004–1015. https://doi.org/10.1002/jobm.201200141

Jeffrey C (2007) Compositae. Introduction with key to tribes. In: Kadereit JW, Jeffrey C (eds) Families and genera of vascular Plants, flowering plants Asterales. Springer-Verlag, Eudicots, pp 61–87

Jeun YC, Park KS, Kim CH, Fowler WD, Kloepper JW (2004) Cytological observations of cucumber plants during induced resistance elicited by rhizobacteria. Biol Cont 29:34–42. https://doi.org/10.1016/S1049-9644(03)00082-3

Jeun YC, Kim KW, Kim KD, Hyun JW (2007) Comparative ultrastructure of cucumbers pretreated with plant growth-promoting rhizobacteria, dl-3-aminobutyric acid or amino salicylic acid after inoculation with *Colletotrichum orbiculare*. J Phytopathol 155:416–425. https://doi.org/10.1111/j.1439-0434.2007.01252.x

Jiménez-Gómez A, Flores-Félix JD, García-Fraile P, Mateos PF, Menéndez E, Velázquez E, Rivas R (2018) Probiotic activities of *Rhizobium laguerreae* on growth and quality of spinach. Scintific Rep 8:295. https://doi.org/10.1038/s41598-017-18632-z

Jung BK, Kim SD, Khan AR, Lim JH, An CH, Kim YH, Song JH, Hong SJ, Shin JH (2015) Rhizobacterial Communities and red pepper (*Capsicum annum*) yield under different cropping systems. Int J Agric Biol 17:734–740. https://doi.org/10.17957/IJAB/14.0010

Kang SM, Hamayun M, Joo GJ et al (2010) Effect of *Burkholderia* sp. KCTC 11096BP on some physiochemical attributes of cucumber. Eur J Soil Biol 46:264–268. https://doi.org/10.1016/j.ejsobi.2010.03.002

Kang SM, Khan AL, Waqas M et al (2014a) Plant growth-promoting rhizobacteria reduce adverse effects of salinity and osmotic stress by regulating phytohormones and antioxidants in *Cucumis sativus*. J Plant Interact 9:673–682. https://doi.org/10.1080/17429145.2014.894587

Kang SM, Khan AL, You YH, Kim JG, Kamran M, Lee IJ (2014b) Gibberellin production by newly isolated strain *Leifsonia soli* SE134 and its potential to promote plant growth. J Microbiol Biotechnol 24:106–112. https://doi.org/10.4014/jmb.1304.04015

Kang SM, Radhakrishnan R, You YH, Khan AL, Park JM, Lee SM, Lee IJ (2015) Cucumber performance is improved by inoculation with plant growth-promoting microorganisms. Acta Agric Scand B Soil Plant Sci 65:36–44. https://doi.org/10.1080/09064710.2014.960889

Khabbaz SE, Zhang L, Cáceres LA, Sumarah M, Wang A, Abbasi PA (2015) Characterisation of antagonistic *Bacillus* and *Pseudomonas* strains for biocontrol potential and suppression of damping-off and root rot diseases. Ann Appl Biol 166:456–471. https://doi.org/10.1111/aab.12196

Khan FM, Inamul H, Tariq S, Muhammad F, Ohia CM, Tauseef A (2018) Isolation, characterization and ıdentification of plant growth promoting rhizobacteria from cauliflower (*Brassica oleracea*). Arch Bas App Med 6:55–60

Kloepper JW, Schroth MN (1978) Plant growth-promoting rhizobacteria on radishes. In: Proceedings of the Fourth International Conference on Plant Pathogen Bacteria, vol 2. INRA, p 879–882

Kaymak HÇ, Güvenç İ, Yarali F, Dönmez MF (2009) The effects of bio-priming with PGPR on germination of radish (*Raphanus sativus* L.) seeds under saline conditions. Turk J Agric For 33:173–179. https://doi.org/10.3906/tar-0806-30

Kaymak HC, Güvenç İ (2010) The influence of vernalization time and day length on flower induction of radish (*Raphanus sativus* L.) under controlled and field conditions. Turk J Agric For 34:401–413. https://doi.org/10.3906/tar-0901-14

Kaymak HC (2010) Potential of PGPR in agricultural innovations. In: Maheshwari DK (ed) Plant growth and health promoting bacteria. Springer, Berlin, Heidelberg, pp 45–79

Kaymak HC, Dönmez MF, Çakmakçı R (2013) N_2-fixing plant growth promoting rhizobacteria: as a potential application to increase yield, growth and element contents of leaves in *Mentha piperita* L. Euro J Plant Sci Biotechnol "Vegetable Science and Biotechnology in Turkey", 7 (Special Issue 1):38–42

Kaushal M, Kaushal R, Thakur BS, Spehia RS (2011) Effect of plant growth-promoting rhizobacteria at varying levels of N and P fertilizers on growth and yield of cauliflower in mid hills of Himachal Pradesh. J Farm Sci 1:19–26

Kaushal M, Kaushal R, Mandyal P (2013) Impact of integrated nutrient management systems on cauliflower (*Brassica oleracea* var *botrytis*) yield and soil nutrient status. Indian J Agr Sci 83:1013–1016

Koh RH, Song HG (2007) Effects of application of *Rhodopseudomonas* sp. on seed germination and growth of tomato under axenic conditions. J Microbiol Biotechnol 17:1805–1810

Kohler J, Caravaca F, Carrasco L, Roldán A (2006) Contribution of *Pseudomonas mendocina* and *Glomus intraradices* to aggregate stabilization and promotion of biological fertility in rhizosphere soil of lettuce plants under field conditions. Soil Use Manag 22:298–304. https://doi.org/10.1111/j.1475-2743.2006.00041.x

Kokalis-Burelle N, Vavrina CS, Rosskopf EN, Shelby RA (2002) Field evaluation of plant growth-promoting Rhizobacteria amended transplant mixes and soil solarization for tomato and pepper production in Florida. Plant Soil 238:257–266

Kokalis-Burelle N, Vavrina CS, Reddy MS, Kloepper JW (2003) Amendment of muskmelon and watermelon transplant media with plant growth-promoting rhizobacteria: effects on seedling quality, disease, and nematode resistance. Horttechnology 13:476–482

Kokalis-Burelle N (2004) Development of multi-component transplant mixes for plant growth-promotion and disease suppression. Proceeding of the Fourth International Congress of Nematology, Nematology Monographs and Perspectives 2:187–194

Krężel J, Kołota E (2014) Source of nitrogen affects the yield and quality of spinach cultivars grown for autumn harvest. Acta Agric Scand B Soil Plant Sci 64:583–589. https://doi.org/10.1080/09064710.2014.934273

Kumar V, Narula N (1999) Solubilization of inorganic phosphates and growth emergence of wheat as affected by *Azotobacter chrococcum*. Biol Fertil Soils 28:301–305

Kushwaha A, Baily SB, Maxton A, Ram GD (2013) Isolation and characterization of PGPR associated with cauliflower roots and its effect on plant growth. The Bioscan 8:95–99

Lawrence BM (2007) Mint: the genus mentha. Medicinal and aromatic plants-industrial profiles. CRC Press/Taylor & Francis, Boca Raton

Lee KH, Koh RH, Song HG (2008a) Enhancement of growth and yield of tomato by *Rhodopseudomonas* sp. under greenhouse conditions. J Microbiol 46:641–646. https://doi.org/10.1007/s12275-008-0159-2

Lee KJ, Kamala-Kannan S, Sub HS, Seong CK, Lee GW (2008b) Biological control of *Phytophthora* blight in red pepper (*Capsicum annuum* L.) using *Bacillus subtilis*. World J Microbiol Biotechnol 24:1139–1145. https://doi.org/10.1007/s11274-007-9585-2

Lee KE, Radhakrishnan R, Kang SM, You YH et al (2015) *Enterococcus faecium* LKE12 cell-free extract accelerates host plant growth via gibberellin and indole-3-acetic acid secretion. J Microbiol Biotechnol 25:1467–1475. https://doi.org/10.4014/jmb.1502.02011

Li CY, Hu WC, Pan B, Liu Y, Yuan SF, Ding YY, Li R, Zheng XY, Shen B, Shen QR (2017) Rhizobacterium *Bacillus amyloliquefaciens* strain SQRT3-mediated induced systemic resistance controls bacterial wilt of tomato. Pedosphere 27:1135–1146. https://doi.org/10.1016/S1002-0160(17)60406-5

Lim JH, Kim SD (2013) Induction of drought stress resistance by multi-functional PGPR *Bacillus licheniformis* K11 in pepper. Plant Pathol J 29:201–208. https://doi.org/10.5423/PPJ.SI.02.2013.0021

Liu K, Garrett C, Fadamiro H, Kloepper JW (2016a) Antagonism of black rot in cabbage by mixtures of plant growth-promoting rhizobacteria (PGPR). Biocontrol 61:605–613. https://doi.org/10.1007/s10526-016-9742-3

Liu K, Garrett C, Fadamiro H, Kloepper JW (2016b) Induction of systemic resistance in Chinese cabbage against black rot by plant growth-promoting rhizobacteria. Biol Cont 99:8–13. https://doi.org/10.1016/j.biocontrol.2016.04.007

Lucas JA, García-Cristobal J, Bonilla A, Ramos B, Gutierrez-Mañero J (2014) Beneficial rhizobacteria from rice rhizosphere confers high protection against biotic and abiotic stress inducing systemic resistance in rice seedlings. Plant Physiol Biochem 82:44–53. https://doi.org/10.1016/j.plaphy.2014.05.007

Mangmang JS, Deaker R, Rogers G (2015) Early seedling growth response of lettuce, tomato and cucumber to *Azospirillum brasilense* inoculated by soaking and drenching. Hort Sci 42:37–46. https://doi.org/10.17221/159/2014-HORTSCI

Mardani-Talaee M, Razmjou J, Nouri-Ganbalani G, Hassanpour M, Naseri B (2017) Impact of Chemical, Organic and Bio-Fertilizers Application on Bell Pepper, *Capsicum annuum* L. and Biological Parameters of *Myzus persicae* (Sulzer) (Hem.: Aphididae). Neotrop Entomol 46:578–586. https://doi.org/10.1007/s13744-017-0494-2

Medeiros FH, Moraes IS, da Silva Neto EB, Silveira EB, Rosa de Lima RM (2009) Management of melon bacterial blotch by plant beneficial bacteria. Phytoparasitica 37:453–460. https://doi.org/10.1007/s12600-009-0063-2

Mehta P, Walia A, Kulshrestha S, Chauhan A, Shirkot CK (2015) Efficiency of plant growth-promoting P-solubilizing *Bacillus circulans* CB7 for enhancement of tomato growth under net house conditions. J Basic Microbiol 55:33–44. https://doi.org/10.1002/jobm.201300562

Mena-Violante HG, Olalde-Portugal V (2007) Alteration of tomato fruit quality by root inoculation with plant growth-promoting rhizobacteria (PGPR): *bacillus subtilis* BEB-13bs. Sci Hort 113:103–106. https://doi.org/10.1016/j.scienta.2007.01.031

Nishihara E, Inoue M, Kondo K, Takahashi K, Nakata N (2001) Spinach yield and nutritional quality affected by controlled soil water matric head. Agric Water Manag 51:217229

Nga NTT, Giau NT, Long NT et al (2010) Rhizobacterially induced protection of watermelon against *Didymella bryoniae*. J Appl Microbiol 109:567–582. https://doi.org/10.1111/j.1365-2672.2010.04685.x

Nosheen A, Bano A (2014) Potential of plant growth promoting rhizobacteria and chemical fertilizers on soil enzymes and plant growth. Pak J Bot 46:1521–1530

Olivares FL, Aguiar NO, Rosa RCC, Canellas LP (2015) Substrate biofortification in combination with foliar sprays of plant growth promoting bacteria and humic substances boosts production of organic tomatoes. Sci Hort 183:100–108. https://doi.org/10.1016/j.scienta.2014.11.012

Ongena M, Daayf F, Jacques P, Thonart P, Benhamou N, Paulitz TC, Bélanger RR (2000) Systemic induction of phytoalexins in cucumber in response to treatments with fluorescent pseudomonads. Plant Pathol 49:523–530. https://doi.org/10.1046/j.13653059.2000.00468.x

Ordookhani K, Khavazi K, Moezzi A, Rejali F (2010) Influence of PGPR and AMF on antioxidant activity, lycopene and potassium contents in tomato. Afr J Agric Res 5:1108–1116. https://doi.org/10.5897/AJAR09.183

Orhan E, Esitken A, Ercisli S, Turan M, Sahin F (2006) Effects of plant growth promoting rhizobacteria (PGPR) on yield, growth and nutrient contents in organically growing raspberry. Sci Hortic 111:38–43. https://doi.org/10.1016/j.scienta.2006.09.002

Palada MC, Kalb TJ, Lumpkin TA (2006) The role of AVRDC-The world vegetable center in enhancing and promoting vegetable production in the tropics. HortScience 41:556–560

Park K, Park JW, Lee SW, Balaraju K (2013) Disease suppression and growth promotion in cucumbers induced by integrating PGPR agent *Bacillus subtilis* strain B4 and chemical elicitor ASM. Crop Prot 54:199–205. https://doi.org/10.1016/j.cropro.2013.08.017

Pastor N, Rosas S, Luna V, Rovera M (2014) Inoculation with *Pseudomonas putida* PCI2, a phosphate solubilizing rhizobacterium, stimulates the growth of tomato plants. Symbiosis 62:157–167. https://doi.org/10.1007/s13199-014-0281-3

Pastor N, Masciarelli O, Fischer S, Luna V, Rovera M (2016) Potential of *Pseudomonas putida* PCI2 for the protection of tomato plants against fungal pathogens. Curr Microbiol 73:346–353. https://doi.org/10.1007/s00284-016-1068-y

Pastor-Bueis R, Mulas R, Gomez X, Gonzalez-Andres F (2017) Innovative liquid formulation of digestates for producing a biofertilizer based on *Bacillus siamensis*: field testing on sweet pepper. J Plant Nut. Soil Sci 180:748–758. https://doi.org/10.1002/jpln.201700200

Patakioutas G, Dimou D, Kostoula O, Yfanti P, Paraskevopoulos A, Ntatsi G, Savvas D (2015) Inoculation of tomato roots with beneficial microorganisms as a means to control *Fusarium oxysporum* f. sp *lycopersici* and improve nutrient uptake and yield. Acta Hortic 1107:141–148. https://doi.org/10.17660/ActaHortic.2015.1107.19

Peng D, Luo K, Jiang H, Deng Y, Bai L, Zhou X (2017) Combined use of Bacillus subtilis strain B-001 and bactericide for the control of tomato bacterial wilt. Pest Manag Sci 73:1253–1257. https://doi.org/10.1002/ps.4453

Peralta IE, Spooner DM (2007) History, origin and early cultivation of tomato (Solanaceae). In: Razdan MK, Mattoo AK (eds) Genetic Improvement of Solanaceous Crops, vol 2. CRC Press. Taylor and Francis, Boca Raton pp, pp 1–27

Pii Y, Penn A, Terzano R, Crecchio C, Mimmo T, Cesco S (2015) Plant-microorganism-soil interactions influence the Fe availability in the rhizosphere of cucumber plants. Plant Physiol Biochem 87:45–52. https://doi.org/10.1016/j.plaphy.2014.12.014

Podile AR, Kishore GK (2006) Plant growth-promoting rhizobacteria. In: Gnanamanickam SS (ed) Plant-Associated Bacteria: rhizosphere bacteria. Springer, Netherlands, pp 195–230

Radhakrishnan R, Lee IJ (2013) Regulation of salicylic acid, jasmonic acid and fatty acids in cucumber (*Cucumis sativus* L.)by spermidine promotes plant growth against salt stress. Acta Physiol Plant 35:3315–3322. https://doi.org/10.1007/s11738-013-1364-0

Ramesh R, Joshi AA, Ghanekar MP (2009) Pseudomonads: major antagonistic endophytic bacteria to suppress bacterial wilt pathogen, *Ralstonia solanacearum* in the eggplant (*Solanum melongena* L.). World J Microbiol Biotechnol 25:47–55. https://doi.org/10.1007/s11274-008-9859-3

Ramesh R, Phadke GS (2012) Rhizosphere and endophytic bacteria for the suppression of eggplant wilt caused by *Ralstonia solanacearum*. Crop Protec 37:35-41. https://doi.org/10.1016/j.cropro.2012.02.008

Rimmer SR, Shattuck VI, Buchwaldt L (2007) Compendium of Brassica disease. American phytopathological society, St. Paul, Minnesota

Robacer M, Canali S, Kristensen HL, Bavec F, Mlakar SG, Jakop M, Bavec M (2016) Cover crops in organic field vegetable production. Sci Hortic 208:104–110. https://doi.org/10.1016/j.scienta.2015.12.029

Rodriguez H, Gonzalez T, Goire I, Bashan Y (2004) Gluconic acid production and phosphate solubilization by the plant growth-promoting bacterium *Azospirillum* spp. Naturewissenschaften 91:552–555. https://doi.org/10.1007/s00114-004-0566-0

Rodriguez Mendoza MDLN, San Miguel-Chavez R, Garcia Cue JL, Benavides Mendoza A (2013) Inoculatıon of growth-promotıng bacterıa ın melon (*Cucumis melo*). Interciencia 38:857–862

Samancioglu A, Yildirim E, Sahin U (2016a) Effect of seedlings development, some physiological and biochemical properties of cabbage seedlings grown at different irrigation levels of the plant growth promoting rhizobacteria application. Kahramanmaras Sutcu Imam Unıversity Journal of Natural Sciences 19:332–338

Samancioglu A, Yildirim E, Turan M, Kotan R, Sahin U, Kul R (2016b) Amelioration of drought stress adverse effect and mediating biochemical content of cabbage seedlings by plant growth promoting rhizobacteria. Int J Agric Bio 18:948–956. https://doi.org/10.17957/IJAB/15.0195

Samaniego-Gamez BY, Garruna R, Tun-Suarez JM, Kantun-Can J, Reyes-Ramirez A, Cervantes-Diaz L (2016) *Bacillus* spp. inoculation improves photosystem II efficiency and enhances photosynthesis in pepper plants. Chil J Agr Res 76:4. https://doi.org/10.4067/s0718-58392016000040000

Sang MK, Jeong JJ, Kim J, Kim KD (2018) Growth promotion and root colonisation in pepper plants by phosphate-solubilising *Chryseobacterium* sp. strain ISE14 that suppresses *Phytophthora* blight. Ann Appl Biol 172:208–223. https://doi.org/10.1111/aab.12413

Sang MK, Kim KD (2012) Plant growth-promoting rhizobacteria suppressive to *Phytophthora* blight affect microbial activities and communities in the rhizosphere of pepper (*Capsicum annuum* L.) in the field. Appl Soil Ecol 62:88–97. https://doi.org/10.1016/j.apsoil.2012.08.001

Santoro MV, Zygadlo J, Giordano W, Banchio E (2011) Volatile organic compounds from rhizobacteria increase biosynthesis of essential oils and growth parameters in peppermint (*Mentha piperita*). Plant Physiol Biochem 49:1177–1182. https://doi.org/10.1016/j.plaphy.2011.07.016

Santoro MV, Cappellari LR, Giordano W, Banchio E (2015) Plant growth-promoting effects of native Pseudomonas strains on *Mentha piperita* (peppermint): an in vitro study. Plant Biol 17:1218–1226. https://doi.org/10.1111/plb.12351

Sharma IP, Sharma AK (2017) Physiological and biochemical changes in tomato cultivar PT-3 with dual inoculation of mycorrhiza and PGPR against root-knot nematode. Symbiosis 71:175–183. https://doi.org/10.1007/s13199-016-0423-x

Sharma SP, Leskovar DI, Crosby KM. Volder A (2017) Root growth dynamics and fruit yield of melon (*Cucumis melo* L) genotypes at two locations with sandy loam and clay soils. Soil Tillage Res 168:50–62. http://dx.doi.org/10.1016/j.still.2016.12.0060167-1987/

Shen M, Kang YJ, Wang HL, Zhang XS, Zhao QX (2012) Effect of plant growth-promoting rhizobacteria (PGPRs) on plant growth, yield, and quality of tomato (*Lycopersicon esculentum* Mill.) under simulated seawater irrigation. J Gen Appl Microbiol 58:253–262. https://doi.org/10.2323/jgam.58.253

Shrestha A, Kim EC, Lim CK, Cho SY, Hur JH, Park DH (2009) Biological control of soft rot on chinese cabbage using beneficial bacterial agents in greenhouse and field. Korean J Pestic Sci 13:325–331

Siddikee MA, Chauhan PS, Sa T (2012) Regulation of ethylene biosynthesis under salt stress in red pepper (*Capsicum annuum* L.) by 1-Aminocyclopropane-1Carboxylic Acid (ACC) deaminase-producing halotolerant bacteria. J Plant Growth Regul 31:265–272. https://doi.org/10.1007/s00344-011-9236-6

Singh N, Raina S, Singh D, Ghosh M, Helfish AIAI (2017) Exploitation of promising native strains of bacillus subtilis with antagonistic properties against fungal pathogens and their PGPR characteristic. J Plant Pathol 99:27–35. https://doi.org/10.4454/jpp.v99i1.3809

Son JS, Sumayo M, Hwang YJ, Kim BS, Ghim SY (2014) Screening of plant growth-promoting rhizobacteria as elicitor ofsystemic resistance against gray leaf spot disease in pepper. Appl Soil Ecol 73:1–8. https://doi.org/10.1016/j.apsoil.2013.07.016

Sottero AN, Freitas SD, de Melo AMT, Trani PE (2006) Rhizobacteria and lettuce: root colonization, plant growth promotion and biological control. Rev Bras Cienc Solo 30:225–234

Sundara B, Natarajan V, Hari K (2002) Influence of phosphorus solubilizing bacteria on the changes in soil available phosphorus and sugarcane and sugar yields. Field Crop Res 77:43–49. https://doi.org/10.1016/S0378-4290(02)00048-5

Supanjani, Han HS, Jung JS, Lee KD (2006) Rock phosphate-potassium and rock-solubilising bacteria as alternative, sustainable fertilisers. Agron Sustain Dev 26:233–240. https://doi.org/10.1051/agro:2006020

Swaider JM, George W, McCollum JP (1992) Producing vegetable crops. Interstate Printers and Publishers, Danville, Illinois

Şahin F, Çakmakçi R, Kantar F (2004) Sugar beet and barley yields in relation to inoculation with N2-fixing and phosphate solubilizing bacteria. Plant Soil 265:123–129

Tailor AJ, Joshi BH (2014) Harnessing plant growth promoting rhizobacteria beyond nature: a review. J Plant Nutr 37:1534–1571. https://doi.org/10.1080/01904167.2014.911319

Tallapragada P, Dikshit R, Seshagiri S (2016) Influence of *Rhizophagus* spp. and *Burkholderia seminalis* on the growth of tomato (*Lycopersicon esculatum*) and bell pepper (*Capsicum annuum*) under drought stress. Commun Soil Sci Plant Anal 47(17):1975–1984. https://dx.doi.org/10.1080/00103624.2016.1216561

Tanwar A, Aggarwal A, Parkash V (2014) Effect of bioinoculants and superphosphate fertilizer on the growth and yield of broccoli (*Brassica oleracea* L. var. *italica* Plenck). New Zeal J Crop Hort 42:288–302. https://doi.org/10.1080/01140671.2014.924537

Tenuta M (2004) Plant PGPR. Prospects for increasing nutrient acquisition and disease control. Available via DIALOG. http://www.umanitoba.ca/faculties/afs/MAC_proceedings/2003/pdf/tenuta_rhizobacteria.pdf. Accessed 08 June 2018

Thakur J, Kumar P, Mohit (2018) Studies on conjoint application of nutrient sources and PGPR on growth, yield, quality, and economics of cauliflower (*Brassica oleracea* var. botrytis L.). J Plant Nutr 41:1862–1867. https://doi.org/10.1080/01904167.2018.1463382

Thanh DT, Tarn LTT, Hanh NT, Tuyen NH, Bharathkumar S, Lee SV, Park KS (2009) Biological control of soilborne diseases on tomato, potato and black pepper by selected PGPR in the greenhouse and field in Vietnam. Plant Pathol J 25:263–269. https://doi.org/10.5423/PPJ.2009.25.3.263

Tripti, Kumar A, Usmani Z, Kumar V, Anshumali (2017) Biochar and flyash inoculated with plant growth promoting rhizobacteria act as potential biofertilizer for luxuriant growth and yield of tomato plant. J Environ Manage 190:20–27. https://doi.org/10.1016/j.jenvman.2016.11.060

Turan M, Ekinci M, Yildirim E, Gunes A, Karagoz K, Kotan R, Dursun A (2014) Plant growth-promoting rhizobacteria improved growth, nutrient, and hormone content of cabbage (*Brassica oleracea*) seedlings. Turk J Agric For 38:327–333. https://doi.org/10.3906/tar-1308-62

Umesha S, Roohie RK (2017) Role of *Pseudomonas fluorescens* and INA against black rot of cabbage. J Phytopathol 165:265–275. https://doi.org/10.1111/jph.12558

Utkhede RS, Koch CA (1999) Rhizobacterial growth and yield promotion of cucumber plants inoculated with *Pythium aphanidermatum*. Can J Plant Pathol 21:265–271. https://doi.org/10.1080/07060669909501189

Urashima Y, Suga Y, Hori K (2005) Growth promotion of spinach by fluorescent *Pseudomonas* strains under application of organic materials. J Soil Sci Plant Nutr 51:841–847. https://doi.org/10.1111/j.1747-0765.2005.tb00119.x

van Loon LC, Bakker PAHM, Pieterse CMJ (1998) Systemic resistance induced by rhizosphere bacteria. Annu Rev Phytopathol 36:453–483. https://doi.org/10.1146/annurev.phyto.36.1.453

Vessey JK (2003) Plant growth promoting rhizobacteria as biofertilizers. Plant Soil 255:571–586

Vural H, Eiyok D, Duman I (2000) Cultural vegetables (Vegetable growing). Bornova, İzmir

Wan T, Zhao H, Wang W (2018) Effects of the biocontrol agent Bacillus amyloliquefaciens SN16-1 on the rhizosphere bacterial community and growth of tomato. J Phytopathol 166:324–332. https://doi.org/10.1111/jph.12690

Welbaum GE (2015) Vegetable production and practices. CABI International, Boston

Widnyana IK, Javandira C (2016) Activities *Pseudomonas* spp. and *Bacillus* sp. to stimulate germination and seedling growth of tomato plants. Agric Agric Sci Procedia 9:419–423. https://doi.org/10.1016/j.aaspro.2016.02.158

Whitelaw MA (2000) Growth promotion of plants inoculated with phosphate-solubilizing fungi. Adv Agron 69:99–151

Wu Y, Zhao CY, Farmer J, Sun JD (2015) Effects of bio-organic fertilizer on pepper growth and *Fusarium* wilt biocontrol. Sci Hortic 193:114–120. https://doi.org/10.1016/j.scienta.2015.06.039

Xu SJ, Kim BS (2016) Evaluation of *Paenibacillus polymyxa* strain SC09-21 for biocontrol of *Phytophthora* blight and growth stimulation in pepper plants. Trop Plant Pathol 41:162–168. https://doi.org/10.1007/s40858-016-0077-5

Yan JM, Smith MD, Glick BR, Liang Y (2014) Effects of ACC deaminase containing rhizobacteria on plant growth and expression of Toc GTPases in tomato (*Solanum lycopersicum*) under salt stress. Botany 92:775–781. https://doi.org/10.1139/cjb-2014-0038

Yaoyao E, Yuan J, Yang F et al (2017) PGPR strain *Paenibacillus polymyxa* SQR-21 potentially benefits watermelon growth by re-shaping root protein expression. AMB Express 7:104. https://doi.org/10.1186/s13568-017-0403-4

Yildirim E, Karlidag H, Turan M, Dursun A, Goktepe F (2011) Growth, nutrient uptake, and yield promotion of broccoli by plant growth promoting rhizobacteriawith manure. HortScience 46:932–936

Yildirim E, Turan M, Ekinci M, Dursun A, Gunes A, Donmez MF (2015) Growth and mineral content of cabbage seedlings in response to nitrogen fixing rhizobacteria treatment. Rom Biotechnol Lett 20:10929–10935

Yildirim E, Turan M, Dursun A, Ekinci M, Kul R, Karagoz FP, Donmez MF, Kitir N (2016) Integrated use of nitrogen fertilization and microbial inoculation: change in the growth and chemical composition of white cabbage. Commun Soil Sci Plant Anal 47:2245–2260. https://doi.org/10.1080/00103624.2016.1228955

Yildiz HN, Altinok HH, Dikilitas M (2012) Screening of Rhizobacteria against *Fusarium oxysporum* f. sp. *melongenae*, the Causal Agent of Wilt Disease of Eggplant. Afr J Microbiol Res 6:3700–3706. https://doi.org/10.5897/AJMR12.307

Zohary D, Hopf M, Weiss E (2012) Domestication of plants in the old world. Oxford University Press, Oxford

Chapter 10
Problem of Mercury Toxicity in Crop Plants: Can Plant Growth Promoting Microbes (PGPM) Be an Effective Solution?

Swapnil Sapre, Reena Deshmukh, Iti Gontia-Mishra and Sharad Tiwari

Abstract Mercury is ranked as the most toxic heavy metals. It enters into the environment due to some natural processes and anthropogenic activities. It has a property of bioaccumulation into the food chain through uptake by crop plants from the contaminated agricultural lands, leading to detrimental impact on human health. Mercury has the toxic effect on plants as it disturbs many biological processes, including photosynthesis, respiration, transpiration, cell division and so on. Phytoremediation involves several plant species which have the ability to accumulate or degrade contaminants, including heavy metals. Another important strategy is the utilization of transgenic plants transformed with bacterial *mer* genes to increase phytoremediation of mercury. The mercury-resistant plant growth promoting microbes (PGPM) enhance plant growth under mercury stress as well as increase the mercury uptake by plants. This chapter summarizes the present understanding toward the mercury toxicity and their molecular responses in plants. It also illustrates the plethora of mechanism adapted by PGPM for plant growth promotion and detoxification of mercury. It also highlights the paradigms for synergistic use of PGPM for improved phytoremediation of mercury from agricultural lands.

S. Sapre · R. Deshmukh · I. Gontia-Mishra (✉)
Biotechnology Centre, Jawaharlal Nehru Agriculture University,
Jabalpur 482004, Madhya Pradesh, India
e-mail: itigontia@gmail.com

S. Sapre
e-mail: swapnil.spr@gmail.com

R. Deshmukh
e-mail: reenad1109@gmail.com

S. Tiwari
Department of Plant Breeding and Genetics, Jawaharlal Nehru Agriculture University,
Jabalpur 482004, Madhya Pradesh, India
e-mail: shtiwari@gmail.com

D. K. Maheshwari and S. Dheeman (eds.), *Field Crops: Sustainable Management by PGPR*, Sustainable Development and Biodiversity 23,
https://doi.org/10.1007/978-3-030-30926-8_10

Keywords Mercury · Mercury hyperaccumulators · *Mer* genes · Plant growth promotion · PGPM-assisted phytoremediation

10.1 Introduction

The contamination of heavy metals in soil by various activities is a major environmental concern. The highly toxic heavy metals such as arsenic (As), lead (Pb), cadmium (Cd) and mercury (Hg) can enter the food chain through uptake by crop plants from the contaminated agricultural lands, leading to detrimental impact on human health (Puglisi et al. 2012). Among the heavy metals, mercury is ranked as the most toxic one and it is present in the environment due to some natural processes and anthropogenic activities. The anthropogenic activities which introduce mercury into the environment are production chemicals (paints, paper and pulp, disinfectants, microbicidal agents, agrochemicals, etc.), use of fossil-fuels in power plants and industrial waste incineration and disposal (Frossard et al. 2017). Mercury is mostly present in the environment in elemental form (Hg^0), which is readily oxidized to the ionic form, Hg^{2+} (Selin 2009). The ionic form of mercury (Hg^{2+}) makes complexes with many inorganic compounds to produce HgS, $ClHg_2$, Hg_2Cl_2 or other organomercurial compounds such as CH_3–Hg (Tiwari and Lata 2018). The agricultural soil has the ionic form (Hg^{2+}) in majority, through adsorption onto sulfides, clay particles and organic matters (Azevedo and Rodriguez 2012; Gontia-Mishra et al. 2016; Beckers and Rinklebe 2017). Interestingly, mercury is a bioaccumulative toxin as it remains in the environment for longer periods (Schroeder and Munthe 1998). Plants growing under mercury contamination are severely affected during all stages of their development. It is observed that plants exhibit poor germination, impaired photosynthesis, compromised biomass, and so on, under mercury stress (Patra and Sharma 2000). It is highly phytotoxic to plant cells, leading to various physiological abnormalities. At cellular level, mercury interferes with the proper water uptake (stomata closure), functioning of mitochondria and generates oxidative stress leading to lipid peroxidation, enzyme inactivation and DNA and membrane damage (Chen et al. 2014).

Hence, it is apparently the need of the hour to select an appropriate methodology for mercury remediation from the environment. There are physiochemical methods, including precipitation, conventional coagulation, adsorption, ion exchange, and so on, and biological methods (bioremediation and phytoremediation) for removal of this hazardous metal from the environment (Dash and Das 2012). The physiochemical methods have a very high operational cost, as well as are not suitable for large areas; therefore, the focus is rapidly shifting toward the cost-effective and eco-friendly biological alternatives for remediation of mercury (Oves et al. 2013). In this context, the use of plant growth promoting microbes (PGPM) for remediation of mercury contaminated site is an attractive and economical option. There are several reports which advocate the role of plant growth promoting rhizobacteria (PGPR) and fungi in enhancement of plant growth and endurance under heavy metal stress condition (Arshad et al. 2007; Gontia-Mishra et al. 2016; Pietro-Souza et al. 2017). They also possess the ability to biotransform toxic ions to non-toxic forms. This chapter summarizes the present understanding toward the mercury toxicity and their molecular responses in plants. It also highlights the paradigms for synergistic use of PGPM for improved phytoremediation of mercury from agricultural lands.

10.2 Mercury Pollution in Agricultural Lands

There are various sources of mercury pollution to the environment. These sources are natural as well as due to human activities. The natural activities include ubiquitous breaking of mercury-containing rocks in the outer layer of the earth. The volcanic explosion and geothermal activity can contribute to the mercury pollution to some extent (AMAP/UNEP 2013). This kind of pollution depends on several factors, like presence of mercury containing rocks and period of geothermal activities (Ferrara 1998, 2000; Pirrone et al. 2001). Global mercury emissions are approximately 80–600 ton/year to the environment from such natural activities (Mason et al. 2012).

Since ancient times, mercury is being used for several purposes, like explosives, preservatives and treatments of various skin diseases (Azevedo and Rodriguez 2012). Anthropogenic activities are the major sources for mercury pollution in the soil and water. These activities include mining, coal burning, cement production, mercury containing pesticides and disposal of mercury containing waste in the environment (AMAP/UNEP 2013; Mason et al. 2012). Electricity generation by burning of coal and dependence of several industries on coal energy emit huge amounts of mercury to the atmosphere.

Some frequently used materials, like batteries, fluorescent lamps and medical instruments such as leftover thermometers and sphygmomanometer, also contribute to Hg emissions. Mercury is also used by dentists as dental amalgams and in various devices like barometers, manometers, float valves, mercury switches, mercury relays and various research activities. Although these devices are useful in weather stations, airports and air fields, wind tunnels and engine manufacturing, as well as in installations offshore or on ships, they also contribute to release of mercury in the atmosphere (Hutchison 2003). Mercury pollution also emerges from agricultural practices, like use of pesticides and fertilizers (Hseu et al. 2010). It was conventionally used in the form of fungicide or pesticide in agriculture. The amount of mercury present in the soil and its uptake by plants rely on various factors, like pH of soil, species of plant, presence of microorganisms, type of mercury ion and so on. It was observed that the mercury uptake can be minimized using high soil pH and/or high amount of lime and salts (Patra and Sharma 2000; Patra et al. 2004). It has been reported that mercury accumulation is much more persistent in soil as compared to aquatic and other biomes (Padmavathiamma and Li 2007; Tangahu et al. 2011).

10.3 Mercury Toxicity in Plants and Their Molecular Responses

The organic form of mercury severely affects plants as they are more toxic than inorganic (Hg^{2+}) counterparts (Patra and Sharma 2000). Mercury has its toxic effect on most of the crop species, beyond the tolerance limit. It tends to amass in the roots; hence the phytotoxic symptoms are also noticed in roots (Chen et al. 2014). The excess mercury in the soil is taken up by plants, causing disturbance and malfunction to many of the biological processes, including photosynthesis, respiration, transpiration and cell division (Fig. 10.1). The plausible mechanism of mercury toxicity is its ability to react with the sulfydryl (SH) groups of proteins and enzymes; similarly, it has high affinity for the phosphate groups of lipids, energy-rich molecules like ATPs and nucleotides. It is also noted that it also substitutes the essential ions such as Mg^{2+} ion in chlorophyll (Azevedo and Rodriguez 2012). Mercury also messes up with the aqua-porins (water channels), causing impaired transpiration and subsequent water uptake via vascular tissues (Zhou et al. 2008). It deliberately disrupts the plant antioxidant defense enzymes, especially glutathione reductase (GR), superoxide dismutase (SOD), catalase and ascorbate peroxidase (APX). Besides, it also affects the other antioxidant entities such as glutathione (GSH) and non-protein thiols (Israr et al. 2006; Zhou et al. 2008).

The plants can tolerate the effect of mercury toxicity to some extent by the interplay of various physiological and molecular mechanisms. First, when plants come into contact with mercury ions, they prohibit or reduce the uptake of mercury into the roots by either complexing them to cell wall or root exudates; if it enters the root cell, the metal ion is restricted to the apoplasts. But if still the mercury ions gain entry into the plant cell, they are countered by detoxification through compartmentalization into vacuoles or complexation with amino acids, organic acids, chelation by phytochelatins (PC) and metallothioneins (MT). Further, some non-enzyme antioxidants such as α-tocopherol and GSH also aid in combating mercury toxicity (Kalaivanan and Ganeshamurthy 2016). These processes mostly

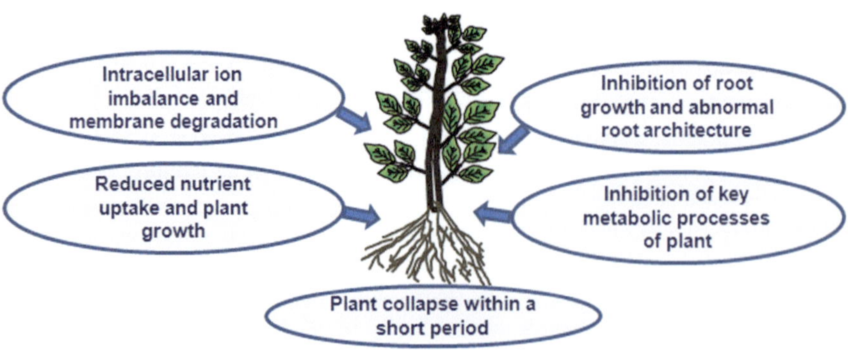

Fig. 10.1 Effects of mercury toxicity on plants

put a check on translocation of mercury ions to the leaf tissues and thereby shielding the photosynthesis from detrimental effect of mercury (Rascio and Navari-Izzo 2011). Finally, plants resort the mercury toxicity by induction of oxidative stress enzymes such as SOD, APX, catalase, glyoxalase and GR. They also trigger the stress-responsive proteins and hormones. Various signaling cascades are stimulated by encountering heavy metal ions, namely calcium-dependent signaling and mitogen-activated protein kinase (MAPK) signaling (Tiwari and Lata 2018). Recently, it was observed that mercury toxicity activates the biosynthesis of aromatic amino acids (tryptophan and phenylalanine), calcium accumulation and also stimulates MAPK in rice (Chen et al. 2014).

10.4 Phytoremediation

As already discussed, the physical and chemical methods of remediation of heavy metals are costly and time-consuming. The focus has shifted to phytoremediation. Phytoremediation is a process utilizing plants to eliminate, detoxify, volatilization or reduce the concentration of environmental contaminants, including heavy metals (Wood et al. 2016). In contrast, this technology is widely accepted as it is an economical, efficient and environment-friendly way to remove metals from contaminated soils. Phytoremediation, in general, involves several plant species with the ability to accumulate or degrade contaminants and in turn the biomass generated through this process can be used for other purposes. Phytoremediation occurs by the use of one or all the following mechanisms, such as phytoextraction, rhizofiltration, phytodegradation and phytovolatilization. In the context of mercury contamination, phytovolatilization is of great importance because Hg^{2+} ion is taken from soil and volatilizes it to non-toxic Hg^0 form from the foliage into the atmosphere (Kumar et al. 2017). Phytoremediation is adopted by using hyperaccumulator plants which have capacity to take up exceedingly enormous amounts of heavy metals from the soil in comparison to other plants (non-hyperaccumulators). Additionally, in hyperaccumulators the heavy metal ions are translocated to the shoots, and interestingly, no visible signs of phytotoxicity are observed in hyperaccumulators (Rascio and Navari-Izzo 2011).

In general, plants belonging to the family Brassicaceae (*Thlaspi* and *Alyssum*, *Brassica*, *Noccaea*), Crassulaceae (*Sedum alfredii*) and Pteridaceae (*Pteris vittata*) have been reported as proficient heavy metal hyperaccumulators (Reeves and Baker 2000). However, the reports for mercury hyperaccumulators are scanty and include plant species *Brassica juncea*, *Polypogon monospeliensis* and *Pteris vittata* (Su et al. 2007). Similarly, plants *Lindernia crustacea*, *Paspalum conjugatum* and *Cyperus kyllingia* have been used for potential phytoextraction of mercury for contaminated sites. It was also noted that these plant species efficiently take up and translocate mercury from roots to shoots (Muddarisna et al. 2013). In another study, *Medicago sativa* and *Dittrichia viscosa* were found suitable plants for phytoremediation of the site contaminated with mercury (Shehu et al. 2014). Hence, there is an ever-increasing demand to search new mercury hyperaccumulators plant for their utilization in phytoremediation.

10.5 Plant Growth Promoting Rhizobacteria/Microbes (PGPR/PGPM)-Assisted Phytoremediation

PGPR are a dynamic group of bacteria which are associated with the roots or present in vicinity of roots, that is, in the rhizosphere and tend to benefit plants through plethora of mechanisms (Gontia-Mishra et al. 2017a). Microbes are well known to take part in the biogeochemical cycling of toxic metals and also in remediating heavy metals from the contaminated sites. The phytoremediation of heavy metals mostly relies on the availability of metal ions. The interference of heavy metal-resistant PGPR addresses this problem by increasing the bioavailability of metal ion by modifying the soil texture and properties (Mishra et al. 2017). PGPR play a major role in plant growth promotions as well as bestow with the ability to reduce the noxious effect of heavy metals on plants. PGPR are also equipped to detoxify the heavy metals. There are ample of reports which advocate that the use of PGPM proposes a better strategy by improving the phytoextraction and phytoremediation of heavy metals, particularly mercury from soil (Glick 2010; Quiñones et al. 2013; Gontia-Mishra et al. 2016; Ashraf et al. 2017; Etesami 2018). The comparative account of mercury detoxification via phytoremediation and PGPM-assisted phytoremediation is represented in Fig. 10.2.

10.5.1 Detoxification of Mercury via Mer *Operon*

With special context to bioremediation of mercury, PGPM can facilitate the process by direct interaction with the metal ion, through volatilization (reduction of Hg^{2+} to Hg^{0}), transformation and rhizodegradation (Gadd 2010). There are many studies which provide the details of mercury-resistant PGPR (Ruiz-Diez et al. 2012; Quiñones et al. 2013; Gontia-Mishra et al. 2016). These mercury-resistant bacteria have evolved numerous ways to combat the mercury toxicity, namely through volatilization, exopolysaccharide (EPS) sequestration, metal complexation and enzymatic detoxification (Dash and Das 2012). Mercury-resistant bacteria are also known to produce thiol compounds which have high affinity for Hg^{2+} ions, further lowering the metal toxicity. Another important mechanism adapted by bacteria to alleviate metal toxicity is bioaccumulation. Bioaccumulation is largely carried out by biosorption, which depends on various processes such as complexation (sulfur, phosphate and bicarbonate groups of cell wall), chelation (siderophore and PC), ion exchange and entrapment in intracellular spaces (Mishra et al. 2017).

Furthermore, many mercury-resistant bacteria also harbor mercury resistance genes, that is, the *mer* operon, either on plasmid or in the chromosome, transposons

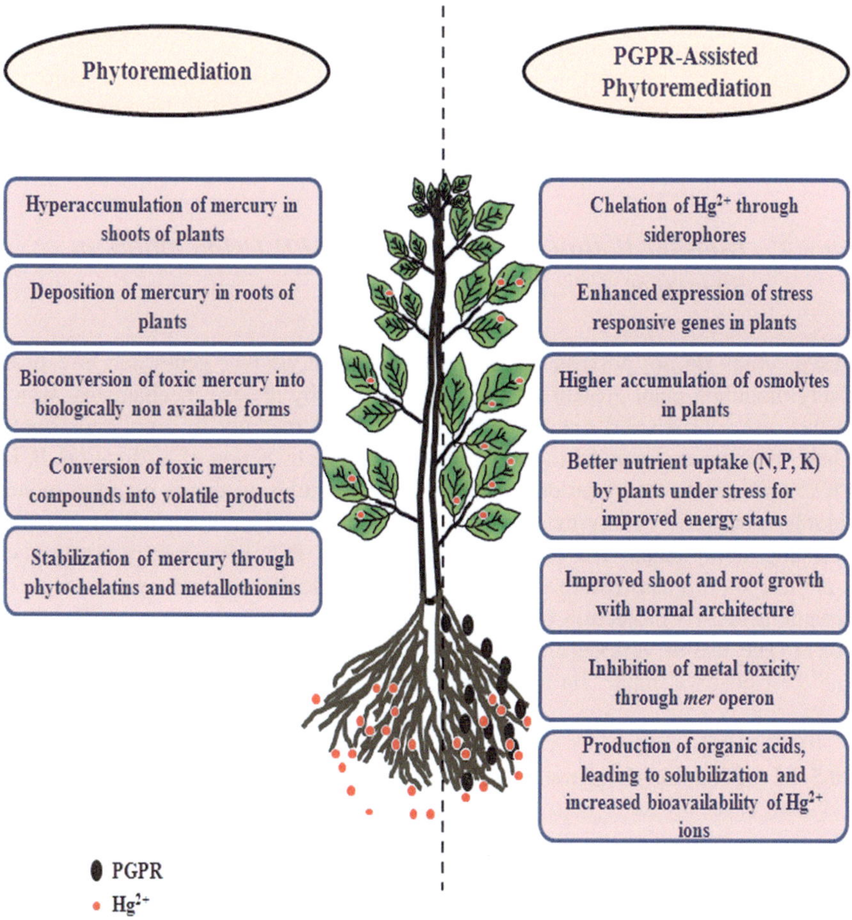

Fig. 10.2 The comparative account of mercury detoxification *via* phytoremediation and PGPM-assisted phytoremediation

or integrons for reduction from Hg^{2+} to Hg^0, the less reactive form (Mathema et al. 2011; Dash et al. 2017a). This is the most efficient mechanism of mercury detoxification widespread in the Gram-positive and Gram-negative bacteria (Chatziefthimiou et al. 2007). The *mer* operon consists of a series of operator, promoter, regulator genes and the functional genes. The functional genes comprise *merP*, *merT*, *merD*, *merA*, *merF*, *merC* and occasionally *merB*. Each gene codes for a functional protein participate in a specific way to detoxify mercury. The detoxification of mercury is generally achieved by the uptake and transport of Hg^{2+} by the periplasmic protein MerP, a mercuric ion transport protein MerF and the inner membrane protein MerT (Powlowski and Sahlman 1999; Mathew et al. 2015). Further, the final process of reduction of Hg^{2+} to volatile metallic mercury is carried out by mercuric ion reductase encoded by the most important *merA* gene

(Ní Chadhain et al. 2006; Dash et al. 2017a). Moreover, some bacteria also possess *merB* gene coding for organomercurial lyase (Reniero et al. 1995) which aid in cleavage of Hg^{2+} from organomercurial compounds. The *mer* operon is positively regulated by regulator protein merR in the presence of Hg^{2+} in the surroundings (Huang et al. 2002).

10.5.2 *Bioremediation of Mercury by PGPR Using Different Strategies*

PGPR which do not possess the mercury resistance via *mer* genes can also positively modulate plant growth under mercury stress by diverse mechanisms, which in turn aid in improved uptake and assimilation of mercury in roots and shoots. These mechanisms include: (a) better nutrient uptake by action of N_2 fixation, P, K, Zn solubilization; (b) by producing plant growth regulators, that is phytohormones [indole acetic acid (IAA), cytokinin and gibberellic acid]; (c) production of metal chelating compounds, for example, siderophores, PC, EPS; (d) production of 1-aminocyclopropane-1-carboxylic acid (ACC) deaminase; (e) production of organic acids; (f) protection of plants from deleterious plant pathogens (Jha et al. 2012). The use of various mercury-resistant PGPR for alleviating mercury stress and their positive impact on host plants is shown in Table 10.1.

10.5.2.1 Rhizobia–Legume Interaction in Mercury Bioremediation

The most important symbiotic interaction between rhizobia and legume which offer nutrients, especially nitrogen to plants, has been extensively exploited to improve the phytoremediation of mercury (Nonnoi et al. 2012; Lebrazi and Fikri-Benbrahim 2018). It is a well-known fact that rhizobia enter the roots of legumes to develop root nodules, where atmospheric N_2 is fixed to ammonia, which is assimilated by the plants. The nitrogen is the major nutrient element for plant growth, and legume–rhizobia interaction enhances plant productivity; hence rhizobia has been recognized as an important plant growth promoter (Teng et al. 2015). Recently, it is noted that many species of rhizobia have mercury resistance which promotes them to be used as excellent candidates for rehabilitating contaminated soils (Ruiz-Diez et al. 2012). Besides, the rhizobia–legume partnership also enhances soil fertility, promotes plant growth and provides ecological benefits. A recent study reports the isolation of mercury-resistant rhizobial strains *Ensifer medicae* and *Rhizobium leguminosarum* bv. *trifolii* from the root nodules of *Medicago* spp. and *Trifolium* spp. plants growing in Hg-contaminated soils. In addition to nitrogen fixation, these isolates also exhibited other plant growth promotion traits such as phosphate solubilization and siderophore production, which assists in increasing the bioavailability of metal ion in the soil through modulation of soil pH (Nonnoi et al. 2012). Similarly, mercury-resistant rhizobial strains *R. leguminosarum*, *Rhizobium*

Table 10.1 List of mercury-resistant PGPM used to enhance plant growth promotion and alleviation of mercury stress in plants

S. no.	Mercury-tolerant PGPRs	Isolation site	Interaction with plant	Level of Hg tolerance	Plant growth promoting traits	References
1	*Pseudomonas fluorescens*	Rhizosphere of soybean and pea	Soybean	147 µM	IAA, siderophore production and P solubilization	Gupta et al. (2005)
2	*Ensifer medicae* and *Rhizobium leguminosarum* bv. *trifolii*	Root nodules of *Medicago* spp. and *Trifolium* spp.	–	250 µM	P solubilization and siderophore production	Nonnoi et al. (2012)
3	*Rhizobium radiobacter*, *R. leguminosarum*, *R. gallicum*, *Bradyrhizobium canariense*, *Ensifer medicae* and *E. meliloti*	Nodules of legumes (*Medicago*, *Trifolium*, *Vicia*, *Lupinus*, *Phaseolus*, and *Retama*)	–	3–30 µM	N_2 fixation and ability to re-nodulate the host plants	Ruiz-Diez et al. (2012)
4	*Bradyrhizobium canariense*	Nodules of *Lupinus albus*	*Lupinus albus*	12.5 µM	N_2 fixation and nodulation at high concentration of Hg	Quiñones et al. (2013)
5	*Halobacillus* sp., *Halomonas* sp. and *Halobacillus* sp.	Mangrove rhizosphere	*Sesuvium portulacastrum*	0.04 mM	Salinity and other metal tolerance with IAA production and P solubilization.	Desale et al. (2014)
6	*Enterobacter, Cronobacter* and *Pseudomonas*	Nodules of different plants	–	–	N_2 fixation	Tariq and Latif (2014)
7	*Photobacterium* spp.	Rhizosphere soil of reed (*Phragmites australis*)	Long yard bean	33 $mg^{-1\ kg}$	IAA producing with multiple metal tolerance	Mathew et al. (2015)
8	*Cronobacter* sp., *Pseudomonas* sp. and *Bacillus* sp.	Root nodules of *Sesbania concolor*, *Trifolium alexandrinum*, *Iresine herbstii* and *Dahlia* sp.		50 µg of $HgCl_2$	N_2 fixation and H_2S producing	Rafique et al. (2015)

(continued)

Table 10.1 (continued)

S. no.	Mercury-tolerant PGPRs	Isolation site	Interaction with plant	Level of Hg tolerance	Plant growth promoting traits	References
9.	*Pseudomonas* sp.	Polluted water, rhizospheric soil and root nodules of different plant species	–	10–20 μg^{-1ml} $HgCl_2$	N_2 fixation and H_2S producing	Tariq et al. (2015)
10	*Enterobacter ludwigii*, *E. cloacae* and *Klebsiella pneumoniae*	Rhizosphere of the weeds *Alternanthera sessilis* and *Cyperus esculentus*	Wheat	500 μM	NH_3, siderophore, IAA, EPS production, ACC deaminase activity, Zn, P and K solubilization	Gontia-Mishra et al. (2016)
11	*Enterobacter* sp.	Rose (root section), Lemon (stem section) and Mosaami (stem section)	–	20–50 mg^{-1L}	N_2 fixation, HCN and H_2S production	Mobeen and Latif (2016)
12	*Bacillus* sp., *Bacillus cereus*, and *Enterobacter cloacae*	Rhizosphere soils near to districts Kasur and Sheikhupura, Pakistan	Chickpea	20 μg^{-1ml}	IAA production	Amin and Latif (2017)
13	*Pseudomonas* spp., *Bacillus* spp., *Paenibacillus sinopodophylli*, *Cupriavidus gilardii* and *Paenarthrobacter* sp.	Rhizosphere of *Lepidium meyenii*	*Trifolium pratense*	0.01 mM	IAA production and P solubilization	Ortiz-Ojeda et al. (2017)
14	*Azotobacter* sp.	Gold mining area at Buru Regency, Maluku	–	15–20 mg^{-1L}	EPS production	Hindersah et al. (2017)
15	*Azotobacter* sp. and *A. chroococcum*	Wild kale rhizosphere grown in tailings of Waekerta Village, Buru Regency, Maluku	Groundnut	0.64 $mg^{-1\ kg}$	EPS production	Hindersah et al. (2018)

radiobacter, *E. medicae* and *Bradyrhizobium canariense* were isolated from the root nodules of leguminous plants *Medicago*, *Trifolium*, *Vicia*, *Lupinus*, *Phaseolus* and *Retama* growing in the mercury-contaminated areas of Spain (Ruiz-Diez et al. 2012). The mercury-resistant *B. canariense* was used in interaction with *Lupinus albus* for phytoremediation of mercury from the contaminated soils (Quiñones et al. 2013). Another advantage with rhizobia–legume system is that it is a stable association over other free-living PGPR; hence it will definitely improve the performance of host plants under heavy metal stress (Li et al. 2012). Thus, it can be suggested that rhizobia-assisted phytoremediation is a feasible option to revitalize mercury-contaminated soils.

10.5.2.2 Production of Phytohormones

Phytohormones mainly comprise auxin, cytokinin (CK), gibberellic acid (GA), abscisic acid and ethylene. These phytohormones are the key players in metal stress mitigation (Bücker-Neto et al. 2017). Almost 80% of PGPR isolated from the rhizosphere are capable of producing IAA, which has selective impact of the root growth and development (Patten and Glick 2002). The phytohormone IAA is mostly associated with cell division and differentiation, formation of vascular bundles and also plays a vital role in nodule formation (Gopalakrishnan et al. 2015). Plants with better root system have greater access to the nutrient and heavy metal uptake further results in improved phytoremediation (Ma et al. 2016). PGPR bear the ability to produce IAA and, therefore, enhance the root surface area of plant for better adsorption of heavy metals from the soil (Etesami 2018). Additionally, IAA is also implicated with diverse metabolic processes in plants, namely, stimulation of plant defense system and function as a cell–cell signaling molecule (Spaepen et al. 2007; Ma et al. 2018). Krishnamurthy and Rathinasabapathi (2013) have revealed an affirmative role of auxin transport through *AUX1* on tolerance of plant to heavy metal stress via ROS-mediated signaling. In a study, *Halobacillus* sp. and *Halomonas* sp. produced IAA and indole butyric acid (IBA) and subsequently increased the root length and root dry weight of *Sesuvium portulacastrum* under mercury stress (Desale et al. 2014). In a similar study, consortium of mercury-resistant PGPR (IAA producing) *Bacillus* sp. and *Enterobacter cloacae* were inoculated with chickpea and showed improved growth of plant under mercury stress (Amin and Latif 2017). In a recent report, inoculation of mercury-resistant IAA producing fungi *Aspergillus* sp. and *Massariosphaeria* sp. was beneficial to host plants *Polygonum acuminatum* and *Aeschynomene fluminensis* in providing defense against the adverse effects of mercury (Pietro-Souza et al. 2017). These studies provide evidences that mercury-tolerant PGPR which produce IAA have capability to positively modulate plant growth under stressful conditions (Table 10.1). The role of other phytohormones such as CK, GA and salicylic acid (SA) is not known in mercury stress. In contrast, the involvement of CK, GA and SA is largely recognized in alleviating other heavy metals such as Mn, As and Cd (Gangwar et al. 2010; Zhu et al. 2012; Gondor et al. 2016).

10.5.2.3 Production of Organic Acids

The PGPR are extensively documented for increasing the bioavailability of essential nutrients such as P, K and Zn (Gontia-Mishra et al. 2017a). The chemical forms P and Zn applied as superphosphates and zinc sulfate, respectively, are readily converted into insoluble forms and are inaccessible for plant uptake (Richardson 2001; Gontia-Mishra et al. 2017b). The phosphate solubilizing bacteria (PSB) and zinc solubilizing bacteria (ZSB) render the insoluble phosphates and zinc into soluble form through the process of acidification (release of organic acids), chelation and exchange reactions (Kim et al. 1998; Kamran et al. 2017). The most important mechanism of P and Zn solubilization is through the action of organic acids. Several workers provided evidences that low molecular weight organic acids like gluconic acid, 2-ketogluconic acid, 5-ketogluconic acid pentanoic acids citric acid, lactic acid, succinic acid and propionic acid are produced during P and Zn solubilization (Chen et al. 2006; Saravanan et al. 2007). These organic acids lower the pH of soil and help in solubilization of P and Zn. The organic acids, such as citrate, malate, oxalate, malonate, tend to form strong bonds with heavy metal ions through metal chelation with carboxyl groups and these complexes are less toxic to plants than the free metal ions (Kavita et al. 2008; Osmolovskaya et al. 2018). Another important fact is that the organic acids produced by microbes have greater affinity for chelating heavy metals than essential nutrients (Gadd 2010). The organic acids also lower the pH of soil; hence they also help in solubilization of heavy metals and increase the mobilization of heavy metals in rhizosphere for uptake by the plants (Etesami 2018). There are reports on P and Zn solubilization by several mercury-resistant PGPR that can aid in alleviation of mercury stress (Gupta et al. 2005; Nonnoi et al. 2012; Desale et al. 2014; Gontia-Mishra et al. 2016). Consequently, organic acids producing PGPM can indirectly contribute to mitigation of heavy metal stress in plant.

10.5.2.4 Action of ACC Deaminase Enzyme

Ethylene is produced in plants subjected to variety of abiotic and biotic stresses encompassing exposure to salt, drought, flooding, heavy metals, organic and inorganic chemicals, attack of nematodes, phytopathogens and so on. (Gontia-Mishra et al. 2014). Hence the ethylene produced during such stress conditions is termed as "stress ethylene" (Glick 2014). Consequently, elevated levels of ethylene can impair root growth; overall plant growth is retarded and induction of senescence occurs in plants (Han et al. 2015). Interestingly, it is noted that 1-aminocyclopropane-1-carboxylic acid (ACC) works as precursor for ethylene biosynthesis. Moreover, several PGPR have potential to substantially lower the ethylene concentration in plants. This is attributed to the enzyme ACC deaminase (present in PGPR), which catalyzes the conversion of ACC to ammonia and α-ketobutyrate, and subsequently reduced ethylene concentration in stressed plants (Glick et al. 1999). Various researchers have documented the application of ACC deaminase producing PGPR for improvement of plant growth under heavy metal stress (Ma et al. 2015; Han et al. 2015; Rizvi and Khan 2017). In context to mercury stress, Gontia-Mishra et al. (2016) have reported the application of mercury resistance PGPR with ACC deaminase activity for combating the effect on mercury in wheat plants.

10.5.2.5 Production of Siderophore, EPS, PC and MT

Iron is a vital nutrient for microorganisms. Fe^{3+} ion usually makes complex insoluble with hydroxide and oxyhydroxide under aerobic conditions and renders it unavailable to microbial uptake (Storey et al. 2006). Mostly, bacteria acquire iron by the secretion of low-molecular weight ferric-ion-specific chelating agents known as siderophores (Neilands 1995). Siderophores are produced by PGPR, both symbiotic and free-living bacteria and fungi, growing under low iron concentrations. Moreover, siderophore has registered a remarkable role in detoxification of various heavy metals (Nonnoi et al. 2012; Hesse et al. 2018). Nevertheless, the undeniable function of siderophore is iron acquisition, but they can bind with other heavy metals like Al, Cd, Cu, Pb, Hg and Zn (Neubauer et al. 2000; Braud et al. 2009). It is also noted that besides iron, other metals can stimulate siderophore production (Hesse et al. 2018). The siderophore–metal complex can increase the soluble metal concentration for uptake by plants (Schalk et al. 2011). The siderophore-producing PGPR provide dual benefit to plants under metal stress first, by improving iron availability, and secondly, by lowering the free metal ion concentration in the rhizosphere (as siderophore forms complex with heavy metals) (Dimkpa et al. 2008). Additionally, siderophores have also displayed an efficient role in biological control against phytopathogens (Kloepper et al. 1980). The siderophore-producing PGPM has canonical function to promote the plant growth via iron acquisition and for control of plant diseases but their use in PGPR-assisted phytoremediation has gained much attention (Rajkumar et al. 2010). A large number of mercury-resistant PGPR and endophytic fungi have been reported to increase the survival and growth of plants in mercury-contaminated soils by ameliorating the metal toxicity (Gupta et al. 2005; Nonnoi et al. 2012; Gontia-Mishra et al. 2016; Pietro-Souza et al. 2017).

Many PGPR have the unique ability to produce exopolysaccharide (EPS)/extracellular polymeric substances. The EPS has multifarious function in bacterial cells ranging from quorum-sensing signals, biofilm formation, development, survival and host colonization (Nocelli et al. 2016). The EPS largely constitutes high molecular weight macromolecules like polysaccharide along with smaller proportions of protein, lipids and uronic acid (Gupta and Diwan 2017). EPS shields the bacterial cells against various environmental stresses, such as heavy metal toxicity, drought and salinity. They have several anionic functional groups (e.g., sulfhydryl, carboxyl, hydroxyl, sulfonate, amine and amide), which binds with metal ions and decreases their mobility in the soils and their accessibility for plants (Rajkumar et al. 2012). In fact, *S. meliloti* produces two kinds of EPS: succinoglycan and galactoglucan that are associated with developing symbiotic relation with host plants. In a study, wild-type and mutant *S. meliloti* (deficient in EPS I and EPS II synthesis) was treated with different concentrations of mercury. The mutant strains could not withstand mercury toxicity; hence it could be proposed that EPS production plays a pivotal role in combating mercury toxicity (Nocelli et al. 2016). EPS secretion is an extensively recognized phenomenon for plant growth promotion. The PGPR which are capable of EPS secretion has been widely known to decrease mercury toxicity during their interaction with host plants (Gontia-Mishra et al. 2016; Hindersah et al. 2018).

Recent studies investigating the role of PGPM in heavy metal uptake by plants have demonstrated that microbial PC and MT have the ability to chelate heavy metals. PC is produced by arbuscular mycorrhizal fungi as well as by plants in response to heavy metal stress (Miransari 2011). However, rhizobacteria do not produce PC; hence phytochelatein synthase gene from *Schizosaccharomyces pombe* (yeast) was transformed to *Pseudomonas putida*, which is induced by several heavy metals such as Cu, Cd, Pb and Hg and this recombinant bacteria improved germination and plant growth in wheat under multiple metal stress condition (Yong et al. 2014). These microbial PCs are cysteine-rich peptides which bind to heavy metals with great affinity and enhance phytoremediation of heavy metals by plants (Kang et al. 2007). In order to sequester, heavy metals such as Cd, Zn, Hg, Cu and Ag PGPR and mycorrhizal fungi produce MT and are also cysteine-rich polypeptides that bind to heavy metals (Ullah et al. 2015). A recent study reported the characterization of MT genes, from the ectomycorrhizal fungus *Laccaria bicolor* under heavy metal stress (Reddy et al. 2014). Thus, it could be predicted that PGPR and symbiotic fungi has capability to produce PC and MT, and can enhance metal tolerance in host plants. The role of different bacterial species as potential bioremediation candidates to control mercury pollution and their mode of action for detoxification of mercury are listed in Table 10.2.

10.6 Transgenic Plants Overexpressing *Mer* Genes

Phytoremediation through naturally occurring plants capable of alleviating heavy metal stress can be difficult to manage and achieve, except for hyperaccumulators. Usually, such plants fail to survive in diverse environment or even if they survive, they fail to alleviate the heavy metal stress, either due to change in soil type or due to excessively high concentration of phytotoxic compounds (Doty 2008). Therefore, development of transgenic has recently been employed in several important plant species, particularly those capable of growing in diverse climatic conditions. Transgenic technology for improving or inducing the phytoremediation uses the genes responsible for either metabolizing phytotoxic compounds or uptake or their translocation (Kawahigashi et al. 2002; Lee et al. 2003; Chowdhury et al. 2015). There have been three basic strategies used to develop transgenic plants for phytoremediation to extract genes, viz., from bacteria or non-plant sources, or transforming with genes derived from different plant species or overexpressing of detoxification genes in the same plant species (Maestri and Marmiroli 2011).

Table 10.2 Bioremediation of mercury using mercury resistant bacteria

Bacterial species	Genera	Mercury toxicity	Isolation site	Mode of action for detoxification of mercury	References
Sphingobium sp.	α-Proteobacteria	5.1 $mg.l^{-1}$	New South Wales, Australia	Volatilization of mercury by the action of mercuric reductase	Mahbub et al. (2016a)
Vibrio fluvialis, *Serratia marcescens* and *Bacillus megaterium*	γ-Proteobacteria Fermicutes		Tagus Estuary (Barreiro and Cala do Norte), Portugal	Biotransformation of mercury to methyl mercury in 48 h	Figueiredo et al. (2016)
Pseudarthrobacter oxydans and *Pseudomonas frederiksbergensis*	Actinobacteria γ-Proteobacteria	2.0 and 1.5 ppm, respectively	Tundra ecosystem of Ny-Ålesund, Svalbard	Biosorption of mercury in cell pellets	Balan et al. (2018)
Alcaligenes faecalis	β-Proteobacteria		Fatehgarh Sahib, Punjab, India	Biosorption of mercury followed by biotransformation of sequestered mercuric ions into monovalent mercury (Hg_2Cl_2).	Gupta and Nirwan (2015)
Bacillus tequilensis, *Bacillus pumilus*, *Bacillus subtilis* and *Achromobacter xylosoxidans*	Fermicutes β-Proteobacteria	0.3 mM mercuric chloride	Mysore, Karnataka, India,	–	Sunil et al. (2015)
Bacillus thuringiensis	Fermicutes γ-Proteobacteria	50 ppm of Hg as $HgCl_2$	Odisha, India	Volatilization of mercury at different environmental parameters, i.e. pH, temperature and salinity	Dash et al. (2014)
Vibrio fluvialis	γ-Proteobacteria	250 $\mu g.ml^{-1}$	Uppanar estuary, Tamil Nadu, India	Solubilization of mercury to non-toxic form	Saranya et al. (2017)
Afifella marina and *Rhodovulum sulfidophilum*	α-Proteobacteria	398.84 $\mu g.l^{-1}$ $HgCl_2$	Thai Peninsular coastal areas (Andaman Sea and	Volatilization of mercury via mercuric reductase	Mukkata et al. (2015)

(continued)

Table 10.2 (continued)

Bacterial species	Genera	Mercury toxicity	Isolation site	Mode of action for detoxification of mercury	References
			Gulf of Thailand), Thailand		
Bacillus thuringiensis	Fermicutes	500 ppm of mercury	Mithi River, Mumbai, India	Solubilization of mercury to non-toxic form	Pushkar et al. (2018)
Pseudoxanthomonas sp.	γ-Proteobacteria	1.7 $mg.l^{-1}$ in low nutrition medium	New South Wales, Australia	Mercury volatilization potential via mercuric reductase and the presence of *merA* gene	Mahbub et al. (2016b)
Pseudomonas sp.	γ-Proteobacteria	250 μM	Brazil	Via action of mercuric reductase	Giovanella et al. (2016)
Bacillus cereus	Fermicutes	50 ppm	–	Through the action of biofilm and EPS trapping of mercury and finally volatilization	Dash et al. (2017b)

Phytoremediation of toxic metals like mercury offers an economic and non-accumulative mode of remediation. With increasing mercury pollutions, transgenics have been widely developed for phytoremediation of mercury toxicity. Most of the transgenics designed for mercurial phytoremediation introgress two major classes of bacteria-derived genes *merA* and *merB* (Rugh et al. 1996; Meagher and Heaton 2005). These transformants are achieved via nuclear or chloroplast genome transfer of *mer* genes. According to the latest reports, mercuric reductase is the only class of enzyme known to metabolize phytotoxic Hg^{2+} ions. The *merA* gene product, a chief enzyme of *mer* operon, codes for mercuric reductase enzyme responsible for conversion of Hg^{2+} to elemental Hg^0.

A similar activity has been initially reported to be associated with the plasmids of few bacteria (Clark et al. 1977; Schottel 1978). However, presence of *merA* genes is observed within the conjugative plasmids and their presence in diverse taxonomic subgroups indicate a possible horizontal transfer (Møller et al. 2014). The detoxification efficiency of *merA* genes varies between genera of bacteria which could probably be correlated with the wide sequence variations within the *merA* sequences (Dash et al. 2017a). Within different bacterial genomes, a 431 bp region is known to be conserved which can be effectively exploited to develop molecular markers (Sotero-Martins et al. 2008).

Volatilization of Hg^{2+} ions through *merA* genes essentially requires the formation of *merA*-NADPH-Flavo redox complex (Dash et al. 2017a), where the NADPH probably acts as an electron donor. The Hg poisoning is usually observed in chloroplast, due to which transformants with chloroplast genome transfer are favored. As the expression of *merA* requires NADPH, the gene is predicted to function efficiently in chloroplast as it has abundant NADPH (Ruiz and Daniell 2009).

To obtain higher volatilization efficiency, plants with higher biomass is usually preferred with intense root system for translocation or uptake of Hg^{2+} ions. Therefore, few forest tree species, capable of transformation, have been exploited to develop *merA* transgenics which can be utilized for large-scale in situ phytoremediation of mercury-polluted soils. The *merA* gene-transformed yellow poplar plantlets were capable of surviving in media containing normal toxic concentration of mercury (Rugh et al. 1998). Similarly, transgenic eastern cottonwood trees expressing *merA9* and *merA18* genes could metabolize 25 μM Hg^{2+} ions. However, wild-type plantlets were killed at this high mercury concentration (Che et al. 2003). Further transgenic studies have demonstrated a successful generation of tobacco transformants via nuclear genome, expressing *merA* gene derived from *E. coli*, metabolizing Hg^{2+} ions. These tobacco transgenics were tolerant to higher levels of mercury concentrations of 80–140 μM Hg^{2+} ions which are approximately five to seven times more than the normal levels (Haque et al. 2010). Such transformation has also been applied to rice, where transgenic rice produced through particle gun bombardment harboring *merA* gene resistant to 250 μM of $HgCl_2$ (Heaton et al. 2003).

Usually the broad-spectrum mercury tolerance is achieved through the compound integration of *merA* and *merB* gene products. Nonetheless, the reduction of

Hg^{2+} ions is accomplished by MerA enzyme, presence of additional *merB* facilitates the breakdown of organomercurial compounds like methylmercury chloride, ethylmercury chloride, and so on (Chien et al. 2010). The *merB* codes for organomercurial lyase, an enzyme that hydrolyzes the breakdown of C–Hg bond followed by reduction of Hg^{2+} ions with *merA* (Weiss et al. 1977; Nascimento and Chartone-Souza 2003; Lafrance-Vanasse et al. 2009). In addition to *merB* gene, *merG* gene is also associated with broad-spectrum mercury tolerance, located between *merA* and *merB* in a mer operon (Kiyono and Pan-Hou 1999). The model plant *A. thaliana* introgressed with *merB* could survive at varying concentrations of monomethylmercuric chloride and phenylmercuric acetate (PMA). The growth of control plants was severely hampered in the presence of Hg at same concentration (Bizily et al. 1999).

Improved detoxification of inorganic and organomercurial can be accomplished using an integrated *mer* gene complex, *merA* and *merB*. A nuclear genome integrated *merAB* complex was first introduced in *A. thaliana* which could survive at 5 μM PMA and 10 μM CH_3Hg. This resistance is a fivefold increase in tolerance to organic Hg in comparison to the plant expressing only *merB* genes (Bizily et al. 2000). The transgenic populus species have been successfully developed, expressing both the genes. The transformants could tolerate 50 and 2 μM of $HgCl_2$ and CH_3HgCl, respectively (Im Choi et al. 2007). A *merAB* complexed system when transformed to chloroplast genome in tobacco, demonstrated an enhanced uptake of inorganic and organomercurial compounds with simultaneous rapid Hg volatilization (Ruiz et al. 2003; Hussein et al. 2007).

Grasses have also been utilized to develop *mer* gene transgenics as they often act as natural pollutant remediators. Their transformation with *mer* genes can improve their phytoremediation capabilities (Czakó et al. 2005). *Spartina alterniflora* has been successfully transformed with *merAB* system which enhanced their resistance to PMA and $HgCl_2$ (Czako et al. 2006). These bacterial-derived *mer* genes functions sufficiently well in plants; however, they have to be modified with an additional requirement of plant promoters. Using genetic engineering to modify plant genome incorporating mercury-resistant genes can potentially help in reducing heavy metal pollution. However, such techniques may or may not offer complete and stable inheritance. Hence, few other alternative modes of phytoremediation should also be simultaneously employed.

10.7 Concluding Remarks and Future Strategy

The contamination of mercury in agriculture land is widespread. As stated, mercury can induce noxious effects on plant growth and development. Hence, it is the need of the hour to quest for the effectual solution to overcome the problem of mercury contamination Phytoremediation is one of the effective, environment-friendly and comparatively cheaper than the conventional strategies for remediation of mercury. The application of PGPM has attained ample attention to mitigate various environmental stresses, especially heavy metals. The PGPM have been proficiently associated with plant growth promotion and simultaneously aid in combating mercury toxicity by varied mechanisms such as directly by utilization of mer operon and indirectly by production of siderphores, organic acids, PC, MT, phytohormones and ACC deaminase. Hence, in recent years, there is a profound upgradation in the role of PGPM from "nutrient provider" to "stress alleviator". In this respect, PGPM-assisted phytoremediation has been proved to be better solution and even stimulates the proficiency of mercury detoxification. In the parallel, several transgenic plants overexpressing *mer* genes from microbial origins have been developed. The transgenic plants overexpressing *mer* genes either individually or along with mercury-resistant PGPM could be used as integrative approach for phytoremediation of mercury-contaminated sites. In the present scenario, serious and constant efforts are to be made to increase the number and diversity of effective and competitive mercury resistance PGPM from the mercury-contaminated sites and investigating their role in phytoremediation of mercury with a suitable host plant. Despite several findings, studies are still needed to explore the underlying molecular aspects of interplay between plant-PGPM in soil, which can hasten the phytoremediation process of under mercury stress. In this context, researchers should investigate and develop an efficient technology using the consortia of mercury-resistant PGPM for preservation and protection of environment against mercury contamination.

Acknowledgment The author I. Gontia-Mishra acknowledges the funding provided by Science and Engineering Research Board, New Delhi, India, grant number PDF/2017/001001.

References

AMAP/UNEP (2013) AMAP/UNEP geospatially distributed mercury emissions dataset 2010v1. http://www.amap.no/mercury-emissions/datasets

Amin A, Latif Z (2017) Screening of mercury-resistant and indole-3-acetic acid producing bacterial-consortium for growth promotion of *Cicer arietinum* L. J Basic Microbiol 57:204–217

Arshad M, Saleem M, Hussain S (2007) Perspectives of bacterial ACC deaminase in phytoremediation. Trends Biotechnol 25:356–362

Ashraf MA, Hussain I, Rasheed R, Iqbal M, Riaz M, Arif MS (2017) Advances in microbe-assisted reclamation of heavy metal contaminated soils over the last decade: a review. J Environ Manag 198:132–143

Azevedo R, Rodriguez E (2012) Phytotoxicity of mercury in plants: a review. J Bot 2012:848614

Balan BM, Shini S, Krishnan KP, Mohan M (2018) Mercury tolerance and biosorption in bacteria isolated from Ny-Alesund, Svalbard, Arctic. J Basic Microbiol 58:286–295

Beckers F, Rinklebe J (2017) Cycling of mercury in the environment: sources, fate, and human health implications: a review. Crit Rev Environ Sci Technol 47:693–794

Bizily SP, Rugh CL, Summers AO, Meagher RB (1999) Phytoremediation of methylmercury pollution: *merB* expression in *Arabidopsis thaliana* confers resistance to organomercurials. Proc Natl Acad Sci USA 96:6808–6813

Bizily SP, Rugh CL, Meagher RB (2000) Phytodetoxification of hazardous organomercurials by genetically engineered plants. Nat Biotechnol 18:213

Braud A, Hoegy F, Jezequel K, Lebeau T, Schalk IJ (2009) New insights into the metal specificity of the Pseudomonas aeruginosa pyoverdine-iron uptake pathway. Environ Microbiol 11: 1079–1091

Bücker-Neto L, Paiva AL, Machado RD, Arenhart RA, Margis-Pinheiro M (2017) Interactions between plant hormones and heavy metals responses. Genet Mol Biol 40:373–386

Chatziefthimiou AD, Crespo-Medina M, Wang Y, Vetriani C, Barkay T (2007) The isolation and initial characterization of mercury resistant chemolithotrophic thermophilic bacteria from mercury rich geothermal springs. Extremophiles 11:469–479

Che D, Meagher RB, Heaton AC, Lima A, Rugh CL, Merkle SA (2003) Expression of mercuric ion reductase in Eastern cottonwood (*Populus deltoides*) confers mercuric ion reduction and resistance. Plant Biotechnol J 1:311–319

Chen YP, Rekha PD, Arun AB, Shen FT, Lai WA, Young CC (2006) Phosphate solubilizing bacteria from subtropical soil and their tricalcium phosphate solubilizing abilities. Appl Soil Ecol 34:33–41

Chen YA, Chi WC, Trinh NN, Huang LY, Chen YC, Cheng KT et al (2014) Transcriptome profiling and physiological studies reveal a major role for aromatic amino acids in mercury stress tolerance in rice seedlings. PLoS One 9:e95163

Chien MF, Narita M, Lin KH, Matsui K, Huang CC, Endo G (2010) Organomercurials removal by heterogeneous *merB* genes harboring bacterial strains. J Biosci Bioeng 110:94–98

Chowdhury AS, Das P, Sarkar I, Islam R, Aksharin L, Parvin F, Islam Z, Faris M, Shaekh MP (2015) Phytoremediation of heavy metals (Ar, Cd, Pb) using transgenic rice plants-an overview. Int J Sci Eng Res 6:878

Clark D, Weiss AA, Silver S (1977) Mercury and organomercurial volatilization activities associated with plasmids in *Pseudomonas aeruginosa* and *Pseudomonas putida*. J Bacteriol 132:186–196

Czako M, Feng X, He Y, Liang D, Marton L (2006) Transgenic *Spartina alterniflora* for phytoremediation. Environ Geochem Health 28:103–110

Czakó M, Feng X, He Y, Liang D, Márton L (2005) Genetic modification of wetland grasses for phytoremediation. Zeitschrift für Naturforschung C 60:285–291

Dash HR, Das S (2012) Bioremediation of mercury and the importance of bacterial *mer* genes. Int Biodeterior Biodegrad 75:207–214

Dash HR, Mangwani N, Das S (2014) Characterization and potential application in mercury bioremediation of highly mercury-resistant marine bacterium *Bacillus thuringiensis* PW-05. Environ Sci Pollut Res 21:2642–2653

Dash HR, Sahu M, Mallick B, Das S (2017a) Functional efficiency of MerA protein among diverse mercury resistant bacteria for efficient use in bioremediation of inorganic mercury. Biochimie 142:207–215

Dash HR, Basu S, Das S (2017b) Evidence of mercury trapping in biofilm-EPS and *mer* operon-based volatilization of inorganic mercury in a marine bacterium *Bacillus cereus* BW-201B. Arch Microbiol 199:445–455

Desale P, Patel B, Singh S, Malhotra A, Nawani N (2014) Plant growth promoting properties of *Halobacillus* sp. and *Halomonas* sp. in presence of salinity and heavy metals. J Basic Microbiol 54:781–791

Dimkpa C, Svatoš A, Merten D, Büchel G, Kothe E (2008) Hydroxamate siderophores produced by *Streptomyces acidiscabies* E13 bind nickel and promote growth in cowpea (*Vigna unguiculata* L.) under nickel stress. Can J Microbiol 54:163–172

Doty SL (2008) Enhancing phytoremediation through the use of transgenics and endophytes. New Phytol 179:318–333

Etesami H (2018) Bacterial mediated alleviation of heavy metal stress and decreased accumulation of metals in plant tissues: mechanisms and future prospects. Ecotoxicol Environ Safety 147:175–191

Ferrara R (1998) Atmospheric mercury sources in the Mt. Amiata area, Italy. Sci Total Environ 213:13–23

Ferrara R (2000) Volcanoes as emission sources of atmospheric mercury in the Mediterranean basin. Sci Total Environ 259:115–121

Figueiredo NL, Canário J, O'Driscoll NJ, Duarte A, Carvalho C (2016) Aerobic Mercury-resistant bacteria alter mercury speciation and retention in the Tagus Estuary (Portugal). Ecotoxicol Environ Safety 124:60–67

Frossard A, Hartmann M, Frey B (2017) Tolerance of the forest soil microbiome to increasing mercury concentrations. Soil Biol Biochem 105:162–176

Gadd GM (2010) Metals: minerals and microbes: geomicrobiology and bioremediation. Microbiology 156:609–643

Gangwar S, Singh VP, Prasad SM, Maurya JN (2010) Modulation of manganese toxicity in *Pisum sativum* L. seedlings by kinetin. Sci Hortic 126:467–474

Giovanella P, Cabral L, Bento FM, Gianello C, Camargo FAO (2016) Mercury (II) removal by resistant bacterial isolates and mercuric (II) reductase activity in a new strain of *Pseudomonas* sp. B50A. New Biotechnol 33:216–223

Glick BR (2010) Using soil bacteria to facilitate phytoremediation. Biotechnol Adv 28:367–374

Glick BR (2014) Bacteria with ACC deaminase can promote plant growth and help to feed the world. Microbiol Res 169:30–39

Glick BR, Patten CL, Holguin G et al (1999) Biochemical and genetic mechanisms used by plant growth promoting bacteria. Imperial College Press, London

Gondor OK, Pál M, Darkó É, Janda T, Szalai G (2016) Salicylic acid and sodium salicylate alleviate cadmium toxicity to different extents in maize (*Zea mays* L.). PLoS One 11:e0160157

Gontia-Mishra I, Sasidharan S, Tiwari S (2014) Recent developments in use of 1-amino cyclopropane-1-carboxylate (ACC) deaminase for conferring tolerance to biotic and abiotic stress. Biotechnol Lett 36:889–898

Gontia-Mishra I, Sapre S, Sharma A, Tiwari S (2016) Alleviation of mercury toxicity in wheat by the interaction of mercury-tolerant plant growth-promoting rhizobacteria. J Plant Grow Regul 35:1000–1012

Gontia-Mishra I, Sapre S, Kachare S, Tiwari S (2017a) Molecular diversity of 1-aminocyclopropane-1-carboxylate (ACC) deaminase producing PGPR from wheat (*Triticum aestivum* L.) rhizosphere. Plant Soil 414:213–227

Gontia-Mishra I, Sapre S, Tiwari S (2017b) Zinc solubilizing bacteria from the rhizosphere of rice as prospective modulator of zinc biofortification in rice. Rhizosphere 3:185–190

Gopalakrishnan S, Sathya A, Vijayabharathi R, Varshney RK, Gowda CL, Krishnamurthy L (2015) Plant growth promoting rhizobia: challenges and opportunities. 3 Biotech 5:355–377

Gupta P, Diwan B (2017) Bacterial exopolysaccharide mediated heavy metal removal: a Review on biosynthesis, mechanism and remediation strategies. Biotechnol Rep 13:58–71

Gupta S, Nirwan J (2015) Evaluation of mercury biotransformation by heavy metal-tolerant A*lcaligenes* strain isolated from industrial sludge. Int J Environ Sci Technol 12:995–1002

Gupta A, Rai V, Bagdwal N, Goel R (2005) In situ characterization of mercury-resistant growth-promoting fluorescent pseudomonads. Microbiol Res 160:385–388

Han Y, Wang R, Yang Z, Zhan Y, Ma Y, Ping S, Zhang L, Lin M, Yan Y (2015) 1-aminocyclopropane-1-carboxylate deaminase from *Pseudomonas stutzeri* A1501 facilitates the growth of rice in the presence of salt or heavy metals. J Microbiol Biotechnol 25:1119–1128

Haque S, Zeyaullah M, Nabi G, Srivastava PS, Ali A (2010) Transgenic tobacco plant expressing environmental E. coli *merA* gene for enhanced volatilization of ionic mercury. J Microbiol Biotechnol 20:917–924

Heaton AC, Rugh CC, Kim T, Meagher RB (2003) Toward detoxifying mercury-polluted aquatic sediments with rice genetically engineered for mercury resistance. Environ Toxicol Chem 22:2940–2947

Hesse E, O'Brien S, Tromas N, Bayer F, Luján AM, van Veen EM, Hodgson DJ, Buckling A (2018) Ecological selection of siderophore-producing microbial taxa in response to heavy metal contamination. Ecol Lett 21:117–127
Hindersah R, Mulyani O, Osok R (2017) Proliferation and exopolysaccharide production of *Azotobacter* in the presence of mercury. Biodivers J 8:21–26
Hindersah R, Handyman Z, Indriani FN, Suryatmana P, Nurlaeny N (2018) *Azotobacter* population, soil nitrogen and groundnut growth in mercury-contaminated tailing inoculated with *Azotobacter*. J Degraded Mining Lands Manag 5:1269–1274
Hseu ZY, Su SW, Lai HY, Guo HY, Chen TC, Chen ZS (2010) Remediation techniques and heavy metal uptake by different rice varieties in metal-contaminated soils of Taiwan: new aspects for food safety regulation and sustainable agriculture. Soil Sci Plant Nutr 56:31–52
Huang CC, Narita M, Yamagata T, Endo G, Silver S (2002) Characterization of two regulatory genes of the mercury resistance determinants from *TnMERI1* by luciferase-based examination. Gene 301:13–20
Hussein HS, Ruiz ON, Terry N, Daniell H (2007) Phytoremediation of mercury and organomercurials in chloroplast transgenic plants: enhanced root uptake, translocation to shoots, and volatilization. Environ Sci Technol 41:8439–8446
Hutchison AR (2003) Mercury pollution and remediation: the chemist's response to a global crisis. J Chem Crystallogr 33:631–645
Im Choi Y, Noh EW, Lee HS, Han MS, Lee JS, Choi KS (2007) Mercury-tolerant transgenic poplars expressing two bacterial mercury-metabolizing genes. J Plant Biol 50:658
Israr M, Sahi S, Datta R, Sarkar D (2006) Bioaccumulation and physiological effects of mercury in *Sesbania drummonii*. Chemosphere 65:591–598
Jha B, Gontia I, Hartmann A (2012) The roots of the halophyte *Salicornia brachiata* are a source of new halotolerant diazotrophic bacteria with plant growth promoting potential. Plant Soil 356:265–277
Kalaivanan D, Ganeshamurthy AN (2016) Mechanisms of heavy metal toxicity in plants. In: Rao N, Shivashankara K, Laxman R (eds) Abiotic stress physiology of horticultural crops. Springer, New Delhi, pp 85–102
Kamran S, Shahid I, Baig DN, Rizwan M, Malik KA, Mehnaz S (2017) Contribution of zinc solubilizing bacteria in growth promotion and zinc content of wheat. Front Microbiol 8:2593
Kang SH, Singh S, Kim JY, Lee W, Mulchandani A, Chen W (2007) Bacteria metabolically engineered for enhanced phytochelatin production and cadmium accumulation. Appl Environ Microbiol 73:6317–6320
Kavita B, Shukla S, Kumar GN, Archana G (2008) Amelioration of phytotoxic effects of Cd on mung bean seedlings by gluconic acid secreting rhizobacterium *Enterobacter asburiae* PSI3 and implication of role of organic acid. World J Microbiol Biotechnol 24:2965–2972
Kawahigashi H, Hirose S, Hayashi E, Ohkawa H, Ohkawa Y (2002) Phytotoxicity and metabolism of ethofumesate in transgenic rice plants expressing the human CYP2B6 gene. Pesticide Biochem Physiol 74:139–147
Kim KY, Jordan D, McDonald GA (1998) Effect of phosphate solubilizing bacteria and vesicular-arbuscular mycorrhizae on tomato growth and soil microbial activity. Biol Fertil Soils 26:79–87
Kiyono M, Pan-Hou H (1999) The *merG* gene product is involved in phenylmercury resistance in *Pseudomonas* strain K-62. J Bacteriol 181:726–730
Kloepper JW, Leong J, Teintze M, Schiroth MN (1980) Enhanced plant growth by siderophores produced by plant growth promoting rhizobacteria. Nature 286:885–886
Krishnamurthy A, Rathinasabapathi B (2013) Auxin and its transport play a role in plant tolerance to arsenite-induced oxidative stress in *Arabidopsis thaliana*. Plant Cell Environ 36:1838–1849
Kumar B, Smita K, Flores LC (2017) Plant mediated detoxification of mercury and lead. Arab J Chem 10:S2335–S2342
Lafrance-Vanasse J, Lefebvre M, Di Lello P, Sygusch J, Omichinski JG (2009) Crystal structures of the organomercurial lyase merb in its free and mercury-bound forms insights into the mechanism of methylmercury degradation. J Biol Chem 284:938–944

Lebrazi S, Fikri-Benbrahim K (2018) Rhizobium-legume symbiosis: heavy metal effects and principal approaches for bioremediation of contaminated soil. In: Meena R, Das A, Yadav G, Lal R (eds) Legumes for soil health and sustainable management. Springer, Singapore, pp 205–233

Lee J, Bae H, Jeong J, Lee JY, Yang YY, Hwang I, Martinoia E, Lee Y (2003) Functional expression of a bacterial heavy metal transporter in Arabidopsis enhances resistance to and decreases uptake of heavy metals. Plant Physiol 133:589–596

Li HY, Wei DQ, Shen M, Zhou ZP (2012) Endophytes and their role in phytoremediation. Fungal Divers 54:11–18

Ma Y, Oliveira RS, Wu L, Luo Y, Rajkumar M, Rocha I, Freitas H (2015) Inoculation with metal-mobilizing plant-growth-promoting rhizobacterium *Bacillus* sp. SC2b and its role in rhizoremediation. J Toxicol Environ Health Part A 78:931–944

Ma Y, Rajkumar M, Zhang C, Freitas H (2016) Beneficial role of bacterial endophytes in heavy metal phytoremediation. J Environ Manag 174:14–25

Ma Q, Grones P, Robert S (2018) Auxin signaling: a big question to be addressed by small molecules. J Exp Bot 69:313–328

Maestri E, Marmiroli N (2011) Transgenic plants for phytoremediation. Int J Phytoremediation 13:264–279

Mahbub KR, Krishnan K, Megharaj M, Naidu R (2016a) Bioremediation potential of a highly mercury resistant bacterial strain *Sphingobium* SA2 isolated from contaminated soil. Chemosphere 144:330–337

Mahbub KR, Krishnan K, Naidu R, Megharaj M (2016b) Mercury resistance and volatilization by *Pseudoxanthomonas* sp. SE1 isolated from soil. Environ Technol Innov 6:94–104

Mason RP, Choi AL, FitzgeraldWF Hammerschmidt CR, Lamborg CH, Soerensen AL, Sunderland EM (2012) Mercury biogeochemical cycling in the ocean and policy implications. Environ Res 119:101–117

Mathema VB, Thakuri BC, Sillanpää M (2011) Bacterial mer operon-mediated detoxification of mercurial compounds: a short review. Arch Microbiol 193:837–844

Mathew DC, Ho YN, Gicana RG, Mathew GM, Chien MC, Huang CC (2015) A rhizosphere-associated symbiont, *Photobacterium* spp. strain MELD1, and its targeted synergistic activity for phytoprotection against mercury. PLoS One 10:e0121178

Meagher RB, Heaton AC (2005) Strategies for the engineered phytoremediation of toxic element pollution: mercury and arsenic. J Ind Microbiol Biotechnol 32:502–513

Miransari M (2011) Hyperaccumulators, arbuscular mycorrhizal fungi and stress of heavy metals. Biotechnol Adv 29:645–653

Mishra J, Singh R, Arora NK (2017) Alleviation of heavy metal stress in plants and remediation of soil by rhizosphere microorganisms. Front Microbiol 8:1706

Mobeen N, Latif Z (2016) Characterization of mercury resistant and growth promoting *Enterobacter* sp. from rhizosphere to use as a biofertilizer. Adv Life Sci 3:36–41

Møller AK, Barkay T, Hansen MA, Norman A, Hansen LH, Sørensen SJ, Boyd ES, Kroer N (2014) Mercuric reductase genes (*merA*) and mercury resistance plasmids in High Arctic snow, freshwater and sea-ice brine. FEMS Microbiol Ecol 87:52–63

Muddarisna N, Krisnayanti BD, Utami SR, Handayanto E (2013) Phytoremediation of mercury-contaminated soil using three wild plant species and its effect on maize growth. Appl Ecol Environ Sci 1:27–32

Mukkata K, Kantachote D, Wittayaweerasak B, Techkarnjanaruk S, Mallavarapu M, Naidu R (2015) Distribution of mercury in shrimp ponds and volatilization of Hg by isolated resistant purple non-sulfur bacteria. Water Air Soil Pollut 226:148

Nascimento AM, Chartone-Souza E (2003) Operon mer: bacterial resistance to mercury and potential for bioremediation of contaminated environments. Genet Mol Res 2:92–101

Neilands J (1995) Siderophores: structure and function of microbial iron transport compounds. J Biol Chem 45:26723–26726

Neubauer U, Furrer G, Kayser A, Schulin R (2000) Siderophores, NTA, and citrate: potential soil amendments to enhance heavy metal mobility in phytoremediation. Int J Phytoremediation 2:353–368

Ní Chadhain SM, Schaefer JK, Crane S, Zylstra GJ, Barkay T (2006) Analysis of mercuric reductase (merA) gene diversity in an anaerobic mercury-contaminated sediment enrichment. Environ Microbiol 8:1746–1752

Nocelli N, Bogino PC, Banchio E, Giordano W (2016) Roles of extracellular polysaccharides and biofilm formation in heavy metal resistance of rhizobia. Materials 9:418

Nonnoi F, Chinnaswamy A, García de la Torre VS, Coba de la Peña T, Lucas MM, Pueyo JJ (2012) Metal tolerance of rhizobial strains isolated from nodules of herbaceous legumes *Medicago* spp. and *Trifolium* spp. growing in mercury contaminated soils. Appl Soil Ecol 61:49–59

Ortiz-Ojeda P, Ogata-Gutiérrez K, Zúñiga-Dávila D (2017) Evaluation of plant growth promoting activity and heavy metal tolerance of psychrotrophic bacteria associated with maca (*Lepidium meyenii* Walp.) rhizosphere. AIMS Microbiology 3:279–292

Osmolovskaya N, Vu D, Kuchaeva L (2018) The role of organic acids in heavy metal tolerance in plants. Biol Comm 63:9–16

Oves M, Khan MS, Zaidi A (2013) Chromium reducing and plant growth promoting novel strain *Pseudomonas aeruginosa* OSG41 enhance chickpea growth in chromium amended soils. Eur J Soil Biol 56:72–83

Padmavathiamma PK, Li LY (2007) Phytoremediation technology: hyperaccumulation metals in plants. Water Air Soil Pollut 184(1–4):105–126

Patra M, Sharma A (2000) Mercury toxicity in plants. Bot Rev 66:379–422

Patra M, Bhowmik N, Bandopadhyay B, Sharma A (2004) Comparison of mercury, lead and arsenic with respect to genotoxic effects on plant systems and the development of genetic tolerance. Environ Exp Bot 52:199–223

Patten CL, Glick BR (2002) Role of *Pseudomonas putida* indole acetic acid in development of host plant root system. Appl Environ Microbiol 48:3795–3801

Pietro-Souza W, Mello IS, Vendruscullo SJ, da Silva GF, da Cunha CN, White JF, Soares MA (2017) Endophytic fungal communities of *Polygonum acuminatum* and *Aeschynomene fluminensis* are influenced by soil mercury contamination. PLoS One 12:e0182017

Pirrone N, Costa P, Pacyna JM, Ferrara R (2001) Mercury emissions to the atmosphere from natural and anthropogenic sources in the Mediterranean region. Atmos Environ 35:2997–3006

Powlowski J, Sahlman L (1999) Reactivity of the two essential cysteine residues of the periplasmic mercuric ion-binding protein, MerP. J Biol Chem 274:33320–33326

Puglisi I, Faedda R, Sanzaro V, Piero ARL, Petrone G, Cacciola SO (2012) Identification of differentially expressed genes in response to mercury I and II stress in *Trichoderma harzianum*. Gene 506:325–330

Pushkar B, Sevak P, Sounderajan S (2018) Assessment of the bioremediation efficacy of the mercury resistant bacterium isolated from the Mithi River. Water Sci Technol Water Supply, ws2018064. https://doi.org/10.2166/ws.2018.064

Quiñones MA, Ruiz-Díez B, Fajardo S, López-Berdonces MA, Higueras PL, Fernández-Pascual M (2013) *Lupinus albus* plants acquire mercury tolerance when inoculated with an Hg-resistant *Bradyrhizobium* strain. Plant Physiol Biochem 73:168–175

Rafique A, Amin A, Latif Z (2015) Screening and characterization of mercury-resistant nitrogen fixing bacteria and their use as biofertilizers and for mercury bioremediation. Pakistan J Zool 47:1271–1277

Rajkumar M, Ae N, Prasad MN, Freitas H (2010) Potential of siderophore-producing bacteria for improving heavy metal phytoextraction. Trends Biotechnol 28:142–149

Rajkumar M, Sandhya S, Prasad MNV, Freitas H (2012) Perspectives of plant-associated microbes in heavy metal phytoremediation. Biotechnol Adv 30:1562–1574

Rascio N, Navari-Izzo F (2011) Heavy metal hyperaccumulating plants: how and why do they do it? And what makes them so interesting? Plant Sci 180:169–181

Reddy MS, Prasanna L, Marmeisse R, Fraissinet-Tachet L (2014) Differential expression of metallothioneins in response to heavy metals and their involvement in metal tolerance in the symbiotic basidiomycete *Laccaria bicolor*. Microbiology 160:2235–2242

Reeves RD, Baker AJM (2000) Metal-accumulating plants. In: Raskin I, Ensley BD (eds) Phytoremediation of toxic metals: using plants to clean up the environment. Wiley, pp 193–229

Reniero D, Galli E, Barbieri P (1995) Cloning and comparison of mercury and organomercurial resistance determinants from a *Pseudomonas stutzeri* plasmid. Gene 166:77–82

Richardson AE (2001) Prospects for using soil microorganisms to improve the acquisition of phosphorus by plants. Aust J Plant Physiol 28:897–906

Rizvi A, Khan MS (2017) Biotoxic impact of heavy metals on growth, oxidative stress and morphological changes in root structure of wheat (*Triticum aestivum* L.) and stress alleviation by *Pseudomonas aeruginosa* strain CPSB1. Chemosphere 185:942–952

Rugh CL, Wilde HD, Stack NM, Thompson DM, Summers AO, Meagher RB (1996) Mercuric ion reduction and resistance in transgenic *Arabidopsis thaliana* plants expressing a modified bacterial *merA* gene. Proc Natl Acad Sci USA 93:3182–3187

Rugh CL, Senecoff JF, Meagher RB, Merkle SA (1998) Development of transgenic yellow poplar for mercury phytoremediation. Nat Biotechnol 16:925

Ruiz ON, Daniell H (2009) Genetic engineering to enhance mercury phytoremediation. Curr Opin Biotechnol 20:213–219

Ruiz ON, Hussein HS, Terry N, Daniell H (2003) Phytoremediation of organomercurial compounds via chloroplast genetic engineering. Plant Physiol 132:1344–1352

Ruiz-Diez B, Quinones MA, Fajardo S, Lopez MA, Higueras P, Fernandez-Pascual M (2012) Mercury-resistant rhizobial bacteria isolated from nodules of leguminous plants growing in high Hg-contaminated soils. Appl Microbiol Biotechnol 96:543–554

Saranya K, Sundaramanickam A, Shekhar S, Swaminathan S, Balasubramanian T (2017) Bioremediation of mercury by *Vibrio fluvialis* screened from industrial effluents. Biomed Res Int 2017:6509648

Saravanan VS, Madhaiyan M, Thangaraju M (2007) Solubilization of zinc compounds by the diazotrophic, plant growth promoting bacterium *Gluconacetobacter diazotrophicus*. Chemosphere 66:1794–1798

Schalk IJ, Hannauer M, Braud A (2011) New roles for bacterial siderophores in metal transport and tolerance. Environ Microbiol 13:2844–2854

Schottel JL (1978) The mercury and organomercurial detoxifying enzymes from a plasmid-bearing strain of *Escherichia coli*. J Biol Chem 253:4341–4349

Schroeder WH, Munthe J (1998) Atmospheric mercury—an overview. Atmospheric Environ 32:809–822

Selin NE (2009) Global biogeochemical cycling of mercury: a review. Ann Rev Environ Res 34:43

Shehu J, Imeri A, Kupe L, Dodona E, Shehu A, Mullaj A (2014) Hyperaccumulators of mercury in the industrial area of a PVC factory in Vlora (Albania). Arch Biol Sci Belgrade 66:1457–1464

Sotero-Martins A, Jesus MS, Lacerda M, Moreira JC, Filgueiras AL, Barrocas PR (2008) A conservative region of the mercuric reductase gene (*merA*) as a molecular marker of bacterial mercury resistance. Braz J Microbiol 39:307–310

Spaepen S, Vanderleyden J, Remans R (2007) Indole-3-acetic acid in microbial and microorganism-plant signaling. FEMS Microbiol Rev 31:425–448

Storey EP, Boghozian R, Little JL, Lowman DW, Chakraborty R (2006) Characterization of 'Schizokinen'; a dihydroxamate-type siderophore produced by *Rhizobium leguminosarum* IARI 917. Biometals 19:637–649

Su Y, Han F, Shiyab S, Monts DL (2007) Phytoextraction and accumulation of mercury in selected plant species grown in soil contaminated with different mercury compounds. In: WM'07 Conference February 25–March 1, 2007, Tucson, Arizon, USA

Sunil KCR, Swati K, Bhavya G, Nandhini M, Veedashree M, Prakash HS, Kini KR, Geetha N (2015) *Streptomyces flavomacrosporus*, a multi-metal tolerant potential bioremediation candidate isolated from paddy field irrigated with industrial effluents. Int J Life Sci 3:9–15

Tangahu BV, Sheikh Abdullah SR, Basri H, Idris M, Anuar N, Mukhlisin M (2011) A review on heavy metals (As, Pb, and Hg) uptake by plants through phytoremediation. Int J Chem Eng 2011:939161

Tariq A, Latif Z (2014) Bioremediation of mercury compounds by using immobilized nitrogen-fixing bacteria. Inter J Agric Biol 16:1129–1134

Tariq S, Amin A, Latif Z (2015) PCR based DNA fingerprinting of mercury resistant and nitrogen fixing *Pseudomonas* spp. Pure Appl Biol 4:129–136

Teng Y, Wang X, Li L, Li Z, Luo Y (2015) Rhizobia and their bio-partners as novel drivers for functional remediation in contaminated soils. Front Plant Sci 6:32

Tiwari S, Lata C (2018) Heavy metal stress, signaling, and tolerance due to plant-associated microbes: an overview. Front Plant Sci 9:452

Ullah A, Heng S, Munis MFH, Fahad S, Yang X (2015) Phytoremediation of heavy metals assisted by plant growth promoting (PGP) bacteria: a review. Environ Exp Bot 117:28–40

Weiss AA, Murphy SD, Silver S (1977) Mercury and organomercurial resistances determined by plasmids in *Staphylococcus aureus*. J Bacteriol 132:197–208

Wood JL, Tang C, Franks AE (2016) Microbial associated plant growth and heavy metal accumulation to improve phytoextraction of contaminated soils. Soil Biol Biochem 103:131–137

Yong X, Chen Y, Liu W, Xu L, Zhou J, Wang S, Chen P, Ouyang P, Zheng T (2014) Enhanced cadmium resistance and accumulation in *Pseudomonas putida* KT2440 expressing the phytochelatin synthase gene of *Schizosaccharomyces pombe*. Lett Appl Microbiol 58:255–261

Zhou ZS, Wang SJ, Yang ZM (2008) Biological detection and analysis of mercury toxicity to alfalfa (*Medicago sativa*) plants. Chemosphere 70:1500–1509

Zhu XF, Jiang T, Wang ZW, Lei GJ, Shi YZ, Li GX, Zheng SJ (2012) Gibberellic acid alleviates cadmium toxicity by reducing nitric oxide accumulation and expression of IRT1 in *Arabidopsis thaliana*. J Hazard Mater 239:302–307

Chapter 11
Regulatory Role of Rhizobacteria to Induce Drought and Salt Stress Tolerance in Plants

Humaira Yasmin, Asia Nosheen, Rabia Naz, Rumana Keyani and Seemab Anjum

Abstract This chapter summarizes the role of rhizosphere dwelling beneficial bacteria for the induction of tolerance against drought and salt stresses in plants. A vast proportion of world's agricultural land is rendered less productive or completely unproductive due to different factors including water scarcity and salinity. Drought can be due to insufficient rainfall, dry spells or changes in rainfall patterns whereas salinity is because of excessive amount of salts in soil or water. This salinity can be primary (arise due to natural phenomena) or it can be secondary (anthropogenic in origin). Plants respond to drought and salinity via morphological, physiological and biochemical mechanisms. To overcome devastating effects of these stresses in plants, different strategies developed along with the traditional agricultural practices. An emerging strategy to overcome drought and salinity is the use of plant growth-promoting rhizobacteria (PGPR), which enable plants to combat these stresses by various direct and indirect mechanisms. Rhizobacteria are under extensive research for their beneficial effects, uncomplicated and cost-effective application methods and their environment-friendly behaviors. Now also serve as best alternatives to chemical and traditional methods so as to overcome to tolerate and ameliorate harmful effects in plants.

Keywords Drought · Salinity · PGPR · Induced systemic resistance · ACC deaminase · Phytohormone · Proline

H. Yasmin (✉) · A. Nosheen · R. Naz · R. Keyani · S. Anjum
COMSATS University Islamabad, Park Road, Chak Shahazad, Islamabad, Pakistan
e-mail: humaira.yasmin@comsats.edu.pk

A. Nosheen
e-mail: asia.nosheen@comsats.edu.pk

R. Naz
e-mail: rabia.naz@comsats.edu.pk

R. Keyani
e-mail: rumana.keyani@comsats.edu.pk

S. Anjum
e-mail: seemabanjum786@yahoo.com

D. K. Maheshwari and S. Dheeman (eds.), *Field Crops: Sustainable Management by PGPR*, Sustainable Development and Biodiversity 23,
https://doi.org/10.1007/978-3-030-30926-8_11

11.1 Introduction

Plants are the creatures of prime importance to Earth, as all other organisms are directly or indirectly dependent on them. Most plants are well adapted to their environment but there exists a huge range of biotic and abiotic stress factors that are strong enough to breach the protective barriers of plants to cause diseases and other adverse conditions in plants. The major biotic stresses for plants include bacteria, viruses, fungi, nematodes, insects, weeds, etc. whereas major abiotic stresses which are also equally or in some cases, more devastating for plants include drought, salinity, water logging, flooding, heat, cold, frost, heavy metal toxicity, etc. All these stress factors contribute to plant growth reduction and huge yield losses worldwide. When a plant faces abiotic stress, it also becomes vulnerable to other abiotic stresses. Abiotic stresses are a great threat to food security due to continuous climate changes and worsening of natural environment by human activities.

Abiotic stresses faced by the largest proportion of plants are drought and salinity. Drought refers to low water availability to plants to carry out their normal physiological processes. Drought is a result of change in rainfall patterns, less rainfalls, extreme high or low temperatures, faster evaporation rates than water replenishing to soils and saline conditions, etc. Salinity is a consequence of deposition of oceanic salts through rain and winds, inadequate agricultural practices, use of saline water for irrigation, insufficient rainfall, elevated evaporation rates, weathering of rocks. In a broader context, water stress is referred to both drought and salinity.

Plants respond to abiotic stresses by altering their functioning on different levels and also by exploiting mechanisms to avoid them. These responses are multiplexed and potent with the involvement of various changes on physiological, morphological, molecular and biochemical levels to ensure the maintenance of normal functions and survival. Plant responses can be reversible or irreversible. The type of plant responses to stress also depends on the intensity and time of exposure to stress.

Classical breeding methods helped to induce tolerance in plants against abiotic stresses, but the success rate is not satisfactory owing to various reasons such as inefficient criteria for selection, complex quantitative traits and less genetic variability in plants under stress. Genetic engineering has proved to be a valuable advancement in developing stress-tolerant plant varieties. An emerging research area to induce tolerance in plants against these stresses is the use of microorganisms specifically bacteria. Various bacterial strains are under extensive research for their potential to make plants resistant to stresses and to enable plants to survive and produce good yield if they encounter stress.

Bacteria are the most common microorganisms present in soil, the possible reason for their great numbers is their rapid growth rates and ability to exploit vast range of materials as carbon or nitrogen sources. Mostly bacteria remain bound to soil particles, a large number of bacteria are closely associated with plant roots. Plant rhizosphere dwelling bacteria that exert positive effects on plants are known as plant growth-promoting rhizobacteria (PGPR). PGPR opened an infinitely broad new chapter about plant stress tolerance.

The criteria for soil bacteria to be called as PGPR are that they strictly colonize plants roots, they must be able to sustain themselves and reproduce in the microenvironment close to root surfaces and compete with the other soil-dwelling microbes for resources and they must promote plant growth either directly or indirectly.

PGPR are well studied for their abilities to increase seed viability and germination rates, enhancement of root proliferation, increased nutrient uptake and use of versatile mechanisms to help plants to survive abiotic and biotic stresses. PGPR have many advantages over classical methods for induction of stress tolerance such as they can be applied easily and by using a number of techniques at any stage of plant life, they are non-pollutants for the environment and do not accumulate in soil, water or on crops, etc. Implementation of PGPR on large scale of crops is an agreeable idea to make crops stress resistant.

11.2 Drought Stress: A Global Problem

Drought stress or water stress is defined as the condition in which the availability of water supply is short as compared to its demand (Wrathall et al. 2018). Recently, Nawrotzki and Bakhtsiyarava (2017) reported that the effects of drought stress which operate in the form of water scarcity change in rainfall dry spell, etc., on agricultural crops leads to the increased migration of communities. Various factors are responsible for an increase in water crisis problem, among those, climate change is the major water limiting factor which is produced due to increased level of atmospheric carbon dioxide, increase in temperature, decrease in rainfall, extreme environmental events, and population growth (Torres and Henry 2016).

According to estimation (United Nations FAO 2013), it was reported that about one-third of the world population is living in the water-scarce areas. The climate change will further increase the water shortage intensity in subtropical areas of Asia and Africa. Further it was reported that by 2035, the glaciers of the Himalayas which are feeding largest rivers of Asia such as Indus, Yangtze, Ganges, Salween, Brahmaputra, Mekong, Yellow, etc., may vanish due to increase in temperature and about 1.8 million people will live in the regions or countries which are suffering from water scarcity. Drought has severely affected the Central Asian countries and they are using the water resources by treaties or by bargaining (Athar and Ashraf 2009).

About 60% of the world is composed of arid and semi-arid regions. The water shortage severely affects the crops yield specifically in Asian countries (Swain et al. 2017). Drought is considered as a major global limiting factor of crop productivity, which hinders the growth of the plants and leads toward yield losses (Zhang et al. 2018). For example, wheat is considered as the major food in many countries of the world. It provides 20% of the daily calories and also acts as a source of protein for

4.5 billion peoples of the world (Shiferaw et al. 2013). It has been reported that the yield of wheat has been decreased by 4.1–6.4% for every 1 °C increase in temperature due to the global climate change (Liu et al. 2016). To meet the ever-increasing population needs the consumption of wheat is expected to rise in the next 40 years (Weigand 2011) and by 2050, the wheat production yet to increase 858 million tons to fulfil the predicted global food demand (Alexandratos and Bruinsma 2012).

11.3 Responses of Crops/Plants to Drought Stress

Drought stress may become worse if the control measures are not taken into consideration. It is one of the major environmental factors, which restrain the plant growth and yield (Butt et al. 2017). The water availability reduced to the plants, hence hindering the growth, development, efficiency, and water relations to several terrestrial plants. The maximum yield losses occur in the crops, which are prone to drought stress. In fact, under drought stress, plants tend to increase their immune response by osmoregulation in the tissues, activation of antioxidant defense system and modulation in hormonal balance (Butt et al. 2017).

The survival of the plants during the early growth stages is quite critical in drought stress areas. However, plants have adapted an array of physiological, morphological, and biochemical strategies to endure the drought stress condition (Basu et al. 2016). There are two major mechanisms by which plants survive under drought stress (i) either by escape from drought or (ii) they must avoid the stress. The prompt maturity and early completion of the life cycle is the phenological phenomenon, which is categorized under drought escape strategy. While to avoid the detrimental effects of drought by maintaining a higher level of water potential is the drought avoidance mechanism of the plants (Athar and Ashraf 2009). Peng and Ismail (2004) reported that dense and deep root system, increased stomatal conductance, increased root penetration ability, avoidance of rolling of leaves, increased water potential during pre-dawn, and greater cuticular resistance in order to reduce or prevent the water loss are some of the characteristics which plants adapt to avoid drought stress conditions.

Under natural dry conditions, a plant tends to mature early and produce seeds before the onset of the dry season. For example, *Escholtzia californica* (California poppy) completes its life cycle earlier before the start of the drought stress. Some xerophyte plants such as *Agave deserti* store water in their stem, leaves, and buds to survive in water stress conditions (Athar and Ashraf 2009). Some plants are characterized as having a deep root system (such as *Cicer arietinum* L., *Vigna aconitifolia, Brassica campestris,* etc.), which provides drought tolerance (Kumar 2005).

11.4 Plant Adaptations to Drought Stress

Plants have developed various mechanisms to maintain their vigor during different levels of water scarcity. Plants respond to drought stress of different intensities through a range of changes from morphological to molecular levels.

11.4.1 Root Morphology

Drought stress is firstly perceived by the roots wherein, shoot growth is significantly reduced. Primary roots are not affected but the lateral roots show a reduction in growth due to the suppression of lateral root meristems (Chen et al. 2012). It was reported that plant microRNA miR393 played a significant role in the root-mediated adaptations caused by reduction of auxin signaling under drought stress (Chen et al. 2012). The small roots in addition to lateral roots are also considered an adaptive response of the plants to absorb more water by increasing the surface area. Similarly, the presence of suberized exodermis, reduction in the number of cortical layers and rhizodermis are the adaptive strategies under drought stress conditions. On the other hand, drought stress, i.e., hydrotropism which is increased by the degradation of columella cells of roots and amyloplast. Hormonal cross-talk is another adaptive strategy to modify root architecture under drought stress condition. (Blilou et al. 2005).

11.4.2 Photosynthesis and Gaseous Exchange

Due to metabolic impairment, stomatal closure and leaf area reduction occur, the rate of photosynthesis decreased under drought stress (Basu et al. 2016). Photosynthesis is also reduced because the intercellular carbon dioxide concentration decreases resulting in the reduction of the components of electron transporter (ET). The reduction in ET leads toward the decreased concentration of molecular oxygen corresponded to the production of reactive oxygen species (ROS). Reactive oxygen species cause damage to photosynthetic apparatus and other macromolecules (Demmig-Adams and Adams 2006). Various photosynthetic pigment such as xanthophyll, light-harvesting complexes from the reaction centers and thermal dissipation of light energy causes some of the adaptive responses of the plants to drought stress. Further, the ribulose-1,5-bisphosphate carboxylase/oxygenase (RuBisCO) activity and regeneration of ribulose-1,5-bisphosphate adversely affect the photosynthetic biochemical efficacy. (Lawlor 2002; Chaves et al. 2009).

11.4.3 Transpiration and Stomatal Conductance

Due to drought, stomatal closure in the leaves is the prompt response which reduced the loss of water, uptake of nutrients, etc., which modulate the metabolic pathways (Xiong and Zhu 2002). The phenomena of leaf shedding, decrease in the number of leaves, leaf size, and branching as well as xeromorphic characteristics reduce the rate of transpiration under drought stress. Sclerophylly is among one of the adaptations of the plants under drought, where the plants do not suffer from wilting and permanent damage due to the formation of hard leaves and resume their activity under normal conditions (De Micco and Aronne 2009).

Reduction in the size and number of stomata is another adaptation of plants for survival under drought stress. Reduction in the surface of chloroplast and a decrease in stomatal conductance due to the reduced expression of *aquaporin* genes leads to stress avoidance (Tosens et al. 2012). Chloroplast differentiation and mesophyll modulation affect the light availability, developmental stage of leaf, and ultimately photosynthesis (Tosens et al. 2012). These adaptations reduce the damaging effects of drought stress on plants and improve water-use efficiency to ultimately increase yield (Blum 2005).

11.4.4 Regulation of Phytohormones

Plants contains ethylene (ET), abscisic acid (ABA), gibberellic acid (GA), cytokinin (CK), and auxin, which are meant for the regulation of physiology of plants. (Wilkinson et al. 2012). When plants face the harsh condition of drought stress, ABA is synthesized in the roots and its translocation is carried out to the leaves, where it causes stomatal closure and reduction in plant growth (Wilkinson and Davies 2010). The ABA signaling genes such as *DSM2*, *OsNAP,* and *OsNAC5* improved the yield of the crops under drought stress conditions (Du et al. 2010; Chen et al. 2014; Liang et al. 2014). If drought occurs at the reproductive stage of plant, ABA-induced senescence and stomatal closure due to the unintended reduction in carbon supply occur. Thus, it seems a great challenge towards the ABA-induced drought adaptation for better growth and yield (Ji et al. 2011). Similarly, another adaptive trait is the increased endogenous production of CK through the expression of CK biosynthetic gene *isopentenyltransferase (IPT)* which delays the senescence and death of premature leaves (Peleg et al. 2011).

Interestingly, auxins negatively regulate the drought adaptation in the plants. Upregulation of a gene that encodes late embryogenesis abundant (LEA) proteins resulted in a decrease of auxin contents under drought stress (Zhang et al. 2009). It has also been reported that auxins negatively regulate the *DRO1*gene (*DEEPER ROOTING 1*), which controls the root growth. Increased expression of *DRO1*gene results in the phenomenon of drought avoidance in shallow rooting in rice cultivars resulting improved the yield under drought (Uga et al. 2013).

Ethylene has also been reported as a negative regulator of the response to drought stress. Ethylene responses under drought caused overall effect on root growth, leaf expansion, photosynthesis, grain development, and leaf senescence (Fukao et al. 2006; Perata and Voesenek 2007). Ethylene enhanced the abortion of embryo and grain and directly affects the crop yield, a rapid reduction in the level of GA occurred due to drought stress which acts as positive regulator of plants adaptation to stress (Wang et al. 2008). Along with these, other less explored hormones such as jasmonic acid (JA), brassinosteroids, strigolactone, and salicylic acid (SA) have also been reported to provide adaptation to the plants under drought stress. Thus, hormones interact with each other and modulate each other's biosynthesis and responses, therefore, the drought stress responses are regulated by a balance among the phytohormones which promote and those that inhibit the traits.

11.4.5 Osmotic Adjustment

The phenomenon of Osmotic adjustment (OA) is the mechanism, which helps in the maintaining of cell turgor during which the solutes are accumulated in the drying cell under reduced water potential (Chaves and Oliveira 2004). During drought stress conditions, the leaf water volume, stomatal conductance, photosynthesis, and plant growth are observed to be maintained by osmotic adjustment (Chaves and Oliveira 2004). The cell enlargement is believed to be inhibited due to osmotic adjustment during drought stress (Serraj and Sinclair 2002). Compatible solutes (i.e., proline, glycine betaine, etc.) are accumulated by OA and help the plant to cope with the detrimental effects of drought stress (Ashraf and Foolad 2007). Some enzymes like pyrroline-5-carboxylate reductase, betaine aldehyde dehydrogenase and ornithine δ-aminotransferase have been reported to play an important role in osmotic adjustment under drought stress.

11.5 Combat Drought Stress

Drought stress caused adverse impacts on the growth and output of the crops. Plants also have natural mechanisms to adapt under drought stress conditions, however, the tolerance mechanisms partially regulated by a variety of measures such as the use of plant mineral nutrients, plant growth regulators, genetic modifications, developing tolerant plant genotypes, compatible solutes, seed treatments, etc.

11.5.1 Role of Plant Growth-Promoting Rhizobacteria (PGPR)

Various authors have investigated the role of PGPR and other microorganisms in relation to drought tolerance. PGPR inoculation to crop plants is also an important strategy to enhance plant growth and yield under water scarcity conditions. The PGPR can increase plant growth by regulating the nutritional and hormonal balance to solubilizing nutrients and provide plant growth regulators under stress conditions. The plant–microbe interaction is very important to regulate plant growth and provide protection against drought stress (Dar et al. 2018). Application of plant growth-promoting rhizobacteria for plant growth and drought stress tolerance is an environment friendly, sustainable and cost-effective approach as compared to other measures.

PGPR are the group of bacteria that have the capability to colonize the root system of plants and improve the growth and yield of plants. A wide range of bacteria actinobacteria, fungi, etc., belong to this category. A total of 2–5% of soil microorganisms have PGPR properties (Saharan and Nehra 2011). Various genera play a significant role to improve drought stress tolerance of crop plants.

11.5.2 Engineering Plants for Drought Tolerance

Another strategy to combat with drought stress is the engineering of drought-tolerant crop plants, which involves the manipulation of functional and regulatory genes in plants which are not drought stress tolerant (Butt et al. 2017). Due to complete availability of the genome sequence, *Arabidopsis* is considered as a model plant to study the drought tolerance mechanism. For the identification of the target gene, the microarray gene expression technique is used. In alfalfa (*Medicago tranculata*), the engineered gene *APETALAL2* transcription factor induces drought tolerance through the process of wax production (Zhang et al. 2005). Insertion of *Mannitol dehydrogenase (MtlD)* gene in wheat improved the drought tolerance capability (Abebe et al. 2003). Similarly, in tobacco, the drought tolerance is enhanced by the overexpression of *inositol Methyl Transferase gene (IMTI)* gene isolated from the ice plants (Sheveleva et al. 1997).

11.5.3 Molecular and Functional Genomics Approaches

Though, conventional breeding methods have been used for a long time for drought tolerance. Recently, encroachments in genomics and molecular breeding techniques have a prominent role in the development of drought-tolerant cultivars (Kumar et al. 2014). Several genes that show response to the drought have been identified

(Yadav 2010; Todaka et al. 2015; Ali et al. 2017; Kaur and Asthir 2017). The drought tolerance mechanism at molecular level involves the regulation of gene expression and the identification of transcription factors. It has been reported that transcription factors (*DREB2A* and *DREB2B*) are involved in the expression of different genes that confer drought tolerance to plants (Ali et al. 2017). Aquaporins can be an important target for the development of plant genotypes, which are drought tolerant because aquaporins play an important role in the regulation of plant–water relations (Afzal et al. 2016). Kaur and Asthir (2017) observed that ABA-responsive elements-binding proteins respond at both levels, i.e., transcriptional level and posttranscriptional level and also important role in determining the capabilities of drought tolerance in plant.

11.5.4 Selection and Breeding Strategies

In order to develop plant genotypes tolerant to drought stress, different approaches such as conventional, omic-based, molecular proved to be successful (Maqbool et al. 2017). Conventional breeding techniques have been adopted to produce crop varieties with improved growth and yield under drought stress (Ahmad et al. 2014). In plant breeding techniques, drought tolerance is induced in crop plants by manipulating the genetic make of the crops to attain desired characteristics (Maqbool et al. 2017). However, in order to screen traits that are associated with drought tolerance, marker-assisted selection proved to be a better option rather than classical breeding technique (Ahmad et al. 2014). Segregation mapping and quantitative trait loci (QTL) analysis are the molecular basis of drought tolerance and QTL is important for the marker-assisted selection of plants with desired traits (Ali et al. 2017).

11.5.5 Application of Compatible Solutes

Compatible solutes such as proline, glycine betaine, organic acids, soluble sugars, trehalose, sugar, alcohols, etc., played important roles by providing protection against the detrimental consequences of osmotic stress on the macromolecules, membranes, and enzymes (Kiani et al. 2007). Proline, sugar alcohols and soluble sugars served as cryo-protectants and osmoregulators in plants and help in the scavenging of ROS under drought stress (Ruelland et al. 2009; Van den Ende and Valluru 2009). Application of glycine betaine exogenously played essential roles in modulating the osmotic adjustment, antioxidant activities, and detoxification of ROS (Farooq et al. 2008).

11.6 Growth Promotion Mechanisms of PGPR

Numerous studies have reported the mechanisms of growth promotion and drought tolerance of plants mediated by PGPR.

11.6.1 Direct Mechanisms

PGPR increase the drought tolerance ability of plants by directly enhancing the availability of nutrients to them in order to stabilize the structures of their biomolecules and to maintain their biomass production.

11.6.1.1 Nitrogen Fixation

Nitrogen is the most important nutrient, essential for plant growth, development, and productivity. It is a fundamental part of essential biomolecules such as nucleic acids and proteins. In the process of biological nitrogen fixation, the unavailable form of nitrogen is converted into ammonia or nitrate ions which can be readily taken up by the plants for growth promotion process. PGPR group of bacterial genera carry out the process of biological nitrogen fixation (Kim and Rees 1994). Both symbiotic and non-symbiotic or free-living microorganisms including few endophytes fix nitrogen that benefits both plant—soil microorganisms. (Bhattacharyya and Jha 2012).

11.6.1.2 Phosphorus Solubilization

Phosphorus (organic and inorganic forms) is among the main macronutrient necessary for plant growth and development. It is present in unavailable or insoluble form in soil but made available with the help of Phosphobacteria solubilized the soil inhibited P which produces phosphatase enzyme This enzyme efficiently hydrolyze the organic phosphate into a soluble form (Nosheen et al. 2018). Few low molecular weight organic acids also released by phosphate solubilizing bacteria. These compounds utilize their carboxyl and hydroxyl groups for the chelation of cations, which are bounded to the phosphate and convert the insoluble phosphorus into soluble form (Patel and Minocheherhomji 2018).

11.6.1.3 Potassium Solubilization

Potassium also is an important macronutrient, required for plant growth, development, and metabolic processes. It is usually absorbed as cationic form. The imbalanced use of chemical fertilizers leads to the deficiency of potassium in the soil hence, hinders the plant growth and development (Sindhu et al. 2014). Therefore, it is pertinent to have alternative sources for the provision and uptake of potassium. The soil microorganisms have essential roles to recycle natural potassium resources. These microorganisms are called as *potassium solubilizing rhizobacteria* which provide potassium to the plants (Basak and Biswas 2010).

Recent investigations have shown that organic exudates of some bacteria play a key role in releasing otherwise unavailable K from K-bearing minerals. K-solubilization could be attributed to excreting organic acids, which either directly dissolves rock K or chelate silicon ions to bring K into solution (Prajapati et al. 2013).

Some bacteria produce organic exudates that exert central effects in the release of unavailable potassium from potassium-containing minerals. These organic acids may directly dissolve potassium in rocks or may cause chelation of silicon ions to cause release of potassium (Prajapati et al. 2013). Wheat inoculation with *Bacillus* sp. or *Pseudomonas* sp. cause a prominent elevation in the uptake of potassium, magnesium, and calcium from calcareous soil without supplementation with fertilizer (Öğüt et al. 2011).

11.6.1.4 Zinc Solubilization

Zinc is an imperious micronutrient, essential for the optimum growth of plants (Goteti et al. 2013). There are two types of zinc fertilizers based on their different solubilization properties, i.e., zinc sulfate and zinc oxide. Zinc sulfate is considered as highly soluble in water as compared to zinc oxide but it can go back to insoluble form (Hafeez et al. 2013). The application of zinc-solubilizing bacteria plays an important role and carries out conversion of insoluble form of zinc to soluble form using several mechanisms such as proton extrusion, excretion of organic acids, chelating agents, and gluconic acids production (Goteti et al. 2013).Different diverse genera, namely, *Pseudomonas* sp., *E. cloacae* and *Pantoea agglomerans* showed increased zinc contents in shoot and roots of inoculated plants (Kamran et al. 2017). A recent study indicated that *ZnSB2* strain increased the zinc solubilization in the soil and can serve as a potential zinc solubilizer to reduce the need of application of chemical zinc fertilizers (Dinesh et al. 2018).

11.6.1.5 Phytohormones Production

Phytohormones are molecules, which play an important role in plant growth and development. Various group of hormones, viz., gibberellins, auxins (Indole-3-Acetic Acid/IAA), abscisic acid, cytokinins, and ethylene are the major plant hormones (Egamberdieva 2013). Glick and Pasternak (2003) reported that PGPR play an important role in the growth and division of plant cells and provide tolerance against the environmental stresses through the biosynthesis of phytohormones. Inoculation of PGPR to the plant species produce phytohormone, which increases the lateral roots and root hairs, thus increasing the water and nutrient uptake efficiency (Dimkpa et al. 2009) and provide tolerance to water-deficit conditions. Han et al. (2018) reported that bacterial strains such as *Streptomyces sundarbansensis* and *rochei* improved growth and development of plants through the production of IAA. Cytokinins, which are considered as purine derivative compounds, are involved in the cell division and differentiation process during growth and development. Recently, *Bacillus megaterium* has been reported to play an important role in the growth promotion by the production of cytokinin (Numan et al. 2018).

11.6.2 Indirect Mechanisms

PGPR exerts beneficial effects on plants against abiotic stresses by acting indirectly such as by enhancement of the activity of ROS-scavenging enzymes and water-use efficiency. PGPR are also known to produce enzymes and other molecules such as exopolysaccharides against phytopathogens. Role of PGPR in the induction of systemic resistance in plants is also well elaborated.

11.6.2.1 Stress Management

Any factor that causes a negative impact on plant growth and development is called as stress (Foyer et al. 2016). During stressed conditions many harmful molecules are formed which are called as reactive oxygen species (ROS), i.e., H_2O_2, OH, O^{2-} and free radicals which damage the photosynthetic machinery, proteins, membranes and nucleic acids (Ramegowda and Kumar 2015). The role of PGPR in stress management is imperative as reported by several authors. The mitigation of stress by the inoculation of plants with PGPR in leaf water potential under stress condition has been improved (Ahmad et al. 2013a; Naveed et al. 2014). Various studies suggested the role of PGPR to mitigate drought stress in soybean, wheat, chickpea etc. (Ngumbi and Kloepper 2016). Habib et al. (2016) reported the salt stress tolerance in *Abelmoschus esculentus* (okra) by inoculation of PGPR which improved the water-use efficiency and activity of ROS-scavenging enzymes.

11.6.2.2 Biocontrol Agents

Use of microorganisms as antagonists against the plant pathogens is called as biocontrol and it is considered safe as compared to synthetic chemical pesticides. Various genera including *Bacillus*, *Pseudomonas* sp. etc. produce antibiotics which play an important role in inhibiting the phytopathogens (Ulloa-Ogaz et al. 2015). A variety of antifungal and antiviral metabolites such as phenazines, phenazine-1-carboxamide, pyoluteorin, cepaciamide A, 2,4 diacetylphloroglucinol, and azomycin, pseudomonic acid, karalicine, etc., have been produced by *Pseudomonas* species (Ramadan et al. 2016). *Bacillus* also reported to produce a large variety of antibiotics such as bacillomycin, iturins, and surfactin and can be used as biocontrol agents (Wang et al. 2015).

Bacillus pumilus and *Pseudomonas* sp. isolates from maize and rice rhizosphere respectively, growing in water-stressed areas increased the production of ABA in maize plants in combination with addition of l-tryptophan under drought stress (Yasmin et al. 2017). *Pseudomonas* sp. and *Proteus* sp. isolates from maize and rice rhizosphere, respectively, grown in water-deficient conditions elevated the concentrations of indole 3-acetic acid and gibberellic acid in maize plants under drought along with l-tryptophan addition. *B. pumilus* produced the most satisfactory results for helping maize plants against drought stress (Yasmin et al. 2017).

Glucanase-producing *Bacillus amyloliquefaciens* and *Bacillus subtilis* successfully inhibited important fungal pathogens of sugarcane including *Fusarium moniliforme* and *Colletotrichum falcatum* in addition to suppression of other damaging fungal plant pathogens such as *Fusarium oxysporum*, *Rhizoctonia solani* and *Macrophomina phaseolina*. These glucanase-producing plant growth-promoting bacteria also increased the activities of antioxidant enzymes (Zia et al. 2018).

Treatment of rice plants with *Bacillus* sp. strains *KFP-5*, *KFP-7*, *KFP-17* markedly elevated the activities of rice antioxidant enzymes against *Pyricularia oryzae* infection and reduced the incidence of blast disease (Rais et al. 2017).

11.6.2.3 Protective Enzymes

It has been reported that PGPR produce enzymes such as chitinase, β-1,3-glucanase, ACC-deaminase, etc., which induce lysis of the cell wall of pathogens (Goswami et al. 2016). The release of protective enzymes by the PGPR improve plant growth by controlling the phytopathogenic agents (Meena et al. 2016). The most catastrophic pathogens (*Rhizoctonia solani* and *Phytophthora capsice*) in the world can be controlled and their growth can be inhibited by PGPR (Islam et al. 2016). Fluorescent *Pseudomonas* isolate GRC_3 is reported to efficiently suppress the growth of fungal phytopathogens *Phytophthora capsici* and *Rhizoctonia solani* by producing antifungal enzymes, i.e., β-1,3-glucanase and chitinase along with antifungal metabolites of nonenzymatic nature (Arora et al. 2007).

PGPR strains *Sinorhizobium meliloti RMP1* and *Pseudomonas aeruginosa GRC2* successfully enhanced the yield from *Brassica juncea* plant. Both strains were tolerant to urea and diammonium phosphate (DAP) and colonized the plant rhizosphere successfully. Low concentrations of urea and DAP were beneficial for PGPR strains (Maheshwari et al. 2010). It was accessed that plant growth and yield parameters were increased when the strains were co-inoculated. Variants of *Sinorhizobium meliloti RMP1* and *Pseudomonas aeruginosa GRC2* that were not tolerant to urea and DAP showed less growth in presence of urea and DAP. Both strains exhibited urease activities (Maheshwari et al. 2010). Ramadan et al.

(2016) reported that *Sinorhizobium fredii* and *P. fluorescens* release chitinase and β-glucanases which can inhibit *Fusarium oxysporum* and *Fusarium udum* wilt. *Fusarium oxysporum* responsible for wilt and *Rhizoctonia solani* causing root rot of *Fagopyrum esculentum* were antagonized by biocontrol agent *Bacillus pumilus MSUA3* by employing the activities of chitinolytic enzymes and surfactin which is a heat-stable antibiotic (Agarwal et al. 2017).

11.6.2.4 Exopolysaccharide Production

Exopolysaccharides (EPSs) are biosynthesized by plants, algae, and bacteria. These are high molecular weight biodegradable polymers which are formed by the residues of monosaccharides and their derivatives (Sanalibaba and Çakmak 2016). The EPSs play an important role in aggregating soil particles, maintaining water potential, sustaining the host under stress conditions, maintaining the contact between rhizobacteria and roots of host plant (Pawar et al. 2017).

EPS producing PGPR strains are well elaborated for their potential to reduce salinity stress in plants. EPS produced by PGPR can bind to the cations in soil including sodium ions, as a result, the amount of salts present in the soil becomes restricted for uptake by plant. In this way, the salt stress on a plant can be reduced. EPS-producing PGPR alleviated salt stress and promoted the growth of maize and soybean. The PGPR such as *Rhizobium* sp., *Azotobacter*, *Enterobacter cloacae, Bacillus, Agrobacterium* sp., *Xanthomonas* sp., etc., produce EPSs and play a substantial role in improving the soil fertility and taking part in sustainable agriculture (Mahmood et al. 2016).

11.6.2.5 Siderophore Production

Majority of siderophores are produced under iron-limiting conditions by microorganisms to enhance the capacity of iron uptake. These are small organic molecules, which extract the iron metal ions (Saha et al. 2016). *Pseudomonas putida* uses the siderophores which are produced by other microbes and augment the amount of iron available in the natural habitat (Rathore 2014). PGPR are increased iron uptake capacity of plants, where the availability of iron is low. Hence, production of siderophores is an important mechanism used by PGPR to enhance growth and development of plants and induce stress tolerance under nonavailability of iron in their soil environment.

11.6.2.6 Hydrogen Cyanide (HCN) Production

HCN is a secondary metabolite produced by PGPR and overpowers the harmful pathogens development. An enzyme called HCN synthase (associated with the rhizobacterial plasma membrane) is responsible for the synthesis of HCN from glycine.

Many bacteria are capable of producing HCN such as *Pseudomonas*, *Alcaligenes*, *Rhizobium*, *Bacillus*, and *Aeromonas* (Ahmad et al. 2008). HCN produced by PGPR is reported to increase the availability of phosphates to plants (Rijavec and Lapanje 2016). HCN-producing PGPR can be utilized as biopesticides/weedicides to eliminate weeds from vicinity of a crop (Kamei et al. 2014).

11.6.2.7 Induced Systemic Resistance (ISR)

Induced systemic resistance (ISR) is the state of enhanced defensive capacity of the plant under environmental stress. PGPR have the capability to induce systemic resistance in plants under stress conditions (Prathap and Kumari 2015). When pathogen invades a plant, specific signals are produced through vascular system, as a result of these signals, enzymes such as polyphenol oxidase (PPO), catalase (CAT), chitinase, β-1,3-glucanase, ascorbate peroxidase (APX), phenylalanine ammonia-lyase (PAL), superoxide dismutase (SOD), and peroxidase are produced which provide tolerance to the plants against the stress. The ISR helps the plant to combat various diseases and it is not specific against a particular pathogen (Kamal et al. 2014).

PGPR consortium consisting of *Pseudomonas putida CRN-09* and *Bacillus subtilis CRN-16* increased the seed germination rate of *Vigna radiata* and also boosted the levels of antioxidants, i.e., peroxidase (PO), polyphenol oxidase (PPO), phenylalanine ammonia-lyase (PAL), β-1,3-glucanase, and chitinase for the induction of systemic resistance in *Vigna radiata* against pathogen *Macrophomina phaseolina* (Sharma et al. 2018). Majority of PGPR induces ISR in plants and revolutionized the research in agriculture sector (Gouda et al. 2017).

11.7 Improvements in Physiological and Morphological Processes

PGPR positively improves the physiology and morphology of plants to eliminate the deleterious effects of stresses such as changes in root and shoot growth patterns, modification of plant water contents and hormones concentrations and osmolytes production.

11.7.1 Enhanced Root Architecture for Stimulating Water Uptake

It has been seen through different studies that PGPR-treated plants exhibited improved root growth along with changes in root architecture (Kloepper et al. 2004; Ngumbi 2011; Yasmin et al. 2017). Further studies suggested that alterations in roots induced by bacteria lead to overall increase in root surface area that directly improved nutrients and water uptake (Somers et al. 2004; Timmusk et al. 2014). For maize study, in growth chamber tests, the seeds were treated with strain *Alcaligenes faecalis (AF3)* (Naseem and Bano 2014). Planting after 3 weeks, an increase in root length by 10% was observed in PGPR-treated plants than non-inoculated control plants under drought stress (Table 11.1). PGPR-treated plants can tolerate drought stress, due to developed root system that allows increased water uptake. Significant increase in root biomass of in *Mazurka* and *Kaleo* cultivars appeared when maize plants were inoculated with *Burkholderia phytofirmans* strain *PsJN* (Naveed et al. 2014). Similarly, when inoculated by *Enterobacter* sp. strain, an increase in root length was observed in *Mazurka* and *Kaleo* cultivars, under drought stress.

Under drought stress, maize plants have enhanced root length, when inoculated with PGPR isolate *9K* (Yasmin et al. 2013). Similar effects were also observed in wheat under drought stress when treated with *thuringiensis AZP2*. Longer root hairs and intense growth of lateral roots were visible in wheat root inoculated with BT. The alteration in root architecture also helps the plants to tolerate drought stress (Timmusk et al. 2014) (Table 11.1) enhancement and alteration of root parameters. Yet, numerous studies are required for better understanding of the direct signaling between drought tolerance and bacterial-induced root architecture. More elaborated studies are required to trace the ideal root traits that could provide tolerance against drought stress.

11.7.2 Shoot Length

Shoot growth is one so as of the major responses to drought stress, that benefits plants by reducing the available leaf area to decrease the evaporative loss of limited water reserves (Sinclair and Muchow 2001; Wang and Yamauchi 2006; Neumann 2008; Skirycz and Inzé 2010). However, essential solutes are diverted from plant growth requirements to stress-related housekeeping functions such as osmotic adjustment as a result of resisted shoot growth. To tolerate drought stress, inhibition of shoot growth is considered a classical adaptive response (Neumann 1995, 2008; Achard et al. 2006). The limited yield potential of a plant declines severely because of drought tolerance but increase the chances of plants survival (Sinclair and Muchow 2001; Neumann 2008; Claeys and Inzé 2013). Therefore, when plants are exposed to moderate stress and inhibition of shoot growth appears as a counter-response, Hence, in such examples, plant survival is not a question but the main concern is about limited yield production of plants under drought stress paired with reduction in shoot growth.

This advantageous strategy could be in the improvement in crop varieties that sustain an ability to maintain near to normal shoot growth during drought stress (Neumann 2008). PGPR-treated plants showed prominently increased shoot growth. Because plants under drought stress are capable to maintain near to normal shoot growth, resulting in increasing crop productivity. A study demonstrated that corn plants had improved shoot growth when inoculated with *Bacillus*

Table 11.1 PGPR in Mitigation of Drought Stress

Plant	Means of applying drought	PGPR	References
Arabidopsis thaliana	Drought induction by removing the covers of culture dishes for 3 days	*Paenibacillus polymyxa*	Timmusk and Wagner (1999)
thaliana	Drought induction by terminating irrigation. Drought started with the emergence of first two true leaves	*Phyllobacterium brassicacearum*	Bresson et al. (2013)
thaliana	Drought-induced when plants were 30 days old by stopping irrigation until symptoms of temporary wilting were visually observed (ca. water was with held after 10 days)	*Azospirillum brasilense*	Cohen et al. (2015)
Capsicum annuum	Progressive drought by suppressing water supply for 15 d. Drought introduced after 5 d of transplantation	*Bacillus licheniformis*	Lim and Kim (2013)
Cucumis sativa	Drought applied by stopping irrigation for 13 days. Drought started after 15 days of transplanting 15 days old cucumber seedlings	*Bacillus cereus*, *B. subtilis*, *Serratia* sp.	Wang et al. (2012)
Helianthus annuus	Drought applied by watering with polyethylene glycol (PEG) 6000 at a concentration sufficient to produce ψ_a = 2.03 MPa	*Achromobacter xylosoxidans Bacillus pumilus*	Castillo et al. (2013)
Hyoscyamus niger	Continuous drought by restraining irrigation for 60 days. Drought introduced when plants were 45 days old	*Pseudomonas putida Pseudomonas fluorescens*	Ghorbanpour et al. (2013)
Pisum sativum	Drought induced by stopping watering when plants were at the stage of vegetative growth (ca. 3 weeks after germination) or at flowering stage (ca. 7 weeks after germination) or at the pod formation stage (ca.8 weeks after germination). Plants were re-watered when symptoms of wilting were visualized	*Pseudomonas* spp.	Arshad et al. (2008)

(continued)

Table 11.1 (continued)

Plant	Means of applying drought	PGPR	References
lycopersicum annuum	Induction of drought by stopping irrigation after 2 weeks of seeds transplantation. Plants were re-watered after 7 or 12 days	*Achromobacter piechaudii*	Mayak et al. (2004b)
tuberosum	Drought application by watering plants with 10% PEG. Drought started after 2 weeks of inoculation of plants with PGPR	*Bacillus pumilus Bacillus firmus*	Gururani et al. (2013)
Sorghum bicolor	Continuous drought by withholding irrigation for 5 days. Drought started when plants were 27 days	*Bacillus* sp.	Grover et al. (2014)
aestivum	Application of drought by stopping water supply for 6 days or 12, 18, and 24 days. Drought started after 50 days of planting. Drought had three intensities: control, moderate and high drought. After drought stress, irrigation was restored	*Azospirillum lipoferum A. lipoferum A. lipoferum*	Arzanesh et al. (2011)
aestivum	Continued drought by withholding irrigation for 4, 5, or 7 days. Drought was introduced when plants were 12 days	*Bacillus amyloliquefaciens*, *Azospirillum brasilense*	Kasim et al. (2013)
aestivum	Progressive drought by withholding irrigation for 10 days (growing plants in sand soil) or 14 days (growing plants in sand soil added with 10% greenhouse soil). Drought introduced after 10 days of seed germination	*Bacillus thuringiensis*, *Paenibacillus polymyxa B*	Timmusk et al. (2014)
Vigna radiata	Progressive drought by 6 d of withholding water. Drought was introduced when plants were 30 days old	*Pseudomonas fluorescens* strain *Bacillus subtilis*	Saravanakumar et al. (2011)
Vigna radiata	Drought applied by withholding watering 10 days after planting	*Pseudomonas aeruginosa* strain	Sarma and Saikia (2014)
Zea mays	Progressive drought by stopping irrigation for 6 days. Drought introduced after 21 days of seed germination	*Pseudomonas entomophila*, *P. stutzeri*, *P. putida*, *P. syringae*, *P. monteilli*	Sandhya et al. (2010)

(continued)

Table 11.1 (continued)

Plant	Means of applying drought	PGPR	References
Z. mays	Drought induced at the vegetative stage by conserving the moisture content of the soil at 15 ± 1%	*Azospirillum lipoferum* strain	Bano et al. (2013)
Z. mays	Drought introduced by withholding watering for 6 days after 7 days of germination	PGPR	Yasmin et al. (2013)
Z. mays	Progressive drought introduced after 1 week of seed germination by stopping irrigation for 10 days	*Proteus penneri* strain, *Pseudomonas aeruginosa*, *Alcaligenes faecalis*	Naseem and Bano, (2014)
Z. mays	Progressive drought introduced when plants were 45 days old by withholding watering and observing for wilting signs	*Burkholderia phytofirmans*, *Enterobacter* sp.	Naveed et al. (2014)
Z. mays	Introduction of continuous drought when plants were 21 d old by stopping irrigation for 6 days	*Bacillus amyloliquefaciens*, *B. licheniformis, B. thuringiensis*, *Paenibacillus favisporus*, *B. subtilis*	Vardharajula et al. (2011)
Z. mays	Drought introduced when plants reached the 4 leaves stage by withholding watering for 10 days	*Burkholderia* sp. strain	Fan et al. (2015)
Z. mays	Induction of drought by terminating watering for 7 days when plants reached the 4 leaves stage	*Pseudomonas* sp. *Bacillus pumilus Proteus* sp. *Pseudomonas* sp. *Bacillus cereus*	Yasmin et al. (2017)
C. arietinum	Plants were subjected to drought stress after well-watering till 1 month, and then by withholding water for 1, 3, and 7 days. Plants were re-watered for recovery for 3 days	*Pseudomonas putida*	Tiwari et al. (2016)
Oryza sativa	Drought stress applied by withholding water at the panicle initiation stage of reproduction for 15 days	*Pseudomonas fluorescens*	Saakre et al. (2017)
Z. mays	Drought stress introduced by stopping water after 14 days of planting	*Pseudomonas putida*	SkZ et al. (2018)

sp. (Vardharajula et al. 2011). All plants inoculated with *Bacillus* sp. under stress conditions showed relatively higher shoot growth and dry biomass as compared to non-inoculated plants. PGPR-treated wheat plants confirmed their potential in enhancement of plant performance by showing higher biomass (Timmusk et al. 2014). Likewise, under drought stress pepper plants were inoculated with *Bacillus licheniformis K11* increased shoot length biomass, etc. (Lim and Kim 2013).

Improved plant growth and increment in shoot growth during drought stress have been reported in several other crops treated with PGPR in sorghum (*Sorghum bicolor* L.) (Grover et al. 2014), mung bean (*Vigna radiata* L.) (Sarma and Saikia 2014), wheat (*Triticum aestivum*) (Arzanesh et al. 2011; Kasim et al. 2013), sunflower (*Helianthus annuus* L.) (Castillo et al. 2013), green gram (*Vigna radiata* L.) (Saravanakumar et al. 2011) and maize (*Zea mays*) (Sandhya et al. 2010; Naseem and Bano 2014; Naveed et al. 2014; Yasmin et al. 2017) Given in Table 11.1.

11.7.3 Relative Water Content (RWC)

One of the major criteria to measure plant water status called relative waste content (RWC) In such cases, leaf tissues exhibited limited cell expansion as a result of a decrease in RWC that occur due to loss of turgor, leading to a decline in plants growth (Ashraf 2010; Lu et al. 2010; Castillo et al. 2013). On the other hand, the high RWC was observed in species that were better adapted to dry environments (Jarvis and Jarvis 1963). Therefore, an important drought tolerance enhancement strategy could be the increment in RWC.

For instance, *Bacillus* sp. strain *129 KB*-treated sorghum, showed an increase in sorghum (Grover et al. 2014) and maize (Sandhya et al. 2010; Vardharajula et al. 2011; Bano et al. 2013; Naveed et al. 2014; Naseem and Bano 2014, Yasmin et al. 2017). In fact, oxidative and osmotic stresses caused by drought stress can be reverted back to higher RWC, contributing to greater productivity of. Maize plants treated with *brasilense BR11005* sp. showed high RWC as a result of bacterial mediated abscisic acid (ABA) that in return induced closing of stomata and mitigate drought stress (Casanovas et al. 2002). Alterations of the sensitivity of physiological processes such as stomatal closure may be the reason for increase in RWC. In light of contrasting views, it is yet to investigate the mechanism involved in bacterial-induced drought tolerance.

11.7.4 Role of Phytohormones

Another mechanism used by rhizobacteria to impart drought tolerance in plants is alteration in their phytohormones contents. Indole-3-acetic acid (IAA), Cytokinins (CK), abscisic acid (ABA), and ethylene (ET) have been reported to be involved in ensuring plant tolerance to drought stress.

11.7.4.1 Indole-3-Acetic Acid (IAA)

PGPR synthesizes indole-3-acetic acid (IAA), is adsorbed on the root surface or seeds. Some of the newly synthesized IAA is taken up by the plant from root exudates, which facilitate plant cell proliferation and elongation in combination with the endogenous nature of plant IAA. Meanwhile, IAA stimulates the activity of the enzyme 1-aminocyclopropane-1-carboxylate (ACC) synthase to convert *S*-adenosyl methionine (SAM) into ACC (Naveed et al. 2014). A bacterial cell surviving under drought stress conditions also needs to cope up with the osmotic stress, bacteria help to tolerate osmotic stress generated by drought conditions in plants (Boiero et al. 2007).

Second most important function of bacterial IAA is its involvement in altering the root architecture in terms of root area and tips extension, which help associated host plants to assimilate nutrients from soil more efficiently underwater limiting conditions (Mantelin and Touraine 2004). Plants inoculated with IAA-producing bacteria showed more tolerance to drought stress (Yasmin et al. 2017, Marulanda et al. 2009). The volatile organic compounds were observed to enhance the production of IAA by upregulating IAA regulating transcripts (Zhang et al. 2007). Higher production of IAA was observed in *lipoferum* due to upregulation of indole-3-pyruvate decarboxylase gene. Re-inoculation studies showed that strain-induced morphological changes in coleoptile xylem of wheat seedlings grown under osmotic stress conditions (Pereyra et al. 2012).

11.7.4.2 Gibberellins

Several types of GAs secreted in diverse genera of PGPR *Pseudomonas putida H-2-3* induces physiological alterations and cause better growth of soybean under drought stress conditions by secretion of gibberellins (GAs) (Sang-Mo et al. 2014). Cohen et al. (2009) found that PGPR strains producing ABA and gibberellic acid (GA) remediated the consequences of drought stress in maize plants.

11.7.4.3 Abscisic Acid (ABA)

ABA has been considered as stress hormone as its secretion is stimulated by stress signals under drought stress conditions. Yamaguchi et al. (1994) found that ABA is involved in stomatal regulation to prevent water loss under drought stress conditions. *Azospirillum brasilense (Sp245)* treated *Arabidopsis* spp. accumulated higher levels of ABA as compared to non- treated plants (Cohen et al. 2008). Bresson et al. (2013) found that *Arabidopsis* plants inoculated with *Phyllobacterium brassicacearum* strain *STM196*. showed a reduction in transpiration and osmotic tolerance by producing higher contents of ABA. On the other hand, cytokinin-producing *Bacillus subtilis* ameliorate the growth retarding effects of drought stress on *Platycladus orientalis* seedlings (Liu et al. 2013).

11.7.4.4 Ethylene (ET)

Ethylene usually regulates the growth and functions of plants through various mechanisms. Physiological indicators induced by ethylene aid in shoot and root growth differentiation, release of dormancy, adventitious root formation, induction of flowering and increased femaleness in dioecious plants, leaf and fruit abscission, fruit ripening, and flower and leaf senescence (Davies 2013). It also helps in overabundance of ethylene which leads to abnormal growth of roots, imparting a visible dent on plant growth and development. Both biotic and abiotic stresses accelerated ethylene production in plant roots. The increase in ethylene concentrations has inhibitory effects on root growth crucial to regulate production in the vicinity of plant roots for normal growth and development (Naveed et al. 2014).

11.7.4.5 Influence of 1-Aminocyclopropane-1-Carboxylate (ACC) Deaminase

The enzyme ACC deaminase metabolizes ACC into ammonia and α-ketobutyrate and checks ethylene production, which inhibits plant growth (Glick 2007). The plants treated with bacteria having ACC deaminase may have comparatively increased root growth due to decreased ethylene production and can resist various stresses more effectively (Glick 2007).

The uptake and hydrolysis of ACC by microorganisms lessen the levels of ACC outside the plant. On the other side, the equilibrium between the external and internal ACC levels is continued through the diffusion of more ACC into the rhizosphere (Shaharoona et al. 2006a). Microbial communities of soil having ACC deaminase activity increased biosynthesis of ACC than the plant needs and stimulate exudation of ACC from roots provide microorganisms with a unique nitrogen source (ACC), and hence, the growth of ACC deaminase containing microbes is accelerated in the close vicinities of plant roots in comparison to the other soil microorganisms Not only the ACC levels decreased within the plant but also inhibits the biosynthesis of the stress hormone ethylene (Shaharoona et al. 2006b). A plant inoculated with ACC deaminase containing bacteria exhibited more root growth. In various studies, inoculation with ACC deaminase containing PGPR has been unequivocally shown to modify the endogenous levels of ethylene, which results in changes in plant growth (Shaharoona et al. 2006a, b; Madhaiyan et al. 2006; Shahzad et al. 2013)

A considerable increase in fresh and dry biomass of pepper and tomato was observed when treated with ACC producing *Achromobacter piechaudii (ARV8)* against water-deficit conditions (Mayak et al. 2004b). Lim and Kim (2013) observed that ACC deaminase-producing *licheniformis K11* impart significant increase in drought tolerance of pepper plant. Similarly, ACC deaminase producing *Pseudomonas* sp. treated P. pea showed significant root elongation which subsequently enhanced uptake of nutrients and water under water scarcity conditions (Zahir et al. 2008). A significant decrease in transcription of stress-responsive genes and upregulation of genes related to cell division was observed in ACC deaminase producing strain *Enterobacter cloacae UW4* treated canola plants (Hontzeas et al. 2004)

11.7.5 Effect of Osmolytes and Secondary Metabolites

The great concern lies in of osmoregulation in plants under drought stress conditions by the production and accumulation of certain secondary metabolites. The most commonly acclimated osmolytes in water-stressed plants and bacteria includes proline, trehalose, and glycine betaine (Kaushal and Wani 2016; Chen et al. 2007; Rodriguez et al. 2009).

11.7.5.1 Amino Acids

Rhizobacteria induce drought tolerance to the plants. The quantity of amino acids sorghum, pepper and wheat have been observed to get increased underwater deficit conditions due to protein breakdown in response to fluctuation in osmotic pressure (Zhu 2002; Yadav et al. 2005; Kaushal and Wani 2016) as stated below.

11.7.5.2 Proline

Accumulation of proline is a common metabolic response of higher plants to drought and salinity stress (Rhodes et al. 1999; Yasmin et al. 2017). Drought-affected cellular machinery responds by proline accumulation via osmoregulation, ROS-scavenging mechanisms, and stabilization of cytoplasmic organelles. Generally, proline protects plants from stress through different means such as detoxification of reactive oxygen species, contribution toward osmotic adjustment, stabilization of native structures of enzymes, proteins, and membranes (Binzel et al. 1987; Kaushal and Wani 2016).

Under water-limiting conditions bacteria produce osmolytes in soil, which are readily taken up by the plants (Kaushal and Wani 2016; Yasmin et al. 2017). PGPR *Bacillus subtilis*, *Pseudomonas* sp. isolated from water-deficient habitats showed more proline production in inoculated maize plants as compared to un-inoculated plants (Yasmin et al. 2017). Long back, Yoshiba et al. (1997) showed significant upregulation in proline synthesis gene *P5CS* and downregulate the proline metabolism gene *ProDH*. Moreover, transgenic *thaliana* plants with *ProBA* genes from *Bacillus subtilis* produced higher amount of proline (Chen et al. 2007).

A significant higher accumulation of osmolytes proline, sugars, and free amino acids in *Bacillus* sp. treated maize plants produced higher biomass, improved relative water content, and overall plant health status under drought stress condition (Vardharajula et al. 2011). Ansary et al. (2012) observed significant increase in accumulation of proline in *P. fluorescens* inoculated maize plants. On the other hand, *lipoferum* showed increase in soluble sugars and free amino acids during drought stress (Qudsaia et al. 2013).

11.7.5.3 Glycine Betaine (GlyBet)

Salt stress leads to water stress because of excess solute concentration. GlyBet amount elevates in both saline and drought conditions. It is a quaternary ammonium compound with osmoprotecting functions, protects the plants by stabilization of both the highly ordered quaternary structure of membranes and proteins. Excess ions may disturb the proper structures and functions of enzymes and other proteins, so GlyBet acts as molecular chaperons and helps in refolding of enzymes and in regaining of proteins functions. This metabolite is synthesized via two-step oxidation (Chen and Murata 2008). *Pseudomonas pseudoalcaligenes* inoculation in rice plants stimulated the synthesis and accumulation of GlyBet (Jha et al. 2011). According to Zhang et al. (2010) volatile organic compounds produced by *subtilis GB03* induced systemic drought tolerance in *Arabidopsis* plants. But, *xipotl* mutant of *Arabidopsis* showed less accumulation of choline with reduced drought tolerance when treated with *GB03*.

11.7.6 Antioxidants Defense

Reactive oxygen species (ROS) are generated at a low level in the organelles chloroplast, mitochondria, etc., under normal growth conditions. However, when the plants encounter any of the abiotic stresses, the dramatic acclimation of ROS level is observed (Sheteawi 2007). During the stress conditions, CO_2 uptake is reduced due to the stress-induced stomatal closure and reduction in leaf area. This indirectly favors the photorespiration in plants. Photorespiration is responsible for the overproduction of H_2O_2 in the peroxisomes. Moreover, the level of H_2O_2 or singlet oxygen is upregulated by over reduction in the photosynthetic electron transport chain (ETC) (Rejeb et al. 2014).

Fortunately, plants possess the defense mechanism that is activated upon the upregulation of ROS than the optimal level and scavenge the activated oxygen species (AOS) (Das and Roychoudhury 2014). The defense response is actually the generation of antioxidant enzymes. There are different antioxidant enzymes like catalase (CAT), superoxide dismutase (SOD), glutathione reductase (GR), glutathione peroxidase (GPX) and ascorbate peroxidase (APX) (Cho and Park 2000). The protective response also involves some other defense enzymes such as polyphenol oxidase (PPO), peroxidase (PO), phenylalanine ammonia-lyase (PAL), and tyrosine ammonia-lyase (TAL) (Valifard et al. 2015). These enzymes not only have antioxidant capacity but also have role in defense against pathogens attack in a variety of crops. Apart from their other normal functions, these enzymes are also involved in the formation of defense barrier for reinforcement of cellular structures (Gao et al. 2008).

Plants have to regularize the production and scavenging of ROS to prevent themselves from cell injury and for normal functions under drought stress conditions. A visible decrease in antioxidant enzymes activity of plants has been observed under water-stressed condition (Yasmin et al. 2017). However, PGPR inoculation showed significant positive effects on ROS-scavenging enzyme machinery to mitigate the adverse effects of drought stress (Bindu et al. 2018). Vardharajula et al. (2011) and Yasmin et al. (2017) observed that *Bacillus* sp. treated maize plants showed obvious reduction in ROS accumulation and decrease in the activities of ROS-scavenging enzymes.

11.8 Salt Stress/Salinity

Soil salinity is a threat and is defined as excessive salt accumulation in the soil causing plant growth inhibition that ultimately leads to plant death. On a world scale, salt is the most toxic substance among other substances that restricts plant growth. that significantly reduces crop yield (Qadir et al. 2014). Salinity poses an increasing threat to agriculture (Gunes et al. 2007). Among multiple sources of salt stress, the combined effect of irrigation with poor drainage is a common and most serious threat, as it causes losses to the productivity of agricultural land.

In comparison to primary salt stress in the seashore, the reason for this secondary salinization is simple, i.e., salts remain in the soil after water evaporates and adversely affect crop growth and yield (Ghassemi et al. 1995). The stresses caused by excessive salt accumulation in the soil are twofold; First, most of the salt ions causing toxicity to plant cells when exposed to high salt concentrations internally or externally. Characteristically, NaCl comprises of majority of the salts. Na^+ ions are toxic to most of the plants, and growth of some plants is also inhibited by excessive Cl^- ion accumulations. Second, salt stress causes a decrease in osmotic potential of the soil leading to osmotic stress (Gilroy et al. 2014; Roy et al. 2014). The objectives are to understand the control of ion homeostasis and osmotic regulation, and to use the knowledge to engineer crop plants with enhanced salt tolerance.

Among various abiotic stresses, salt stress is one of the major limiting factors in crop productivity, particularly in arid and semiarid areas. It has been stated that about 7% of the total land and 20% of the total arable lands are affected by salt stress (Zhu 2001; Shrivastava and Kumar 2015). Soil reclamation is essential as the cultivable lands are being greatly affected by salinity (Gunes et al. 2005). Salt stress/salinity is a complex trait to study as it hinders designing and interpreting the experiments but current OMICS-driven research has made it comparatively easier to study (Negrão et al. 2017). Plant growth could be recovered when salt stress is relieved but the plants primarily respond to salinity by reducing the leaf surface area followed by cessation of expansion as the stress intensifies (Parida and Das 2005).

Salinity causes ionic and osmotic stress that ultimately leads to oxidative stress (secondary stress) affecting different metabolic processes that include protein synthesis, photosynthesis and lipid metabolisms (Jampeetong and Brix 2009).

Salt stress is one of the major issues affecting agricultural lands worldwide. *Food and Agriculture Organization (FAO)* states that about 397 million hectares (ha) of land has been salinized globally. It also stated that out of 230 million ha of irrigated land, 45 million ha (19.5%) have been affected by salinity and in 1500 million hectares of dryland cultivation, 32 million ha (2.1%) have been under salt stress.

Secondary salinization, which is the result of mismanaged irrigation, is also a vital issue with regards to the world's food production. Though irrigation accounts for 17% of the world's cultivable land, it provides over 30% of its agricultural yields (Pitman and Läuchli 2002). Salt stress is a pervasive issue in irrigated regions of dry and hot territories, especially in countries of Africa and Asia like India, Egypt, and Iran. According to estimates, 33% of irrigated land is salt-affected in which 20% is due to secondary salinization (Shrivastava and Kumar 2015). About 75% of secondary salinization in dry, subhumid, arid, and semiarid regions occurs in the Asian-Pacific region; during the mid-1980s, about 50% (30 million ha) of irrigated land damaged by salinization was found in Pakistan, India, and China (Pitman and Läuchli 2002).

11.9 Effects of Salinity in Pakistan

In Pakistan, salinity is among the most serious abiotic stresses. It is increasing day by day affecting both quality and quantity of crops. The major cause of salinity is due to imbalance of entrance and exit of salt in the soil. Every year, around 120 million tons of salt is added to the land from canal and brackish water and from this only one-fifth of this salt can go to the seawater. Remaining salt leaches in the soil and causes reduction in growth of plants (Elnaggar and Noller 2009). About 2.5 million hectares of irrigated land is being adversely affected by severe saline conditions in Pakistan. Out of this, 18% of this land is affected in Sindh while NWFP and Punjab contribute 2% and 3%, respectively. Areas affected by moderate saline conditions are 10%, 4%, and 2% in Sindh, Punjab, and NWFP, respectively (Ali et al. 2005).

Crop yield losses in Pakistan are estimated to reach to feared amount of Rs. 880 million ($ 28.5 million) this year due to water logging and salt stress, while total economic loss is around $ 300 million. As for as wheat is concerned, the yield losses in moderate saline areas are about 65% (El-Hendawy et al. 2005). Yield losses in important crops due to salt stress are given in Table 11.2.

Other than crop yields, salinity also affects photosynthesis, ion regulation, and water relations. It also affects plant physiology at adult plant stages as well as at cellular levels through ionic and osmotic stress (Parida et al. 2004). It has been observed that various physiological processes that are severely affected by salt stress include mineral distribution, plant growth inhibition, membrane instability

Table 11.2 Yield losses of different crops caused by salinity

Crop	Yield loss (%)	References
Wheat	60–65	El-Hendawy et al. (2005)
Rice	30–50	Joseph and Mohanan (2013)
Sugarcane	40–50	Rao and Shaw (1985)
Pepper	8–15	Navarro et al. (2002); Chartzoulakis and Klapaki 2000)

as a result of calcium displacement by sodium, membrane permeability, and drastic reduction in rate of photosynthesis and Na^+ and K^+ discrimination (Shafi et al. 2010).

11.10 Strategies Used to Overcome Salinity

Supply of sufficient water along with proper drainage and irrigation management are also good but comparatively expensive practices to control salinity problem (Cominelli et al. 2013; Mwadzingeni et al. 2016).

The most significant step to control salinity issue is developing salt-tolerant plant varieties. Improvements in the field of genetic engineering have greatly reduced the risk of salinity by the production of salt-tolerant plants. However, the progress in this field is still slow because of the contradiction in views among plant physiologists, plant breeders, and plant molecular biologists (Munns 2002; Blum 2014; Flowers 2004; Ashraf et al. 2008; Hasegawa et al. 2000; Mittler and Blumwald 2010; Hasegawa 2013; Munns and Gilliham 2015; Yeo 1998; Zhang et al. 2014). But still several salt-tolerant plants have been developed through genetic engineering technique (Zhang et al. 2001; Zhu 2001).

Forests are the environmental buffers which keep the soil in place, deforestation leads to soil erosion and ultimately increase the risk of salinity and water logging so planting more trees to renovate the custom of the green revolution is the need of the hour. Acid rain also causes leaching of important nutrients from the soil such as calcium and potassium. To reduce the risk of acid rain, use of biofilters in industries is emphasized so that soil could be protected from grave problem of salinity and soil erosion.

In irrigated soil, reclamation of soil involves the replacement of sodium ions with calcium ions. The released sodium ions are finally percolated out of the root zone by using excess water which is later carried out of the fields in the drainage water. Because of continuous process of evaporation, salt concentration increases in the top soil surface which is then scraped and transported out of the field. Pre sowing irrigation with good quality water is a good strategy to remove salt from the soil surface. This ensures better seed germination and establishment of seedling.

Mulching (a practice that uses crop residue, straw) reduces water loss from the top soil causing reduction in uptake of salts. Reduced evaporation also improves water-use efficiency. As a result, lesser salts are accumulated. Mixing organic matter with crop residues or green-manure crops also improves soil surface, supply structure, and improves water infiltration which prevents adverse effects of salinity. Halophytes being native flora of saline environment complete their life in salinity which also plays a positive role in reduction of salt stress.

11.11 Concerns with Strategies

Currently, numerous crop improvement approaches are being used against salt stress. In the past, selection breeding using the mechanism of stress tolerance has led to smaller crop improvements. For example, salt stress causes 40–70% reduction in yield and crop yield improvement is only 5–20% through conventional or molecular breeding techniques (Ashraf et al. 2012; Ashraf and Foolad 2013; Munns et al. 2012; Negrão et al. 2011, 2017). There are several transgenic approaches available (more than three thousand reports of improved salt tolerance in wheat, rice, maize, sorghum, canola, etc.) but unfortunately none is being used by farmers.

Salinity has been reported as one of the major limiting factors in crop productivity in Pakistan. In the past, several agricultural regions have significantly lost their productivity due to soil stress. The saline conditions are found in various parts of the country but the most affected lands as mentioned earlier are found in Sindh, Punjab and few areas of Khyber Pakhtunkhwa. Farmers do not pay proper attention to salinity. It indicates that farmers did not adopt any consistent strategy to fight against this serious issue. No significant pattern in farmers' response toward salinity has been noticed. Farmers were just preferring to buy lands with minimal soil salinity and increasing production of existing agricultural lands. Instead of adopting strategies to get rid of salinity, farmers were demanding larger government actions to install a proper drainage system for successful reclamation. Suggestions of farmers for land reclamation (Tanwir et al. 2003) are included in Table 11.3.

The above table suggests that farmers were unable to form any opinion which is quite alarming. Salt stress has emerged as a serious issue that not only causes a loss in crop productivity but also puts far-reaching impacts on the livelihood strategies of small farmers. The problem has been intensified to such an extent that it has made it very difficult for the farmers to fight against the severity of the problem. A joint step by government, NGOs, and the farmers is promptly required to control the alarming situation (Tanwir et al. 2003).

Table 11.3 Farmers' suggestions for reclamation of land

Suggestions	Percentage, n = 150
Laying of tile drain	19.8
Installation of tube well	21.5
Cleaning of drainage	11.2
Tile drain, tube well and cleaning	9.1
No opinion	38.4

11.12 PGPR-Mediated Amelioration of Salt Stress

Over the past century, large volumes of publications have appeared on the beneficial interactions between microbes and plants. Microbial communities dwelling in the nutrient-enriched plant rhizosphere play an essential role in plant functioning by influencing their growth, physiology, and other developmental processes (Mendes et al. 2013). PGPRs residing as free-living in the rhizosphere and carry out useful activities for plants. Few bacteria are facultative endophytes invading the host plant tissues to mutually establish a beneficial interaction. Most of the PGPR inhabit the roots by colonization and proliferate in rhizodermal spaces and root hairs, while few species reside without any physical contact with roots (Gray and Smith 2005). Both rhizospheric, as well as endophytic bacteria, promote plant growth through various direct as well as indirect mechanisms. In such mechanisms, bacteria directly enhance the plant growth efficiently by improving plant nutrition through phytohormone production and solubilizing several important minerals (Chauhan et al. 2016; Vaishnav et al. 2016; Maheshwari et al. 2014).

For regulation and rhizospheric signal communication, root secretions play the most important role and are involved in the plant–microbe interaction. In the field of agriculture, the main focus is to remove the unadorned stress conditions and adverse outcomes, through exogenous application of inoculum of these microbial organisms (Shrivastava and Kumar 2015). Before introducing commercially, the PGPR that are isolated from the rhizospheric soil are screened in vitro for testing the characteristics of PGP and some other valuable effects in field trails and greenhouse experiments. Plant growth is promoted by PGPR through the vast chain of signaling and mechanisms like efficient uptake of nutrients by fixing nitrogen biologically, iron chelation and solubilization of phosphorus as well as pathogen control through competition for survival and antagonistic relationships (Compant et al. 2005; Steenhoudt and Vanderleyden 2000; Maheshwari 2010; Beneduzi et al. 2012; Sharma et al. 2013; Jin et al. 2013; Chowdhury et al. 2015; Kuan et al. 2016), soil decontamination by degrading several organic contaminants and reduction of heavy metals from the soil, and assistance in phytoremediation (Nie et al. 2011; Divya and Kumar 2011; Janssen et al. 2015).

PGPR inoculation is known to induce the abiotic stress regulatory mechanisms through direct and indirect systemic induced resistance pathways (Yang et al. 2009; Maheshwari 2010). The mechanism involved in the elevation of salt stress, maintenance of ion homeostasis, improvement of photosynthesis and water plant relations. A complex channel of plant–microbe interaction and cascade of signaling mechanism is implemented in the removal of salinity stress (Smith et al. 2017). The plant growth promotion by the application of different salt-resistant PGPR genera including *Bacillus*, *Serratia, Pseudomonas*, *Rhizobium*, *Azospirillum* and *Acetobacter* under salt stress conditions is given in Table 11.4.

Table 11.4 Plant growth-promoting rhizobacteria mediating salt tolerance in plants

Plant	PGPR specie	Effects	References
Glycine max	*Rhizobium*	Increased nodulation	Ma et al. (2002)
Zea mays	*Azospirillum*	Obstructed Na^+ uptake, increased uptake of K^+ and Ca^{2+}, and increased nitrogenase and nitrate reductase activity	Hamdia et al. (2004)
Lycopersicon esculentum	*Achromobacter piechaudii*	Reduced ethylene and improved plant growth	Mayak et al. (2004a)
Triticum aestivum	*Aeromonas hydrophila/ Caviae, Bacillus insolitus, Bacillus* sp.	Exopolysaccharide production	Ashraf et al. (2004)
Triticum aestivum	*Azospirillum*	Improved water relations and yield	Creus et al. (2004)
Vitis vinifera	*Burkholderia phytofirmans*	Increased chilling resistance	Barka et al. (2006)
Arachis hypogea	*Pseudomonas fluorescens*	Increased ACC deaminase activity	Saravanakumar and Samiyappan (2007)
Zea mays	*Pseudomonas syringae, Pseudomonas fluorescens, Enterobacter aerogenes*	ACC deaminase activity increased	Nadeem et al. (2007)
Arabidopsis thaliana	*Bacillus subtilis*	Regulation of tissue-specific sodium transporter HKT1	Zhang et al. (2008)
Zea mays	*Rhizobium, Pseudomonas*	Electrolyte leakage decreased, increase in proline, relative water content and K uptake	Bano and Fatima (2009)
Lactuca sativa cv. *Tafalla*	*Pseudomonas mendocina*	ACC deaminase activity, improved nutrient uptake	Kohler et al. (2009)
Gossypium hirsutum	*Pseudomonas putida* Rs-198	Increased K^+, Mg^{2+}, and Ca^{2+} absorption, decrease Na^{2+} uptake from soil	Yao et al. (2010)
Zea mays	*Bacillus megaterium*	Upregulated expression of *ZmPIP* two isoforms	Marulanda et al. (2010)
Oryza sativa	*Pseudomonas pseudoalcaligenes, Bacillus pumilus*	Increased glycine betaine concentration	Jha et al. (2011)
Zea mays	*Micrococcus luteus*	Increased IAA and HCN production	Raza and Faisal (2013)
Fragaria ananassa	*Kocuria erythromyxa (EY43), Bacillus atrophaeus (EY6), Staphylococcus kloosii (EY37)*	Increased yield and nutrient uptake	Karlidag et al. (2013)

(continued)

Table 11.4 (continued)

Plant	PGPR specie	Effects	References
Triticum aestivum	*Pseudomonas putida, Pseudomonas fluorescens, Serratia ficaria, Enterobacter cloacae*	Increased germination percentage, rate and index, enhanced nutrient status	Nadeem et al. (2013)
Oryza sativa	*Pseudomonas pseudoalcaligenes, Bacillus Pumilus,*	Increased antioxidants, plant growth, and nutrient uptake	Jha and Subramanian (2013)
Vigna radiata	*Rhizobium* and *Pseudomonas*	IAA production and ACC deaminase activity	Ahmad et al. (2013b)
Solanum lycopersicum	*Pseudomonas putida (UW4)*	Increased shoot growth and expression of *Toc-GTPase*	Yan et al. (2014)
Hordeum vulgare and *Avena sativa*	*Acinetobacter* sp., *Pseudomonas* sp.	Improved IAA and ACC deaminase production	Chang et al. (2014)
Oryza sativa (GJ-17)	*Bacillus pumilus, Pseudomonas pseudoalcaligenes*	Decreased superoxide dismutase and lipid peroxidation activity	Jha and Subramanian (2014)
Solanum lycopersicum 'Micro tom'	*Streptomyces* sp. strain PGPA39	Enhanced IAA production, ACC deaminase activity, phosphate solubilization	Palaniyandi et al. (2014)
Glycine max	*Pseudomonas simiae (AU)*	Upregulating the expression of storage vegetative proteins, RuBisCO large unit proteins. Increased chlorophyll and proline contents and decreased root *NaCl* accumulation	Vaishnav et al. (2015)
Abelmoschus esculentus	Enterobacter sp. *(UPMR18)*	Improved antioxidant activities, upregulation of ROS specific pathway genes	Habib et al. (2016)
Triticum aestivum	*Dietzia natronolimnaea*	Protection by modulating the expression of stress-responsive genes	Bharti et al. (2016)
Zea mays	*Bacillus amyloliquefaciens (SQR9)*	Improved photosynthesis and nutrient uptake	Chen et al. (2016)
Glycine max	*Bacillus thuringiensis (NEB17)*	Upregulation of the expression of RuBisCo-oxygenase, PEP carboxylase, proteins and pyruvate kinase, antioxidants, photosystems I and II, Glutathione-S transferase and isocitrate lyase.	Subramanian et al. (2016)

(continued)

Table 11.4 (continued)

Plant	PGPR specie	Effects	References
Capsicum annum	*Microbacterium oleivorans* (*KNUC7074*), *Brevibacterium iodinum* (*KNUC7183*), *Rhizobium massiliae* (*KNUC7586*)	Greater plant height, fresh and dry weight, enhanced chlorophyll content	Hahm et al. (2017)
Triticum aestivum	*Arthrobacter protophormiae (SA3)*, *Dietzia natronolimnaea (STR1)*	Enhanced photosynthetic efficiency	Barnawal et al. (2017)
Panax ginseng	*Paenibacillus yonginensis (DCY84T)*	Induction of the defense-related systems, for instance, ROS-scavenging enzymes, ion transport, proline content, biosynthetic ABA genes, total sugars and induction of genes involved proliferation of root hairs	Sukweenadhi et al. (2018)
Lolium perenne	*M30-35* strain (novel bacterium)	Improved chlorophyll contents, shoot fresh and dry weights, root activity, root volume, soluble sugar, catalase activity, and proline content	He et al. (2018)

11.12.1 PGPR-Mediated Induced Systemic Tolerance

It is important to develop the knowledge about the vibrant functions of PGPR by using different standardized techniques and understanding the basic mechanisms and functions of PGPR regarding the conductance of stomata, transport of ions, phytohormonal status, antioxidant enzymes, carbohydrate metabolism, and proteins signal transduction. On improvement of salt stress different PGPR traits have been reported, e.g., ACC deaminase, phosphate, and potassium solubilization, exopolysaccharides, volatiles, and IAA production, etc. In this context, Dodd and Pérez-Alfocea (2012) reported that the hormonal root to shoot signaling in plants is altered by different rhizospheric microbes. Secretion of flavonoids by bean plants stimulated by IAA-producing PGPR regulates the nutrient uptake, nitrogen fixation and reduces unfavorable results of salt stress. The ACC-deaminase (ACC-D), is produced by PGPR that reduces the ethylene level under salt stress conditions through hydrolysis of ethylene precursor ACC (Choudhary et al. 2015).

Further, during saline conditions reduction of the chelation of excessive Na^+ and decrease in its bioavailability to the plants is maintained by the microbial secreted exopolysaccharides (EPSs) (Choudhary et al. 2015). PGPR increase the bioavailability of nutrients like phosphorus, potassium, zinc, and iron through secretion of different enzymes in soil for enhanced nutrients uptake by plants. In the past few years, the volatile organic compounds produced by specific bacterial strains were proved as advanced way of communal signaling between plants and PGPR, enhancing the plant growth through regulation of different biological processes including nutrient uptake, osmoprotectant biosynthesis, ion balance and distribution of hormones (Liu and Zhang 2015).

11.13 Mechanisms of PGPR for Salinity Stress Tolerance

Mechanisms employed by microbes which help in the improvement of plant growth and reduction of negative effects of salt stress are given below:

11.13.1 Osmotic Balance

Osmotic stress affects the plant growth that is partially recovered by osmolytes accumulation during salt stress. Salt ions gathering around the root creates water stress which results in an osmotic imbalance in plants. Photosynthetic structures and osmotic balance are dire to the salt stress so their maintenance proved crucial for the removal of unsafe effects of salinity on plant growth (Iqbal et al. 2014). PGPR not only improve plant–water relations through accumulation of osmolytes in the plants but enhanced synthesis of proline in the under abiotic stress through application of *Burkholderia*, *Arthrobacter*and *Bacillus* (Choudhary 2012). Vardharajula et al. (2011) reported the higher proportion of proline in the salt-stressed plants inoculated with *Bacillus* sp. Kumari et al. (2015) further confirmed that roots maintain osmotic balance because of accumulation of higher proline content, resulting in enhanced water uptake into the roots. Increased proline, as well as water content, was observed in the *Z. mays* inoculated with *Rhizobium* and *Pseudomonas* strains (Bano and Fatima 2009). Osmotic potential, antioxidants, relative water content, proline and endogenous phytohormones ABA and salicylic acid (SA) were found to be increased in *H. annus* due to PGPR under salt stress (Naz and Bano 2013; Naz and Bano 2015). The relation between accumulated proline and *pyrroline-5-carboxylate synthase (P5CS)* gene expression has been determined by the PGPR applications. The bacterial applications resulted in free proline accumulation in plant roots by upregulating *P5CS* gene expression (Kumari et al. 2015). In addition to proline, total soluble sugars increased in PGPR inoculated salt-stressed plants (Shukla et al. 2012).

Under salinity, water conductance from soil to the plant through roots is maintained by the rhizobacteria (Marulanda et al. 2007). During salt stress, curtailment of the relative water content indicated water stress. Bacterial exopolysaccharides play a primary role in providing resistance to the plants against drought stress in addition to other activities (Vardharajula et al. 2009).

Inoculation with *Pseudomonas putida* sp. *GAP-P45* in sunflower increased the rate of survival as well as its biomass and root adherence to the soil under water scarcity (Sandhya, et al. 2009). Earlier, Mhadhbi et al. (2004) reported *Sinorhizobium medicae, Mesorhizobium mediterraneum,* and *Mesorhizobium ciceri* potential to cope with salt stress and provide tolerance to the plants in salinity through increasing POD enzyme activity. *Lactuca sativa* seedlings were exposed to drought stress at different levels after being inoculated with *P. mendocina,* showed an increase in photosynthetic activity, biomass content and relative water content occurred due to increase in peroxidase (POD) and catalase (CAT) activities (Kohler et al. 2008). Rice plants were inoculated with salt-resistant *B. amyloliquefaciens NBRISN13 (SN13)* reported increase in the chlorophyll content, plant–water relations, root shoot lengths, and biomass by upregulating the catalase activity and proline content as a defensive mechanism (Nautiyal et al. 2013).

PGPR helps in the maintenance of transpiration rate and water conductance so as to regulate plant stomatal openings and water potential. In saline conditions, maize plants inoculated with *B. megaterium* exhibited an increase root water conductance in relation to the upregulation of two isoforms of plasma membrane aquaporin protein *(ZmPIP)* (Marulanda et al. 2010). PGPR genera supported osmolytes and different phytohormones storage after salinization to cope with initial osmotic shock. Genetically engineered *proBA* genes that were derived from *B.subtilis* in *A. thaliana* provided salt stress tolerance by upregulating the synthesis of proline (Chen et al. 2007). When the rice plants inoculated with salt-tolerant *B.amyloliquefaciens (SN13)* exposed to the saline environment in hydroponic and saline soil, the plants tolerance against stress was increased due to the altered expression of 14 genes, among them four genes *(NADP-malic enzyme NADP-Me2*, *somatic embryogenesis receptor-like kinase SERK1*, *ethylene responsive element binding proteins EREBP* and *SOS1*) were upregulated and the expression of 2 genes *(glucose insensitive growth GIG* and *serinethreonine protein kinase SAPK4*) was downregulated in hydroponic setup, while in greenhouse experiment only *MAPK5* was upregulated. *SN13* inoculation enhanced the gene synthesis that is involved in the osmotic and salt stress tolerance (Nautiyal et al. 2013). Carbohydrate synthesis and transport is enhanced by the beneficial microbes directly associated with the relationship between source and sink, rate of photosynthesis, biomass accumulation, and plant growth rate. Wheat seeds inoculated with *B. aquimaris* have shown an increase in the total soluble sugar content and reducing sugars resulting in higher biomass, sodium reduction and NPK accumulation (Upadhyay and Singh 2015).

In another study, it was observed that the inoculation of pepper plants with *Pantoea dispersa* and *A. brasilense* following salt stress resulted in increased dry mass storage in relation to stomatal conductance and photosynthesis, but no change in the chlorophyll content and photosynthesis was observed (del Amor and Cuadra-Crespo 2012).

Large amount of osmoprotectants used to store in the cytosol of microbes exhibiting in osmolality fluctuating environment (Kempf and Bremer 1998). The osmolytes synthesis and accumulation by PGPR under such conditions became faster. Under oxidative stress, the absorbance of compatible solutes through roots assisted them to impart resistance under salinity. Increased nodulation and better plant growth were observed in the bean (*Phaseolus vulgaris*) inoculated with *P. polymyxa* and *R. tropici* due to upregulation of *trehalose-6-phosphate* gene synthesis. Excessive trehalose synthesis acts as osmoprotectant and provide salt stress tolerance (Figueiredo et al. 2008).

11.13.2 Ion Homeostasis in Abiotic Stress

To control the ion uptake at the initial stages rather than spending a large amount of energy in damage repair and clearing excessive salts, plants cells have adaptive mechanisms to control the excessive salt by storing them in the large membrane-bound vacuoles. The most important factors limiting the plant growth and adolescence occurred due to the accumulation of excessive salts like Na^+ and Cl^- ions in the plant tissues. Glycophytes are known to have ion exclusion strategy as a tolerance mechanism in the salt stress. A group of rhizobacteria is known to remove the toxic ionic stress and improved development and growth under saline conditions (ref.). These microbes supported tolerance to salt stress by altering the ion transporters expression and by secretion of specialized biomacromolecules around roots, i.e., EPS that acts as a physical barrier between the roots and excessive salt. It was reported that the wheat plants inoculated with *Aeromonas hydrophila/caviae* and *Bacillus* sp. reduced uptake of Na^+ because of EPS secretion that clogs Na^+ by binding to the plant roots and preventing its transport to the leaves (Ashraf et al. 2004). In addition, the EPS producing *B. circulans* and *B. polymyxa* have potential in higher dry matter accumulation in wheat roots and shoots during the salt stress. Khodair et al. (2008) evide by due to the Bacterial volatile organic compounds (VOCs) also involved in Na^+ exchange pathways in PGPR-based ion mediated homeostasis. Bacterial VOCs are distant signaling low molecular weight compounds that help in signal transduction between two microorganisms. Exposure of *Arabidopsis* to *B. subtilis* volatile organic compounds (VOCs) exhibited salinity tolerance as compared to the unexposed plants by downregulating *AtHKT1* expression in roots and increasing expression in the shoots to facilitate root to root recirculation (Shkolnik-Inbar et al. 2013). The vegetative storage protein expression was increased through *P. simiae* AU-mediated VOCs expression in soybean leaves in relation to reduced Na^+ uptake under salinity stress by altering the sodium transport channels (Vaishnav et al. 2015).

Bacteria have ability to reduce the salt uptake by trapping cations with the help of rhizosheath in the EPS matrix and altering the ion exchange channels. Macro and micronutrients imbalance and mineral nutrients exchange caused by the sodium and chloride ions is controlled with the help of PGPR. Siderophore production by microbes, nutrients circulation and pH changes in the rhizospheric area facilitate bioavailability of minerals in the soil (Lugtenberg et al. 2013). In plant shoot, the Na/K ratio and ionic balance is maintained with the help of PGPR by decreasing the Na and Cl buildup in the leaves and increasing the Na removal from roots along with increasing the affinity K transporter channels. In maize plants K uptake and Na exclusion enhanced by inoculating with *Azotobacter* strains *C5* (auxin producing) and *C9*. Plant stress response observed to enhance increase in proline, chlorophyll and polyphenol contents in leaves after PGPR inoculation (Rojas-Tapias et al. 2012). Similarly, *B. subtilis GB03* inoculation in *P. tenuiflora* (a halophyte grass) showed less Na salt accumulation because of increased expression of *PtHKT1* and *PtSOS1* genes in roots but under salt stress the gene *PtHKT2* in roots was down-regulated (Niu et al. 2016).

11.13.3 Phytohormone Signaling and Salt Tolerance

Many soil bacteria secreted, modulated and altered the status of endogenous phytohormones by releasing exogenous enzymes, metabolites and hormones, contributed to enhance salinity tolerance. Plant–microbe associations are known to initiate signaling events to enhance the endogenous production of phytohormones and several other metabolites under stress conditions (Dodd et al. 2010).

A number of studies elaborated the roles of abscisic acid (ABA), auxin (IAA), cytokinins (CK), ethylene and gibberellic acid (GA) under salinity stress. These hormones alter morphology, metabolism, water uptake, and nutrient acquisition efficiency of plants and consequently result in the larger, healthier and sturdy plants.

11.13.3.1 Abscisic Acid (ABA)

Abscisic acid (ABA) mainly cuts off the leaf and shoot growth of the plant, but recent research evidenced that the elevated concentrations of ABA in stressed plants suppress the excess production of ethylene. Mainly the raised ABA level generate an adaptive response that is fundamental for plant endurance (Pliego et al. 2011).

The aspects of exogenously applied ABA in plant–microbe associations under salinity stress and the role of the bacterial ABA. In salt-stressed plants, PGPR regulated the synthesis of ABA and its signaling pathways that resulted in the heightened plant growth. Halotolerant *Dietzia natronolimnaea* (*STR1*) persuaded the salt stress (150 *mM NaCl*) resistance system in wheat plants through regulation of ABA signaling course via upregulation of *ABA-reactive gene (TaABARE)* and *12-oxophytodienoate reductase 1 (TaOPR1)* activating *TaWRKY* and *TaMYB* input signal, resulting trailed due to expression of stress-sensitive genes including upregulation of *TaST* (a salt stress instigated gene). In PGPR-inoculated plants, the enhanced antioxidant enzyme gene expression and proline content provided maximum resistance against salt stress (Bharti et al. 2016). Under salinity *Cucumis sativus* plants inoculated with *Acinetobacter calcoaceticus* (*SE370*), *Promicromonospora* sp. *SE188* and *Burkholderia cepacia* (*SE4*) resulted in greater biomass. PGPR reduced electrolyte leakage and increased water potential and, also caused ABA upregulation and decrease in salicylic acid and gibberellin *GA4* contents (Kang et al. 2014).

Decreased ABA accumulation and increased plant biomass in salinized soil were observed in cottonseed inoculated with *P. putida* (*Rs-198*). However, the enhanced salt resistance likewise is credited to control ionic balance and enhanced endogenous IAA content (Yao et al. 2010). Under salinity stress conditions, wheat inoculation with *B. subtilis (LDR2)* and *Arthrobacter protophormiae* (*SA3*) resulted in enhanced IAA content and decreased ACC and ABA contents. The upregulation of *Serine/Threonine protein kinase–ethylene responsive* (*TaCTR1*) and *drought responsive element (TaDRE2)* genes exerted an improved impact and was additionally approved (Barnawal et al. 2017).

11.13.3.2 Auxin (Indole-3-Acetic Acid)

Biosynthesis of auxin occurs by means of complex pathways in rhizobacteria involved utilization of tryptophan secretion from root exudates and its into IAA (Spaepen et al. 2007). Upgraded plant cell growth occurred due to activated auxin signaling pathway and endogenous IAA pool in plants. In plant–microbe associations, the bacterial signaling molecule is IAA produced by PGPR. Under ideal conditions of growth, gaining of bacterial IAA in triggered, repressed or impartial plant growth (Dodd et al. 2010; Spaepen and Vanderleyden 2011). Salinity in soil influenced cell elongation due to the accumulation of IAA in roots. This accumulated IAA is likewise expected to hinder the synthesis and transport of cytokinin from root to shoot and brings in increased root lengthening under pressure conditions (Dodd et al. 2005). Enhanced salt stress tolerance was observed in maize when inoculated with *B. amyloliquefaciens* (*SQR9*). It was supported by increased total soluble sugar and chlorophyll contents, enhanced catalase and peroxidase activity, improved KC/NaC ratio, and glutathione content. The negative impacts of 50 *mM NaCl* on *Phaseolus vulgaris* grown under hydroponic conditions were found to be alleviated by *Azospirillum brasilense* application because of higher accumulation of flavonoids as well as improved proliferation of roots (Dardanelli et al. 2008). Improved germination and development of corn and soybean was found when an IAA-producing co-inoculations of *Bradyrhizobium japonicum* (*E109*) and *Azospirillum brasilense* (*Az39*) were done (Cassán et al. 2009, 2014).

Research revealed that plants displayed higher root and leaf development inoculated with IAA-producing bacterial strains and considered as an adaptive response to salinity (Albacete et al. 2008). Under hydroponic conditions, PGPR strains producing IAA documented for enhanced efficiency of nutrient uptake (Shukla et al. 2012). Recent research avowed that GB03 activated the plant growth by cell wall releasing enzymes and auxin homeostasis as well (Zhang et al. 2007). Interestingly, the reduced effects of salt stress in the PGPR inoculated seedlings were affirmed by the upregulation of genes of *RuBisCo* encoding subunits *(RBCS, RBCL)*, *HC encoding pumping pyrophosphatase (H(C)-Ppase)*, *NHX2*, *NHX3*, *NHX1*, and *HKT1*, and downregulation of gene encoding *9-cisepoxycarotenoid dioxygenase* (NCED expression) (Chen et al. 2016).

11.13.3.3 Ethylene

In plants ethylene, a stress hormone synthesized under stress conditions. Rhizobial nodulation of legumes is also inhibited by ethylene. However, low production under normal condition soccurred due to adjustment of some physiological reactions, for instance, breaking of seed dormancy (Dodd et al. 2005). Yang cycle in plants is involved in ethylene synthesis. In this process ACC oxidase catalyze the conversion of ACC into ethylene. In stress conditions, transportation of ACC to a specific stressed organ resulted in ethylene production in that particular organ (Yoon and Kieber 2013). Under salt stress, growth of tomato is reduced due to the Na^+ build up and it is associated with elevated foliar ethylene (Mayak et al. 2004a). In another study, Albacete et al. (2008) reported that the application of salt, elevated the ACC levels along with the Na^+ accumulation in the leaf, xylem sap, and root. This phenomenon is associated with initiating the oxidative stress and decrease in photosynthetic capacity indicating ethylene's role in foliar senescence. In several studies, there is a decrease in ethylene quantity in stressed plants as numerous rhizobacteria contain ACC-deaminase enzyme, which splits ACC into ammonia and α-ketobutyrate. As stated in the reported model, plant roots are known to release and then hydrolyze ACC via ACC-D mechanism as ACC-D producing bacteria gets attached to the plant roots and absorb the released ACC (Glick et al. 1998). Consequently, more ACC is exuded from roots to keep the balanced ACC level inside the plant cell. Few reports revealed that PGPR containing ACC-D reduce salt stress (Wu et al. 2012). *Achromobacter piechaudii (ARV8)* inoculated strain under salt stress in tomato produce ACC-D which in turn has remarkably increased the efficacy of nutrient uptake and plant weights (Mayak et al. 2004a).

Researchers reported that improved nodulation in mung bean can be done by applying the ACC-D co-inoculated with *Bradyrhizobium* as it can bring down ethylene synthesis compared to the *Bradyrhizobium* alone (Shaharoona et al. 2006a). Red pepper under 150 *mM NaCl* stress increased the ethylene production that has been accounted for to decrease by ACC-D producing *B. iodinum (RS16)*, *Z. alba (RS111)* and *B. licheniformis (RS656)*. (Siddikee et al. 2011). Additionally, plants treated with PGPR strains those have capability to produce ACC-D showed even more root nodules, and improved plant growth and productivity under oxidative stress conditions (Zafar-ul-Hye et al. 2013).

Under saline conditions, up to 90 *mM NaCl* tomato seedlings inoculated with *P. putida (UW4)* exhibited enhanced shoot development in *Pisum sativum* cv. Alderman. Inoculation with *Variovorax paradoxus (5C-2)* reported to have positive effects under salt stress conditions. In saline soil, ACC-D producing *Enterobacter* sp. and *P. fluorescens* enhanced maize productivity. In inoculated plants under salinity stress elevated KC/NaCl proportion and NPK uptake were also recorded (Nadeem et al. 2009). *Pantoea dispersa (PSB3),* local bacterium remarkable improvement in pod number, biomass, seed weight, seed number and pod weight under salt stress. The enhanced salt resilience was related to critical decrease in electrolyte leakage and Na^+ uptake while increase in KC uptake, chlorophyll and relative water contents (Panwar et al. 2016).

11.13.3.4 Cytokinins

Cell division and differentiation are critically regulated by auxin and cytokinin proportion. Some PGP bacteria synthesize rhizobacteria is cytokinin generation (Dodd et al. 2010; García de Salamone et al. 2001; Aslantaş et al. 2007; Kudoyarova et al. 2014). The biomass content within the lettuce plant shoot was found to be raised by the application of CK-mediated *B. subtilis* under drought stress (Arkhipova et al. 2007). The role of CK signaling pathway in the plant development was anticipated by observing the increased CK levels in *A. thaliana* roots as a result of the application of cytokinin-producing bacterium *B. megaterium* (Ortíz-Castro et al. 2008). Giraud et al. (2007) reported that in the absence of nod factor in soybean plants the nodulation process occurred with the help of CKs after inoculating plants with PGPR strain *Bradyrhizobium*.

11.13.3.5 Gibberellins

Inoculation of wheat and soybean by GA-producing PGP strains of various genera involved in their growth and Gibberellins development under salt stress (Cassán et al. 2014). According to another study, tomato plant growth was enhanced by GA-producing *Promicromonospora* sp. *SE188*, revealed in the form of higher biomass and shoot length (Kang et al. 2012). Kang et al. (2014) also reported that GA-producing *P. putida H-2-3* essentially improved the plant weight, length, and chlorophyll substance in GA-deficient mutant soybean plants.

It is reported that PGPR isolates *B. cepacia*, *A. calcoaceticus*, and *Promicromonospora* sp. exerted positive effects on cucumber plants growth under salinity and drought conditions induced by sodium chloride and polyethylene glycol treatments (Kang et al. 2014). *Enterobacter cloacae*, P6 and *Bacillus drentensis*, P16 were used in combination with foliar application of silicon to maintain yield and growth of mung bean plants under saline conditions. The combination produced good results (Mahmood et al. 2016).

11.14 Impact of PGPR on Crop Growth Under Salinity

Researchers observed that PGPR enhance plant growth and have ability to promote their salinity tolerance levels, Besides the improvement in physiological responses and antioxidant potential. Ability of PGPR activated indole-3-acetic acid (IAA), gibberelins (GA), cytokinins (CK), abscisic acid (ABA), ethylene, and cofactor pyrroloquinoline quinone (PQQ), helpful in promoting the plant growth (Perrig et al. 2007). When spermidine introduced against saline stress as a new defender molecule produced by *S. rhizophile*, induced plant–microbe interaction process (Alavi et al. 2013). In soybean seedlings, IAA level decreases during stress conditions but by PGPR inoculation, major improvement was observed in the level of hormones increased. The morphological changes in the roots improved the growth of shoots and ultimately increasing the yield (Asim et al. 2013). Isolation of halotolerant bacteria and their characterization by a high proline production with other plant hormones like trans-zeatin riboside, ABA, IAA and GA3 and the bacterial inoculation into the soybean plant improved length of roots shoots, and dry mass under salinity stress.

An osmotolerant PGPR named as *A. brasilense NH* isolated from the salt-affected soil in Algeria near Mediterranean coast, positively enhanced the growth of durum wheat in salinity effected soil (Nabti et al. 2007, 2010). To produce IAA at very high salt concentration, the strain was found effective. IAA played important role in providing resistance to plants against salt stress and in growth promotion of plant (Kang et al. 2014; Khalid et al. 2013; Kaya et al. 2010, 2013).

A variety of halophilic bacterial strains belong to the genus *Halomonas* isolated from the root zone of Salicornia plants, *Halomonas* spp. exhibited PGP characters under high salt concentration specifically due to IAA production(Mapelli et al. 2013). The effect of *B. cereus* and *Pseudomonas* sp., in wheat cultivation, revealed that osmotolerant bacteria have a direct and indirect effect on seed germination of paddy under saline conditions (Jha and Subramanian 2013). Thus may be due to important role of bacteria in upregulating IAA that helps in seed germination. PGPR through the enhancement of IAA under salinity stress cause reduction in toxin uptake results in the improved plant growth (Chakraborty et al. 2011; Zhang et al. 2008).

Hormones such as ABA, IAA, and GA3 ameliorated high concentration of salt stress after treatment with halotolerant strains like *P. extremorientalis* and *P. chlororaphis* resulted in improved growth of staple crops such as common bean (*Phaseolus vulgaris*) (Jha and Subramanian 2013).

11.15 Conclusions

Drought and salinity are major problems faced by the agricultural world. PGPR are the best alternatives to the chemical and traditional methods used against these problems. PGPR are proved to be efficient in increasing plant growth, yield and stress tolerance levels. They are non-pollutants and methods for their application are not costly, moreover, PGPR can be applied on plants at different life stages such as treatment of seeds with PGPR before planting or after emergence. Development of the bioinoculum is beneficial to promote economically important crops grown in arid and semiarid and salt-affected regions of the world. Extensive research on PGPR is needed to explore mechanisms of their beneficial effects on different cash and food crops along with searching for suitable methods of PGPR application according to crop types.

References

Abebe T, Guenzi AC, Martin B, Cushman JC (2003) Tolerance of mannitol-accumulating transgenic wheat to water stress and salinity. Plant Physiol 131(4)L:1748–1755

Achard P, Cheng H, De Grauwe L, Decat J, Schoutteten H, Moritz T (2006) Integration of plant responses to environmentally activated phytohormonal signals. Science 311(57):91–94

Afzal Z, Howton TC, Sun Y, Mukhtar MS (2016) The roles of aquaporins in plant stress responses. J Dev Biol 1:9

Agarwal M, Dheeman S, Dubey RC, Kumar P, Maheshwari DK, Bajpai VK (2017) Differential antagonistic responses of *Bacillus pumilus* MSUA3 against *Rhizoctonia solani* and *Fusarium oxysporum* causing fungal diseases in *Fagopyrum esculentum* Moench. Microbiol Res 205:40–47

Ahmad F, Ahmad I, Khan MS (2008) Screening of free-living rhizospheric bacteria for their multiple plant growth promoting activities. Microbiol Res 163(2):173–181

Ahmad M, Zahir ZA, Khalid M, Nazli F, Arshad M (2013a) Efficacy of *Rhizobium* and *Pseudomonas* strains to improve physiology, ionic balance and quality of mung bean under salt-affected conditions on farmer's fields. Plant Physiol Biochem 63:170–176

Ahmad M, Zahir ZA, Nazli F, Akram F, Arshad M, Khalid M (2013b) Effectiveness of halo-tolerant, auxin producing *Pseudomonas* and *Rhizobium* strains to improve osmotic stress tolerance in mung bean (*Vigna radiata* L.). Braz J Microbiol 44(4):1341–1348

Ahmad P, Wani MR, Azooz MM, Tran LSP (2014) Improvement of crops in the era of climatic changes, vol 2. Springer Internation Publishing, p 397

Alavi P, Starcher M, Zachow C, Müller H, Berg G (2013) Root-microbe systems: the effect and mode of interaction of Stress Protecting Agent (SPA) *Stenotrophomonas rhizophila* DSM14405T. Front Plant Sci 4:141

Albacete A, Ghanem ME, Martínez-Andújar C, Acosta M, Sánchez-Bravo J, Martínez V, Lutts S, Dodd IC, Pérez-Alfocea F (2008) Hormonal changes in relation to biomass partitioning and shoot growth impairment in salinized tomato (*Solanum lycopersicum* L.) plants. J Exp Bot 59(15):4119–4131

Alexandratos N, Bruinsma J (2012) World agriculture towards 2030/2050: the 2012 revision, vol 12, no. 3. ESA Working paper, Rome, FAO

Ali J, Xu JL, Gao YM, Ma XF, Meng LJ, Wang Y (2017) Harnessing the hidden genetic diversity for improving multiple abiotic stress tolerance in rice (*Oryza sativa* L.). Plos One 12(3): e0172515

Ali Y, Aslam Z, Sarwar G, Hussain F (2005) Genotypic and environmental interaction in advanced lines of wheat under salt-affected soils environment of Punjab. Int J Environ Sci Technol 2 (3):223–228

Ansary MH, Rahmani HA, Ardakani MR, Paknejad F, Habibi D, Mafakheri S (2012) Effect of Pseudomonas fluorescens on proline and phytohormonal status of maize (*Zea mays* L.) under water deficit stress. Ann Biol Res 3:1054–1062

Arkhipova TN, Prinsen E, Veselov SU, Martinenko EV, Melentiev AI, Kudoyarova GR (2007) Cytokinin producing bacteria enhance plant growth in drying soil. Plant Soil 292(1–2):305–315

Arora NK, Kim MJ, Kang SC, Maheshwari DK (2007) Role of chitinase and β-1, 3-glucanase activities produced by a fluorescent pseudomonad and in vitro inhibition of *Phytophthora capsici* and *Rhizoctonia solani*. Can J Microbiol 53(2):207–212

Arshad M, Shaharoona B, Mahmood T (2008) Inoculation with *Pseudomonas* spp. containing ACC-deaminase partially eliminates the effects of drought stress on growth, yield, and ripening of pea (*Pisum sativum* L.). Pedosphere 18(5):611–620

Arzanesh MH, Alikhani HA, Khavazi K, Rahimian HA, Miransari M (2011) Wheat (*Triticum aestivum* L.) growth enhancement by *Azospirillum* sp. under drought stress. World J Microbiol Biotechnol 27(2):197–205

Ashraf M, Berge SH, Mahmood OT (2004) Inoculating wheat seedling with exopolysaccharides-producing bacteria restrict sodium uptake and stimulates plant growth under salt stress. Biol Fertil Soil 40:157–162

Ashraf M (2010) Inducing drought tolerance in plants: recent advances. Biotechnol Adv 28(1): 169–183
Ashraf MA, Ashraf M, Shahbaz M (2012) Growth stage-based modulation in antioxidant defense system and proline accumulation in two hexaploid wheat (*Triticum aestivum* L.) cultivars differing in salinity tolerance. Flora-Morphol Distrib Func Ecol Plant 207(5):388–397
Ashraf M, Foolad MR (2007) Roles of glycine betaine and proline in improving plant abiotic stress resistance. Environ Exp Bot 59(2):206–216
Ashraf M, Foolad MR (2013) Crop breeding for salt tolerance in the era of molecular markers and marker-assisted selection. Plant Breed 132(1):10–20
Ashraf M, Athar HR, Harris PJC, Kwon TR (2008) Some prospective strategies for improving crop salt tolerance. Adv Agron 97:45–110
Asim M, Aslam M, Bano A, Munir M, Majeed A, Abbas SH (2013) Role of phytohormones in root nodulation and yield of soybean under salt stress. Am J Res Commun 1:191–208
Aslantaş R, Cakmakçi R, Şahin F (2007) Effect of plant growth promoting rhizobacteria on young apple tree growth and fruit yield under orchard conditions. Sci Horticul 111(4):371–377
Athar HR, Ashraf M (2009) Strategies for crop improvement against salinity and drought stress: an overview. Salinity and water stress. Springer, Dordrecht, pp 1–16
Bano A, Fatima M (2009) Salt tolerance in *Zea mays* (L.) following inoculation with *Rhizobium* and *Pseudomonas*. Biol Fertil Soils 45:405–413
Bano Q, Ilyas N, Bano A, Zafar N, Akram A, Hassan F (2013) Effect of *Azospirillum* inoculation on maize (*Zea mays* L.) under drought stress. Pak J Bot 45(S1):13–20
Barka EA, Nowak J, Clément C (2006) Enhancement of chilling resistance of inoculated grapevine plantlets with a plant growth-promoting rhizobacterium *Burkholderia phytofirmans* strain PsJN. Appl Environ Microbiol 72:7246–7252
Barnawal D, Bharti N, Pandey SS, Pandey A, Chanotiya CS, Kalra A (2017) Plant growth-promoting rhizobacteria enhance wheat salt and drought stress tolerance by altering endogenous phytohormone levels and *TaCTR1/TaDREB2* expression. Physiol Plant 161:502–514
Basak BB, Biswas DR (2010) Co-inoculation of potassium solubilizing and nitrogen fixing bacteria on solubilization of waste mica and their effect on growth promotion and nutrient acquisition by a forage crop. Biol Fer Soil 46(6):641–648
Basu S, Ramegowda V, Kumar A, Pereira A (2016) Plant adaptation to drought stress. F1000Res. 5:F1000 Faculty Rev-1554
Beneduzi A, Ambrosini A, Passaglia LM (2012) Plant growth-promoting rhizobacteria (PGPR): their potential as antagonists and biocontrol agents. Genet Mol Biol 35(4):1044–1051
Bharti N, Pandey SS, Barnawal D, Patel VK, Kalra A (2016) Plant growth promoting rhizobacteria Dietzia natronolimnaea modulates the expression of stress responsive genes providing protection of wheat from salinity stress. Sci Rep 6:34768. https://doi.org/10.1038/srep34768
Bhattacharyya PN, Jha DK (2012) Plant growth-promoting rhizobacteria (PGPR): emergence in agriculture. W J Microbiol Biotechnol 28(4):1327–1350
Bindu GH, Selvakuma G, Shivashankara KS, Kumar NS (2018) Osmotolerant plant growth promoting bacterial inoculation enhances the antioxidant enzyme levels of tomato plants under water stress conditions. Int J Curr Microbiol Appl Sci 7(1):2824–2833
Binzel ML, Hasegawa PM, Rhodes D, Handa S, Handa AK, Bressan RA (1987) Solute accumulation in tobacco cells adapted to NaCl. Plant Physiol 84(4):1408–1415
Blilou I, Xu J, Wildwater M, Willemsen V, Paponov I, Friml J, Scheres B (2005) The PIN auxin efflux facilitator network controls growth and patterning in *Arabidopsis* roots. Nature 433 (7021):39
Blum A (2005) Drought resistance, water-use efficiency, and yield potential—are they compatible, dissonant, or mutually exclusive? Aust J Agri Res 56(11):1159–1168
Blum A (2014) Genomics for drought resistance–getting down to earth. Func Plant Biol 41(11): 1191–1198
Boiero L, Perrig D, Masciarelli O, Penna C, Cassán F, Luna V (2007) Phytohormone production by three strains of *Bradyrhizobium japonicum* and possible physiological and technological implications. Appl Microbiol Biotechnol 74(4):874–880

Bresson J, Varoquaux F, Bontpart T, Touraine B, Vile D (2013) The PGPR strain *Phyllobacterium brassicacearum* STM196 induces a reproductive delay and physiological changes that result in improved drought tolerance in *Arabidopsis*. New Phytol 200(2):558–569

Butt HI, Yang Z, Gong Q, Chen E, Wang X, Zhao G, Li F (2017) GaMYB85 an R2R3 MYB gene, in transgenic Arabidopsis plays an important role in drought tolerance. BMC Plant Biol 17(1): 142

Casanovas EM, Barassi C A, Sueldo RJ (2002) *Azospirillum* inoculation mitigates water stress effects in maize seedlings. Cereal Res Commun 343–350

Cassán F, Perrig D, Sgroy V, Masciarelli O, Penna C, Luna V (2009) *Azospirillum brasilense* Az39 and *Bradyrhizobium japonicum* E109, inoculated singly or in combination, promote seed germination and early seedling growth in corn (*Zea mays* L.) and soybean (*Glycine max* L.). Eur J Soil Biol 45(1):28–35

Cassán F, Vanderleyden J, Spaepen S (2014) Physiological and agronomical aspects of phytohormone production by model plant-growth-promoting rhizobacteria (PGPR) belonging to the genus *Azospirillum*. J Plant Growth Regul 33:440–459

Castillo P, Escalante M, Gallardo M, Alemano S, Abdala G (2013) Effects of bacterial single inoculation and co-inoculation on growth and phytohormone production of sunflower seedlings under water stress. Acta Physiol Plant 35(7):2299–2309

Chakraborty U, Roy S, Chakraborty AP, Dey P, Chakraborty B (2011) Plant growth promotion and amelioration of salinity stress in crop plants by a salt-tolerant bacterium. Recent Res Sci Technol 3:11

Chang P, Gerhardt KE, Huang XD, Yu XM, Glick BR, Gerwing PD, Greenberg BM (2014) Plant growth-promoting bacteria facilitate the growth of barley and oats in salt-impacted soil: implications for phytoremediation of saline soils. Int J Phytoremed 16(11):1133–1147

Chartzoulakis K, Klapaki G (2000) Response of two greenhouse pepper hybrids to NaCl salinity during different growth stages. Sci Horticul 86(3):247–260

Chauhan AK, Maheshwari DK, Kim K, Bajpai VK (2016) Termitarium-inhabiting *Bacillus endophyticus* TSH42 and *Bacillus cereus* TSH77 colonizing *Curcuma longa* L.: isolation, characterization, and evaluation of their biocontrol and plant-growth-promoting activities. Can J Microbiol 62(10):880–892

Chaves MM, Oliveira MM (2004) Mechanisms underlying plant resilience to water deficits: prospects for water-saving agriculture. J Exp Bot 55(407):2365–2384

Chaves MM, Flexas J, Pinheiro C (2009) Photosynthesis under drought and salt stress: regulation mechanisms from whole plant to cell. Annal Bot 103(4):551–560

Chen TH, Murata N (2008) Glycinebetaine, an effective protectant against abiotic stress in plants. Trend Plant Sci 13:499–505

Chen H, Li Z, Xiong L (2012) A plant microRNA regulates the adaptation of roots to drought stress. Febs Lett 586(12):1742–1747

Chen L, Liu Y, Wu G, Veronican Njeri K, Shen Q, Zhang N, Zhang R (2016) Induced maize salt tolerance by rhizosphere inoculation of *Bacillus amyloliquefaciens* SQR9. Physiol Plant 158 (1):34–44

Chen M, Wei H, Cao J, Liu R, Wang Y, Zheng C (2007) Expression of *Bacillus subtilis* proBA genes and reduction of feedback inhibition of proline synthesis increases proline production and confers osmotolerance in transgenic *Arabidopsis*. J Biochem Mol Biol 40:396–403

Chen X, Wang Y, Lv B, Li J, Luo L, Lu S, Ming F (2014) The NAC family transcription factor OsNAP confers abiotic stress response through the ABA pathway. Plant Cell Physiol 55(3): 604–619

Cho UH, Park JO (2000) Mercury-induced oxidative stress in tomato seedlings. Plant Sci 156(1): 1–9

Choudhary DK (2012) Microbial rescue to plant under habitat-imposed abiotic and biotic stresses. Appl Microbiol Biotechnol 96(5):1137–1155

Choudhary DK, Kasotia A, Jain S, Vaishnav A, Kumari S, Sharma KP, Varma A (2015) Bacterial-mediated tolerance and resistance to plants under abiotic and biotic stresses. J Plant Growth Regul 35:276–300

Choudhary DK, Kasotia A, Jain S, Vaishnav A, Kumari S, Sharma KP, Varma A (2016) Bacterial-mediated tolerance and resistance to plants under abiotic and biotic stresses. J Plant Growth Regul 35(1):276–300

Chowdhury SP, Hartmann A, Gao X, Borriss R (2015) Biocontrol mechanism by root-associated *Bacillus amyloliquefaciens* FZB42–a review. Front Microbiol 6:780

Claeys H, Inzé D (2013) The agony of choice: how plants balance growth and survival under water-limiting conditions. Plant Physiol 113

Cohen AC, Bottini R, Piccoli PN (2008) *Azospirillum brasilense* Sp 245 produces ABA in chemically-defined culture medium and increases ABA content in arabidopsis plants. Plant Growth Regul 54(2):97–103

Cohen AC, Bottini R, Pontin M, Berli FJ, Moreno D, Boccanlandro H, Piccoli PN (2015) *Azospirillum brasilense* ameliorates the response of *Arabidopsis thaliana* to drought mainly via enhancement of ABA levels. Physiol Plant 153(1):79–90

Cohen AC, Travaglia CN, Bottini R, Piccoli PN (2009) Participation of abscisic acid and gibberellins produced by endophytic *Azospirillum* in the alleviation of drought effects in maize. Bot 87(5):455–462

Cominelli E, Conti L, Tonelli C, Galbiati M (2013) Challenges and perspectives to improve crop drought and salinity tolerance. New Biotechnol 30(4):355–361

Compant S, Reiter B, Sessitsch A, Nowak J, Clément C, Barka EA (2005) Endophytic colonization of *Vitis vinifera* L. by plant growth-promoting bacterium *Burkholderia* sp. strain PsJN. Appl Environ Microbiol 71(4):1685–1693

Creus CM, Sueldo RJ, Barassi CA (2004) Water relations and yield in *Azospirillum*-inoculated wheat exposed to drought in the field. Can J Bot 82:273–281

Dar ZM, Masood A, Mughal AH, Asif M, Malik MA (2018) Review on Drought Tolerance in Plants Induced by Plant Growth Promoting Rhizobacteria. Int J Curr Microbiol Appl Sci 7(5):1. https://www.ijcmas.com/7-5-2018/Zaffar%20Mahdi%20Dar,%20et%20al.pdf

Dardanelli MS, Fernández de Córdoba FJ, Espuny MR, Rodríguez Carvajal MA, Soria Díaz ME, Gil Serrano AM, Okon Y, Megías M (2008) Effect of *Azospirillum brasilense* coinoculated with *Rhizobium* on *Phaseolus vulgaris* flavonoids and Nod factor production under salt stress. Soil Biol Biochem 40:2713–2721

Das K, Roychoudhury A (2014) Reactive oxygen species (ROS) and response of antioxidants as ROS-scavengers during environmental stress in plants. Front Environ Sci 2:53

Davies PJ (2013) Plant hormones: physiology, biochemistry and molecular biology. Springer Science & Business Media, p 796

De Micco V, Aronne G (2009) Seasonal dimorphism in wood anatomy of the Mediterranean *Cistus incanus* L. subsp. incanus. Trees 23(5):981–989

del Amor FM, Cuadra-Crespo P (2012) Plant growth-promoting bacteria as a tool to improve salinity tolerance in sweet pepper. Funct Plant Biol 39:82–90

Demmig-Adams B, Adams WW (2006) Photoprotection in an ecological context: the remarkable complexity of thermal energy dissipation. New Phytol 172(1):11–21

Dimkpa C, Weinand T, Asch F (2009) Plant–rhizobacteria interactions alleviate abiotic stress conditions. Plant, Cell Environ 32(12):1682–1694

Dinesh R, Srinivasan V, Hamza S, Sarathambal C, Anke Gowda SJ, Ganeshamurthy AN, Divya VC (2018) Isolation and characterization of potential Zn solubilizing bacteria from soil and its effects on soil Zn release rates, soil available Zn and plant Zn content. Geoderma 321:173–186

Divya B, Kumar MD (2011) Plant-microbe interaction with enhanced bioremediation. Res J Biotechnol 6(1):72–79

Dodd IC, Belimov AA, Sobeih WY, Safronova VI, Grierson D, Davies WJ (2005) Will modifying plant ethylene status improve plant productivity in water limited environments? In: 4th International crop science congress

Dodd IC, Zinovkina NY, Safronova VI, Belimov AA (2010) Rhizobacterial mediation of plant hormone status. Ann Appl Biol 157(3):361–379

Dodd IC, Pérez-Alfocea F (2012) Microbial amelioration of crop salinity stress. J Exp Bot 63 (9):3415–3428

Du H, Wang N, Cui F, Li X, Xiao J, Xiong L (2010) Characterization of a β-carotene hydroxylase gene DSM2 conferring drought and oxidative stress resistance by increasing xanthophylls and ABA synthesis in rice. Plant Physiol 110

Egamberdieva D (2013) The role of phytohormone producing bacteria in alleviating salt stress in crop plants. Biotechnological techniques of stress tolerance in plants. Biotechnological techniques of stress tolerance in plants. Studium, Houston, TX, pp 21–39

El-Hendawy SE, Hu Y, Yakout GM, Awad AM, Hafiz SE, Schmidhalter U (2005) Evaluating salt tolerance of wheat genotypes using multiple parameters. Eur J Agr 22(3):243–253

Elnaggar AA, Noller JS (2009) Application of remote-sensing data and decision-tree analysis to mapping salt-affected soils over large areas. Remote Sens 2(1):151–165

Fan X, Hu H, Huang G, Huang F, Li Y, Palta J (2015) Soil inoculation with Burkholderia sp. LD-11 has positive effect on water-use efficiency in inbred lines of maize. Plant Soil 390 (1–2): 337–349

Farooq M, Basra SMA, Wahid A, Cheema ZA, Cheema MA, Khaliq A (2008) Physiological role of exogenously applied glycinebetaine to improve drought tolerance in fine grain aromatic rice (Oryza sativa L.). J Agr Crop Sci 194(5):325–333

Figueiredo MVB, Burity HA, Martinez CR, Chanway CP (2008) Alleviation of drought stress in the common bean (*Phaseolus vulgaris* L.) by coinoculation with *Paenibacillus polymyxa* and *Rhizobium tropici*. Appl Soil Ecol 40:182–188

Flowers TJ (2004) Improving crop salt tolerance. J Exp Botany 55(396):307–319

Foyer CH, Rasool B, Davey JW, Hancock RD (2016) Cross-tolerance to biotic and abiotic stresses in plants: a focus on resistance to aphid infestation. J Exp Bot 67(7):2025–2037

Fukao T, Xu K, Ronald PC, Bailey-Serres J (2006) A variable cluster of ethylene response factor–like genes regulates metabolic and developmental acclimation responses to submergence in rice. Plant Cell 18(8):2021–2034

Gao S, Ouyang C, Wang S, Xu Y, Tang L, Chen F (2008) Effects of salt stress on growth, antioxidant enzyme and phenylalanine ammonia-lyase activities in *Jatropha curcas* L. seedlings. Plant Soil Environ 54(9):374–381

García de Salamone IE, Hynes RK, Nelson LM (2001) Cytokinin production by plant growth promoting rhizobacteria and selected mutants. Can J Microbiol 47(5):404–411

Ghassemi F, Jakeman AJ, Nix HA (1995) Salinisation of land and water resources: human causes, extent, management and case studies. CAB International, p 544

Ghorbanpour M, Hatami M, Khavazi K (2013) Role of plant growth promoting rhizobacteria on antioxidant enzyme activities and tropane alkaloid production of Hyoscyamus niger under water deficit stress. Turkish J Biol 37(3):350–360

Gilroy S, Suzuki N, Miller G, Choi WG, Toyota M, Devireddy AR, Mittler R (2014) A tidal wave of signals: calcium and ROS at the forefront of rapid systemic signaling. Trend Plant Sci 19 (10):623–630

Giraud E, Moulin L, Vallenet D, Barbe V, Cytryn E, Avarre JC, Bena G (2007) Legumes symbioses: absence of Nod genes in photosynthetic bradyrhizobia. Science 316(5829):1307–1312

Glick BR (2007) Promotion of plant growth by bacterial ACC deaminase. Crit Rev Plant Sci 26:227–242

Glick BR, Pasternak JJ (2003) Plant growth promoting bacteria. In: Molecular biology-principles and applications of recombinant DNA, pp 436–454

Glick BR, Penrose DM, Li J (1998) A model for the lowering of plant ethylene concentrations by plant growth-promoting bacteria. J Theor Biol 190(1):63–68

Goswami D, Thakker JN, Dhandhukia PC (2016) Portraying mechanics of plant growth promoting rhizobacteria (PGPR): a review. Cogent Food Agri 2(1):1127500

Goteti PK, Emmanuel LDA, Desai S, Shaik MHA (2013) Prospective zinc solubilising bacteria for enhanced nutrient uptake and growth promotion in maize (*Zea mays* L.). Int J Microbiol 2013: Article ID 869697

Gouda S, Kerry RG, Das G, Paramithiotis S, Shin HS, Patra JK (2017) Revitalization of plant growth promoting rhizobacteria for sustainable development in agriculture. Microbio Res 206:131–140

Gray EJ, Smith DL (2005) Intracellular and extracellular PGPR: commonalities and distinctions in the plant–bacterium signaling processes. Soil Biol Biochem 37(3):395–412

Grover M, Madhubala R, Ali SZ, Yadav SK, Venkateswarlu B (2014) Influence of *Bacillus* spp. strains on seedling growth and physiological parameters of sorghum under moisture stress conditions. J Basic Microbiol 54(9):951–961

Gunes A, Inal A, Alpaslan M, Cicek N, Guneri E, Eraslan F, Guzelordu T (2005) Effects of exogenously applied salicylic acid on the induction of multiple stress tolerance and mineral nutrition in maize (*Zea mays* L.). Arch Agr Soil Sci 51(6):687–695

Gunes A, Inal A, Alpaslan M, Eraslan F, Bagci EG, Cicek N (2007) Salicylic acid induced changes on some physiological parameters symptomatic for oxidative stress and mineral nutrition in maize (*Zea mays* L.) grown under salinity. J Plant Physiol 164(6):728–736

Gururani MA, Upadhyaya CP, Baskar V, Venkatesh J, Nookaraju A, Park SW (2013) Plant growth-promoting rhizobacteria enhance abiotic stress tolerance in *Solanum tuberosum* through inducing changes in the expression of ROS-scavenging enzymes and improved photosynthetic performance. J Plant Growth Reg 32(2):245–258

Habib SH, Kausar H, Saud HM (2016) Plant growth-promoting rhizobacteria enhance salinity stress tolerance in okra through ROS-scavenging enzymes. BioMed Res Int 2016: Article ID 6284547

Hafeez B, Khanif YM, Saleem M (2013) Role of zinc in plant nutrition-a review. Am J Exp Agri 3 (2):374

Hahm MS, Son JS, Hwang YJ, Kwon DK, and Ghim SY (2017) Alleviation of salt stress in pepper (Capsicum annum L.) plants by plant growth-promoting rhizobacteria. J Microbiol Biotechnol 27(10):1790–1797

Hamdia MAES, Shaddad MAK, Doaa MM (2004) Mechanisms of salt tolerance and interactive effects of Azospirillum brasilense inoculation on maize cultivars grown under salt stress conditions. Plant Growth Reg 44(2):165–174

Han D, Wang L, Luo Y (2018) Isolation, identification, and the growth promoting effects of two antagonistic actinomycete strains from the rhizosphere of *Mikania micrantha* Kunth. Microbiol Res 208:1–11

Hasegawa PM (2013) Sodium (Na+) homeostasis and salt tolerance of plants. Environ Exp Bot 92:19–31

Hasegawa PM, Bressan RA, Zhu JK, Bohnert HJ (2000) Plant cellular and molecular responses to high salinity. Ann Rev Plant Biol 51(1):463–499

He AL, Niu SQ, Zhao Q, Li YS, Gou JY, Gao HJ, Zhang JL (2018) Induced salt tolerance of perennial ryegrass by a novel bacterium strain from the rhizosphere of a desert shrub *Haloxylon ammodendron*. Int J Mole Sci 19(2):469

Hontzeas N, Saleh SS, Glick BR (2004) Changes in gene expression in canola roots induced by ACC-deaminase-containing plant-growth-promoting bacteria. Mol Plant Microb Inter 17 (8):865–871

Iqbal N, Umar S, Nazar R (2014) Manipulating osmolytes for breeding salinity-tolerant plants. In: Ahmad P, Rasool S (eds) Emerging technologies and management of crop stress tolerance a sustainable approach. Elsevier Inc., UK. ISBN: 978-0-12-800875-1

Islam S, Akanda AM, Prova A, Islam MT, Hossain MM (2016) Isolation and identification of plant growth promoting rhizobacteria from *cucumber* rhizosphere and their effect on plant growth promotion and disease suppression. Front Microbiol 6:1360

Jampeetong A, Brix H (2009) Effects of NaCl salinity on growth, morphology, photosynthesis and proline accumulation of *Salvinia natans*. Aquat Bot 91(3):181–186

Janssen J, Weyens N, Croes S, Beckers B, Meiresonne L, Van Peteghem P, Vangronsveld J (2015) Phytoremediation of metal contaminated soil using willow: exploiting plant-associated bacteria to improve biomass production and metal uptake. Int J Phytorem 17(11):1123–1136
Jarvis PG, Jarvis MS (1963) The water relations of tree seedlings: IV. Some aspects of the tissue water relations and drought resistance. Physiol Plant 16(3):501–516
Jha Y, Subramanian RB, Patel S (2011) Combination of endophytic and rhizospheric plant growth promoting rhizobacteria in *Oryza sativa* shows higher accumulation of osmoprotectant against saline stress. Acta Physiol Plant 33:797–802
Jha Y, Subramanian RB (2013) Paddy plants inoculated with PGPR show better growth physiology and nutrient content under saline conditions. Chil J Agric Res 73(3):213–219
Jha Y, Subramanian RB (2014) PGPR regulate caspase-like activity, programmed cell death, and antioxidant enzyme activity in paddy under salinity. Physiol Mole Biol Plant 20(2):201–207
Ji X, Dong B, Shiran B, Talbot MJ, Edlington JE, Hughes T, Dolferus R (2011) Control of abscisic acid catabolism and abscisic acid homeostasis is important for reproductive stage stress tolerance in cereals. Plant Physiol 156(2):647–662
Jin CW, Ye YQ, Zheng SJ (2013) An underground tale: contribution of microbial activity to plant iron acquisition via ecological processes. Ann Bot 113(1):7–18
Joseph EA, Mohanan KV (2013) A study on the effect of salinity stress on the growth and yield of some native rice cultivars of Kerala state of India. Agric Fores Fisher 2(3):141–150
Kamal R, Gusain YS, Kumar V (2014) Interaction and symbiosis of AM fungi, *Actinomycetes* and Plant Growth Promoting Rhizobacteria with plants: strategies for the improvement of plants health and defense system. Int J Curr Microbiol Appl Sci 3(7):564–585
Kamei A, Dolai AK, Kamei A (2014) Role of hydrogen cyanide secondary metabolite of plant growth promoting rhizobacteria as biopesticides of weeds. Global J Sci Front Res 14(6):109–112
Kamran S, Shahid I, Baig DN, Rizwan M, Malik KA, Mehnaz S (2017) Contribution of zinc solubilizing bacteria in growth promotion and zinc content of wheat. Front Microbiol 8:2593
Kang SM, Radhakrishnan R, Khan AL, Kim M-J, Park JM, Kim BR, Shin D-H, Lee IJ (2014a) Gibberellin secreting rhizobacterium, *Pseudomonas putida* H-2-3 modulates the hormonal and stress physiology of soybean to improve the plant growth under saline and drought conditions. Plant Physiol Biochem 84:115–124
Kang SM, Khan AL, Hamayun M, Hussain J, Joo GJ, You YH, Lee IJ (2012) Gibberellin-producing *Promicromonospora* sp. SE188 improves *Solanum lycopersicum* plant growth and influences endogenous plant hormones. J Microbiol 50(6):902–909
Kang SM, Khan AL, Waqas M, You YH, Kim JH, Kim JG, Lee IJ (2014b) Plant growth-promoting rhizobacteria reduce adverse effects of salinity and osmotic stress by regulating phytohormones and antioxidants in *Cucumis sativus*. J Plant Inter 9(1):673–682
Karlidag H, Ertan Y, Metin T, Mucahit P, Figen D (2013) Plant growth-promoting rhizobacteria mitigate deleterious effects of salt stress on strawberry plants (*Fragaria ananassa*). HortScience 48(5):563–567
Kasim WA, Osman ME, Omar MN, El-Daim IAA, Bejai S, Meijer J (2013) Control of drought stress in wheat using plant-growth-promoting bacteria. J Plant Growth Reg 32(1):122–130
Kaur G, Asthir B (2017) Molecular responses to drought stress in plants. Biol Plant 61(2):201–209
Kaushal M, Wani SP (2016) Plant-growth-promoting rhizobacteria: drought stress alleviators to ameliorate crop production in drylands. Ann Microbiol 66(1):35–42
Kaya C, Ashraf M, Dikilitas M, Tuna AL (2013) Alleviation of salt stress-induced adverse effects on maize plants by exogenous application of indoleacetic acid (IAA) and inorganic nutrients-a field trial. Aust J Crop Sci 7(2):249
Kaya C, Tuna AL, Okant AM (2010) Effect of foliar applied kinetin and indole acetic acid on maize plants grown under saline conditions. Turk J Agric For 34(6):529–538
Kempf B, Bremer E (1998) Uptake and synthesis of compatible solutes as microbial stress responses to high-osmolality environments. Arch Microbiol 170:319–330
Khalid S, Parvaiz M, Nawaz K, Hussain H, Arshad A, Shawaka S, Sarfaraz Z, Waheed T (2013) Effect of Indole Acetic Acid (IAA) on morphological: biochemical and chemical attributes of Two varieties of maize (*Zea mays* L.) under salt stress. World Appl Sci J 26:1150–1159

Khodair TA, Galal GF, El-Tayeb TS (2008) Effect of inoculating wheat seedlings with exopolysaccharide-producing bacteria in saline soil. J Appl Sci Res 4:2065–2070

Kiani SP, Talia P, Maury P, Grieu P, Heinz R, Perrault A, Sarrafi A (2007) Genetic analysis of plant water status and osmotic adjustment in recombinant inbred lines of sunflower under two water treatments. Plant Sci 172(4):773–787

Kim J, Rees DC (1994) Nitrogenase and biological nitrogen fixation. Biochem 33(2):389–397

Kloepper JW, Reddy MS, Kenney DS, Kokalis-Burelle N, Martinez-Ochoa N, Vavrina CS (2004) Theory and application for rhizobacteria in transplant production and yield enhancement. Acta Horticul 631:219–229

Kohler J, Hernández JA, Caravacaa F, Roldána A (2008) Plant-growth-promoting rhizobacteria and arbuscular mycorrhizal fungi modify alleviation biochemical mechanisms in water-stressed plants. Funct Plant Biol 35:141–151

Kohler J, Hernández JA, Caravaca F, Roldán A (2009) Induction of antioxidant enzymes is involved in the greater effectiveness of a PGPR versus AM fungi with respect to increasing the tolerance of lettuce to severe salt stress. Environ Exp Botany 65(2–3):245–252

Kuan KB, Othman R, Rahim KA, Shamsuddin ZH (2016) Plant growth-promoting rhizobacteria inoculation to enhance vegetative growth, nitrogen fixation and nitrogen remobilisation of maize under greenhouse conditions. PLoS One 11(3):e0152478

Kudoyarova GR, Melentiev AI, Martynenko EV, Timergalina LN, Arkhipova TN, Shendel GV, Veselov SY (2014) Cytokinin producing bacteria stimulate amino acid deposition by wheat roots. Plant Physiol Biochem 83:285–291

Kumar A, Dixit S, Ram T, Yadaw RB, Mishra KK, Mandal NP (2014) Breeding high-yielding drought-tolerant rice: genetic variations and conventional and molecular approaches. J Exp Bot 65(21):6265–6278

Kumar D (2005) Breeding for drought resistance. In: Abiotic stresses. CRC Press, pp 167–198

Kumari S, Vaishnav A, Jain S, Varma A, Choudhary DK (2015) Bacterial-mediated induction of systemic tolerance to salinity with expression of stress alleviating enzymes in soybean (*Glycine max* L. Merrill). J Plant Growth Regul 34:558–573

Lawlor DW (2002) Carbon and nitrogen assimilation in relation to yield: mechanisms are the key to understanding production systems. J Exp Bot 53(370):773–787

Liang C, Wang Y, Zhu Y, Tang J, Hu B, Liu L, Chu C (2014) OsNAP connects abscisic acid and leaf senescence by fine-tuning abscisic acid biosynthesis and directly targeting senescence-associated genes in rice. Proc Natl Acad Sci 111(27):10013–10018

Lim JH, Kim SD (2013) Induction of drought stress resistance by multi-functional PGPR *Bacillus licheniformis* K11 in pepper. Plant Pathol J 29(2):201

Liu XM, Zhang H (2015) The effects of bacterial volatile emissions on plant abiotic stress tolerance. Front Plant Sci 6:774

Liu B, Asseng S, Müller C, Ewert F, Elliott J, Lobell DB, Rosenzweig C (2016) Similar estimates of temperature impacts on global wheat yield by three independent methods. Nat Climat Chang 6(12):1130

Liu J, Xia Z, Wang M, Zhang X, Yang T, Wu J (2013) Overexpression of a maize E3 ubiquitin ligase gene enhances drought tolerance through regulating stomatal aperture and antioxidant system in transgenic tobacco. Plant Physio Biochem 73:114–120

Lu GH, Ren DL, Wang XQ, Wu JK, Zhao MS (2010) Evaluation on drought tolerance of maize hybrids in China. J Maize Sci 3:20–24

Lugtenberg BJ, Malfanova N, Kamilova F, Berg G (2013) Plant growth promotion by microbes. Mole Microbial Ecol Rhizosphere 2:561–573

Ma W, Penrose DM, Glick BR (2002) Strategies used by rhizobia to lower plant ethylene levels and increase nodulation. Can J Microbiol 48(11):947–954

Madhaiyan M, Poonguzhali S, Ryu J, Sa T (2006) Regulation of ethylene levels in canola (*Brassica campestris*) by 1-aminocyclopropane-1-carboxylate deaminase-containing *Methylobacterium fujisawaense*. Planta 224(2):268–278

Maheshwari DK, Aeron A, Dubey RC, Agarwal M, Dheeman S, Shukla S (2014) Multifaceted beneficial associations with *Pseudomonas* and rhizobia on growth promotion of *Mucuna pruriens* L. J Pure Appl Microbiol 8(6):4657–4667
Maheshwari DK, Kumar S, Kumar B, Pandey P (2010) Co-inoculation of urea and DAP tolerant *Sinorhizobium meliloti* and *Pseudomonas aeruginosa* as integrated approach for growth enhancement of *Brassica juncea*. Ind J Microbiol 50(4):425–431
Mahmood S, Daur I, Al-Solaimani SG, Ahmad S, Madkour MH, Yasir M, Ali Z (2016) Plant growth promoting rhizobacteria and silicon synergistically enhance salinity tolerance of mung bean. Front Plant Sci 7:876
Mantelin S, Touraine B (2004) Plant growth-promoting bacteria and nitrate availability: impacts on root development and nitrate uptake. J Exp Bot 55(394):27–34
Mapelli F, Marasco R, Rolli E, Barbato M, Cherif H, Guesmi A, Ouzari I, Daffonchio D, Borin S (2013) Potential for plant growth promotion of rhizobacteria associated with Salicornia growing in Tunisian hypersaline soils. Biomed Res Int 2013:248078
Maqbool MA, Aslam M, Ali H (2017) Breeding for improved drought tolerance in Chickpea (*Cicer arietinum* L.). Plant Breed 136(3):300–318
Marulanda A, Porcel R, Barea JM, Azcón R (2007) Drought tolerance and antioxidant activities in lavender plants colonized by native drought-tolerant or drought-sensitive *Glomus* species. Microb Ecol 54:543–552
Marulanda A, Azcon R, Chaumont F, Ruiz-Lozano JM, Aroca R (2010) Regulation of plasma membrane aquaporins by inoculation with a *Bacillus megaterium* strain in maize (*Zea mays* L.) plants under unstressed and salt-stressed conditions. Planta 232:533–543
Marulanda A, Barea JM, Azcón R (2009) Stimulation of plant growth and drought tolerance by native microorganisms (AM fungi and bacteria) from dry environments: mechanisms related to bacterial effectiveness. J Plant Growth Reg 28(2):115–124
Mayak S, Tirosh T, Glick BR (2004a) Plant growth-promoting bacteria confer resistance in tomato plants to salt stress. Plant Physiol Biochem 42:565–572
Mayak S, Tirosh T, Glick BR (2004b) Plant growth-promoting bacteria that confer resistance to water stress in tomatoes and peppers. Plant Sci 166(2):525–530
Meena MK, Gupta S, Datta S (2016) Antifungal potential of PGPR, their growth promoting activity on seed germination and seedling growth of winter wheat and genetic variabilities among bacterial isolates. Int J Cur Microbiol Appl Sci 5(1):235–243
Mendes R, Garbeva P, Raaijmakers JM (2013) The rhizosphere microbiome: significance of plant beneficial, plant pathogenic, and human pathogenic microorganisms. FEMS Microbiol Rev 37 (5):634–663
Mhadhbi H, Jebara M, Limam F, Aouani ME (2004) Rhizobial strain involvement in plant growth, nodule protein composition and antioxidant enzyme activities of chickpea-rhizobia symbioses: modulation by salt stress. Plant Physiol Biochem 42(9):717–722
Maheshwari DK (ed) (2010) Microbiology monographs (V-18). Plant growth and health promoting bacteria. Springer, Heidelberg, Germany. ISBN: 978-3-642-13611-5
Minerdi D, Bossi S, Maffei ME, Gullino ML, Garibaldi A (2011) *Fusarium oxysporum* and its bacterial consortium promote lettuce growth and expansin A5 gene expression through microbial volatile organic compound (MVOC) emission. FEMS Microbiol Ecol 76(2):342–351
Minocheherhomji A, Yasmin H, Naz R, Bano A, Keyani R, Hussain I (2018) *Pseudomonas putida* improved soil enzyme activity and growth of kasumbha under low input of mineral fertilizers. Soil Sci Plant Nut 1–6
Mittler R, Blumwald E (2010) Genetic engineering for modern agriculture: challenges and perspectives. Ann Rev Plant Biol 61:443–462
Munns R (2002) Salinity, growth and phytohormones. In: Salinity: environment-plants-molecules, Springer Dordrecht 271–290
Munns R, Gilliham M (2015) Salinity tolerance of crops–what is the cost? New Phytol 208 (3):668–673
Munns R, Tester M (2008) Mechanisms of salinity tolerance. Annu Rev Plant Biol 59:651–681

Munns R, James RA, Xu B, Athman A, Conn SJ, Jordans C, Plett D (2012) Wheat grain yield on saline soils is improved by an ancestral Na+ transporter gene. Nat Biotechnol 30(4):360
Mwadzingeni L, Shimelis H, Dube E, Laing MD, Tsilo TJ (2016) Breeding wheat for drought tolerance: progress and technologies. J Integ Agric 15(5):935–943
Nabti E, Sahnoune M, Adjrad S, Van Dommelen A, Ghoul M, Schmid M, Hartmann A (2007) A halophilic and osmotolerant *Azospirillum brasilense* strain from Algerian soil restores wheat growth under saline conditions. Eng Life Sci 7(4):354–360
Nabti E, Sahnoune M, Ghoul M, Fischer D, Hofmann A, Rothballer M, Schmid M, Hartmann A, (2010) Restoration of growth of durum wheat (*Triticum durum* var. waha) under saline conditions due to inoculation with the rhizosphere bacterium *Azospirillum brasilense* NH and extracts of the marine alga *Ulva lactuca*. J Plant Growth Regul 29 (1):6–22
Nadeem SM, Zahir ZA, Naveed M, Arshad M (2007) Preliminary investigations on inducing salt tolerance in maize through inoculation with rhizobacteria containing ACC deaminase activity. Can J Microbiol 53(10):1141–1149
Nadeem SM, Zahir ZA, Naveed M, Arshad M (2009) Rhizobacteria containing ACC-deaminase confer salt tolerance in maize grown on salt-affected fields. Can J Microbiol 55(11):1302–1309
Nadeem SM, Zahir ZA, Naveed M, Nawaz S (2013) Mitigation of salinity-induced negative impact on the growth and yield of wheat by plant growth-promoting rhizobacteria in naturally saline conditions. Anna Microbiol 63(1):225–232
Narula N, Kothe E, Behl RK (2009) Role of root exudates in plant-microbe interactions. J Appl Bot Food Qual 82(2):122–130
Naseem H, Bano A (2014) Role of plant growth-promoting rhizobacteria and their exopolysaccharide in drought tolerance of maize. J Plant Interact 9(1):689–701
Nautiyal CS, Srivastava S, Chauhan PS, Seem K, Mishra A, Sopory SK (2013) Plant growth-promoting bacteria *Bacillus amyloliquefaciens* NBRISN13 modulates gene expression profile of leaf and rhizosphere community in rice during salt stress. Plant Physiol Biochem 66:1–9
Naveed M, Mitter B, Reichenauer TG, Wieczorek K, Sessitsch A (2014) Increased drought stress resilience of maize through endophytic colonization by *Burkholderia phytofirmans* PsJN and Enterobacter sp. FD17. Environ Exp Bot 97:30–39
Nawrotzki RJ, Bakhtsiyarava M (2017) International climate migration: evidence for the climate inhibitor mechanism and the agricultural pathway. Popul Space Place 23(4):e2033
Naz R, Bano A (2013) Influence of exogenously applied SA and PGPR inoculation on the growth and physiology of sunflower (*Helianthus annus* L.) under salt stress. Pak J Bot 45(2):367–373
Naz R, Bano A (2015) Molecular and Physiological responses of Sunflower (*Helianthus annus* L.) to PGPR and SA under Salt Stress. Pak J Bot 47(1):35–42
Naz I, Bano A, Ul-Hassan T (2009) Isolation of phytohormones producing plant growth promoting rhizobacteria from weeds growing in Khewra salt range, Pakistan and their implication in providing salt tolerance to *Glycine max* L. Afr J Biotechnol 8:21
Negrão S, Courtois B, Ahmadi N, Abreu I, Saibo N, Oliveira MM (2011) Recent updates on salinity stress in rice: from physiological to molecular responses. Crit Rev Plant Sci 30(4):329–377
Negrão S, Schmöckel SM, Tester M (2017) Evaluating physiological responses of plants to salinity stress. Ann Bot 119(1):1–11
Neumann PM (1995) The role of cell wall adjustments in plant resistance to water deficits. Crop Sci 35(5):1258–1266
Neumann PM (2008) Coping mechanisms for crop plants in drought-prone environments. Anna Bot 101(7):901–907
Ngumbi EN (2011) Mechanisms of olfaction in parasitic wasps: analytical and behavioral studies of response of a specialist (Microplitis croceipes) and a generalist (Cotesia marginiventris) parasitoid to host-related odor. PhD thesis, Auburn University, Albama USA
Ngumbi E, Kloepper J (2016) Bacterial-mediated drought tolerance: current and future prospects. Appl Soil Ecol 105:109–125

Nie M, Wang Y, Yu J, Xiao M, Jiang L, Yang J, Li B (2011) Understanding plant-microbe interactions for phytoremediation of petroleum-polluted soil. PLoS One 6(3):e17961

Niu SQ, Li HR, Paré PW, Aziz M, Wang SM, Shi H, Guo Q (2016) Induced growth promotion and higher salt tolerance in the halophyte grass *Puccinellia tenuiflora* by beneficial rhizobacteria. Plant Soil 407(1–2):217–230

Nosheen A, Yasmin H, Naz R, Bano A, Keyani R, Hussain I (2018) improved soil enzyme activity and growth of kasumbha under low input of mineral fertilizers. Soil Sci Plant Nut 1–6

Numan M, Bashir S, Khan Y, Mumtaz R, Shinwari ZK, Khan AL, Ahmed AH (2018) Plant growth promoting bacteria as an alternative strategy for salt tolerance in plants: a review. Microbiol Res 209:21–32

Öğüt M, Er F, Neumann G (2011) Increased proton extrusion of wheat roots by inoculation with phosphorus solubilising microorganims. Plant Soil 339(1–2):285–297

Ortíz-Castro R, Valencia-Cantero E, López-Bucio J (2008) Plant growth promotion by *Bacillus megaterium* involves cytokinin signaling. Plant Sig Beh 3(4):263–265

Palaniyandi SA, Damodharan K, Yang SH, Suh JW (2014) *Streptomyces* sp. strain PGPA39 alleviates salt stress and promotes growth of 'Micro Tom' tomato plants. J Appl Microbiol 117 (3):766–773

Panwar M, Tewari R, Gulati A, Nayyar H (2016) Indigenous salt-tolerant rhizobacterium *Pantoea dispersa* (PSB3) reduces sodium uptake and mitigates the effects of salt stress on growth and yield of chickpea. Acta Physiol Plant 38:278

Parida AK, Das AB (2005) Salt tolerance and salinity effects on plants: a review. Ecotoxicol Environ Saf 60(3):324–349

Parida AK, Das AB, Mittra B (2004) Effects of salt on growth, ion accumulation, photosynthesis and leaf anatomy of the mangrove, *Bruguiera parviflora*. Trees 18(2):167–174

Patel TS, Minocheherhomji FP (2018) Plant growth promoting rhizobacteria: blessing to agriculture. Int J Pure Appl Biosci 6(2):481–492

Pawar ST, Bhosale AA, Gawade TB, Nale TR (2017) Isolation, screening and optimization of exopolysaccharide producing bacterium from saline soil. J Microbiol Biotechnol Res 3(3): 24–31

Peleg Z, Reguera M, Tumimbang E, Walia H, Blumwald E (2011) Cytokinin-mediated source/sink modifications improve drought tolerance and increase grain yield in rice under water-stress. Plant Biotechnol J 9(7):747–758

Peng S, Ismail AM (2004) Physiological basis of yield and environmental adaptation in rice. Physiol Biotechnol Integ Plant Breed 83–140

Perata P, Voesenek LA (2007) Submergence tolerance in rice requires Sub1A, an ethylene-response-factor-like gene. Trend Plant Sci 12(2):43–46

Pereyra MA, Garcia P, Colabelli MN, Barassi CA, Creus CM (2012) A better water status in wheat seedlings induced by *Azospirillum* under osmotic stress is related to morphological changes in xylem vessels of the coleoptile. Appl Soil Ecol 53:94–97

Perrig D, Boiero M, Masciarelli O, Penna C, Ruiz O, Cassán F, Luna M (2007) Plant-growth-promoting compounds produced by two agronomically important strains of *Azospirillum brasilense*, and implications for inoculant formulation. Appl Microbiol Biotechno 75(5):1143–1150

Pitman MG, Läuchli A (2002) Global impact of salinity and agricultural ecosystems. In: Salinity: environment-plants-molecules. Springer, Dordrecht, pp 3–20

Pliego C, Kamilova F, Lugtenberg B (2011) Plant growth-promoting bacteria: fundamentals and exploitation. In: Bacteria in Agrobiology: crop ecosystems, Springer, Berlin, pp 295–343

Prajapati K, Sharma MC, Modi HA (2013) Growth promoting effect of potassium solubilizing microorganisms on *Abelmoscus esculantus*. Int J Agric Sci 3(1):181–188

Prathap M, Kumari BR (2015) A critical review on plant growth promoting rhizobacteria. J Plant Pathol Microbiol 6(4):1

Qadir M, Quillérou E, Nangia V, Murtaza G, Singh M, Thomas RJ, Noble AD (2014) Economics of salt-induced land degradation and restoration. Nat Resour Forum 38(4):282–295

Qudsaia B, Noshinil Y, Asghari B, Nadia Z, Abida A, Fayazul H (2013) Effect of *Azospirillum* inoculation on maize (*Zea mays* L.) under drought stress. Pak J Bot 45:13–20
Rais A, Jabeen Z, Shair F, Hafeez FY, Hassan MN (2017) *Bacillus* spp., a bio-control agent enhances the activity of antioxidant defense enzymes in rice against *Pyricularia oryzae*. Plos One 12(11):e0187412
Ramadan EM, AbdelHafez AA, Hassan EA, Saber FM (2016) Plant growth promoting rhizobacteria and their potential for biocontrol of phytopathogens. Afri J Microbiol Res 10 (15):486–504
Ramegowda V, Kumar S (2015) The interactive effects of simultaneous biotic and abiotic stresses on plants: mechanistic understanding from drought and pathogen combination. J Plant Physiol 176:47–54
Rao SM, Shaw MEA (1985) A review of research on sugar cane soils of Jamaica. In: 23. Sugar technologists' conference, (Trinidad and Tobago), 4–8 Mar 1985
Rathore P (2014) A review on approaches to develop plant growth promoting rhizobacteria. Inter J Rec Sci Res 5:403–407
Raza FA, Faisal M (2013) Growth promotion of maize by desiccation tolerant *Micrococcus luteus*-chp37 isolated from Cholistan desert. Pak Aust J Crop Sci 7(11):1693
Rejeb KB, Abdelly C, Savouré A (2014) How reactive oxygen species and proline face stress together. Plant Physiol Biochem 80:278–284
Rhodes D, Verslues PE, Sharp RE (1999) Role of amino acids in abiotic stress resistance. In: Singh BK (ed) Plant amino acids: biochem biotechnol marcel dekker NY, 319–356
Rhodes D, Handa S (1989) Amino acid metabolism in relation to osmotic adjustment in plant cells. In: Environmen stress plant. Springer, Berlin 41–62
Richards DE, King KE, Ait-ali T, Harberd NP (2001) How gibberellin regulates plant growth and development: a molecular genetic analysis of gibberellin signaling. Annu Rev Plant Physiol Plant Mol Biol 52:67–88
Rijavec T, Lapanje A (2016) Hydrogen cyanide in the rhizosphere: not suppressing plant pathogens, but rather regulating availability of phosphate. Front Microbiol 7:1785
Rodriguez-Salazar J, Suarez R, Caballero-Mellado J, Itturiaga G (2009) Trehalose accumulation in *Azospirillum brasilense* improves drought tolerance and biomass in maize plants. FEMS Microbiol Lett 296:52–59
Rojas-Tapias D, Moreno-Galván A, Pardo-Díaz S, Obando M, Rivera D, Bonilla R (2012) Effect of inoculation with plant growth-promoting bacteria (PGPB) on amelioration of saline stress in maize (*Zea mays*). Appl Soil Ecol 61:264–272
Rowley G (1993) Multinational and national competition for water in the Middle East: Towards the deepening crisis. J Environ Manag 39(3):187–197
Roy SJ, Negrão S, Tester M (2014) Salt resistant crop plants. Curr Opin Biotechnol 26:115–124
Ruelland E, Vaultier MN, Zachowski A, Hurry V (2009) Cold signalling and cold acclimation in plants. Adv Bot Res 49:35–150
Saakre M, Baburao TM, Salim AP, Ffancies RM, Achuthan VP, Thomas G, Sivarajan SR (2017) Identification and characterization of genes responsible for drought tolerance in rice mediated by *Pseudomonas fluorescens*. Rice Sci 24(5):291–298
Saha M, Sarkar S, Sarkar B, Sharma BK, Bhattacharjee S, Tribedi P (2016) Microbial siderophores and their potential applications: a review. Environ Sci Poll Res 23(5):3984–3999
Saharan BS, Nehra V (2011) Plant growth promoting rhizobacteria: a critical review. Life Sci Med Res 21(1):30
Sanalibaba P, Çakmak GA (2016) Exopolysaccharides production by lactic acid bacteria. Appl Microbiol 2:115
Sandhya VSKZ, Ali SZ, Grover M, Reddy G, Venkateswarlu B (2010) Effect of plant growth promoting *Pseudomonas* spp. on compatible solutes, antioxidant status and plant growth of maize under drought stress. Plant Growth Reg 62(1):21–30
Sandhya VZAS, Grover M, Reddy G, Venkateswarlu B (2009) Alleviation of drought stress effects in sunflower seedlings by the exopolysaccharides producing *Pseudomonas putida* strain GAP-P45. Biol Fer Soil 46(1):17–26

Saravanakumar D, Samiyappan R (2007) ACC deaminase from Pseudomonas fluorescens mediated saline resistance in groundnut (*Arachis hypogea*) plants. J Appl Microbiol 102 (5):1283–1292

Saravanakumar D, Kavino M, Raguchander T, Subbian P, Samiyappan R (2011) Plant growth promoting bacteria enhance water stress resistance in green gram plants. Acta Physiol Plant 33 (1):203–209

Sarma RK, Saikia R (2014) Alleviation of drought stress in mung bean by strain *Pseudomonas aeruginosa* GGRJ21. Plant Soil 377(1–2):111–126

Serraj R, Sinclair TR (2002) Osmolyte accumulation: can it really help increase crop yield under drought conditions? Plant Cell Environ 25(2):333–341

Shafi M, Bakht J, Khan MJ, Khan MA, Anwar S (2010) Effect of salinity on yield and ion accumulation of wheat genotypes. Pak J Bot 4113–4121

Shaharoona B, Arshad M, Zahir ZA (2006a) Effect of plant growth promoting rhizobacteria containing ACC-deaminase on maize (*Zea mays* L.) growth under axenic conditions and on nodulation in mung bean (*Vigna radiata* L.). Lett Appl Microbiol 42(2):155–159

Shaharoona B, Arshad M, Zahir ZA, Khalid A (2006b) Performance of *Pseudomonas* spp. containing ACC-deaminase for improving growth and yield of maize (Zea mays L.) in the presence of nitrogenous fertilizer. Soil Biol Biochem 38(9):2971–2975

Shahzad SM, Arif MS, Riaz M, Iqbal Z, Ashraf M (2013) PGPR with varied ACC-deaminase activity induced different growth and yield response in maize (*Zea mays* L.) under fertilized conditions. Eur J Soil Biol 57:27–34

Sharma A, Shankhdhar D, Shankhdhar SC (2013) Enhancing grain iron content of rice by the application of plant growth promoting rhizobacteria. Plant Soil Environ 59(2):89–94

Sharma CK, Vishnoi VK, Dubey RC, Maheshwari DK (2018) A twin rhizospheric bacterial consortium induces systemic resistance to a phytopathogen *Macrophomina phaseolina* in mung bean. Rhizosphere 5:71–75

Sheteawi SA (2007) Improving growth and yield of salt-stressed soybean by exogenous application of jasmonic acid and ascobin. Inter J Agric Biol 2007 http://agris.fao.org/agris-search/search.do?recordID=PK2007001114

Sheveleva E, Chmara W, Bohnert HJ, Jensen RG (1997) Increased salt and drought tolerance by D-ononitol production in transgenic *Nicotiana tabacum* L. Plant Physiol 115(3):1211–1219

Shi H, Xiong L, Stevenson B, Lu T, Zhu JK (2002) The *Arabidopsis* salt overly sensitive 4 mutants uncover a critical role for vitamin B6 in plant salt tolerance. Plant Cell 14(3):575–588

Shiferaw B, Smale M, Braun HJ, Duveiller E, Reynolds M, Muricho G (2013) Crops that feed the world 10. Past successes and future challenges to the role played by wheat in global food security. Food Sec 5(3):291–317

Shkolnik-Inbar D, Adler G, Bar-Zvi D (2013) ABI4 downregulates expression of the sodium transporter *HKT1* in *Arabidopsis* roots and affects salt tolerance. Plant J 73:993–1005

Shrivastava P, Kumar R (2015) Soil salinity: a serious environmental issue and plant growth promoting bacteria as one of the tools for its alleviation. Sau J Biol Sci 22(2):123–131

Shukla PS, Agarwal PK, Jha B (2012) Improved salinity tolerance of *Arachis hypogaea* (L.) by the interaction of halotolerant plant growth promoting rhizobacteria. J Plant Growth Regul 31:195–206

Siddikee MA, Glick BR, Chauhan PS, Yim W, Sa T (2011) Enhancement of growth and salt tolerance of red pepper seedlings (*Capsicum annuum* L.) by regulating stress ethylene synthesis with halotolerant bacteria containing 1-aminocyclopropane-1 carboxylic acid deaminase activity. Plant Physiol Biochem 49:427–434

Sinclair TR, Muchow RC (2001) System analysis of plant traits to increase grain yield on limited water supplies. Agric J 93(2):263–270

Sindhu SS, Parmar P, Phour M (2014) Nutrient cycling: potassium solubilization by microorganisms and improvement of crop growth. In: Geomicrobiol Biogeochem. Springer, Berlin, 175–198

Skirycz A, Inzé D (2010) More from less: plant growth under limited water. Curr Opin Biotechnol 21(2):197–203

SkZ A, Vardharajula S, Vurukonda SSKP (2018) Transcriptomic profiling of maize (*Zea mays* L.) seedlings in response to *Pseudomonas putida* stain FBKV2 inoculation under drought stress. Ann Microbiol 68(6):331–349

Smith DL, Gravel V, Yergeau E (2017) Signaling in the phytomicrobiome. Front Plant Sci 8:611

Somers E, Vanderleyden J, Srinivasan M (2004) Rhizosphere bacterial signalling: a love parade beneath our feet. Crit Rev Microbiol 30(4):205–240

Spaepen S, Vanderleyden J, Remans R (2007) Indole-3-acetic acid in microbial and microorganism- plant signaling. FEMS Microbiol Rev 31(4):425–448

Spaepen S, Vanderleyden J (2011) Auxin and plant-microbe interactions. Cold Spring Harb Perspect Biol 3:a001438

Steenhoudt O, Vanderleyden J (2000) *Azospirillum*, a free-living nitrogen-fixing bacterium closely associated with grasses: genetic, biochemical and ecological aspects. FEMS Microbiol Rev 24 (4):487–506

Subramanian S, Ricci E, Souleimanov A, Smith DL (2016) A proteomic approach to lipo-chitooligosaccharide and thuricin 17 effects on soybean germination unstressed and salt stress. PLoS One 11(8):e0160660

Sukweenadhi J, Balusamy SR, Kim YJ, Lee CH, Kim YJ, Koh SC, Yang DC (2018) A Growth Promoting Bacteria, *Paenibacillus yonginensis* DCY84T Enhanced Salt Stress Tolerance by Activating Defense-Related Systems in *Panax ginseng*. Front Plant Sci 9:813

Swain DL, Singh D, Horton DE, Mankin JS, Ballard TC, Diffenbaugh NS (2017) Remote linkages to anomalous winter atmospheric ridging over the northeastern Pacific. J Geophy Res Atmos 22:122

Syvertsen JP, Boman B, Tucker DPH (1989) Salinity in Florida citrus production. Proceed Florida State Horticul Soc 102:61–64

Szabolcs I (1992) Salinization of soil and water and its relation to desertification. Desertification Control Bull 21:32–37

Szabolcs I (1994) Soils and salinisation. In: Handbook of Plant Crop Stress 3–11

Tanji KK (1990) Nature and extent of agricultural salinity. In: Agricultural salinity assessment and management 71–92

Tanwir F, Saboor A, Nawaz N (2003) Soil salinity and the livelihood strategies of small farmers: a case study in Faisalabad district, Punjab. Pak Int J Agric Biol 5:440–441

Timmusk S, Wagner EGH (1999) The plant-growth-promoting rhizobacterium *Paenibacillus polymyxa* induces changes in *Arabidopsis thaliana* gene expression: a possible connection between biotic and abiotic stress responses. Mole Plant Microbe Interact 12(11):951–959

Timmusk S, El-Daim IAA, Copolovici L, Tanilas T, Kännaste A, Behers L, Niinemets Ü (2014) Drought-tolerance of wheat improved by rhizosphere bacteria from harsh environments: enhanced biomass production and reduced emissions of stress volatiles. PLoS One 9(5):e96086

Tiwari S, Lata C, Chauhan PS, Nautiyal CS (2016) *Pseudomonas putida* attunes morphophysiological, biochemical and molecular responses in *Cicer arietinum* L. during drought stress and recovery. Plant Physiol Biochem 99:108–117

Todaka D, Shinozaki K, Yamaguchi-Shinozaki K (2015) Recent advances in the dissection of drought-stress regulatory networks and strategies for development of drought-tolerant transgenic rice plants. Front Plant Sci 6:84

Torres RO, Henry A (2016) Yield stability of selected rice breeding lines and donors across conditions of mild to moderately severe drought stress. Field Crop Res 220:37–45

Tosens T, Niinemets U, Vislap V, Eichelmann H, Castro Diez P (2012) Developmental changes in mesophyll diffusion conductance and photosynthetic capacity under different light and water availabilities in *Populus tremula*: how structure constrains function. Plant Cell Environ 35 (5):839–856

Uga Y, Sugimoto K, Ogawa S, Rane J, Ishitani M, Hara N, Inoue H (2013) Control of root system architecture by deeper rooting 1 increases rice yield under drought conditions. Nat Gen 45(9): 1097

Ulloa-Ogaz AL, Muñoz-Castellanos LN, Nevárez-Moorillón GV (2015) Biocontrol of phytopathogens: antibiotic production as mechanism of control. In: The battle against microbial pathogens: basic science, technological advances and educational programes. Formatex Research Center, Spain, pp 305–309

Upadhyay SK, Singh DP (2015) Effect of salt-tolerant plant growth promoting rhizobacteria on wheat plants and soil health in a saline environment. Plant Biol 17:288–293

Vaishnav A, Kumari S, Jain S, Varma A, Choudhary DK (2015) Putative bacterial volatile-mediated growth in soybean (Glycine max L. Merrill) and expression of induced proteins under salt stress. J Appl Microbiol 119:539–551

Vaishnav A, Kumari S, Jain S, Varma A, Tuteja N, Choudhary DK (2016) PGPR-mediated expression of salt tolerance gene in soybean through volatiles under sodium nitroprusside. J Basic Microbiol 56(11):1274–1288

Valifard M, Mohsenzadeh S, Niazi A, Moghadam A (2015) Phenylalanine ammonia lyase isolation and functional analysis of phenylpropanoid pathway under salinity stress in 'Salvia' species. Aus J Crop Sci 9(7):656

Van den Ende W, Valluru R (2009) Sucrose, sucrosyl oligosaccharides, and oxidative stress: scavenging and salvaging? J Exp Bot 60(1):9–18

Vardharajula S, Ali SKZ, Grover M, Reddy G, Bandi V (2009) Alleviation of drought stress effects in sunflower seedlings by the exopolysaccharides producing *Pseudomonas putida* strain GAP-P45. Biol Fertil Soil 46:17–26

Vardharajula S, Zulfikar Ali S, Grover M, Reddy G, Bandi V (2011) Drought-tolerant plant growth promoting *Bacillus* spp.: effect on growth, osmolytes, and antioxidant status of maize under drought stress. J Plant Interact 6(1):1–14

Vendruscolo ECG, Schuster I, Pileggi M, Scapim CA, Molinari HBC, Marur CJ, Vieira LGE (2007) Stress-induced synthesis of proline confers tolerance to water deficit in transgenic wheat. J Plant Physiol 164(10):1367–1376

Wang H, Yamauchi A (2006) Growth and function of rootsunder abiotic stress soils. In: Huang B (ed) Plant–environment interactions, 3rd edn. CRC, Taylor and Francis Group, LLC, New York, pp 271–320

Wang CJ, Yang W, Wang C, Gu C, Niu DD, Liu HX, Guo JH (2012) Induction of drought tolerance in cucumber plants by a consortium of three plant growth-promoting rhizobacterium strains. PLoS One 7(12):e52565

Wang C, Yang A, Yin H, Zhang J (2008) Influence of water stress on endogenous hormone contents and cell damage of maize seedlings. J Integ Plant Biol 50(4):427–434

Wang Q, Dodd IC, Belimov AA, Jiang F (2016) Rhizosphere bacteria containing 1-aminocyclopropane-1-carboxylate deaminase increase growth and photosynthesis of pea plants under salt stress by limiting Na + accumulation. Funct Plant Biol 43(2):161–172

Wang X, Mavrodi DV, Ke L, Mavrodi OV, Yang M, Thomashow LS, Zhang J (2015) Biocontrol and plant growth-promoting activity of rhizobacteria from C hinese fields with contaminated soils. Microb Biotechnol 8(3):404–418

Weigand C (2011) Wheat import projections towards 2050. US Wheat Associates, USA

Wilkinson S, Davies WJ (2010) Drought, ozone, ABA and ethylene: new insights from cell to plant to community. Plant Cell Environ 33(4):510–525

Wilkinson S, Kudoyarova GR, Veselov DS, Arkhipova TN, Davies WJ (2012) Plant hormone interactions: innovative targets for crop breeding and management. J Exp Bot 63(9):3499–3509

Wrathall DJ, Van Den Hoek J, Walters A, Devenish A (2018) Water stress and human migration: a global, georeferenced review of empirical research. Land Water Discus Paper 11

Wu Z, Yue H, Lu J, Li C (2012) Characterization of rhizobacterial strain Rs-2 with ACC deaminase activity and its performance in promoting cotton growth under salinity stress. World J Microbiol Biotechnol 28(6):2383–2393

Xiong L, Zhu JK (2002) Molecular and genetic aspects of plant responses to osmotic stress. Plant Cell Environ 25(2):131–139

Xuemei J, Dong B, Shiran B, Talbot MJ, Edlington JE, Trijntje H, Dolferus R (2011) Control of ABA catabolism and ABA homeostasis is important for reproductive stage stress tolerance in cereals. Plant Physiol 111

Yadav SK, Jyothi Lakshmi N, Maheswari M, Vanaja M, Venkateswarlu B (2005) Influence of water deficit at vegetative, anthesis and grain filling stages on water relation and grain yield in Sorghum. Indian J Plant Physiol 10:20–22

Yadav OP (2010) Drought response of pearl millet landrace-based populations and their crosses with elite composites. Field Crop Res 118(1):51–56

Yamaguchi-Shinozaki K, Shinozaki K (1994) A novel cis-acting element in an *Arabidopsis* gene is involved in responsiveness to drought, low-temperature, or high-salt stress. Plant Cell 6 (2):251–264

Yan J, Smith MD, Glick BR, Liang Y (2014) Effects of ACC deaminase containing rhizobacteria on plant growth and expression of Toc GTPases in tomato (*Solanum lycopersicum*) under salt stress. Botany 92(11):775–781

Yang J, Kloepper JW, Ryu CM (2009) Rhizosphere bacteria help plants tolerate abiotic stress. Trend Plant Sci 14(1):1–4

Yao L, Wu Z, Zheng Y, Kaleem I, Li C (2010) Growth promotion and protection against salt stress by *Pseudomonas putida* Rs-198 on cotton. Eur J Soil Biol 46(1):49–54

Yasmin H, Bano A, Samiullah A (2013) Screening of PGPR isolates from semi-arid region and their implication to alleviate drought stress. Pak J Bot 45:51–58

Yasmin H, Nosheen A, Naz R, Bano A, Keyani R (2017) l-tryptophan-assisted PGPR-mediated induction of drought tolerance in maize (*Zea mays* L.). J Plant Interact 12(1):567–578

Yeo A (1998) Molecular biology of salt tolerance in the context of whole-plant physiology. J Exp Bot 49(323):915–929

Yoon GM, Kieber JJ (2013) 1-Aminocyclopropane-1-carboxylic acid as a signalling molecule in plants. Aob Plant 5:17

Yoshiba Y, Kiyosue T, Nakashima K, Yamaguchi-Shinozaki K, Shinozaki K (1997) Regulation of levels of proline as an osmolyte in plants under water stress. Plant Cell Physiol 38:1095–1102

Zafar-ul-Hye M, Ahmad M, Shahzad SM (2013) Synergistic effect of rhizobia and plant growth promoting rhizobacteria on the growth and nodulation of lentil seedlings under axenic conditions. Soil Environ 32:79–86

Zahir ZA, Munir A, Asghar HN, Shaharoona B, Arshad M (2008) Effectiveness of rhizobacteria containing ACC deaminase for growth promotion of peas (*Pisum sativum*) under drought conditions. J Microbiol Biotechnol 18(5):958–963

Zhang H, Kim MS, Krishnamachari V, Payton P, Sun Y, Grimson M, Farag MA, Ryu CM, Allen R, Melo SI, Paré PW (2007) Rhizobacterial volatile emissions regulate auxin homeostasis and cell expansion in *Arabidopsis*. Planta 226:839–851

Zhang H, Murzello C, Sun Y, Kim MS, Xie X, Jeter RM, Zak JC, Dowd SE, Pare PW (2010) Choline and osmotic-stress tolerance induced in Arabidopsis by the soil microbe *Bacillus subtilis* (GB03). Mol Plant Microbe Interact 23:1097–1104

Zhang HX, Hodson JN, Williams JP, Blumwald E (2001) Engineering salt-tolerant Brassica plants: characterization of yield and seed oil quality in transgenic plants with increased vacuolar sodium accumulation. Proceed Nat Acad Sci 98(22):12832–12836

Zhang H, Kim MS, Sun Y, Dowd SE, Shi H, Paré PW (2008) Soil bacteria confer plant salt tolerance by tissue-specific regulation of the sodium transporter HKT1. Mol Plant Microbe Interact 21(6):737–744

Zhang JY, Broeckling CD, Blancaflor EB, Sledge MK, Sumner LW, Wang ZY (2005) Overexpression of WXP1, a putative Medicago truncatula AP2 domain-containing transcription factor gene, increases cuticular wax accumulation and enhances drought tolerance in transgenic alfalfa (Medicago sativa). Plant J 42(5):689–707

Zhang J, Zhang S, Cheng M, Jiang H, Zhang X, Peng C, Jin J (2018) Effect of drought on agronomic traits of rice and wheat: a meta-analysis. Inter J Environ Res Public Health 5:15

Zhang N, Sun Q, Zhang H, Cao Y, Weeda S, Ren S, Guo YD (2014) Roles of melatonin in abiotic stress resistance in plants. J Exp Bot 66(3):647–656

Zhang SW, Li CH, Cao J, Zhang YC, Zhang SQ, Xia YF, Sun Y (2009) Altered architecture and enhanced drought tolerance in rice via the down-regulation of indole-3-acetic acid by TLD1/OsGH3. 13 activation. Plant Physiol 151(4):1889–1901

Zhu JK (2002) Salt and drought stress signal transduction in plants. Annu Rev Plant Biol 53: 247–273

Zhu JK (2001) Plant salt tolerance. Trend. Plant Sci 6(2):66–71

Zia MA, Yasmin H, Shair F, Jabeen Z, Mumtaz S, Hayat Z, Hassan MN (2018) glucanolytic rhizobacteria produce antifungal metabolites and elicit ROS scavenging system in sugarcane. Sugar Tech 1–12 https://doi.org/10.1007/s12355-018-0654-7

Chapter 12
Growth and Yield of Field Crops Grown Under Drought Stress Condition Is Influenced by the Application of PGPR

Naeem Khan and Asghari Bano

Abstract Water stress causes significant losses to agriculture crops mostly grown in rainfed condition. Water stress affects the plant–water relation that causes specific and nonspecific damages to crop. Water stress is categorized as the dominant abiotic stress that is responsible for secondary stresses including oxidative stress that has hazardous effects on the biomolecules of cell. Plant growth-promoting rhizobacteria (PGPR) inoculated plants grow well under biotic and abiotic stresses. Plant survival in abiotic stresses defends on many adaptations and mitigation strategies. PGPR play dominant role in protecting plants from these stresses either directly or indirectly. PGPR colonize the rhizosphere and impose tolerance by producing different metabolites and other volatile compounds and by regulating gene expression and altering root morphology under water scarcity. PGPR influence physiology of plant in response to drought. Considerable growth in cereals has been noted in response to bacterial inoculation. PGPRs like *Azospirillum*, *Pseudomonas*, *Bacillus* and *Azotobacter* are associated with plant roots, improve shoot and root growth and drought tolerance; however, current works unveiled that PGPRs not have just stimulated the systemic tolerance to abiotic stresses but also improve nutrient uptake.

Keywords Drought · PGPR · EPS · Tolerance

N. Khan
Department of Plant Sciences, Quaid-i-Azam University, Islamabad, Pakistan
e-mail: naeemkhan@bs.qau.edu.pk

A. Bano (✉)
Department of Biosciences, University of Wah, Wah Cantt., Pakistan
e-mail: banoasghari@gmail.com

D. K. Maheshwari and S. Dheeman (eds.), *Field Crops: Sustainable Management by PGPR*, Sustainable Development and Biodiversity 23,
https://doi.org/10.1007/978-3-030-30926-8_12

12.1 Introduction

Crop production is significantly reduced worldwide as a result of drought stress. Crop growth models calculation suggest that this matter will be harsher in the forthcoming days (Khan et al. 2018). Drought damages normal growth interrupts plant–water relation and decreases water-use efficiency. It had been reported that less precipitation and high evapotranspiration result moisture deficient condition (Mishra and Cherkauer 2010). Agriculture drought is the absence of sufficient wetness needed for growth of plants (Manivannan et al. 2008). Drought-induced reduction in growth causes significant decrease in plant biomass. The key effects of drought consist of a reduction in cell division, leaf size; stem growth and nutrient uptake (Li et al. 2009; Farooq et al. 2009). Plants need to adopt different mechanisms against drought (Khan et al. 2017). In plants, changes in response to unfriendly environmental conditions could be used while selecting drought-tolerant germplasms for better growth and productivity even under stressful unfriendly condition (Nam et al. 2001; Martínez et al. 2007). The response of the plant to various abiotic stresses fluctuates from species to species and at different organizational levels which depends on the age of the plant, the intensity of stress, and duration of stress (Chaves et al. 2002; Jaleel et al. 2008). Drought stress tolerance could be understood from plant reactions to lesser water condition (Reddy et al. 2004). Though the responses of plants to various stresses are well known, the performance of the plant under multiple stresses is scrappy. This is why it is important for plants to respond instantaneously to numerous stresses co-occurring at one time. These types of inquiries are typically not expectable from single factor study (Zhou et al. 2007). It is noted that water shortage, high rate of evapotranspiration, and intense sunlight reduces the yield 30–50% (Fathi and Tari 2016) (Fig. 12.1).

Soil inhabitant bacteria have some useful effects on soil properties and plant health. Kloepper and Schroth (1981) were the first to use the term PGPR for these

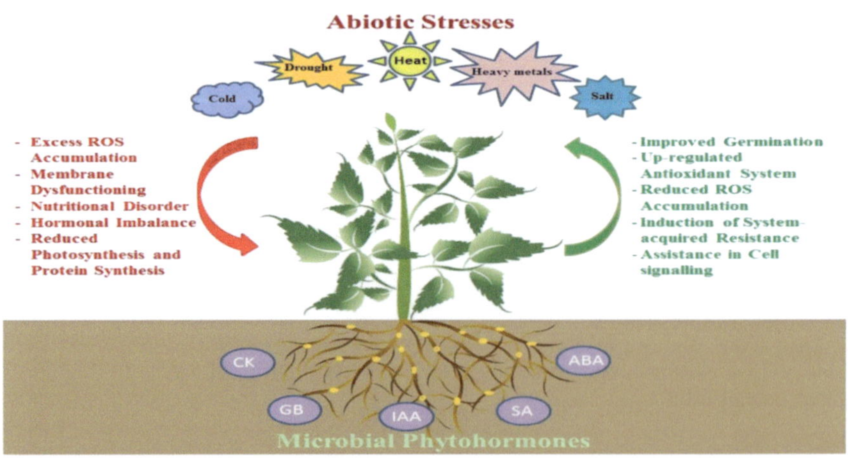

Fig. 12.1 An overview of mechanism in microbial phytohormone-mediated plant stress tolerance (Egamberdieva et al. 2008)

rhizobacteria. Whereas, the rhizosphere is the region in the vicinity of roots and an active zone for microbial activities and is of vital significance for plant health and nutrition (Nihorimbere et al. 2011). The number of rhizobacteria is relatively higher near roots than surrounding soil due to the presence of root exudates in this region (Khan and Bano 2016a). PGPR enhance plant growth and impart drought tolerance by a number of different mechanisms. PGPR improve plant growth by enhancing abiotic stress tolerance in plants, increase uptake of nutrients, production of phytohormones and ACC-deaminase activity that prevent diseases in plants (Saleem et al. 2007). PGPR have been utilized commercially (Dutta and Podile 2010) and hold great potential for sustainable agriculture. Growth-promoting potentials of PGPR have been examined in a large number of cereals (Gray and Smith 2005; Khan and Bano 2016b; Khan et al. 2017).

12.2 Effects of Drought Stress on Plant Growth and Development

Drought affects plants to the extent that impairs normal functions (Siddique et al. 2000) and brings significant changes in plant physiology and morphology (Rahdari et al. 2012). Drought stress induces a decline in growth have been examined in many crops (Samarah 2005; Kamara et al. 2003; Oh et al. 2005; Rampino et al. 2006). Drought stress adversely affects fresh weight and water content of the developing plant (Jaleel et al. 2009). Besides this it affects the accessibility and nutrient movement in soil. Therefore, drought reduces nutrient availability and diffusion (Barber et al. 2000). Free radicals like ROS are induced in plants as a result of water shortage, affecting plant growth by producing oxidative stress. The higher concentration of ROS is harmful to various plant tissues as it initiates lipid peroxidation, protein degradation, and destruction of lipids in the affected plants (Smirnoff 1993; Sgherri et al. 2000; Nair et al. 2008).

Drought coupled with high temperature reduces photosynthesis and is responsible for the continuous production of ROS that may cause plant death (Chaves et al. 2009). A surge in drought leads to the ionic condition in roots, resulting in osmotic stress and ion toxicity. In fact, with an increase in droughty period, cell-wall shriveled and gets loose, resulting poor turgor pressure which ultimately leads to decreased growth (Anjum et al. 2011; Farooq et al. 2012; Simões-Araújo et al. 2003). The continuous increase in the period of water stress, increase in the events of heat waves that enhance drought and heat stresses (Sekhon et al. 2011; Wahid et al. 2007). There are three main factors responsible for the decrease in yield by soil water deficit, i.e., reduce in plant canopy, decrease in radiation-use efficiency, and reduction in harvest index (Earl and Davis 2003). A slow pace in illuminating drought tolerance mechanism has weakened both traditional breeding efforts as well as modern genetics approaches for the development of drought-tolerant varieties (Xiong et al. 2006).

Water deficit has dominant effects on the incidence of kernel abortion in maize during pollination (Sangoi 2001). Water stress diminishes normal growth and productivity by affecting grain filling (Morgan et al. 1990; Ober et al. 1991). In cereals, filling of grains is an important step of starch biosynthesis and four enzymes, viz., sucrose synthase, adenosine diphosphate–glucose–pyrophosphorylase, starch synthase, and starch branching enzyme are known to play a key role during this process (Taiz and Zeiger 2006). It is believed that a decrease in sucrose synthetase results in a decrease in the rate of grain growth in water-stressed wheat (Ahmadi and Baker 2001). In pigeon pea and rice, drought stress concurring with the flowering stage causes a reduction by 40–55% in grain yield (Nam et al. 2001; Yang et al. 2009).

Lack of moisture results in a substantial decrease in dry matter of plant organs, though this varies from organ to organ (Asrar and Elhindi 2011). Similarly, drought greatly restricted the biomass of Asian red sage, though roots were more resistant in comparison to shoots (Liu et al. 2011; Farooq et al. 2010; Ge et al. 2006; Blum 2005). New plantation and ill-developed plants are very vulnerable to drought stress because of a limited number of developed roots and a huge mass of stems and leaves in contrast to roots. Different plant species respond differently to drought stress. For instance, maize plant responds by delaying flowering (McMaster et al. 2008), a similar response is also present in quinoa (Geerts et al. 2008) and rice (Fukai et al. 1999), whereas in other crops (soybean, wheat, etc.) drought accelerated flowering and physiological maturity (Desclaux and Roumet 1996). Under drought stress, plants modify their functions in order to curtail adverse effects and enhance existence (Thapa et al. 2011). In reaction to drying soils, plants adjust the expression of entire genome that intricate in drought tolerance, thus impart tolerance to plants experiencing stress condition (Clement et al. 2008).

12.3 PGPR and Their Mode of Action

PGPR have the ability to inhabit the rhizosphere and improve the tolerance of plants to various stresses. This is because of their ability to solubilize inorganic phosphate, produces plant hormones, siderophore, exopolysaccharides (EPS), organic acids (OA), fix atmospheric nitrogen, and control plant diseases (Muleta et al. 2013; Rani et al. 2012). A large number of bacteria species have been studied for this purpose and many of them commercialized (Vejan et al. 2016). However, they are little utilized, this is because of their inconsistency that could influence crop production. Success through PGPR inoculation greatly depends on their survival in the rhizosphere, compatibility with crop and interactions with crop and other soil micro-flora (Compant et al. 2010). Another aspect is that the mode of action, which is diverse and vary from species to species of PGPR (Siddiqui 2005). This weakness limits the advantages of PGPR. Hence, the rivalry between chemical fertilizers and biofertilizer is believed redundant in the face of the worldwide agricultural productivity necessary to nurse the growing world's population (Lelieveld et al. 2015).

PGPR have been categorized based on their actions as biofertilizers, phytostimulators, rhizoremediators, and biopesticides (Antoun and Prévost 2005; Kloepper 2003; Vessey 2003). In general, PGPR may be categorized into extracellular (ePGPR) and intracellular (iPGPR) PGPR (Figueiredo et al. 2016; Bhattacharyya and Jha 2012). Most of the iPGPR are gram-negative rod-shaped bacteria, while some are gram-positive cocci or pleomorphic (Bhattacharyya and Jha 2012). Besides, many actinomycetes are also the dominant inhabitants of rhizosphere microbial community beneficial characters (Bhattacharyya and Jha 2012; Merzaeva and Shirokikh 2010). Among them, *Streptosporangium* sp., *Micromonospora* sp., *Thermobifida* sp., and *Streptomyces* spp., which have presented huge potential as bio-control agents contrary to a variety of root fungal pathogens (Franco-Correa et al. 2010).

12.4 PGPR and Their Mechanism of Action Under Drought Stress

One alternative treatment against drought is the application of PGPR to crops as they boost plant growth and increase drought tolerance either directly or indirectly (Bashan 1998; Cassán et al. 2014). This may be due to their ability for fixation of atmospheric nitrogen, production of different plant hormones including GA, IAA, and cytokinin, solubilization of inorganic phosphate and production of exopolysaccharides and siderophores (Mayak et al. 1999). In addition, PGPR is related to catabolism of molecules linked to stress-signaling in plants (Glick et al. 2007).

The rhizosphere is inhabited by millions of different microbes that form complex interactions with the plant (Berg 2009; Schmidt et al. 2014). Plant-associated PGPR adopt itself to stressful environmental conditions and thus improve the resistant of plants towards a variety of abiotic stressors which also promote plant health and yield (Schmidt et al. 2014; Yin et al. 2018). However, this depends on the distribution of microorganisms in the rhizosphere and endosphere. Differences in the bacterial population were experiential at the rhizosphere and in the soil nearby plant roots due to root exudates (Marasco et al. 2012).

Microbiome related to roots can improve micronutrient uptake or disturb the content of phytohormones or indirectly they arouse the plant immunity against phytopathogens (Balloi et al. 2010; Rolli et al. 2015; Berg et al. 2014). ACCdeaminase producing PGPR species have the ability to lesser stress and impart tolerance against that particular stress (Lucy et al. 2004). Similarly, *Acinetobacter* and *Pseudomonas* species significantly increased the shoot and leaf biomass the photosynthetic activity even in drought-sensitive grapevine plants grown under stress condition (Rolli et al. 2015). Similar results were noted by Khan et al. (2017) for drought-sensitive chickpea genotype. PGPR, *Stenotrophomonas rhizophila* DSM14405 triggered the expression of several functional genes responsible for stress protection, energy production and for motility of cells (Alavi et al. 2013). The extremely eroded region of southern Sonoran desert had been reinstated with leguminous plants by the application of PGPR in combination with mycorrhizal fungi and compost (Bashan et al. 2012).

ACC-deaminase-producing bacteria results in more seed yield and enhance seed size and restore the process of nodulation even under unavailability of water (Dodd et al. 2005). These bacteria effectively reduced the negative effects of moisture stress in plants grown under field and pot condition (Arshad et al. 2008). Inoculation of a pea plant with ACC-deaminase producing *P. fluorescens* caused increase root system and improved nutrient uptake under water scarcity (Zahir et al. 2008; Rossi et al. 2012), whereas *V. paradoxus* augmented growth, yield, and water-use efficiency of droughty peas (Belimov et al. 2009). This PGPR augmented the soil drying-induced surge in the abscisic acid of xylem tissues and lessened soil drying-induced rise of xylem ACC (Dodd et al. 2005; Tiemann and Billings 2011). It has been reported that nitrogen-fixing PGPR prevents drought-induced decrease in nodules formation (Belimov et al. 2009) (Fig. 12.2).

PGPR were found responsive even at the transcriptional level in *Arabidopsis thaliana* where they improved the growth and tolerance (Timmusk et al. 2011). Six newly identified stress-related proteins were reported in plants after treatment with *B. licheniformis*. Among them the genes of Cadhn, VA, sHSP, and CaPR-10 showed 1.5-fold increase in inoculated plants than control (Lim and Kim 2013). Upregulation of stress-responsive genes was noted in the inoculated wheat plants, which also alleviated the harmful effect of drought stress (Kasim et al. 2013;

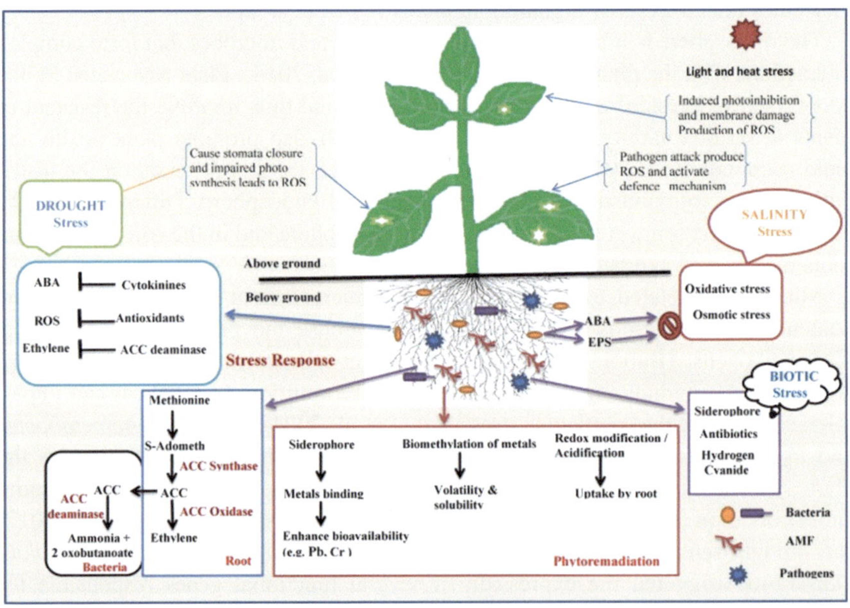

Fig. 12.2 Mechanism of action of ACC-deaminase-producing rhizobacteria for mitigating the adverse effects of drought stress (Kumar and Verma 2017)

Hecker et al. 1996). Similarly, a set of drought signaling responsive genes was downregulated in the *A. thaliana* treated with *P. chlororaphis* compared to those droughty plants lacking bacterial treatment (von Rad et al. 2008). Transcripts of the jasmonic acid (JA) marker genes were upregulated in the PGPR inoculated plants but varied in their reaction to water scarcity (Cho et al. 2013; Schimel et al. 2007).

Rhizobia are well-known mutualistic symbiotic bacteria, could create symbiotic relations with leguminous crops, and help them in fixing atmospheric nitrogen in specific root structures known as nodules. These bacteria are also equally beneficial in non-leguminous plants. Maize plants inoculated with PGPR under drought stress displayed lessen antioxidant enzyme activity as compared to un-inoculated plants (Sandhya et al. 2010; Conlin and Nelson 2007). Similarly, inoculated maize with *Bacillus* species developed resistance to drought by reducing the activities of the antioxidants (Vardharajula et al. 2011; Heidari et al. 2011; Placella et al. 2012). Reduction in the activity of apex, catalase, and GPX was noted in inoculated maize plants with EPS-producing bacteria resulting in stress tolerance in test plants (Naseem and Bano 2014). The effectivity of *B. thuringiensis* in lavender and in common sage under rainfed condition have been studied and was found a promising PGPR against drought stress (Armada et al. 2014; Gusain et al. 2015). These studies provide evidence regarding the advantageous effect of PGPR for drought tolerance in plants by altering the antioxidants activity under water-deficit conditions (Gusain et al. 2015) (Fig. 12.3).

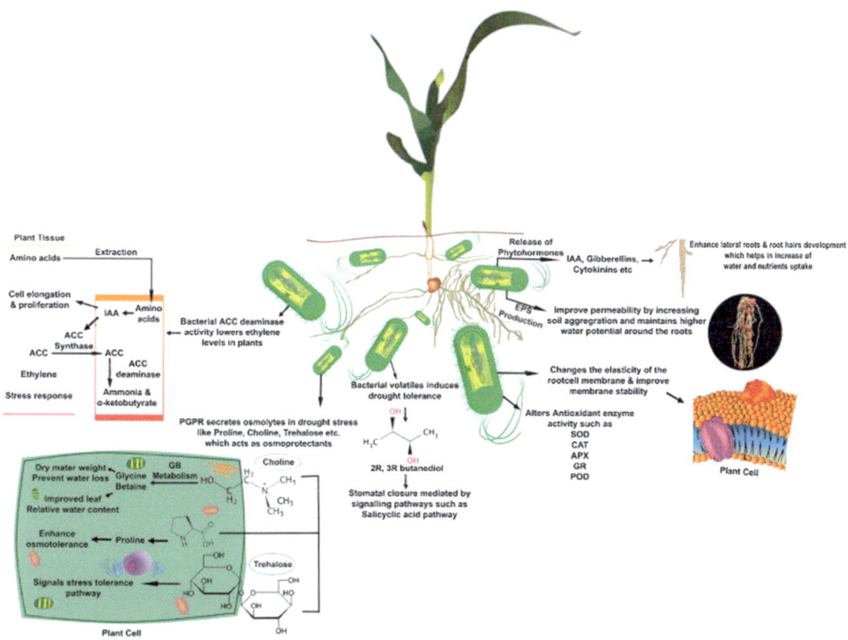

Fig. 12.3 Mechanism of plant drought tolerance induced by plant growth-promoting rhizobacteria (Dubey et al. 2019)

12.5 Conclusion

Water scarcity to the agriculture crops mainly in arid and semiarid regions decreasing crop production. PGPR are playing a significant role in the course of managing plant stresses. Enhanced production of exopolysaccharides, phytohormone, 1-aminocyclopropane- 1-carboxylate (ACC) deaminase, volatile compounds by PGPR underwent considerable increases in growth and yield of important cereals. There is recent urgent to explore the draught resistant mechanisms of PGPR to facilitate this plant protection and phytoremediation properties in order to sustain agriculture.

Conflict of Interest The author(s) have no conflict of interest.

References

Ahmadi A, Baker DA (2001) The effect of water stress on grain filling processes in wheat. J Agric Sci 136(3):257–269

Alavi P, Starcher M, Zachow C, Müller H, Berg G (2013) Root-microbe systems: the effect and mode of interaction of stress protecting agent (SPA) *Stenotrophomonas rhizophila* DSM14405T. Front Plant Sci 4:141

Anjum SA, Xie XY, Wang LC, Saleem MF, Man C, Lei W (2011) Morphological, physiological and biochemical responses of plants to drought stress. Afr J Agric Res 6(9):2026–2032

Antoun H, Prévost D (2005) Ecology of plant growth promoting rhizobacteria. In: Siddqui ZA (ed) PGPR: Biocontrol and biofertilization. Springer, Dordrecht, pp 1–38

Armada E, Roldán A, Azcon R (2014) Differential activity of autochthonous bacteria in controlling drought stress in native *Lavandula* and *Salvia* plants species under drought conditions in natural arid soil. Microb Ecol 67(2):410–420

Arshad M, Shaharoona B, Mahmood T (2008) Inoculation with *Pseudomonas* spp. containing ACC-deaminase partially eliminates the effects of drought stress on growth, yield, and ripening of pea (*Pisum sativum* L.). Pedosphere 18(5):611–620

Asrar AWA, Elhindi KM (2011) Alleviation of drought stress of marigold (*Tagetes erecta*) plants by using arbuscular mycorrhizal fungi. Saudi J Biol Sci 18(1):93–98

Balloi A, Rolli E, Marasco R, Mapelli F, Tamagnini I, Cappitelli F, Borin S, Daffonchio D (2010) The role of microorganisms in bioremediation and phytoremediation of polluted and stressed soils. Agrochimia 54(6):353–369

Barber VA, Juday GP, Finney BP (2000) Reduced growth of Alaskan white spruce in the twentieth century from temperature-induced drought stress. Nature 405(6787):668

Bashan A, Bartsch RP, Kantelhardt JW, Havlin S, Ivanov PC (2012) Network physiology reveals relations between network topology and physiological function. Nat Comm 3:702

Bashan Y (1998) Inoculants of plant growth-promoting bacteria for use in agriculture. Biotechnol Adv 16(4):729–770

Belimov AA, Dodd IC, Hontzeas N, Theobald JC, Safronova VI, Davies WJ (2009) Rhizosphere bacteria containing 1-aminocyclopropane-1-carboxylate deaminase increase yield of plants grown in drying soil via both local and systemic hormone signalling. New Phytol 181(2):413–423

Berg G (2009) Plant–microbe interactions promoting plant growth and health: perspectives for controlled use of microorganisms in agriculture. Appl Microbiol Biotechnol 84(1):11–18

Berg G, Grube M, Schloter M, Smalla K (2014) Unraveling the plant microbiome: looking back and future perspectives. Front Microbiol 5:148
Bhattacharyya PN, Jha DK (2012) Plant growth-promoting rhizobacteria (PGPR): emergence in agriculture. World J Microbiol Biotechnol 28(4):1327–1350
Blum A (2005) Drought resistance, water-use efficiency, and yield potential—are they compatible, dissonant, or mutually exclusive? Aus J Agric Res 56(11):1159–1168
Cassán F, Vanderleyden J, Spaepen S (2014) Physiological and agronomical aspects of phytohormone production by model plant-growth-promoting rhizobacteria (PGPR) belonging to the genus *Azospirillum*. J Plant Growth Regul 33(2):440–459
Chaves MM, Flexas J, Pinheiro C (2009) Photosynthesis under drought and salt stress: regulation mechanisms from whole plant to cell. Annal Bot 103(4):551–560
Chaves MM, Pereira JS, Maroco J, Rodrigues ML, Ricardo CPP, Osório ML, Carvalho I, Faria T, Pinheiro C (2002) How plants cope with water stress in the field? Photosynthesis and growth. Annal Bot 89(7):907–916
Cho SW, Kim S, Kim JM, Kim JS (2013) Targeted genome engineering in human cells with the Cas9 RNA-guided endonuclease. Nat Biotechnol 31(3):230
Clement M, Lambert A, Herouart D, Boncompagni E (2008) Identification of new up-regulated genes under drought stress in soybean nodules. Gene 426(1):15–22
Compant S, Clément C, Sessitsch A (2010) Plant growth-promoting bacteria in the rhizo-and endosphere of plants: their role, colonization, mechanisms involved and prospects for utilization. Soil Biol Biochem 42(5):669–678
Conlin LK, Nelson HC (2007) The natural osmolyte trehalose is a positive regulator of the heat-induced activity of yeast heat shock transcription factor. Mol Cell Biol 27(4):1505–1515
Desclaux D, Roumet P (1996) Impact of drought stress on the phenology of two soybean (*Glycine max* L. Merr) cultivars. Field Crop Res 46(1–3):61–70
Dodd AN, Salathia N, Hall A, Kévei E, Tóth R, Nagy F, Hibberd JM, Millar AJ, Webb AA (2005) Plant circadian clocks increase photosynthesis, growth, survival, and competitive advantage. Science 309(5734):630–633
Dubey A, Kumar A, Abd_Allah EF, Hashem A, Khan ML (2019) Growing more with less: breeding and developing drought resilient soybean to improve food security. Ecol Ind 105:425–437
Dutta S, Podile AR (2010) Plant growth promoting rhizobacteria (PGPR): the bugs to debug the root zone. Crit Rev Microbiol 36(3):232–244
Earl HJ, Davis RF (2003) Effect of drought stress on leaf and whole canopy radiation use efficiency and yield of maize. Agron J 95(3):688–696
Egamberdieva D, Kamilova F, Validov S, Gafurova L, Kucharova Z, Lugtenberg B (2008) High incidence of plant growth-stimulating bacteria associated with the rhizosphere of wheat grown on salinated soil in Uzbekistan. Environ Microbiol 10(1):1–9
Farooq M, Hussain M, Wahid A, Siddique KHM (2012) Drought stress in plants: an overview. In: Ricardo A (ed) Plant responses to drought stress. Springer, Berlin, pp 1–33
Farooq M, Wahid A, Kobayashi N, Fujita D, Basra SMA (2009) Plant drought stress: effects, mechanisms and management. In: Lichtfouse E et al (eds) Sustainable agriculture. Springer, Dordrecht, pp 153–188
Farooq M, Wahid A, Lee DJ, Cheema SA, Aziz T (2010) Drought stress: comparative time course action of the foliar applied glycinebetaine, salicylic acid, nitrous oxide, brassinosteroids and spermine in improving drought resistance of rice. J Agro Crop Sci 196(5):336–345
Fathi A, Tari DB (2016) Effect of drought stress and its mechanism in plants. Int J Life Sci 10 (1):1–6
Figueiredo MDVB, Bonifacio A, Rodrigues AC, de Araujo FF (2016) Plant growth-promoting rhizobacteria: key mechanisms of action. In: Devendra K et al (eds) Microbial-mediated induced systemic resistance in plants. Springer, Singapore, pp 23–37
Franco-Correa M, Quintana A, Duque C, Suarez C, Rodríguez MX, Barea JM (2010) Evaluation of actinomycete strains for key traits related with plant growth promotion and mycorrhiza helping activities. Appl Soil Ecol 45(3):209–217

Fukai S, Pantuwan G, Jongdee B, Cooper M (1999) Screening for drought resistance in rainfed lowland rice. Field Crops Res 64(1–2):61–74

Ge TD, Sui FG, Bai LP, Lu YY, Zhou GS (2006) Effects of water stress on the protective enzyme activities and lipid peroxidation in roots and leaves of summer maize. Agric Sci China 5(4):291

Geerts S, Raes D, Garcia M, Condori O, Mamani J, Miranda R, Cusicanqui J, Taboada C, Yucra E, Vacher J (2008) Could deficit irrigation be a sustainable practice for quinoa (*Chenopodium quinoa* Willd.) in the Southern Bolivian Altiplano? Agric Water Manag 95(8):909–917

Glick BR, Cheng Z, Czarny J, Duan J (2007) Promotion of plant growth by ACC deaminase-producing soil bacteria. Eur J Plant Pathol 119:329–339

Gray EJ, Smith DL (2005) Intracellular and extracellular PGPR: commonalities and distinctions in the plant–bacterium signaling processes. Soil Biol Biochem 37(3):395–412

Gusain YS, Kamal R, Mehta CM, Singh US, Sharma AK (2015) Phosphate solubilizing and indole-3-acetic acid producing bacteria from the soil of Garhwal Himalaya aimed to improve the growth of rice. J Environ Biol 36(1):301

Hecker M, Schumann W, Völker U (1996) Heat-shock and general stress response in *Bacillus subtilis*. Mol Microbiol 19(3):417–428

Heidari M, Mousavinik SM, Golpayegani A (2011) Plant growth promoting rhizobacteria (PGPR) effect on physiological parameters and mineral uptake in basil (*Ociumum basilicm* L.) under water stress. ARPN J Agric Biol Sci 6(5):6–11

Jaleel CA, Gopi R, Sankar B, Gomathinayagam M, Panneerselvam R (2008) Differential responses in water use efficiency in two varieties of *Catharanthus roseus* under drought stress. Comptes Rendus Biol 331(1):42–47

Jaleel CA, Manivannan P, Wahid A, Farooq M, Al-Juburi HJ, Somasundaram R, Panneerselvam R (2009) Drought stress in plants: a review on morphological characteristics and pigments composition. Int J Agric Biol 11(1):100–105

Kamara AY, Menkir A, Badu-Apraku B, Ibikunle O (2003) The influence of drought stress on growth, yield and yield components of selected maize genotypes. J Agric Sci 141(1):43–50

Kasim WA, Osman ME, Omar MN, El-Daim IAA, Bejai S, Meijer J (2013) Control of drought stress in wheat using plant-growth-promoting bacteria. J Plant Growth Regul 32(1):122–130

Khan N, Bano A (2016a) Modulation of phytoremediation and plant growth by the treatment with PGPR, Ag nanoparticle and untreated municipal wastewater. Int J Phytoremed 18(12):1258–1269

Khan N, Bano A (2016b) Role of plant growth promoting rhizobacteria and Ag-nano particle in the bioremediation of heavy metals and maize growth under municipal wastewater irrigation. Int J Phytoremed 18(3):211–221

Khan N, Bano A, Babar MA (2017) The root growth of wheat plants, the water conservation and fertility status of sandy soils influenced by plant growth promoting rhizobacteria. Symbiosis 72 (3):195–205

Khan N, Bano A, Rahman MA, Rathinasabapathi B, Babar MA (2018) UPLC-HRMS-based untargeted metabolic profiling reveals changes in chickpea (*Cicer arietinum*) metabolome following long-term drought stress. Plant Cell Environ 42(1):1–5

Kloepper JW, Schroth MN (1981) Development of a powder formulation of rhizobacteria for inoculation of potato seed pieces. Phytopathol 71(6):590–592

Kloepper JW (2003) A review of mechanisms for plant growth promotion by PGPR. In: 6th international PGPR workshop, October, vol 10, pp 5–10

Kumar A, Verma JP (2017) Does plant—Microbe interaction confer stress tolerance in plants: a review? Microbiol Res 207:41–52

Lelieveld J, Beirle S, Hörmann C, Stenchikov G, Wagner T (2015) Abrupt recent trend changes in atmospheric nitrogen dioxide over the Middle East. Sci Adv 1(7):e1500498

Li Y, Ye W, Wang M, Yan X (2009) Climate change and drought: a risk assessment of crop-yield impacts. Clim Res 39(1):31–46

Lim JH, Kim SD (2013) Induction of drought stress resistance by multi-functional PGPR *Bacillus licheniformis* K11 in pepper. Plant Pathol J 29(2):201

Liu C, Liu Y, Guo K, Fan D, Li G, Zheng Y, Yu L, Yang R (2011) Effect of drought on pigments, osmotic adjustment and antioxidant enzymes in six woody plant species in karst habitats of southwestern China. Environ Exp Bot 71(2):174–183

Lucy M, Reed E, Glick BR (2004) Applications of free living plant growth-promoting rhizobacteria. Antonie Van Leeuwenhoek 86(1):1–25

Manivannan P, Jaleel CA, Somasundaram R, Panneerselvam R (2008) Osmoregulation and antioxidant metabolism in drought-stressed *Helianthus annuus* under triadimefon drenching. Comptes Rendus Biol 331(6):418–425

Marasco R, Rolli E, Ettoumi B, Vigani G, Mapelli F, Borin S, Abou-Hadid AF, El-Behairy UA, Sorlini C, Cherif A, Zocchi G (2012) A drought resistance-promoting microbiome is selected by root system under desert farming. PLoS ONE 7:e48479

Martínez JP, Silva H, Ledent JF, Pinto M (2007) Effect of drought stress on the osmotic adjustment, cell wall elasticity and cell volume of six cultivars of common beans (*Phaseolus vulgaris* L.). Euro J Agron 26(1):30–38

Mayak S, Tirosh T, Glick BR (1999) Effect of wild-type and mutant plant growth-promoting rhizobacteria on the rooting of mung bean cuttings. J Plant Growth Regul 18(2):49–53

McMaster GS, White JW, Weiss A, Baenziger PS, Wilhelm WW, Porter JR, Jamieson PD (2008) Simulating crop phenological responses to water deficits. Response of crops to limited water: understanding and modeling water stress effects on plant growth processes, pp 277–300 (Response of Crops)

Merzaeva OV, Shirokikh IG (2010) The production of auxins by the endophytic bacteria of winter rye. Appl Biochem Microbiol 46(1):44–50

Mishra V, Cherkauer KA (2010) Retrospective droughts in the crop growing season: implications to corn and soybean yield in the Midwestern United States. Agric Forest Meteorol 150(7–8):1030–1045

Morgan PW, He CJ, De Greef JA, Maurice P (1990) Does water deficit stress promote ethylene synthesis by intact plants? Plant Physiol 94(4):1616–1624

Muleta D, Assefa F, Börjesson E, Granhall U (2013) Phosphate-solubilising rhizobacteria associated with *Coffea arabica* L. in natural coffee forests of south western Ethiopia. J Saudi Soc Agric Sci 12(1):73–84

Nair AS, Abraham TK, Jaya DS (2008) Studies on the changes in lipid peroxidation and antioxidants in drought stress induced cowpea (*Vigna unguiculata* L.) varieties. J Environ Biol 29:689–691

Nam NH, Chauhan YS, Johansen C (2001) Effect of timing of drought stress on growth and grain yield of extra-short-duration pigeonpea lines. J Agric Sci 136(2):179–189

Naseem H, Bano A (2014) Role of plant growth-promoting rhizobacteria and their exopolysaccharide in drought tolerance of maize. J Plant Interact 9(1):689–701

Nihorimbere V, Ongena M, Smargiassi M, Thonart P (2011) Beneficial effect of the rhizosphere microbial community for plant growth and health. Biotechnol Agron Soc Environ 15(2):327–337

Ober ES, Setter TL, Madison JT, Thompson JF, Shapiro PS (1991) Influence of water deficit on maize endosperm development: enzyme activities and RNA transcripts of starch and zein synthesis, abscisic acid, and cell division. Plant Physiol 97(1):154–164

Oh SJ, Song SI, Kim YS, Jang HJ, Kim SY, Kim M, Kim YK, Nahm BH, Kim JK (2005) *Arabidopsis* CBF3/DREB1A and ABF3 in transgenic rice increased tolerance to abiotic stress without stunting growth. Plant Physiol 138(1):341–351

Placella SA, Brodie EL, Firestone MK (2012) Rainfall-induced carbon dioxide pulses result from sequential resuscitation of phylogenetically clustered microbial groups. Proc Natl Acad Sci 109 (27):10931–10936

Rahdari P, Hosseini SM, Tavakoli S (2012) The studying effect of drought stress on germination, proline, sugar, lipid, protein and chlorophyll content in purslane (*Portulaca oleracea* L.) leaves. J Med Plants Res 6(9):1539–1547

Rampino P, Pataleo S, Gerardi C, Mita G, Perrotta C (2006) Drought stress response in wheat: physiological and molecular analysis of resistant and sensitive genotypes. Plant Cell Environ 29(12):2143–2152

Rani MU, Arundhathi A, Reddy G (2012) Screening of rhizobacteria containing plant growth promoting (PGPR) traits in rhizosphere soils and their role in enhancing growth of pigeon pea. Afr J Biotechnol 11(32):8085–8091

Reddy AR, Chaitanya KV, Vivekanandan M (2004) Drought-induced responses of photosynthesis and antioxidant metabolism in higher plants. J Plant Physiol 161(11):1189–1202

Rolli E, Marasco R, Vigani G, Ettoumi B, Mapelli F, Deangelis ML, Gandolfi C, Casati E, Previtali F, Gerbino R, Pierotti Cei F (2015) Improved plant resistance to drought is promoted by the root-associated microbiome as a water stress-dependent trait. Environ Microbiol 17 (2):316–331

Rossi F, Potrafka RM, Pichel FG, De Philippis R (2012) The role of the exopolysaccharides in enhancing hydraulic conductivity of biological soil crusts. Soil Biol Biochem 46:33–40

Saleem M, Arshad M, Hussain S, Bhatti AS (2007) Perspective of plant growth promoting rhizobacteria (PGPR) containing ACC deaminase in stress agriculture. J Ind Microbiol Biotechnol 34(10):635–648

Samarah NH (2005) Effects of drought stress on growth and yield of barley. Agron Sustain Dev 25 (1):145–149

Sandhya V, Ali SZ, Grover M, Reddy G, Venkateswarlu B (2010) Effect of plant growth promoting *Pseudomonas* spp. on compatible solutes, antioxidant status and plant growth of maize under drought stress. Plant Growth Regul 62(1):21–30

Sangoi L (2001) Understanding plant density effects on maize growth and development: an important issue to maximize grain yield. Ciência Rural 31(1):159–168

Schimel J, Balser TC, Wallenstein M (2007) Microbial stress-response physiology and its implications for ecosystem function. Ecology 88(6):1386–1394

Schmidt R, Köberl M, Mostafa A, Ramadan EM, Monschein M, Jensen KB, Bauer R, Berg G (2014) Effects of bacterial inoculants on the indigenous microbiome and secondary metabolites of chamomile plants. Front Microbiol 5:64

Sekhon RS, Lin H, Childs KL, Hansey CN, Buell CR, de Leon N, Kaeppler SM (2011) Genome-wide atlas of transcription during maize development. Plant J 66(4):553–563

Sgherri C, Maffei M, Navari-Izzo F (2000) Antioxidative enzymes in wheat subjected to increasing water deficit and rewatering. J Plant Physiol 157(3):273–279

Siddique MRB, Hamid A, Islam MS (2000) Drought stress effects on water relations of wheat. Bot Bull Acad Sinica 41

Siddiqui ZA (2005) PGPR: prospective biocontrol agents of plant pathogens. In: Siddiqui ZA (ed) PGPR: biocontrol and biofertilization. Springer, Dordrecht, pp 111–142

Simões-Araújo JL, Rumjanek NG, Margis-Pinheiro M (2003) Small heat shock proteins genes are differentially expressed in distinct varieties of common bean. Braz J Plant Physiol 15(1):33–41

Smirnoff N (1993) The role of active oxygen in the response of plants to water deficit and desiccation. New Phytol 125(1):27–58

Taiz L, Zeiger E (2006) Water and plant cells. Plant physiology, 4th edn. Sinauer Associates Inc, p 672

Thapa GD, Dey M, Sahoo L, Panda SK (2011) An insight into the drought stress induced alterations in plants. Biol Plant 55(4):603

Tiemann LK, Billings SA (2011) Changes in variability of soil moisture alter microbial community C and N resource use. Soil Biol Biochem 43(9):1837–1847

Timmusk S, Paalme V, Pavlicek T, Bergquist J, Vangala A, Danilas T, Nevo E (2011) Bacterial distribution in the rhizosphere of wild barley under contrasting microclimates. PLoS ONE 6(3): e17968

Vardharajula S, Zulfikar Ali S, Grover M, Reddy G, Bandi V (2011) Drought-tolerant plant growth promoting *Bacillus* spp.: effect on growth, osmolytes, and antioxidant status of maize under drought stress. J Plant Interact 6(1):1–14

Vejan P, Abdullah R, Khadiran T, Ismail S, Nasrulhaq BA (2016) Role of plant growth promoting rhizobacteria in agricultural sustainability—a review. Molecules 21(5):57

Vessey JK (2003) Plant growth promoting rhizobacteria as biofertilizers. Plant Soil 255(2):571–586 von Rad U, Klein I, Dobrev PI, Kottova J, Zazimalova E, Fekete A, Hartmann A, Schmitt-Kopplin P, Durner J (2008) Response of *Arabidopsis thaliana* to N-hexanoyl-DL-homoserine-lactone, a bacterial quorum sensing molecule produced in the rhizosphere. Planta 229(1):73–85

Wahid A, Gelani S, Ashraf M, Foolad MR (2007) Heat tolerance in plants: an overview. Environ Exp Bot 61(3):199–223

Xiong L, Wang RG, Mao G, Koczan JM (2006) Identification of drought tolerance determinants by genetic analysis of root response to drought stress and abscisic acid. Plant Physiol 142 (3):1065–1074

Yang Z, Wu Y, Li Y, Ling HQ, Chu C (2009) OsMT1a, a type 1 metallothionein, plays the pivotal role in zinc homeostasis and drought tolerance in rice. Plant Mol Biol 70(1–2):219–229

Yin A, Jia Y, Qiu T, Gao M, Cheng S, Wang X, Sun Y (2018) Poly-γ-glutamic acid improves the drought resistance of maize seedlings by adjusting the soil moisture and microbial community structure. Appl Soil Ecol 129:128–135

Zahir ZA, Munir A, Asghar HN, Shaharoona B, Arshad M (2008) Effectiveness of rhizobacteria containing ACC deaminase for growth promotion of peas (*Pisum sativum*) under drought conditions. J Microbiol Biotechnol 18(5):958–963

Zhou J, Wang X, Jiao Y, Qin Y, Liu X, He K, Chen C, Ma L, Wang J, Xiong L, Zhang Q (2007) Global genome expression analysis of rice in response to drought and high-salinity stresses in shoot, flag leaf, and panicle. Plant Mol Biol 63(5):591–608

Chapter 13
Plant Growth Promotion and Suppression of Fungal Pathogens in Rice (*Oryza Sativa* L.) by Plant Growth-Promoting Bacteria

Hassan Etesami

Abstract Crop plants play an outstanding function in providing food and energy to humans. Rice (*Oryza sativa* L.) is one of the most important stable crops that have a role in providing the main food to more than half of the world's people. One of the important factors in increasing yield in rice is the balanced nutrition or supply of the required nutrients in the proper form and ratios. Chemical fertilizers are essential components of modern agriculture by providing essential plant nutrients. However, the overuse of these fertilizers causes serious environmental pollution. But threats of plant pathogens on the attack and damages on the crop productivity cannot be ruled out. Therefore, chemical-based pesticides are thought to be an effective and trustworthy agricultural management measure for repressing pests. Nowadays, the use of beneficial microorganisms and biological control agents are proved as good as synthetic pure/chemicals for the increased plant growth and yield. The diminished utilization of chemical fertilizers for the management of plant pathogens is considered as a secure and maintainable strategy for safe and rewarding agricultural productivity. Based on research conducted until this moment, rice-associated bacteria are encouraging alternatives to chemical fertilizers in an eco-friendly manner. In general, the application of plant growth-promoting bacteria (PGPB) could offer a cheaper and cost-effective approach to overcome the environmental problems caused by chemical fertilizers and their use in the form of biofertilizers and biopesticides could decrease our reliance on synthetic agrochemicals. This chapter highlights the importance of PGPB for enhancing sustainable rice production.

Keywords Sustainable agriculture · Biocontrol agents · PGPR · Biofertilizer · Co-inoculation

---

H. Etesami (✉)
Department of Soil Science, Faculty of Agricultural Engineering and Technology, University College of Agriculture and Natural Resources, University of Tehran, Tehran, Iran
e-mail: hassanetesami@ut.ac.ir

D. K. Maheshwari and S. Dheeman (eds.), *Field Crops: Sustainable Management by PGPR*, Sustainable Development and Biodiversity 23,
https://doi.org/10.1007/978-3-030-30926-8_13

13.1 Introduction

Cereals are the main source of nutrition for human beings in the world. Among the cereals, rice (*Oryza sativa* L.) is of great importance. Rice is a staple food in food diet over 40% of the world's population, especially, in Asia (Naureen et al. 2009a). Of the total energy produced by cereals per person per day, more of it is related to rice. This has made rice as the most important food product in developing countries. According to statistics, the world's rice cultivation area in 2009 was 153 million hectares, with a production of 585 million tons, which should increase to 800 million tons in 2025. In other words, in order to meet the food needs of this growing population, an increase of 70% in rice production is needed over the next few years. Rice is mostly produced in countries, whose population are growing rapidly and often are limited in terms of land and resources. Therefore, given the limiting factors of production (including decreasing the quality and quantity of agricultural land, reducing water resources and labor shortage), the only rational solution is to increase the yield of rice per unit area of land cultivation or use of high-yielding rice varieties to meet the demand for rice demanded in 2025 (Mishra et al. 2006). However, the use of these varieties requires extensive application of fertilizers such as nitrogen (N) and phosphorus (P) (Hazell 2010). Some of the main constraints on the growth of this crop can be inadequate fertilizer use, pest infestation, and growing of low-yielding traditional varieties, and paucity of water (Datta et al. 2017). In general, one of the most important factors in increasing the rice yield is the balanced nutrition or the supply of essential nutrients. Low-soil fertility is the most important factor which not only seriously affects the rice production but also reduces the quality of the rice (Vaid et al. 2014).

Chemical fertilizers are essential components of modern agriculture due to the provision of essential plant elements. However, the excessive use of these chemical fertilizers for greater production of crop plants including rice can cause unpredictable environmental impacts including leaching and runoff of nutrients, especially N and P, leading to environmental degradation (deterioration in air and water quality) (Gyaneshwar et al. 2002). In addition to essential nutrients, diseases are also among the most significant limiting factors that affect rice production, causing annual yield losses conservatively estimated at five percent (Song and Goodman 2001).

In agricultural systems, the utilization of plant growth-promoting bacteria (PGPB) is of particular consequence in augmenting crop production and preserving sustainable soil fertility (Bagyaraj and Balakrishna 2012). In the past decade, the use of PGPB as a biofertilizer or biological control agent in agriculture has been considered by many researchers. The growth of different crops by these bacteria has been proved in greenhouse and field experiments. Most studies show that these bacteria could have positive and economic effects on crop plants such as corn, wheat, and rice (Freitas and Germida 1990; Çakmakçı et al. 2007; Etesami et al. 2013, 2014a, c, 2015; Ghorchiani et al. 2018; Etesami and Maheshwari 2018) by mechanisms like increasing the availability of soil mineral elements (i.e., through solubilizing insoluble P compounds and potassium (K)-bearing minerals and releasing P and K), producing plant growth-regulating hormones (i.e., indole-3-acetic acid, gibberellin, and cytokinin) and siderophores (increase in availability of Fe, Zn, etc.), producing ACC (1-aminocyclopropane-1-carboxylate) deaminase (decrease of stress ethylene), controlling pathogenic microorganisms (Etesami et al. 2017; Etesami and Maheshwari 2018), and nitrogen fixation (Bhattacharjee et al. 2008; Saharan and Nehra 2011).

It has been well proven that PGPB could increase plant growth and resistance to environmental stresses (Fig. 13.1) such as salinity (Dimkpa et al. 2009; Egamberdieva and Lugtenberg 2014; Paul and Lade 2014; Choudhary et al. 2016; Qin et al. 2016), drought (Timmusk et al. 2013; Choudhary et al. 2016; Kaushal and Wani 2016; Ngumbi and Kloepper 2016), heavy metal toxicity (Carmen and Roberto 2011; Sessitsch et al. 2013; Ullah et al. 2015), nutritional imbalance (Adesemoye and Kloepper 2009; Yang et al. 2009; Miransari 2013; Chakraborty et al. 2015; Pii et al. 2015; Choudhary et al. 2016), and plant pathogens (bacterium, virus, fungi, etc.) (Compant et al. 2005; Pal and Gardener 2006; Ryan et al. 2008) via miscellaneous mechanisms usually more than one action mechanism (Etesami and Maheshwari 2018). Despite these good reviews, there are a few review studies on PGPB-mediated nutrient availability and biological control of fungal pathogens in rice. Better understanding on interactions of rice with the plant-associated PGPB enhanced nutrient acquisition and controlled fungal rice pathogens is needed for increasing the efficiency of nutrient management and rice disease management in soil and also for promoting eco-friendly low-input sustainable agriculture. Therefore, the aim of this chapter was to reviews advances in research on PGPB

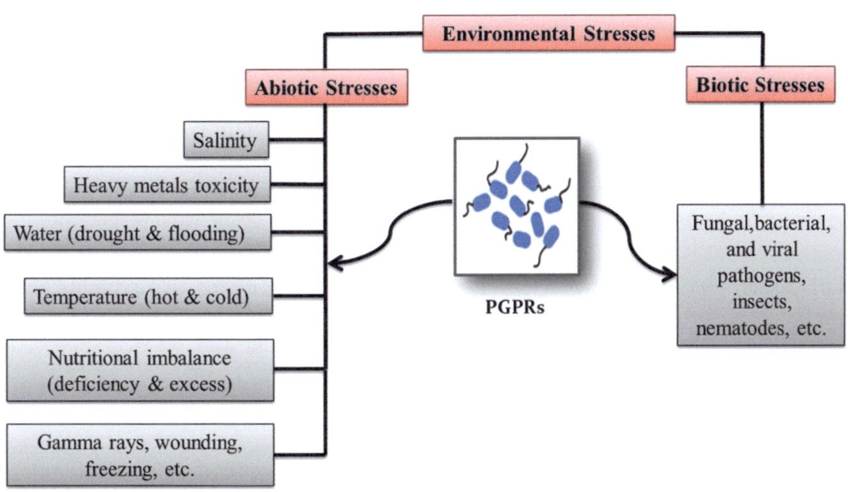

Fig. 13.1 Common abiotic and biotic stresses in agricultural environments alleviated by PGPR

capable of increasing the availability of soil insoluble nutrients, their mechanism of action, and their potential use for biofertilization of nutrients in rice and biological control of fungal rice pathogens (Table 13.1).

13.2 Nitrogen-Fixing Bacteria (NFB) in Nitrogen Nutrition

Nitrogen (N) is an important element in the plant and a component of chlorophyll molecules and therefore plays an important role in photosynthesis and in the production of proteins, nucleic acids, and coenzymes. Chemical N-fertilizers are one of the most influential factors in the production and yield of rice. Without the addition of chemical N-fertilizers, the yield of existing varieties is severely limited (Ladha et al. 1997). The excessive use of the chemical fertilizers for greater production of this crop has caused unpredictable environmental impacts. Currently, most of the N-fertilizers are produced through the Haber–Bosch process at chemical fertilizer factories. This process requires a large amount of energy (natural gas or oil), all of which are nonrenewable sources. It also generates carbon dioxide (CO_2), which is a greenhouse gas. In developing countries, the cost of purchasing N-fertilizers is usually higher than farmers' income, which limits yield potential of their crops. Approximately one-third of the applied N (urea-N or nitrate-N, which is applied as fertilizer) is consumed by the plant; the rest of the N can enter as nitrate form into underground waters and are a potential hazard to environmental health. Excess N can also produce nitrous oxide (N_2O), an effective greenhouse gas. In addition, since rice grows in an environment susceptible to N loss, more than half of the N-fertilizer used in the paddies is lost through denitrification, ammonia volatilization, and leaching/runoff (Ladha et al. 1997).

In general, the agrosystems that require a lot of N-fertilizers are not sustainable systems because they require the use of nonrenewable natural resources and can endanger the health and the environment (Yanni and Dazzo 2010). Reducing the amount of industrial N production for agricultural systems is one of the important goals of agricultural researchers. In the case of sustainable rice production, an important goal is to replace the industrial N fixation to biological N fixation (Yanni et al. 2001). Two basic ways to solve the problem of N-fertilizer loss in paddy fields can be proposed: One is the regulation of N application time based on rice needs, which increases the efficiency of plant use of applied N and another way is to increase the ability of the rice to biological nitrogen fixation. The second approach is a long-term strategy, but it has multiple environmental benefits and also helps low-income farmers. Additionally, farmers can easily adopt a variety of genotypes with useful features rather than conducting soil and crop management operations that are costly (Ladha et al. 1997).

Recent advances in understanding the legume–rhizobium–symbiotic relationships at the molecular level and the ability to introduce new genes into the rice

Table 13.1 Potential of rice-associated bacteria in controlling some important fungal rice pathogens

Biocontrol PGPB	Pathogenic fungi	Effect	References
Streptomyces sp., *Pseudomonas* sp., *Ochrobactrum anthropic, Bacillus firmus*, *Pseudomonas aureofaciens,* and *Kocuria rhizophila*	*Alternaria* sp., *Fusarium oxysporum, Pyricularia oryzae* and *Sclerotium* sp.	In dual-culture technique, the siderophore-producing rhizobacteria showed a strong antagonistic effect against the *Alternaria* (35.4%), *F. oxysporum* (37.5%), *P. oryzae* (31.2%), and *Sclerotium* sp. (10.4%)	Chaiharn and Lumyong (2011)
Streptomyces globisporus	*Magnaporthe oryzae*	*S. globisporus* inhibited mycelial growth of *M. oryzae*, and histological investigations showed that conidial germination and appressorial formation of *M. oryzae* were suppressed on detached rice leaves. This bacterium could suppress disease incidence of rice blast caused by *M. oryzae*	Li et al. (2011)
Streptomyces sindeneusis	*Magnaporthe oryzae*	*S. sindeneusis* resulted in strong inhibition of the pathogen and suppression of leaf symptoms	Zarandi et al. (2009)
Streptomyces flavotricini	*Magnaporthe oryzae*	*S. flavotricini* showed the strongest antifungal activity against *M. oryzae;* the antifungal compound produced by *S. flavotricini* was successfully purified and identified as dihydroxy viridiofungin (C37H58N2O10)	Khalil et al. (2014)
Bacillus methylotrophicus BC79	*Magnaporthe oryzae*	This bacterial strain showed the highest efficiency for *M. oryzae*, with 84.8% biocontrol effect	Shan et al. (2013)

(continued)

Table 13.1 (continued)

Biocontrol PGPB	Pathogenic fungi	Effect	References
Bacillus spp.	*Rhizoctonia solani, Sclerotium rolfsii,* and *Sclerotium oryzae*	All *Bacillus* strains showed strongly inhibited (up to 90%) the growth of phytopathogens, were capable of enhancing the vegetative growth and yield parameters (shoot/root length, biomass, chlorophyll content and yield), and showed significant increase than non-inoculated control plants	Sethi and Mukherjee (2018)
Bacillus mojavensis, B. amyloliquefaciens, B. subtilis, and B. cereus	*Magnaporthe oryzae, M. salvinii, Fusarium verticillioides, F. fujikuroi,* and *F. proliferum*	These bacterial strains exhibited significant antifungal activity against these pathogens, with 70–90% inhibition of mycelial growth	Etesami and Alikhani (2017), Etesami and Alikhani (2016a), Etesami et al. (2014a),
Streptomyces sp. UPMRS4	*Pyricularia oryzae*	This bacterial strain was able to reduce 67.9% of disease severity and able to increase shoot height (15.13%), shoot dry weight (45.75%), leaf surface area (44.6%), root length (48.93), root dry weight (63.25%), number of tillers (42.26%), yield (36.96%), panicle length (15.4%), and the number of spikelet/panicles (29.39%) compared to the control plants at 3 months after inoculation	Awla et al. (2017)
Streptomyces philanthi RM-1-138	*Magnaporthe oryzae*	This bacterial strain exhibited significant antifungal activity against *M. oryzae* (*P. oryzae* PTRRC-18), with 88.73% inhibition of mycelial growth of the rice blast fungus	Boukaew and Prasertsan (2014)

(continued)

Table 13.1 (continued)

Biocontrol PGPB	Pathogenic fungi	Effect	References
Streptomyces philanthi RM-1-138	*Rhizoctonia solani, Pyricularia grisea, Bipolaris oryzae*, and *Fusarium fujikuroi*	Volatile organic compounds (VOCs) produced by *S. philanthi* inhibited mycelial growth of the rice pathogenic fungi. The inhibition was most pronounced on *R. solani* while the least inhibition was against *F. fujikuroi*	Boukaew et al. (2013)
Streptomyces vinaceusdrappus	*Magnaporthe oryzae, Curvularia oryzae, Bipolaris oryzae,* and *Fusarium oxysporum*	*S. vinaceusdrappus* showed maximal antagonistic activity against fungal pathogens *Curvularia oryzae*, *Bipolaris oryzae*, and *Fusarium oxysporum. S. vinaceusdrappus* inhibited the mycelial growth of P. oryzae by 53.5%, which was relatively good since more than 50% of the mycelial growth was inhibited	Ningthoujam et al. (2009)
P. fluorescens Aur6 and *Chryseobacterium balustinum* Aur9	*Magnaporte grisea*	Each strain individually protected rice against rice blast, but the combination of both strains was the most effective treatment (reaching 50% of protection against disease). A relation between protection and increase in rice productivity and quality was found	Lucas et al. (2009)
Streptomyces corchorusii UCR3-16	*Bipolaris oryzae, Pyricularia oryzae, Rhizoctonia solani, Rhizoctonia oryzae-sativae, Fusarium oxysporum,* and *Curvularia oryzae*	This bacterial strain exhibited significant biocontrol potential against rice fungal pathogens showing the highest inhibition against *Rhizoctonia solani*. The strain could significantly enhance the growth and grain yield production of rice plants under pot conditions	Tamreihao et al. (2016)

genome by transformation have made it an excellent opportunity to study the ability of N fixation in rice, although this has remained largely unfinished until now (Dawe 2000). There are such opportunities for cereals including rice. In general, the strategies enabling rice to fix nitrogen are complex and have a long-term nature, but if done, they can increase rice productivity, resource conservation, and environmental security. In addition to the strategies mentioned above, it has been well known that the use of nitrogen fixation technology can reduce the use of N-fertilizers in agricultural land, which can be effective at reducing environmental hazards. Biological nitrogen fixation in rice paddies has significantly contributed to the sustainable yield of these systems. Studies show that biological nitrogen fixation in rice paddies can produce up to 50 kg N per hectare (Elbeltagy et al. 2001). It has been well known that nitrogen fixation through the bacteria associated with rice (associative and free-living bacteria) has a high potential for supply of N for rice. For example, in a study, Mäder et al. (2011) observed an increase of 23% in rice yield obtained upon rice inoculation with N_2-fixing *Pseudomonas* sp. In another study, the co-inoculation of N_2-fixing bacteria (i.e., *Brevundimonas diminuta* PR7, *Anabaena oscillarioides* CR3, and *Ochrobactrum anthropi* PR10) remarkably augmented N, P, and K content and bettered rice yield by 21.2%, as compared to the utilization of recommended quantity of N, P, and K fertilizers (Rana et al. 2015). Due to having a very close relationship with the plant, as compared to other bacteria, endophytic bacteria can offer the fixed N to rice without its loss.

Endophyte bacteria seem to be more effective at supplying rice with N than other bacteria. The bacteria isolated from the internal tissues of the plant or isolated from the plants with sterilized surfaces that do not show any symptoms of the disease are regarded as endophytic bacteria (Di Fiore and Del Gallo 1995). It is well documented that a significant diversity of endophytic bacteria such as *Pantoea, Burkholderia, Azospirillum, Herbaspirillum, Rhizobium, Methylobacterium*, etc., is naturally associated with rice (Carvalho et al. 2014; Mano and Morisaki 2008). Diazotrophs that effectively colonized into rice roots can have a greater potential for N fixation. It has been reported that the contribution of endorhizosphere bacteria to N fixation is much more extensive than the contribution of rhizospheric bacteria because there is no competition in the endorhizosphere with other rhizosphere microorganisms, and carbon sources with low-pressure oxygen oscillations are provided (James et al. 2002). Several endophytic N_2-fixing bacteria have been isolated from various rice species including the genera *Klebsiella, Citrobacter, Enterobacter, Bacillus, Alcaligenes, Azospirillum, Rhizobium, Sphingomonas, Agrobacterium, Corynebacterium, Herbaspirillum, Azoarcus, Penibacillus, Microbacterium,* and *Burkholderia* (Reinhold-Hurek et al. 2007; Prayitno and Rolfe 2010; Yanni and Dazzo 2010; Gupta et al. 2012; Hongrittipun et al. 2014; Ji et al. 2014).

It has been found that the stimulation of growth of the crop plants (such as rice) inoculated with N_2-fixing bacteria may be due to other mechanisms like increasing the availability of soil mineral elements, producing plant growth-regulating hormones, siderophores, and ACC deaminase, and controlling pathogenic microorganisms (Etesami et al. 2017; Etesami and Maheshwari 2018) other than nitrogen fixation (Bhattacharjee et al. 2008; Saharan and Nehra 2011). For example, previous studies have shown that nitrogen accumulation in inoculated non-leguminous plants can be due to either biological N fixation (Elbeltagy et al. 2001; Oliveira et al. 2002) or escalation in nitrogen uptake from soil (Prayitno et al. 1999; Yanni et al. 1997). In another study, Etesami and Alikhani (2016a) showed that bacterial IAA had considerable role in improving use efficiency of N and could increase N content of rice. In other works, Estrada et al. (2013) showed that P-solubilizing diazotrophic bacteria augmented nutrient uptake by rice plants. de Souza et al. (2013) showed that the bacteria (e.g., *Herbaspirillum* sp., *Burkholderia* sp., *Burkholderia* sp., *Pseudacidovorax* sp., and *Rhizobium* sp.) unable to solubilize phosphate in in vitro assay and reduce acetylene (low capacity to reduce acetylene) increased levels of N, P, and K in rice shoots. These observations could indicate that growth promotion mechanisms other than N_2 fixation such as IAA production and improved nutrient uptake balance (de Souza et al. 2013; Ji et al. 2014). The above studies show that if the purpose of rice inoculation with bacteria is to supply nitrogen to the plant, it is better to use nitrogen-fixing bacteria that have other PGP characteristics (such as IAA, ACC deaminase, siderophores, and phosphate solubilization) as well.

13.3 Phosphate-Solubilizing Bacteria (PSB) in Phosphorus Nutrition

After nitrogen (N), phosphorus (P), as a necessary nutrient and a macronutrient, is the most restricting nutrient for the plant (Schachtman et al. 1998; Theodorou and Plaxton 1993). Phosphorus plays several key roles in the plant, including participation in energy transfer reactions, photosynthesis, deformation of sugar into starch, key enzymatic reactions in important metabolic and signaling pathways, and transference of genetic characteristics in plants (Theodorou and Plaxton 1993). There has been an enduring increment in the application of P fertilizers in rice production (Syers et al. 2008) because it is one of the main restricting factors for upland rice production in many regions of the world (Sahrawat et al. 2001). Since water scarcity is becoming a major problem for agriculture, there is a pressing need to cultivate aerobic rice. Aerobic rice requires the same amount of nutrients as flooded rice, but there is a problem of P availability due to its rapid immobilization/fixation with elements such as calcium (Ca^{2+}), iron (Fe^{3+}), and aluminum (Al^{3+}) (Goldstein 1986; Othman and Panhwar 2014). The previous findings also suggest that P deficiency in aerobic crops is quite common (Fageria 2001).

Phosphorus is the most sensitive nutrient to soil pH. The best pH for P uptake by the plant is 6.5. In alkaline condition, P becomes insoluble by reacting with calcium (Ca^{2+}), whereas in acidic soils, it reacts with iron (Fe^{3+}) and aluminum (Al^{3+}) and becomes unavailable to the plants. The amount of P absorbed by the plant in the soil is controlled by several factors such as soil pH, calcium ion concentration, soil organic matter, clay type, and clay amount, root density and exudates, and soil moisture and texture. In order to compensate for the shortage of P, large amounts of P fertilizers are added to the soil annually. The excessive use of P fertilizers and the subsequent accumulation of P in the soil, in addition to increasing costs, have a negative effect on uptake of micronutrients and also contribute to environmental pollution (e.g., eutrophication).

The majority of P fertilizers are adsorbed by solid particles and stored in a solid phase of soil. Most of the P in the fertilizer, after entering the soil, gradually turns into insoluble compounds and is stored as plant unavailable forms in the soil (Dey 1988). It has been reported that the P fertilizer use efficiency in calcareous and alkaline soils does not exceed 20%. The P mobility in the soil is very low and cannot respond to the rapid absorption of the plant. This leads to the emergence and development of phosphate-depleted areas adjacent to the contact surface of roots with soil. Under P-deficient conditions, by modifying root morphology, carbon metabolism; membrane structure, exudation of organic acids, protons, and enzymes; and association with mycorrhizal fungi, and harboring phosphate-solubilizing microorganisms (PSM), some plants have been able to somehow compensate for their lack of P (uptake of adequate P) (Begum and Islam 2005; Islam and Hossain 2012). Among these strategies, secretion of organic acids and association of mycorrhizal fungi are very poor in rice under flooding conditions (Begum et al. 2005; Islam and Hossain 2012). Therefore, the rice plants need an auxiliary system that can easily go beyond these depleted areas and, by developing a wide network around the root system, receive P from an exorbitant volume of adjoining soil.

PGPB such as phosphate-solubilizing bacteria (PSB) are considered to be the most effective plant assistants for the supply of P at the optimal level and seems to be another manner for P nutrition in rice under P-insufficient tropical soils (Islam and Hossain 2012). PSB have been able to dissolve insoluble phosphates through a set of mechanisms such as production of low-molecular-weight organic acids (i.e., gluconic, oxalic, 2-ketogluconic, citric, succinic, lactic, and malic), inorganic acids, siderophores, and exopolysaccharides (EPS), and secretion of hydrolytic enzymes (e.g., phosphatases and phytases, which convert the organic forms of P into P inorganic forms, and thereby increase plant growth under conditions of P deficiency) (Khan et al. 2007, 2014; Sharma et al. 2013). PSB have the ability to dissolve inorganic P in a range of 25–42 μg P ml^{-1} and organic P between 8 and 18 μg P ml^{-1} (Guang-Can et al. 2008). *Agrobacterium, Pseudomonas, Bacillus, Rhizobium, Flavobacterium, Acinetobacter, Micrococcus, Burkholderia, Achromobacter, Erwinia, Pantoea,* and *Streptomyces* are of the most important bacterial genera of solubilizing insoluble phosphates (Khan et al. 2007, 2014; Sharma et al. 2013).

In addition to improving soil P status, members of the bacterial genera such as *Burkholderia, Pseudomonas, Bacillus, Streptomyces,* and *Pantoea* could also suppress soil-borne pathogens (Islam and Hossain 2012; Rodríguez and Fraga 1999). PSB, which form less than one percent total bacterial populations in the soils (Kucey 1983), have been isolated from approximately all agricultural soils (both fertile soils and P-deficient ones) (Oehl et al. 2001). Previous studies show that rhizosphere and endorhiza of rice plants also harbor the bacteria with a good potential for solubilizing insoluble phosphates such as *Bacillus* spp., *Pantoea agglomerans*, *Streptomyces anthocysnicus*, *Pseudomonas pieketti*, *P. aeroginosa*, *Acinetobacter* sp., *Klebsiella* sp., *Acinetobacter* sp., *Enterobacter* sp., *Microbacterium* sp., *Pseudomonas* sp., *B. megaterium*, *B. firmus*, *Erwinia*, *Serratia*, and *Staphylococcus epidermidis* (Etesami et al. 2014a; Islam and Hossain 2012; Naik et al. 2008; Panhwar et al. 2011a; Sapsirisopa et al. 2009; Thakuria et al. 2004; Zeng et al. 2012). Previous studies show that PSB alone or in combination with varying doses of P fertilizers could also increase soil available P and P content in the rice plant tissue (Duarah et al. 2011; Othman and Panhwar 2014; Panhwar et al. 2011a, b, 2013). There are reports that PSB also have the ability to increase the efficiency of P fertilizer and diminish about 25–50% of the required P to plants (Attia et al. 2009; Islam and Hossain 2012; Yildirim et al. 2011). In addition to increasing the efficiency of P fertilizer, PSB also increased total N, K, Ca, S, P, Mg, Fe, Mn, Zn, and Cu contents in plant tissues (Duarah et al. 2011; Gyaneshwar et al. 2002; Islam and Hossain 2012; Yildirim et al. 2011).

It is well known that PSB can increase plant seed germination (Duarah et al. 2011; Sapsirisopa et al. 2009), plant growth and development (i.e., augmented leaf chlorophyll content, leaf area index, tiller numbers, plant height, photosynthesis rate, root morphology, and plant biomass of aerobic rice genotypes) (Duarah et al. 2011; Panhwar et al. 2011a, b), and plant yield and quality (Islam and Hossain 2012), through other mechanisms such as phytohormone production, nitrogen fixation, urease activity, siderophore production, ACC deaminase, and/or antagonisms against phytopathogens, in addition to by solubilizing insoluble phosphates (Islam and Hossain 2012). In general, the above studies show that PSB have been found to have the ability to solubilize P in soil and could reduce fertilizers inputs in rice fields.

13.4 Plant Growth-Promoting Bacteria (PGPB) in Micronutrient Nutrition

Similar to macronutrients, micronutrients are also required for optimum plant growth. Micronutrient deficiencies are omnipresent (Das 2014). For example, 50% of world cereal soils are deficient in zinc (Zn) and 30% of cultivated soils globally are deficient in iron (Fe). Fe deficiency is common in upland, high pH, and aerobic soil due to the low solubility of the oxidized ferric form in aerobic environments (Das 2014; Samaranayake et al. 2012; Zuo and Zhang 2011). Rice is also substantially deficient in Fe (Bouis and Welch 2010). Toxicity of Fe is one of the major constraints to lowland rice production (Das 2014). Manganese (Mn) deficiency is also very common in upland rice (Das 2014). In general, micronutrients-deficient soils hamper the growth of many plants including staple foods such as wheat, rice, corn and sugarcane (Kamran et al. 2017).

Use of micronutrients fertilizers may not be cost-effective in alleviating deficiency of these nutrients and increasing yield of these crop plants. It has been known that bacteria can cause a substantial increase in concentration of micronutrients in crop plants (Etesami and Maheshwari 2018) including in rice grains (Mäder et al. 2011; Pooniya et al. 2012) through various mechanisms such as acidification, production of organic acids/organic anions in soil, which sequester the cations of micronutrients and decrease the pH of the adjacent soil as well as chelate micronutrients and enhance the solubility of these nutrients, and production of siderophores, which mainly form the complexes with Fe(III) (Alexander 1977; Etesami and Maheshwari 2018; Jones and Darrah 1994; Kamran et al. 2017; Saravanan et al. 2007). For example, Zn-solubilizing bacteria such as *Pseudomonas fragi*, *Pantoea dispersa*, *Pantoea agglomerans*, *E. cloacae,* and *Rhizobium* sp. are potential alternatives for Zn supplementation and convert applied inorganic Zn to available forms (Kamran et al. 2017). In a study, Vaid et al. (2014) showed that Zn-solubilizing *Burkholderia* and *Acinetobacter* caused significant increase in productive tillers per plant (15.1%), number of panicles per plant (13.3%), total Zn uptake of rice (52.5%), the mean dry matter-yield per pot (12.9%), yield of straw (12.4%), yield of grain (17.0%), and number of grains per panicle (12.8%) relative to rice plants non-inoculated with the bacterial isolates in a Zn-deficient soil. It was reported that this increment might be due to solubilization of insoluble soil Zn via generating gluconic acid by these bacteria. In another study, co-inoculation of rice with *Providencia* sp. PR3, *Brevundimonas diminuta* PR7, and *Ochrobactrum anthropi* PR10 recorded an increment of 13–16% in Fe, Zn, Cu, and Mn concentrations, respectively, in rice grains (Rana et al. 2015). Adak et al. (2016) also observed 13–46% enhancement of Fe and 15–41% enhancement of Zn in rice grains through the use of cyanobacterial inoculants, under different modes of rice cultivation. The above studies indicate the potential of the PGPB associated with rice to be used as biofertilizer and overcome deficiency of micronutrients.

13.5 Silicate-Solubilizing Bacteria (SSB) in Silicon Nutrition

Silicon (Si) is known as the second most copious element in soils (Epstein 1994). Utilization of Si is known as an ecologically congenial and environmentally friendly technique to augment plant growth, attenuate miscellaneous environmental stresses (i.e., nutritional imbalance, salinity, drought, heavy metal toxicity, and pathogens) in plants, and enhance the plant resistance to multiple stresses (Etesami and Jeong 2018). Despite these benefits, Si is still not classified as an essential element but considered as a beneficial element. This element is useful for some plants such as monocotyledons and *Poaceae* species (Etesami and Jeong 2018; Epstein 1999; Ma et al. 2007). Rice is one of the plants that accumulate this element (a Si-accumulator/a siliceous plant-containing Si up to 10% in shoots on a dry weight basis) and requires high Si content (a high Si-accumulating crop) (Ma and Takahashi 2002). Rice is known that the escalation in its yield per unit area is connected with Si depletion, which is a matter of concern (Savant et al. 1997). Due to being removed from the soil to produce every 100 kg brown rice (about 20 kg/ hm^2 SiO_2) (Ma and Takahashi 2002), being exported from fields by removing straw residues with the harvest by farmers, and being connived the exogenous use of Si in rice cultivation, plant accessible Si in paddy fields is usually low (Cuong et al. 2017; Etesami and Jeong 2018; Ma and Takahashi 2002). This suggests that Si may become a yield-limiting element for rice production and its exogenous application may be necessary to Si-deficient paddy soil for an economic and sustainable rice production system (both high rice yield and disease resistance) (Bocharnikova et al. 2010; Ning et al. 2014). At the present time, Si-fertilizers are exerted in many countries for augmenting rice yield (Guntzer et al. 2012). In previous studies, the positive effects of Si on rice growth and yield have been reported (Detmann et al. 2012; Gerami et al. 2013; Etesami and Jeong 2018; Jawahar and Vaiyapuri 2013; Lavinsky et al. 2016; Liang et al. 1994; Pati et al. 2016; Prakash et al. 2011; Singh et al. 2005). For example, in a study, Cuong et al. (2017) showed that application of Si in combination with the recommended dose of N, P, and K fertilizers positively affected agronomic and yield-related traits, yield and nutrient uptakes of rice. Si had also beneficial effects on disease resistance of rice (i.e., brown spot caused by the fungus *Bipolaris oryzae*, rice blast caused by the fungus *Pyricularia grisea*, and sheath blight caused by *Rhizoctonia solani* Kuhn, which are becoming more severe on rice plants are grown in Si-depleted soils) (Abed-Ashtiani et al. 2012; Ashtiani et al. 2012; Cacique et al. 2012; Dallagnol et al. 2014; Fauteux et al. 2005; Hayasaka et al. 2005; Ning et al. 2014; Prabhu et al. 2001; Rodrigues and Datnoff 2005; Sakr 2016; Song et al. 2016; Van Bockhaven 2014) by various mechanisms such as maintaining mesophyll cells relatively intact, increasing the thickness of silicon layer, enhancing physiological or induced resistance to fungal colonization (Si acts as a modulator of host resistance to pathogen), depositing in host cell walls and papillae sites, which is the first physical barricade for fungal penetration (Ning et al. 2014), and accumulating phenolics and phytoalexins as well as with the activation of some PR-genes (Rodrigues and Datnoff 2005).

There are some bacteria like *Bacillus globisporus, B. mucilaginosus, B. flexus, B. megaterium, Burkholderia eburnean*, and *Pseudomonas fluorescens* that can mobilize K and Si from silicate minerals (i.e., feldspar, muscovite, and biotite) (Kang et al. 2017; Naureen et al. 2015a; Sheng et al. 2008; Vasanthi et al. 2018; Vijayapriya and Muthukkaruppan 2010) by various mechanisms such as producing excess proton, organic ligands, organic acids (i.e., gluconic acid), hydroxyl anion, extracellular EPS, and enzymes (Meena et al. 2014). Inoculation of rice with silicate-solubilizing bacteria (SSB) also caused a significant increase in growth and yield of this plant. In a study (Kang et al. 2017), when combined with silica fertilization, soil inoculation with *Burkholderia eburnean* CS4-2 promoted all rice growth attributes over those of the water-treated (control) and insoluble silica-fertilized plants. In addition to solubilizing Si, K, and P (e.g., by competing with P fixation sites in soil and decreasing the availability of Fe and Mn in plants) (Kannan and Raj 1998; Sahebi et al. 2015), SSB were also capable of controlling the growth of fungal pathogens such as *Magnaporthae grisae, Rhizoctonia solani, Altarnaria alternata,* and *Macrophomina pheasolina* (Naureen et al. 2015a). In addition to the role of Si in increasing rice resistance to pathogens, SSB can antagonize fungal pathogens by the production of hydrolytic enzymes, HCN (hydrogen cyanide), siderophores, and antibiotics (Hassan et al. 2010; Naureen et al. 2015b, 2009b). In a previous study (Vijayapriya and Muthukkaruppan 2010), *B. mucilaginosus*, which was efficient in silicate solubilization, showed antagonistic activity against *Pyricularia oryzae*. The above studies indicate the potential of SSB to be used as biofertilizer for overcoming Si deficiency and as biocontrol agents for controlling fungal rice pathogens.

13.6 Combined Use of PGPB and Chemical Fertilizers for Rice Production

Application of biological fertilizers, in particular, GPGB, combined with the use of fertilizers, is the most important integrated plant nutrition strategy for sustainable agricultural management and increasing their production in a sustainable agricultural system with sufficient input (Bagyaraj and Balakrishna 2012; Etesami and Alikhani 2016b).

Beneficial effects of PGPB in increasing nutrient uptake by rice, including NPK uptake, have been reported in previous studies (Adesemoye and Kloepper 2009; Adesemoye et al. 2009; Biswas et al. 2000; Duarah et al. 2011; Etesami and Alikhani 2016a; Vessey 2003). It has been reported that the PGPB can diminish the exertion of chemical fertilizers without compromising with the growth and yield of rice under nutrient-poor soil conditions (Etesami and Alikhani 2016a; Khan et al. 2017). In a study, Etesami and Alikhani (2016a) showed that co-inoculation with endophytic (*Pseudomonas putida* REN5) and rhizosphere (*Pseudomonas fluorescens* REN1) bacteria can reduce application rates of N-fertilizer up to 25% for rice plant. These researchers showed that the compound application of *P. putida* REN5 and *P. fluorescens* REN1 and nitrogen fertilizer levels (50, 75, and 100% of the recommended N-fertilizer rate) compared to the application of these bacterial isolates with minimum nitrogen fertilizer (25% of the recommended N-fertilizer rate) and or control (25% of the recommended N-fertilizer rate) significantly increased the rice growth indices. It was found that 75% of the recommended fertilizer rate was the minimum level to diminish N-fertilizer. This indicates that nitrogen plays a key role in the growth of the rice plant, and the plant's yield decreases without the presence of nitrogen. de Souza et al. (2013, 2016) showed that rice plants inoculated with bacterial strains (*Herbaspirillum* sp. AG15, *Herbaspirillum* sp. AC32, *Pseudacidovorax* sp. AC32, *Burkholderia* sp. CA21,

Azospirillum sp. UR51, and *Rhizobium* sp. UR51), which were isolated from rice rhizosphere, with 50% of the recommended N-fertilizer rate achieved growth indices (i.e., shoot length and dry matter, the number of panicles, and plant yields) similar to those that received the full-fertilization dose without inoculation. Other researchers also confirmed that single PGPR or combinations of PGPR promoted the growth of rice, increased plant height, dry shoot matter, N, P, and K uptake and grain production even when the recommended amount of nitrogen fertilizer was reduced in half (Biswas et al. 2000; Duarah et al. 2011; Khorshidi et al. 2011; Yanni and Dazzo 2010). In addition, rice plants inoculated with these bacterial strains supplemented with 50% N fertilizer accumulated a higher amount of N and P than those that received 100% of N fertilizer alone (de Souza et al. 2016). Khan et al. (2017) also inoculated rice with *Burkholderia* sp. BRRh-4 and *Pseudomonas aeruginosa* BRRh-5 along with 50% of recommended N, P, and K fertilizers. Both bacterial strains generated equivalent or higher grain yield of rice relative to the control—plants grown with full recommended—fertilizer doses. The above studies show that PGPB can interact with rice plant under different nitrogen fertilizer levels, but this interaction can be much more productive when plants are treated with low levels of chemical N-fertilizers (de Souza et al. 2016).

Generally speaking, it is believed that PGPB are more effectual in augmenting plant growth under restricting nutrient conditions. Besides, the colonization of the plant root by PGPR might have been repressed by the augmenting levels of nutrients (i.e., N) in the growth medium (Egamberdiyeva 2007; Shaharoona et al. 2008). It was also reported that these bacteria can be used as a supplement to chemical fertilizers to reduce the use of fertilizers but cannot replace nitrogen fertilizer in rice (Etesami and Alikhani 2016a). Generally speaking, the PGPB-based inoculation technology should be consumed along with desired levels of fertilization to achieve maximal benefits in terms of fertilizer savings, nutrient uptake, and rice plant growth (de Souza et al. 2016).

13.7 Biological Control of Fungal Rice Pathogens

Pathogenic microorganisms affecting plant fitness are an outstanding and chronic threat to food production and ecosystem steadfastness throughout the world (Compant et al. 2005). Diseases of fungal, bacterial, viral origin, and damage brought about by insects and nematodes can be led to a significant diminution in crop production. Diseases are one of the most important limiting factors affecting rice production, which reduces annual rice yield by about 5% (Song and Goodman 2001). More than 70% of the diseases caused by fungi, bacteria, viruses or nematodes have been reported in rice (Manandhar et al. 1998). In other words, rice is susceptible to diseases. Pathogenic fungi can reduce the quality and quantity of rice grain production (Chaiharn et al. 2009) and affliction with these fungi are among the most niggling of these diseases as it may result in remarkable crop yield losses (Chaiharn et al. 2009; Suprapta 2012). In addition, the consumption of mycotoxins (e.g., aflatoxins, citrinin, ochratoxin A, fumonisins, and zearalenone)-polluted rice can be hazardous to human beings (Almaguer et al. 2012; Ferre 2016).

To control fungal diseases, fungal pathogens-resistant rice cultivars and fungicides are commonly used. But, due to the loss of resistance to pathogens, despite the high variability of disease agents of the pathogen population, the useful life of many pathogen-resistant cultivars is only several years. Use of fungicides is also expensive and environmentally unfriendly and has led to risks to human health, environmental pollution, residual toxicity, development of pesticide resistance, and other beneficial organisms in the soil (Komárek et al. 2010; Suprapta 2012; Yoon et al. 2013). These fungicides also reduced soil fertility and quality and damaged to natural ecosystems (Chaiharn et al. 2009). Furthermore, there are a number of painstaking diseases for which chemical solutions are few, unproductive, or nonexistent (Gerhardson 2002). Biocontrol is thus being considered as an alternative or a supplemental way of diminishing the utilization of chemicals in agricultural land (Compant et al. 2005, 2010; Etesami and Alikhani 2016b, 2016d; Gerhardson 2002; Pal and Gardener 2006; Suprapta 2012; Welbaum et al. 2004). Bacterial biocontrol agents can control plant pathogens including fungal pathogens by various mechanisms (Fig. 13.2). Various suitable nutrient-rich niches on/or inside roots attract a great diversity of microorganisms, including phytopathogens. Competition for the nutrients (root exudates including organic acids, amino acids, specific sugars, etc.) and niches is a underlying mechanism by which PGPB preserve plants from phytopathogens (Compant et al. 2005).

Biocontrol PGPB are aggressive root colonizers and play an important role in the biological control of plant diseases caused by soil-borne fungal pathogens (Chaiharn et al. 2009). Another mechanism of biological control by PGPR is production of allelochemicals like (i) iron(III)-chelating siderophores, which deprive pathogenic fungi of Fe since the fungal siderophores have lower affinity to Fe compared to bacterial siderophores (Loper and Henkels 1999; O'sullivan and O'Gara 1992; Van Loon and Bakker 2005); (ii) production of antibiotics such as amphisin, 2,4-diacetylphloroglucinol (DAPG), rhizoxin, oomycin A, phenazines, tensin, pyoluteorin, pyrrolnitrin, tensin, tropolone, oligomycin A, kanosamine, zwittermicin A, xanthobaccin, viscosinamide, and cyclic lipopeptides (Compant et al. 2005; de Souza et al. 2003; Défago 1993; Hashidoko et al. 1999; Joseph et al. 2012; Kai et al. 2009; Kim et al. 1999; Nain et al. 2012; Nielsen et al. 2002; Pal and Gardener 2006); (iii) biocidal volatiles like HCN and ammonia (NH_3) (Blumer and Haas 2000; Kai et al. 2009; Pal and Gardener 2006; Zou et al. 2007); (iv) lytic enzymes (Chernin and Chet 2002; Sindhu and Dadarwal 2001) such as chitinase (Ordentlich et al. 1988), which inhibits spore germination and germ-tube elongation (Frankowski et al. 2001), laminarinase, which digests and lyses mycelia of some fungi (Lim et al. 1991), β-1,3-glucanase, which lyses fungal cell walls of some fungi (Fridlender et al. 1993; Singh et al. 1999), glucanases, cellulases, and detoxification enzymes (Abbas-Zadeh et al. 2010; Fridlender et al. 1993; Kai et al. 2009; Nain et al. 2012; Pal and Gardener 2006; Sindhu and Dadarwal 2001; Zhao et al. 2010). ISR (induced systemic resistance) is an consequential mechanism by which PGPR in the rhizosphere prime the whole plant body for augmented defense

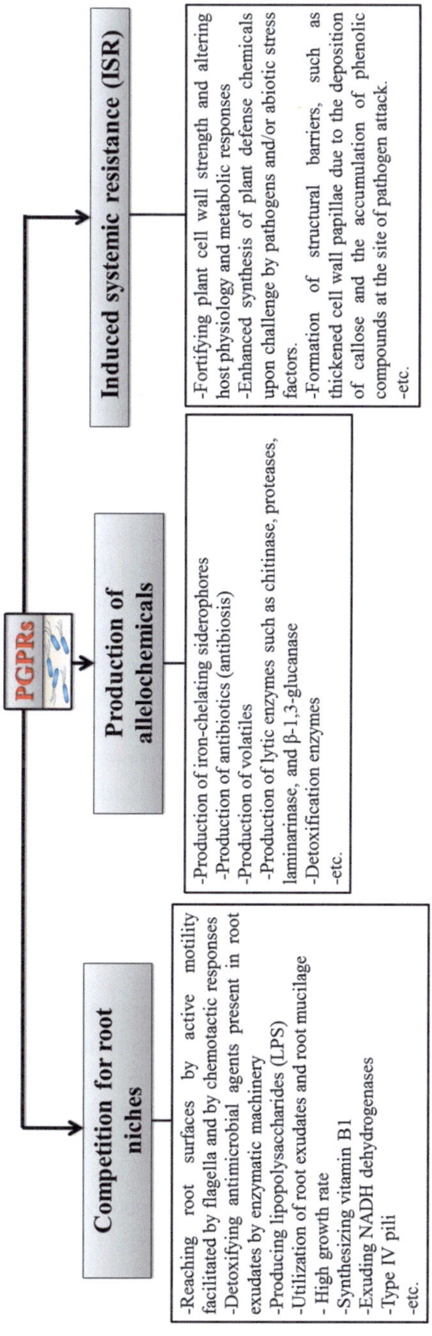

Fig. 13.2 Some of action mechanisms of PGPR in controlling plant pathogens

against a scopious range of pathogens and insect herbivores (Compant et al. 2005; Pal and Gardener 2006; Van Loon and Bakker 2005). Biocontrol PGPR, through different mechanisms such as production of siderophores, lipopolysaccharides (Leeman et al. 1995; Maurhofer et al. 1994; Meziane et al. 2005; Van Loon and Bakker 2005; Van Loon et al. 1998; Van Wees et al. 1997), volatile organic compounds (Ping and Boland 2004; Ryu et al. 2004), cyclic lipopeptides, 2,4-diacetylphloroglucinol, and homoserine lactones (Lugtenberg and Kamilova 2009), sensitize the plant immune system for enhanced defense without directly activating overpriced defenses (Pieterse et al. 2014).

Biocontrol PGPR-mediated control of several bacterial, fungal, and viral plant diseases in plants by this mechanism (ISR) has been reported (Leeman et al. 1995; Pal and Gardener 2006; Park et al. 2009). It has been also known that the ISR contains ethylene and jasmonate intracellular signaling, and these hormones stimulate host plant defense responses against plant diseases (Glick 2012). Biocontrol PGPR-mediated ISR also fortifies plant cell wall strength (Benhamou et al. 1996, 1998), alters host physiology and metabolic responses (Jeun et al. 2004; Park and Kloepper 2000), and increases accumulation of compounds (i.e., phenylalanine ammonia-lyase, peroxidase, phytoalexins, polyphenol oxidase, and/or chalcone synthase) (Chen et al. 2000; Ongena et al. 2000) that augment synthesis of plant defense chemicals upon challenge by plant pathogens (Compant et al. 2005; Nowak and Shulaev 2003; Ramamoorthy et al. 2001). The total of these changes lead to increased plant resistance to diseases. Generally speaking, the most effectual biological control agents (BCAs) studied to date appear to antagonize pathogens using multitudinous mechanisms (Iavicoli et al. 2003; Pal and Gardener 2006).

The ability of biocontrol PGPR to lessen or prevent the deleterious effects of certain fungal rice pathogens has been well documented (Amruta et al. 2018; Awla et al. 2017; Chaiharn et al. 2009; Etesami and Alikhani 2016b, 2016d, 2018; Velusamy and Gnanamanickam 2008; Verma et al. 2018).

Magnaporthe oryzae (anamorph *Pyricularia oryzae*), which causes diseases generically called "blast disease" or "blight disease—the most destructive disease of rice (Chaiharn et al. 2009; Dean et al. 2012) and attacks rice plants at all stages of development and infects the aerial parts of the rice plant—including leaves, nodes, stems, and panicles, bringing about annual losses of approximately 10–30% in miscellaneous rice—producing regions (Law et al. 2017), *Alternaria* sp., which cause leaf spots, *Fusarium* oxysporum, which cause root rot, *Sclerotium* sp., which cause stem rot (Chaiharn et al. 2009), *Bipolaris oryzae*, which causes brown spot disease, *Rhizoctonia solani*, which causes sheath blight disease, *Curvularia oryzae*, which causes leaf spot disease, *Gibberella fujikuroi*, which causes bakanae disease in rice seedlings, and *Rhizoctonia oryzae-sativae*, which causes aggregate sheath blight disease, have been reported as the most consequential fungal pathogen bringing about diseases in rice (Boukaew et al. 2013; Tamreihao et al. 2016). By a combination of different modes of action such as hydrogen ions and gaseous products including ethylene, HCN and NH_3, and siderophore (hydroxamate type), cell wall degrading enzymes (i.e., chitinase, protease, cellulase, β-1,3-glucanase, β-1,4-glucanase, and lipase) and antibiotics, biocontrol PGPB (e.g., *Streptomyces* sp. *S. globisporus, S. sindeneusis, S. flavotricini, S. philanthi, S. vinaceusdrappus, S. corchorusii, Ochrobactrum anthropic, Bacillus* sp., *B. cereus, B. subtilis, B. methylotrophicus, Enterobacter* sp., *Pseudomonas aeruginosa*, and *Pseudomonas* sp.) significantly inhibited the mycelia growth of these fungi (Awla et al. 2017; Boukaew et al. 2013; Boukaew and Prasertsan 2014; Chaiharn et al. 2009; Khalil et al. 2014; Li et al. 2011; Ningthoujam et al. 2009; Prapagdee et al. 2008; Shan et al. 2013; Tamreihao et al. 2016; Tokpah et al. 2016; Zarandi et al. 2009).

In previous studies, Etesami and Alikhani (2016d), (2017), and Etesami et al. (2014b) investigated the potential of antifungal activity of the bacterial isolates isolated from rhizosphere and endorhiza of rice, oilseed rape (*Brassica napus* L.), and berseem clover (*Trifolium alexandrinum* L.), respectively, against five rice pathogenic fungi (*Magnaporthe oryzae, M. salvinii, Fusarium verticillioides, F. fujikuroi,* and *F. proliferum*—the most important pathogenic fungi of rice in Iran) under in vitro conditions. A considerable part of these isolates showed a good percentage of mycelial growth inhibition against all the tested major rice fungal pathogens in dual cultures on solid media (Fig. 13.3) (Etesami and Alikhani 2016d).

Bacillus species (*Bacillus mojavensis, B. amyloliquefaciens, B. subtilis*, and *B. cereus*) were reported as the most propitious bacterial biocontrol agents in rhizosphere and endorhiza of these plants (Etesami and Alikhani 2018). In addition, endophytic bacterial isolates were more effective at mycelial growth inhibition than rhizosphere bacterial isolates. Probably endophytic bacteria use mechanisms similar to PGPR to control plant fungal pathogens. Biocontrol activities of these

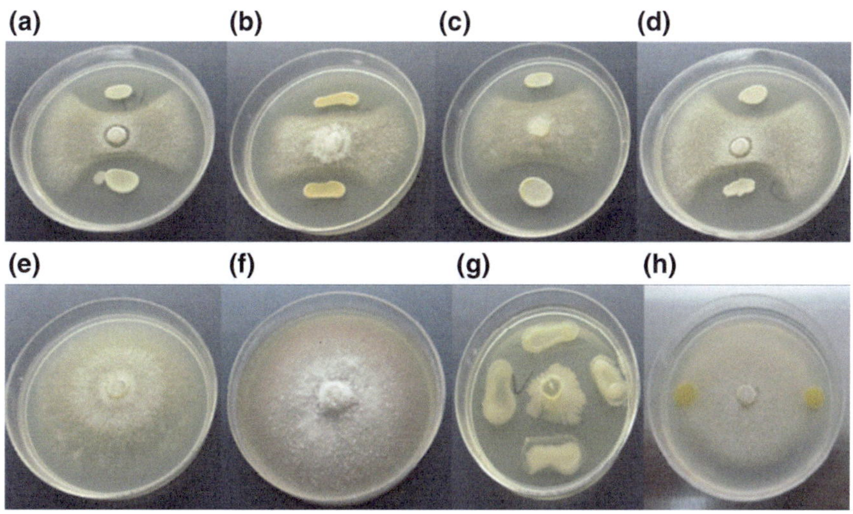

Fig. 13.3 Dual culture assay for in vitro inhibition of mycelia of fungal rice pathogens by the endophytic and rhizoshpere strains grown on PDA agar for 5 days. **a** endophytic strain *B. subtilis* CEN_3; **b** rhizosphere strain *B. cereus* CEN_5; **c** endophytic isolate; **d** rhizosphere isolate; **e** and **f** control (pathogen alone); **g** combination of endophytic and rhizosphere isolates with each other; and **h** rhizosphere isolate resulted in no inhibition zones

bacterial strains may be owing to the creation of antifungal metabolites, volatile organic compounds (VOCs), siderophores, and cell wall degrading enzymes (Etesami and Alikhani 2016c).

Among biocontrol bacteria, spore-forming *Bacillus* bacteria have properties that make them more suitable for development as biocontrol agents, including high resistance to stress, production of various secondary metabolites, induction of ISR in order to reduce the severity of the disease caused by a wide range of pathogens, stimulating plant growth, simplicity in cultivating and maintaining them, as well as use of them as spores on plant or seed inoculation (Alina et al. 2015; Shafi et al. 2017). Besides, *Streptomyces* bacteria also appear to be auspicious biocontrol agents against a wide range of phytopathogenic fungi due to generating various bioactive compounds such as antibiotics (e.g., Blasticidin S, Kasugamycin, Oligomycin A, geldanamycin, and nigericin) or antifungals which can inhibit or kill the pathogen (Copping and Duke 2007; González-Franco and Robles-Hernandez 2009; Law et al. 2017; Tapadar and Jha 2013; Trejo-Estrada et al. 1998; Yang et al. 2010), the release of extracellular lytic enzymes such as chitinases and glucanases, which play consequential roles in ruination of fungal cell walls (El-Tarabily et al. 2000; González-Franco and Robles-Hernandez 2009; Palaniyandi et al. 2013), and their colonization ability, competitive traits, and survival in various types of soil (ability to produce spores which allow them to survive longer and in various extreme conditions) (González-Franco and Robles-Hernandez 2009; Law et al. 2017; Ningthoujam et al. 2009). Under greenhouse conditions, *Streptomyces* could result in up to 88.3% disease diminution of rice blast (Law et al. 2017). Approximately, 75% commercially practicable antibiotics were derived from the genus *Streptomyces* (Kashif et al. 2016). Besides, *Streptomyces* produces spores that help dissemination and confer resistance to many hostile conditions (Goodfellow and Williams 1983). The biocontrol bacteria not only prevent the growth of pathogens, but also improve plant growth. These bacteria were also positive for different PGP traits such as IAA, ACC deaminase, siderophores, and phosphate solubilization, and could significantly enhance the growth and grain yield production of the plants (Alina et al. 2015; Etesami and Alikhani 2016c, 2017; Shafi et al. 2017; Tamreihao et al. 2016).

There are many studies that show bacterial biocontrol agents can be very promising antagonist candidates against plant pathogens which can be developed for sustainable plant diseases management. Despite these studies and the recent interest in bioassays of plant diseases, it is difficult to find examples of commercial use of biological control agents in controlling pathogens. This can be due to inappropriate screening systems that are used. In general, biocontrol PGPB by colonizing the root system of the plant prevent the establishment of harmful rhizospheric microorganisms on the root of the plant. These rhizobacteria must compete with indigenous microorganisms and effectively colonize the rhizosphere. In other words, the biocontrol agents and PGPB are influenced by native microbial communities. Generally, the antagonistic activity of biocontrol bacteria is tested through in vitro inhibition of fungal pathogens in dual cultures on solid media and then confirmed in bioassays on host plants. It has been reported that in vitro evaluations have some limitations (Compant et al. 2005). Many biological control agents effectively control diseases in vitro conditions, but have not been successful in field conditions. The ineffectiveness of biocontrol bacteria in the field is often attributed to their inability to colonize the roots. Rhizosphere competence and colonization are considered as an important factor in controlling fungal pathogens by biocontrol bacteria because both organisms colonize the same ecology niche and use the same nutrient (Compant et al. 2005). Factors such as temperature, soil moisture, soil texture, and environmental stresses affect the survival and establishment of bacteria.

In general, in many studies, a single biological control agent is usually used to control a pathogen under controlled and greenhouse conditions. This can sometimes result in incompatible performance by the biological control agent under natural conditions because a biological control agent cannot be active in all types of soil environments/agricultural ecosystems (Raupach and Kloepper 1998) or against all pathogens that attack the host plant (there is usually more than one pathogen in the soil). Moreover, this may also be due to inadequate colonization, limited resistance to changes in environmental conditions, and fluctuations in the production of antifungal metabolites by this biological control agent (Dowling and O'Gara 1994). Several solutions have been proposed to overcome these problems including the combined use of two or more isolates in biological control (Raupach and Kloepper 1998). Mixtures of biological control agents with different plant colonization patterns or a biological control agent with antifungal activities against several pathogens (formulation of a biocontrol isolate is simpler and cheaper than that of multiple biocontrol isolates) can be useful for controlling the biological diversity of miscellaneous pathogens via assorted mechanisms of repression of the disease. In general, the use of a combination of bacterial antagonists for biological control of pathogens can expand the range of antifungal activities (protection of the plant against a wide range of fungal pathogens), increase the efficiency, sustainability, and effect of biological control agents, and combine different characteristics without applying genetic engineering. In addition, designing a combination of biocontrol isolates and the use of multiple antifungal properties demonstrated by these isolates can be useful in the sense that at least one of the biological control mechanisms among these isolates may exist under unpredictable field conditions. In addition, mixtures of biocontrol microorganisms can increase the genetic diversity of biological control systems that prolong the stay in the rhizosphere and use a spacious range of biological control mechanisms.

A higher efficiency of several isolates from biocontrol agents against plant pathogens has been reported in previous studies (Etesami and Alikhani 2016c; Lucas et al. 2009; Schisler et al. 1997). In addition to controlling the disease, the combination of biocontrol isolates has also increased plant growth in terms of germination, plant height, and yield. It is noteworthy that the compatibility of biocontrol isolates to be inoculated with each other on plant should be considered. The incompatibility of inoculants (biocontrol isolates) can sometimes prevent the growth of each other and target pathogens. Selection of effective biocontrol isolates of bacteria is also very important for the control of pathogens in plants.

The isolation of these bacteria from pathogens repulsive soils can increase the chance of isolating effective isolates. In order to obtain effective isolates, biocontrol bacteria should be isolated from the same environment that they are supposed to be used in it. Formulations and application methods are often of great importance in the effectiveness of biological control, which should pay attention to them. In general, according to the studies conducted on biological control of fungal pathogens in rice, it can be concluded that *Streptomyces* and *Bacillus* bacteria may be taken advantage of as a potential bioinoculant agent for biocontrol as well as rice plant growth promoter.

13.8 Conclusions and Future Prospects

Reviews of literature clearly show that rhizosphere and endorhiza of rice harbor bacteria with a potential in promoting rice growth and controlling fungal rice pathogens. The co-inoculation of rice with the PGPB, as an attractive technique for utilization in commercial inoculant formulations than sole-inoculation of these bacteria, could allow declines in the prevalent high rates of fertilizer and the succeeding environmental problems without making compromise plant productivity under in vitro and greenhouse conditions. One of the major challenges encountered during the selection of biocontrol agents and biofertilizers is that biocontrol agents/biofertilizers that appear efficacious based on in vitro and greenhouse experiments might not be effective at controlling plant diseases and increasing rice growth and yield under field conditions. This inefficiency of bacteria can be owing to the variations in environmental conditions in different locations. Therefore, the environmental factors at the location where biocontrol agents/biofertilizers will be applied should be taken into consideration during the selection of suitable biocontrol agents/biofertilizers. Ideally, the biocontrol agents/biofertilizers should be isolated from and applied to locations with similar environmental factors in order to achieve successful biological control/biofertilizers. Besides, the formulation such as liquid, powder, or granule and the method of use of biocontrol agents/biofertilizers such as seed inoculation, soil inoculation, and vegetative part inoculation should be inspected as they are consequential in specifying the outcomes of field experiments. In general, before PGPB can be regarded for agricultural practices, further studies are essential to evaluate the efficacy of PGPB on rice plants under field conditions where there are a variety of constraints such as soil conditions (i.e., pH, soil nutrients status, nutrients sorption capacity, organic matter, and moisture level of the soil, etc.), environmental stresses, and types of autochthonous microorganisms that can affect the survival and growth promotion activities of PGPB/biocontrol agents.

Acknowledgements I wish to thank the University of Tehran for providing the necessary facilities for this study.

Conflict of Interest The author has no conflict of interest.

References

Abbas-Zadeh P, Saleh-Rastin N, Asadi-Rahmani H, Khavazi K, Soltani A, Shoary-Nejati AR, Miransari M (2010) Plant growth-promoting activities of fluorescent pseudomonads, isolated from the Iranian soils. Acta Physiol Plan 32:281–288

Abed-Ashtiani F, Kadir J-B, Selamat A-B, Hanif AHB-M, Nasehi A (2012) Effect of foliar and root application of silicon against rice blast fungus in MR219 rice variety. Plant Pathol J 28:164–171

Adak A et al (2016) Micronutrient enrichment mediated by plant-microbe interactions and rice cultivation practices. J Plant Nutr 39:1216–1232

Adesemoye AO, Kloepper JW (2009) Plant–microbes interactions in enhanced fertilizer-use efficiency. Appl Microbiol Biotechnol 85:1–12

Adesemoye AO, Torbert HA, Kloepper JW (2009) Plant growth-promoting rhizobacteria allow reduced application rates of chemical fertilizers. Microb Ecol 58:921–929

Alexander M (1977) Introduction to soil microbiology-2

Alina SO, Constantinscu F, Petruţa CC (2015) Biodiversity of *Bacillus subtilis* group and beneficial traits of *Bacillus* species useful in plant protection Romanian. Biotechnol Lett 20:10737

Almaguer M, Rojas TI, Rodríguez-Rajo FJ, Aira MJ (2012) Airborne fungal succession in a rice field of Cuba. Euro J Plant Pathol 133:473–482

Amruta N, Kumar MKP, Puneeth ME, Sarika G, Kandikattu HK, Vishwanath K, Narayanaswamy S (2018) Exploring the potentiality of novel rhizospheric bacterial strains against the rice blast fungus *Magnaporthe oryzae*. Plant Pathol J 34:126

Ashtiani FA, Kadir J, Nasehi A, Rahaghi SRH, Sajili H (2012) Effect of silicon on rice blast disease pertanika. J Trop Agric Sci 35

Attia M, Ahmed MA, El-Sonbaty MR (2009) Use of biotechnologies to increase growth, productivity and fruit quality of *Maghrabi banana* under different rates of phosphorus. World J Agric Sci 5:211–220

Awla HK, Kadir J, Othman R, Rashid TS, Hamid S, Wong M-Y (2017) Plant growth-promoting abilities and biocontrol efficacy of *Streptomyces* sp. UPMRS4 against *Pyricularia oryzae*. Biol Cont 112:55–63

Bagyaraj DJ, Balakrishna AN (2012) Biofertilizers for sustainable agriculture phytotechnology: Emerg Trends:73

Begum HH, Islam MT (2005) Role of synthesis and exudation of organic acids in phosphorus nutrition in plants in tropical soils. Biotechnology 4:333–340

Begum HH, Osaki M, Shinano T, Miyatake H, Wasaki J, Yamamura T, Watanabe T (2005) The function of a maize-derived phosphoenol pyruvate carboxylase (PEPC) in phosphorus-deficient transgenic rice. Soil Sci Plant Nutr 51:497–506

Benhamou N, Belanger RR, Paulitz TC (1996) Induction of differential host responses by *Pseudomonas fluorescens* in Ri T-DNA-transformed pea roots after challenge with *Fusarium oxysporum* f. sp. *pisi* and *Pythium ultimum*. Phytopathology 86:1174–1185

Benhamou N, Kloepper JW, Tuzun S (1998) Induction of resistance against *Fusarium* wilt of tomato by combination of chitosan with an endophytic bacterial strain: ultrastructure and cytochemistry of the host response. Planta 204:153–168

Bhattacharjee RB, Singh A, Mukhopadhyay SN (2008) Use of nitrogen-fixing bacteria as biofertiliser for non-legumes: prospects and challenges. Appl Microbiol Biotechnol 80:199–209

Biswas JC, Ladha JK, Dazzo FB (2000) Rhizobia inoculation improves nutrient uptake and growth of lowland rice

Blumer C, Haas D (2000) Mechanism, regulation, and ecological role of bacterial cyanide biosynthesis. Arch Microbiol 173:170–177

Bocharnikova EA, Loginov SV, Matychenkov VV, Storozhenko PA (2010) Silicon fertilizer efficiency. Russ Agric Sci 36:446–448

Bouis HE, Welch RM (2010) Biofortification—a sustainable agricultural strategy for reducing micronutrient malnutrition in the global south. Crop Sci 50:S-20
Boukaew S, Plubrukam A, Prasertsan P (2013) Effect of volatile substances from *Streptomyces philanthi* RM-1-138 on growth of *Rhizoctonia solani* on rice leaf. Biol Cont 58:471–482
Boukaew S, Prasertsan P (2014) Suppression of rice sheath blight disease using a heat stable culture filtrate from *Streptomyces philanthi*. RM-1-138. Crop Prot 61:1–10
Cacique IS, Domiciano GP, Rodrigues FÁ, Vale FXRd (2012) Silicon and manganese on rice resistance to blast. Bragantia 71:239–244
Çakmakçı R, Erat M, Erdoğan Ü, Dönmez MF (2007) The influence of plant growth–promoting rhizobacteria on growth and enzyme activities in wheat and spinach plants. J Plant Nutr Soil Sci 170:288–295
Carmen B, Roberto D (2011) Soil bacteria support and protect plants against abiotic stresses In: Shaner A (ed) Abiotic stress in plants mechanisms and adaptations, Pub In Tech:143–170
Carvalho TLG, Balsemão-Pires E, Saraiva RM, Ferreira PCG, Hemerly AS (2014) Nitrogen signalling in plant interactions with associative and endophytic diazotrophic bacteria. J Exp Bot 65:5631–5642
Chaiharn M, Chunhaleuchanon S, Lumyong S (2009) Screening siderophore producing bacteria as potential biological control agent for fungal rice pathogens in Thailand. World J Microbiol Biotechnol 25:1919–1928
Chaiharn M, Lumyong S (2011) Screening and optimization of indole-3-acetic acid production and phosphate solubilization from rhizobacteria aimed at improving plant growth. Curr Microbiol 62:173–181
Chakraborty U, Chakraborty B, Dey P, Chakraborty AP (2015) Role of microorganisms in alleviation of abiotic stresses for sustainable agriculture abiotic stresses in crop. Plants:232
Chen C, Belanger RR, Benhamou N, Paulitz TC (2000) Defense enzymes induced in cucumber roots by treatment with plant growth-promoting rhizobacteria (PGPR) and *Pythium aphanidermatum*. Physiol Mol Plant Pathol 56:13–23
Chernin L, Chet I (2002) Microbial enzymes in biocontrol of plant pathogens and pests Enzymes in the environment: activity, ecology, and applications. Marcel Dekker, New York, pp 171–225
Choudhary DK, Kasotia A, Jain S, Vaishnav A, Kumari S, Sharma KP, Varma A (2016) Bacterial-mediated tolerance and resistance to plants under abiotic and biotic stresses. J Plant Growth Regul 35:276–300
Compant S, Clément C, Sessitsch A (2010) Plant growth-promoting bacteria in the rhizo- and endosphere of plants: their role, colonization, mechanisms involved and prospects for utilization. Soil Biol Biochem 42:669–678. https://doi.org/10.1016/j.soilbio.2009.11.024
Compant S, Duffy B, Nowak J, Clément C, Barka EA (2005) Use of plant growth-promoting bacteria for biocontrol of plant diseases: principles, mechanisms of action, and future prospects. Appl Environ Microbiol 71:4951–4959
Copping LG, Duke SO (2007) Natural products that have been used commercially as crop protection agents. Pest Manag Sci 63:524–554
Cuong TX, Ullah H, Datta A, Hanh TC (2017) Effects of silicon-based fertilizer on growth, yield and nutrient uptake of rice in tropical zone of Vietnam rice. Science 24:283–290
Dallagnol LJ, Rodrigues FA, Mielli MVB, Ma JF (2014) Rice grain resistance to brown spot and yield are increased by silicon. Trop Plant Pathol 39:56–63
Das SK (2014) Role of micronutrient in rice cultivation and management strategy in organic agriculture—a reappraisal. Agric Sci 5:765
Datta A, Ullah H, Ferdous Z (2017) Water management in rice. In: Rice production worldwide. Springer, Cham, pp 255–277
Dawe D (2000) The potential role of biological nitrogen fixation in meeting future demand for rice and fertilizer. The quest for nitrogen fixation in rice, pp 1–9
de Souza JT, de Boer M, de Waard P, van Beek TA, Raaijmakers JM (2003) Biochemical, genetic, and zoosporicidal properties of cyclic lipopeptide surfactants produced by *Pseudomonas fluorescens*. Appl Environ Microbiol 69:7161–7172

de Souza R et al (2013) The effect of plant growth-promoting rhizobacteria on the growth of rice (*Oryza sativa* L.) cropped in southern Brazilian fields. Plant Soil 366:585–603
de Souza R, Schoenfeld R, Passaglia LMP (2016) Bacterial inoculants for rice: effects on nutrient uptake and growth promotion. Arch Agron Soil Sci 62:561–569
Dean R et al (2012) The top 10 fungal pathogens in molecular plant pathology. Mol Plant Pathol 13:414–430
Défago G (1993) 2, 4-Diacetylphloroglucinol, a promising compound in biocontrol. Plant Pathol 42:311–312
Detmann KC et al (2012) Silicon nutrition increases grain yield, which, in turn, exerts a feed-forward stimulation of photosynthetic rates via enhanced mesophyll conductance and alters primary metabolism in rice. New Phytol 196:752–762
Dey KB (1988) Phosphate solubilizing organisms in improving fertility status, pp 237–248
Di Fiore S, Del Gallo M (1995) Endophytic bacteria: their possible role in the host plant. In: Azospirillum VI and related microorganisms. Springer, pp 169–187
Dimkpa C, Weinand T, Asch F (2009) Plant–rhizobacteria interactions alleviate abiotic stress conditions. Plant, Cell Environ 32:1682–1694
Dowling DN, O'Gara F (1994) Metabolites of *Pseudomonas* involved in the biocontrol of plant disease. Trend Biotechnol 12:133–141
Duarah I, Deka M, Saikia N, Boruah HPD (2011) Phosphate solubilizers enhance NPK fertilizer use efficiency in rice and legume cultivation. 3 Biotech 1:227–238
Egamberdieva D, Lugtenberg B (2014) Use of plant growth-promoting rhizobacteria to alleviate salinity stress in plants. In: Use of microbes for the alleviation of soil stresses, vol 1. Springer, New York, pp 73–96
Egamberdiyeva D (2007) The effect of plant growth promoting bacteria on growth and nutrient uptake of maize in two different soils. Appl Soil Ecol 36:184–189
El-Tarabily KA, Soliman MH, Nassar AH, Al-Hassani HA, Sivasithamparam K, McKenna F, Hardy GS (2000) Biological control of *Sclerotinia minor* using a chitinolytic bacterium and actinomycetes. Plant Pathol 49:573–583
Elbeltagy A et al (2001) Endophytic colonization and in planta nitrogen fixation by a *Herbaspirillum* sp. isolated from wild rice species. Appl Environ Microbiol 67:5285–5293
Epstein E (1994) The anomaly of silicon in plant biology. Proc Natl Acad Sci 91:11–17
Epstein E (1999) Silicon. Ann Rev Plant Biol 50:641–664
Estrada GA, Baldani VLD, de Oliveira DM, Urquiaga S, Baldani JI (2013) Selection of phosphate-solubilizing diazotrophic *Herbaspirillum* and *Burkholderia* strains and their effect on rice crop yield and nutrient uptake. Plant Soil 369:115–129
Etesami H, Alikhani HA (2016a) Co-inoculation with endophytic and rhizosphere bacteria allows reduced application rates of N-fertilizer for rice plant. Rhizosphere 2:5–12
Etesami H, Alikhani HA (2016b) Evaluation of Gram-positive rhizosphere and endophytic bacteria for biological control of fungal rice (*Oryza sativa* L.) pathogens. Euro J Plant Pathol 1–8
Etesami H, Alikhani HA (2016c) Rhizosphere and endorhiza of oilseed rape (*Brassica napus* L.) plant harbor bacteria with multifaceted beneficial effects. Biol Cont 94:11–24
Etesami H, Alikhani HA (2016d) Suppression of the fungal pathogen Magnaporthe grisea by *Stenotrophomonas maltophilia*, a seed-borne rice (*Oryza sativa* L.) endophytic bacterium Arch Agron Soil Sci 1–14
Etesami H, Alikhani HA (2017) Evaluation of Gram-positive rhizosphere and endophytic bacteria for biological control of fungal rice (*Oryza sativa* L.) pathogens. Euro J Plant Pathol 147:7–14
Etesami H, Alikhani HA (2018) *Bacillus* species as the most promising bacterial biocontrol agents in rhizosphere and endorhiza of plants grown in rotation with each other. Euro J Plant Pathol 150:497–506

Etesami H, Hosseini HM, Alikhani HA (2014a) Bacterial biosynthesis of 1-aminocyclopropane-1-caboxylate (ACC) deaminase, a useful trait to elongation and endophytic colonization of the roots of rice under constant flooded conditions. Physiol Mol Biol Plant 20:425–434
Etesami H, Hosseini HM, Alikhani HA, Mohammadi L (2014b) Bacterial biosynthesis of 1-aminocyclopropane-1-carboxylate (ACC) deaminase and indole-3-acetic acid (IAA) as endophytic preferential selection traits by rice plant seedlings. J Plant Growth Regul 33:654–670
Etesami H, Alikhani HA, Hosseini HM (2015) Indole-3-acetic acid (IAA) production trait, a useful screening to select endophytic and rhizosphere competent bacteria for rice growth promoting agents. MethodsX 2:72–78
Etesami H, Emami S, Alikhani HA (2017) Potassium solubilizing bacteria (KSB): Mechanisms, promotion of plant growth, and future prospects—a review. J Soil Sci Plant Nutr 17:897–911
Etesami H, Jeong BR (2018) Silicon (Si): review and future prospects on the action mechanisms in alleviating biotic and abiotic stresses in plants. Ecotoxicol Environ Safe 147:881–896. https://doi.org/10.1016/j.ecoenv.2017.09.063
Etesami H, Maheshwari DK (2018) Use of plant growth promoting rhizobacteria (PGPRs) with multiple plant growth promoting traits in stress agriculture: action mechanisms and future prospects. Ecotoxicol Environ Safe 156:225–246. https://doi.org/10.1016/j.ecoenv.2018.03.013
Etesami H, Mirsyed Hosseini H, Alikhani HA (2014c) In planta selection of plant growth promoting endophytic bacteria for rice (*Oryza sativa* L.). J Soil Sci Plant Nutr 14:491–503
Etesami H, Mirsyedhosseini H, Alikhani HA (2013) Rapid screening of berseem clover (*Trifolium alexandrinum*) endophytic bacteria for rice plant seedlings growth-promoting agents. ISRN Soil Sci 2013
Fageria NK (2001) Nutrient management for improving upland rice productivity and sustainability. Comm Soil Sci Plant Anal 32:2603–2629
Fauteux F, Rémus-Borel W, Menzies JG, Bélanger RR (2005) Silicon and plant disease resistance against pathogenic fungi. FEMS Microbiol Lett 249:1–6
Ferre FS (2016) Worldwide occurrence of mycotoxins in rice. Food Cont 62:291–298
Frankowski J, Lorito M, Scala F, Schmid R, Berg G, Bahl H (2001) Purification and properties of two chitinolytic enzymes of *Serratia plymuthica* HRO-C48. Arch Microbiol 176:421–426
Freitas JRD, Germida JJ (1990) Plant growth promoting rhizobacteria for winter wheat. Can J Microbiol 36:265–272
Fridlender M, Inbar J, Chet I (1993) Biological control of soilborne plant pathogens by a β-1, 3 glucanase-producing *Pseudomonas cepacia*. Soil Biol Biochem 25:1211–1221
Gerami M, Fallah A, Moghadam MRK (2013) Study of potassium and sodium silicate on the morphological and chlorophyll content on the rice plant in pot experiment (*Oryza sativa* L.). Int J Agric Crop Sci 5:6
Gerhardson B (2002) Biological substitutes for pesticides. Trend Biotechnol 20:338–343
Ghorchiani M, Etesami H, Alikhani HA (2018) Improvement of growth and yield of maize under water stress by co-inoculating an arbuscular mycorrhizal fungus and a plant growth promoting rhizobacterium together with phosphate fertilizers. Agric Ecosys Environ 258:59–70
Glick BR (2012) Plant growth-promoting bacteria: mechanisms and applications. Scientifica
Goldstein AH (1986) Bacterial solubilization of mineral phosphates: historical perspective and future prospects. Am J Alter Agric 1:51–57
González-Franco AC, Robles-Hernandez RY (2009) Actinomycetes as biological control agents of phytopathogenic fungi. Tecno Chihua 3:64–73
Goodfellow M, Williams ST (1983) Ecology of actinomycetes. Ann Rev Microbiol 37:189–216
Guang-Can TAO, Shu-Jun T, Miao-Ying CAI, Guang-Hui XIE (2008) Phosphate-solubilizing and-mineralizing abilities of bacteria isolated from soils. Pedosphere 18:515–523
Guntzer F, Keller C, Meunier J-D (2012) Benefits of plant silicon for crops: a review. Agron Sustain Dev 32:201–213

Gupta G, Panwar J, Akhtar MS, Jha PN (2012) Endophytic nitrogen-fixing bacteria as biofertilizer. In: Sustainable agriculture reviews. Springer, pp 183–221

Gyaneshwar P, Kumar GN, Parekh LJ, Poole PS (2002) Role of soil microorganisms in improving P nutrition of plants. In: Food security in nutrient-stressed environments: exploiting plants' genetic capabilities. Springer, pp 133–143

Hashidoko Y, Nakayama T, Homma Y, Tahara S (1999) Structure elucidation of xanthobaccin A, a new antibiotic produced from *Stenotrophomonas* sp. strain SB-K88. Tetrahed Lett 40:2957–2960

Hassan MN, Afghan S, Hafeez FY (2010) Suppression of red rot caused by *Colletotrichum falcatum* on sugarcane plants using plant growth-promoting rhizobacteria. Biocontrol 55:531–542

Hayasaka T, Fujii H, Namai T (2005) Silicon content in rice seedlings to protect rice blast fungus at the nursery stage. J Gen Plant Pathol 71:169–173

Hazell PBR (2010) Asia's green revolution: past achievements and future challenges rice in the global economy: strategic research and policy issues for food security. International Rice Research Institute, Los Baños, pp 61–92

Hongrittipun P, Youpensuk S, Rerkasem B (2014) Screening of nitrogen fixing endophytic bacteria in *Oryza sativa L.* J Agric Sci 6:66

Iavicoli A, Boutet E, Buchala A, Métraux J-P (2003) Induced systemic resistance in *Arabidopsis thaliana* in response to root inoculation with *Pseudomonas fluorescens* CHA0. Mol Plant-Microbe Interact 16:851–858

Islam MT, Hossain MM (2012) Plant probiotics in phosphorus nutrition in crops, with special reference to rice. In: Maheshwari DK (ed) Bacteria in agrobiology: plant probiotics. Springer, Berlin, pp 325–363

James EK et al (2002) Infection and colonization of rice seedlings by the plant growth-promoting bacterium *Herbaspirillum seropedicae* Z67. Mol Plant-Microbe Interact 15:894–906

Jawahar S, Vaiyapuri V (2013) Effect of sulphur and silicon fertilization on yield, nutrient uptake and economics of rice. Int Res J Chem 3:35–43

Jeun YC, Park KS, Kim CH, Fowler WD, Kloepper JW (2004) Cytological observations of cucumber plants during induced resistance elicited by rhizobacteria. Biol Cont 29:34–42

Ji SH, Gururani MA, Chun S-C (2014) Isolation and characterization of plant growth promoting endophytic diazotrophic bacteria from Korean rice cultivars. Microbiol Res 169:83–98

Jones DL, Darrah PR (1994) Role of root derived organic acids in the mobilization of nutrients from the rhizosphere. Plant Soil 166:247–257

Joseph B, Ranjan Patra R, Lawrence R (2012) Characterization of plant growth promoting rhizobacteria associated with chickpea (*Cicer arietinum* L.). Int J Plant Produc 1:141–152

Kai M, Haustein M, Molina F, Petri A, Scholz B, Piechulla B (2009) Bacterial volatiles and their action potential. Appl Microbiol Biotechnol 81:1001–1012

Kamran S, Shahid I, Baig DN, Rizwan M, Malik KA, Mehnaz S (2017) Contribution of zinc solubilizing bacteria in growth promotion and zinc content of wheat. Fron Microbiol 8:2593

Kang S-M et al (2017) Isolation and characterization of a novel silicate-solubilizing bacterial strain *Burkholderia eburnea* CS4-2 that promotes growth of japonica rice (*Oryza sativa* L. cv. *Dongjin*). Soil Sci Plant Nutr 1–9

Kannan NM, Raj SA (1998) Occurrence of silicate solubilizing bacteria in rice ecosystem. Madras Agric J 85:47–49

Kashif MD, Kumar V, Kalpana VN, Devi Rajeswari V (2016) Phylogenetic diversity and biological activity of actinomycetes isolated from Gulf of Mannar, Tamil Nadu, India. Pharm Lett 8:16–24

Kaushal M, Wani SP (2016) Plant-growth-promoting rhizobacteria: drought stress alleviators to ameliorate crop production in drylands. Ann Microbiol 66:35–42

Khalil MS, Moubasher H, Hasan FF (2014) Biological control of rice blast disease by *Streptomyces flavotricini*. Res J Pharm Biol Chem Sci 5:1453–1461

Khan MMA, Haque E, Paul NC, Khaleque MA, Al-Garni SMS, Rahman M, Islam MT (2017) Enhancement of growth and grain yield of rice in nutrient deficient soils by rice probiotic bacteria. Rice Sci 24:264–273

Khan MS, Zaidi A, Ahmad E (2014) Mechanism of phosphate solubilization and physiological functions of phosphate-solubilizing microorganisms. In: Khan MS (ed) Phosphate solubilizing microorganisms. Springer, Cham, pp 31–62

Khan MS, Zaidi A, Wani PA (2007) Role of phosphate-solubilizing microorganisms in sustainable agriculture—a review. Agron Sustain Dev 27:29–43

Khorshidi YR, Ardakani MR, Ramezanpour MR, Khavazi K, Zargari K (2011) Response of yield and yield components of rice (*Oryza sativa* L.) to *Pseudomonas flouresence* and *Azospirillum lipoferum* under different nitrogen levels. Am -Eurasian J Agric Environ Sci 10:387–395

Kim BS, Moon SS, Hwang BK (1999) Isolation, identification, and antifungal activity of a macrolide antibiotic, oligomycin A, produced by *Streptomyces libani*. Can J Bot 77:850–858

Komárek M, Čadková E, Chrastný V, Bordas F, Bollinger J-C (2010) Contamination of vineyard soils with fungicides: a review of environmental and toxicological aspects. Environ Int 36:138–151

Kucey RMN (1983) Phosphate-solubilizing bacteria and fungi in various cultivated and virgin Alberta soils. Can J Soil Sci 63:671–678

Ladha JK, Barraquio WL, Revilla L (1997) Isolation of endophytic diazotrophic bacteria from wetland rice. In: Opportunities for biological nitrogen fixation in rice and other non-legumes. Springer, Dordrecht, pp 15–24

Lavinsky AO et al (2016) Silicon improves rice grain yield and photosynthesis specifically when supplied during the reproductive growth stage. J Plant Physiol 206:125–132

Law JW-F et al. (2017) The potential of *Streptomyces* as biocontrol agents against the rice blast fungus, *Magnaporthe oryzae* (*Pyricularia oryzae*). Front Microbiol 8:3

Leeman M, Van Pelt JA, Den Ouden FM, Heinsbroek M, Bakker P, Schippers B (1995) Induction of systemic resistance against *Fusarium wilt* of radish by lipopolysaccharides of *Pseudomonas fluorescens*. Phytopathology 85:1021–1027

Li Q et al (2011) Suppression of *Magnaporthe oryzae* by culture filtrates of *Streptomyces globisporus* JK-1. Biol Cont 58:139–148

Liang YC, Ma TS, Li FJ, Feng YJ (1994) Silicon availability and response of rice and wheat to silicon in calcareous soils. Comm Soil Sci Plant Anal 25:2285–2297

Lim H-S, Kim Y-S, Kim S-D (1991) *Pseudomonas stutzeri* YPL-1 genetic transformation and antifungal mechanism against *Fusarium solani*, an agent of plant root rot. Appl Environ Microbiol 57:510–516

Loper JE, Henkels MD (1999) Utilization of heterologous siderophores enhances levels of iron available to *Pseudomonas putida* in the rhizosphere. Appl Environ Microbiol 65:5357–5363

Lucas JA, Solano BR, Montes F, Ojeda J, Megias M, Mañero FJG (2009) Use of two PGPR strains in the integrated management of blast disease in rice (*Oryza sativa*) in Southern Spain. Field Crop Res 114:404–410

Lugtenberg B, Kamilova F (2009) Plant-growth-promoting rhizobacteria. Ann Rev Microbiol 63:541–556

Ma JF, Takahashi E (2002) Soil, fertilizer, and plant silicon research in Japan. Elsevier

Ma JF et al (2007) An efflux transporter of silicon in rice. Nature 448:209

Mäder P et al (2011) Inoculation of root microorganisms for sustainable wheat–rice and wheat–black gram rotations in India. Soil Biol Biochem 43:609–619

Manandhar HK, Lyngs Jørgensen HJ, Mathur SB, Smedegaard-Petersen V (1998) Suppression of rice blast by preinoculation with avirulent *Pyricularia oryzae* and the nonrice pathogen *Bipolaris sorokiniana*. Phytopathology 88:735–739

Mano H, Morisaki H (2008) Endophytic bacteria in the rice plant. Microb Environ 23:109–117

Maurhofer M, Hase C, Meuwly P, Metraux JP, Defago G (1994) Induction of systemic resistance of tobacco to tobacco necrosis virus by the root-colonizing *Pseudomonas fluorescens* strain CHA0: influence of the *gac*A gene and of pyoverdine production. Phytopathology 84:139–146

Meena VS, Maurya BR, Verma JP (2014) Does a rhizospheric microorganism enhance K^+ availability in agricultural soils? Microbiol Res 169:337–347

Meziane H, Van Der Sluis I, Van Loon LC, Höfte M, Bakker PAHM (2005) Determinants of *Pseudomonas putida* WCS358 involved in inducing systemic resistance in plants. Mol Plant Pathol 6:177–185

Miransari M (2013) Soil microbes and the availability of soil nutrients. Acta Physiol Plant 35:3075–3084

Mishra RPN, Singh RK, Jaiswal HK, Kumar V, Maurya S (2006) Rhizobium-mediated induction of phenolics and plant growth promotion in rice (*Oryza sativa* L.). Curr Microbiol 52:383–389

Naik PR, Raman G, Narayanan KB, Sakthivel N (2008) Assessment of genetic and functional diversity of phosphate solubilizing fluorescent pseudomonads isolated from rhizospheric soil. BMC Microbiol 8:230

Nain L, Yadav RC, Saxena J (2012) Characterization of multifaceted *Bacillus* sp. RM-2 for its use as plant growth promoting bioinoculant for crops grown in semi arid deserts. Appl Soil Ecol 59:124–135

Naureen Z et al. (2015a) Isolation and screening of silicate bacteria from various habitats for biological control of phytopathogenic fungi. American J Plant Sci 6:2850

Naureen Z, Hafeez FY, Hussain J, Al Harrasi A, Bouqellah N, Roberts MR (2015b) Suppression of incidence of *Rhizoctonia solani* in rice by siderophore producing rhizobacterial strains based on competition for iron. Euro Scientific J 11

Naureen Z, Price AH, Hafeez FY, Roberts MR (2009a) Identification of rice blast disease-suppressing bacterial strains from the rhizosphere of rice grown in Pakistan. Crop Protec 28:1052–1060

Naureen Z, Price AH, Wilson MJ, Hafeez FY, Roberts MR (2009b) Suppression of rice blast disease by siderophore-producing bioantagonistic bacterial isolates isolated from the rhizosphere of rice grown in Pakistan. Crop Protec 28:1052–1060

Ngumbi E, Kloepper J (2016) Bacterial-mediated drought tolerance: current and future prospects. Appl Soil Ecol 105:109–125

Nielsen TH, Sørensen D, Tobiasen C, Andersen JB, Christophersen C, Givskov M, Sørensen J (2002) Antibiotic and biosurfactant properties of cyclic lipopeptides produced by fluorescent *Pseudomonas* spp. from the sugar beet rhizosphere. Appl Environ Microbiol 68:3416–3423

Ning D, Song A, Fan F, Li Z, Liang Y (2014) Effects of slag-based silicon fertilizer on rice growth and brown-spot resistance. PLoS ONE 9:e102681

Ningthoujam S, Sanasam S, Tamreihao K, Nimaich S (2009) Antagonistic activities of local actinomycete isolates against rice fungal pathogens. Afr J Microbiol Res 3:737–742

Nowak J, Shulaev V (2003) Priming for transplant stress resistance in in vitro propagation. Vitro Cell Dev Biol-Plant 39:107–124

O'sullivan DJ, O'Gara F (1992) Traits of fluorescent *Pseudomonas* spp. involved in suppression of plant root pathogens. Microbiol Rev 56:662–676

Oehl F, Oberson A, Probst M, Fliessbach A, Roth H-R, Frossard E (2001) Kinetics of microbial phosphorus uptake in cultivated soils. Biol Fertil Soil 34:31–41

Oliveira ALMd, Urquiaga S, Döbereiner J, Baldani JI (2002) The effect of inoculating endophytic N_2-fixing bacteria on micropropagated sugarcane plants. Plant Soil 242:205–215

Ongena M, Daayf F, Jacques P, Thonart P, Benhamou N, Paulitz TC, Bélanger RR (2000) Systemic induction of phytoalexins in cucumber in response to treatments with fluorescent pseudomonads. Plant Pathol 49:523–530

Ordentlich A, Elad Y, Chet I (1988) The role of chitinase of *Serratia marcescens* in biocontrol of *Sclerotium rolfsii*. Phytopathology (USA)

Othman R, Panhwar QA (2014) Phosphate-solubilizing bacteria improves nutrient uptake in aerobic rice. In: Khan MS (ed) Phosphate solubilizing microorganisms. Springer, Cham, pp 207–224

Pal KK, Gardener BM (2006) Biological control of plant pathogens. Plant Health Instr 2:1117–1142

Palaniyandi SA, Yang SH, Zhang L, Suh J-W (2013) Effects of actinobacteria on plant disease suppression and growth promotion. Appl Microbiol Biotechnol 97:9621–9636

Panhwar QA, Jusop S, Naher UA, Othman R, Razi MI (2013) Application of potential phosphate-solubilizing bacteria and organic acids on phosphate solubilization from phosphate rock in aerobic rice. Scientific World J

Panhwar QA, Radziah O, Rahman AZ, Sariah M, Razi IM, Naher UA (2011a) Contribution of phosphate-solubilizing bacteria in phosphorus bioavailability and growth enhancement of aerobic rice. Span J Agric Res 9:810–820

Panhwar QA, Radziah O, Zaharah AR, Sariah M, Razi IM (2011b) Role of phosphate solubilizing bacteria on rock phosphate solubility and growth of aerobic rice

Park KS, Kloepper JW (2000) Activation of PR-1a promoter by rhizobacteria that induce systemic resistance in tobacco against *Pseudomonas syringae* pv. tabaci. Biol Cont 18:2–9

Park KS, Paul D, Kim JS, Park JW (2009) L-Alanine augments rhizobacteria-induced systemic resistance in cucumber. Folia Microbiol 54:322–326

Pati S, Pal B, Badole S, Hazra GC, Mandal B (2016) Effect of silicon fertilization on growth, yield, and nutrient uptake of rice. Comm Soil Sci Plant Anal 47:284–290

Paul D, Lade H (2014) Plant-growth-promoting rhizobacteria to improve crop growth in saline soils: a review. Agron Sustain Dev 34:737–752

Pieterse CMJ, Zamioudis C, Berendsen RL, Weller DM, Van Wees SCM, Bakker PAHM (2014) Induced systemic resistance by beneficial microbes. Ann Rev Phytopathol 52:347–375

Pii Y, Mimmo T, Tomasi N, Terzano R, Cesco S, Crecchio C (2015) Microbial interactions in the rhizosphere: beneficial influences of plant growth-promoting rhizobacteria on nutrient acquisition process. Rev Biol Fertil Soil 51:403–415

Ping L, Boland W (2004) Signals from the underground: bacterial volatiles promote growth in Arabidopsis. Trend Plant Sci 9:263–266

Pooniya V, Shivay YS, Rana A, Nain L, Prasanna R (2012) Enhancing soil nutrient dynamics and productivity of Basmati rice through residue incorporation and zinc fertilization. Euro J Agron 41:28–37

Prabhu AS, Barbosa Filho MP, Filippi MC, Datnoff LE, Snyder GH (2001) Silicon from rice disease control perspective in Brazil. In: Studies in plant science, vol 8. Elsevier, pp 293–311

Prakash NB, Chandrashekar N, Mahendra C, Patil SU, Thippeshappa GN, Laane HM (2011) Effect of foliar spray of soluble silicic acid on growth and yield parameters of wetland rice in hilly and coastal zone soils of Karnataka, South India. J Plant Nutr 34:1883–1893

Prapagdee B, Kuekulvong C, Mongkolsuk S (2008) Antifungal potential of extracellular metabolites produced by *Streptomyces hygroscopicus* against phytopathogenic fungi. Int J Biol Sci 4:330

Prayitno J, Rolfe B (2010) Characterization of endophytic diazotroph bacteria isolated from rice HAYATI. J Biosci 17:73–78

Prayitno J et al (1999) Interactions of rice seedlings with bacteria isolated from rice roots. Fun Plant Biol 26:521–535

Qin Y, Druzhinina IS, Pan X, Yuan Z (2016) Microbially mediated plant salt tolerance and microbiome-based solutions for saline agriculture. Biotechnol Adv

Ramamoorthy V, Viswanathan R, Raguchander T, Prakasam V, Samiyappan R (2001) Induction of systemic resistance by plant growth promoting rhizobacteria in crop plants against pests and diseases. Crop Protec 20:1–11

Rana A et al (2015) Prospecting plant growth promoting bacteria and cyanobacteria as options for enrichment of macro-and micronutrients in grains in rice–wheat cropping sequence. Cogent Food Agric 1:1037379

Raupach GS, Kloepper JW (1998) Mixtures of plant growth-promoting rhizobacteria enhance biological control of multiple cucumber pathogens. Phytopathology 88:1158–1164

Reinhold-Hurek B, Krause A, Leyser B, Miché L, Hurek T (2007) The rice apoplast as a habitat for endophytic N 2-Fixing bacteria. In: The apoplast of higher plants: compartment of storage, transport and reactions. Springer, Dordrecht, pp 427–443

Rodrigues FA, Datnoff LE (2005) Silicon and rice disease management. Fitopatol Brasileira 30:457–469

Rodríguez H, Fraga R (1999) Phosphate solubilizing bacteria and their role in plant growth promotion. Biotechnol Adv 17:319–339

Ryan RP, Germaine K, Franks A, Ryan DJ, Dowling DN (2008) Bacterial endophytes: Recent developments and applications. FEMS Microbiol Lett 278:1–9

Ryu C-M, Farag MA, Hu C-H, Reddy MS, Kloepper JW, Paré PW (2004) Bacterial volatiles induce systemic resistance in *Arabidopsis*. Plant Physiol 134:1017–1026

Saharan BS, Nehra V (2011) Plant growth promoting rhizobacteria: a critical review. Life Sci Med Res 21:30

Sahebi M et al. (2015) Importance of silicon and mechanisms of biosilica formation in plants. BioMed Res Int

Sahrawat KL, Abekoe MK, Diatta S (2001) Application of inorganic phosphorus fertilizer. Sustaining soil fertility in West Africa, pp 225–246

Sakr N (2016) The role of silicon (Si) in increasing plant resistance against fungal diseases. Hellenic Plant Protec J 9:1–15

Samaranayake P, Peiris BD, Dssanayake S (2012) Effect of excessive ferrous (fe 2 +) on growth and iron content in rice (*Oryza sativa*). Int J Agric Biol 14

Sapsirisopa S, Chookietwattana K, Maneewan K, Khaengkhan P (2009) Effect of salt-tolerant *Bacillus* inoculum on rice KDML 105 cultivated in saline soil. As J Food Ag-Ind 2:S69–S74

Saravanan VS, Madhaiyan M, Thangaraju M (2007) Solubilization of zinc compounds by the diazotrophic, plant growth promoting bacterium *Gluconacetobacter diazotrophicus*. Chemosphere 66:1794–1798

Savant NK, Datnoff LE, Snyder GH (1997) Depletion of plant-available silicon in soils: a possible cause of declining rice yields. Comm Soil Sci Plant Anal 28:1245–1252

Schachtman DP, Reid RJ, Ayling SM (1998) Phosphorus uptake by plants: from soil to cell. Plant Physiol 116:447–453

Schisler DA, Slininger PJ, Bothast RJ (1997) Effects of antagonist cell concentration and two-strain mixtures on biological control of *Fusarium* dry rot of potatoes. Phytopathology 87:177–183

Sessitsch A, Kuffner M, Kidd P, Vangronsveld J, Wenzel WW, Fallmann K, Puschenreiter M (2013) The role of plant-associated bacteria in the mobilization and phytoextraction of trace elements in contaminated soils. Soil Biol Biochem 60:182–194

Sethi SK, Mukherjee AK (2018) Screening of biocontrol potential of indigenous Bacillus spp. isolated from rice rhizosphere against *R. solani, S. oryzae, S. rolfsii* and response towards growth of rice

Shafi J, Tian H, Ji M (2017) *Bacillus* species as versatile weapons for plant pathogens: a review. Biotechnol Biotechnol Equip 31:446–459

Shaharoona B, Naveed M, Arshad M, Zahir ZA (2008) Fertilizer-dependent efficiency of *Pseudomonads* for improving growth, yield, and nutrient use efficiency of wheat (*Triticum aestivum* L.). Appl Microbiol Biotechnol 79:147–155

Shan H et al (2013) Biocontrol of rice blast by the phenaminomethylacetic acid producer of *Bacillus methylotrophicus* strain BC79. Crop Protec 44:29–37

Sharma SB, Sayyed RZ, Trivedi MH, Gobi TA (2013) Phosphate solubilizing microbes: sustainable approach for managing phosphorus deficiency in agricultural soils. SpringerPlus 2:587

Sheng XF, Zhao F, He LY, Qiu G, Chen L (2008) Isolation and characterization of silicate mineral-solubilizing *Bacillus globisporus* Q12 from the surfaces of weathered feldspar. Can J Microbiol 54:1064–1068

Sindhu SS, Dadarwal KR (2001) Chitinolytic and cellulolytic *Pseudomonas* sp. antagonistic to fungal pathogens enhances nodulation by *Mesorhizobium* sp. Cicer in chickpea. Microbiol Res 156:353–358

Singh AK, Singh R, Singh K (2005) Growth, yield and economics of rice (*Oryza sativa*) as influenced by level and time of silicon application. Indian J Agron 50:190–193

Singh PP, Shin YC, Park CS, Chung YR (1999) Biological control of *Fusarium* wilt of cucumber by chitinolytic bacteria. Phytopathology 89:92–99

Song A et al. (2016) The role of silicon in enhancing resistance to bacterial blight of hydroponic-and soil-cultured rice. Sci Rep 6

Song F, Goodman RM (2001) Molecular biology of disease resistance in rice. Physiol Mol Plant Pathol 59:1–11

Suprapta DN (2012) Potential of microbial antagonists as biocontrol agents against plant fungal pathogens. J ISSAAS 18:1–8

Syers JK, Johnston AE, Curtin D (2008) Efficiency of soil and fertilizer phosphorus use. FAO Fertil Plant Nutrition Bull 18

Tamreihao K, Ningthoujam DS, Nimaichand S, Singh ES, Reena P, Singh SH, Nongthomba U (2016) Biocontrol and plant growth promoting activities of a Streptomyces corchorusii strain UCR208-16 and preparation of powder formulation for application as biofertilizer agents for rice plant. Microbiol Res 192:260–270

Tapadar SA, Jha DK (2013) Disease management in staple crops: a bacteriological approach. In: Maheshwari DK (ed) Bacteria in agrobiology: disease management. Springer, Berlin, pp 111–152

Thakuria D, Talukdar NC, Goswami C, Hazarika S, Boro RC, Khan MR (2004) Characterization and screening of bacteria from rhizosphere of rice grown in acidic soils of Assam. Curr Sci 978–985

Theodorou ME, Plaxton WC (1993) Metabolic adaptations of plant respiration to nutritional phosphate deprivation. Plant Physiol 101:339–344

Timmusk S, Timmusk K, Behers L (2013) Rhizobacterial plant drought stress tolerance enhancement: towards sustainable water resource management and food security. J Food Secur 1:6–9

Tokpah DP, Li H, Wang L, Liu X, Mulbah QS, Liu H (2016) An assessment system for screening effective bacteria as biological control agents against *Magnaporthe grisea* on rice. Biol Cont 103:21–29

Trejo-Estrada SR, Paszczynski A, Crawford DL (1998) Antibiotics and enzymes produced by the biocontrol agent *Streptomyces violaceusniger* YCED-9. J Ind Microbiol Biotechnol 21:81–90

Ullah A, Heng S, Munis MFH, Fahad S, Yang X (2015) Phytoremediation of heavy metals assisted by plant growth promoting (PGP) bacteria: a review. Environ Exp Bot 117:28–40

Vaid SK, Kumar B, Sharma A, Shukla AK, Srivastava PC (2014) Effect of Zn solubilizing bacteria on growth promotion and Zn nutrition of rice. J Soil Sci Plant Nutr 14:889–910

Van Bockhaven J (2014) Silicon-induced resistance in rice (*Oryza sativa* L.) against the brown spot pathogen *Cochliobolus miyabeanus*

Van Loon LC, Bakker P (2005) Induced systemic resistance as a mechanism of disease suppression by rhizobacteria. In: PGPR: biocontrol and biofertilization. Springer, Dordrecht, pp 39–66

Van Loon LC, Bakker P, Pieterse CMJ (1998) Systemic resistance induced by rhizosphere bacteria. Ann Rev Phytopathol 36:453–483

Van Wees SCM, Pieterse CMJ, Trijssenaar A, Van't Westende YAM, Hartog F, Van Loon LC (1997) Differential induction of systemic resistance in *Arabidopsis* by biocontrol bacteria. Mol Plant-Microbe Interact 10:716–724

Vasanthi N, Saleena LM, Raj SA (2018) Silica solubilization potential of certain bacterial species in the presence of different silicate minerals. Silicon 10:267–275

Velusamy P, Gnanamanickam SS (2008) The effect of bacterial secondary metabolites on bacterial and fungal pathogens of rice. In: Secondary metabolites in soil ecology. Springer, Berlin, pp 93–106

Verma SK, Kingsley KL, Bergen MS, Kowalski KP, White JF (2018) fungal disease prevention in seedlings of rice (*Oryza sativa*) and other grasses by growth-promoting seed-associated endophytic bacteria from invasive *Phragmites australis*. Microorganisms 6:21

Vessey JK (2003) Plant growth promoting rhizobacteria as biofertilizers. Plant Soil 255:571–586

Vijayapriya M, Muthukkaruppan SM (2010) Isolation and screening of silicate solubilizing bacteria and its biocontrol nature against *Pyricularia oryzae*. Int J Rec Sci Res 4:87–91

Welbaum GE, Sturz AV, Dong Z, Nowak J (2004) Managing soil microorganisms to improve productivity of agro-ecosystems. Crit Rev Plant Sci 23:175–193

Yang J, Kloepper JW, Ryu C-M (2009) Rhizosphere bacteria help plants tolerate abiotic stress. Trend Plant Sci 14:1–4

Yang PW et al (2010) Oligomycins A and C, major secondary metabolites isolated from the newly isolated strain *Streptomyces diastaticus*. Folia Microbiol 55:10–16

Yanni YG, Dazzo FB (2010) Enhancement of rice production using endophytic strains of *Rhizobium leguminosarum* bv. *trifolii* in extensive field inoculation trials within the Egypt Nile delta. Plant Soil 336:129–142

Yanni YG et al. (1997) Natural endophytic association between *Rhizobium leguminosarum* bv. trifolii and rice roots and assessment of its potential to promote rice growth. In: Opportunities for biological nitrogen fixation in rice and other non-legumes. Springer, Dordrecht, pp 99–114

Yanni YG et al (2001) The beneficial plant growth-promoting association of *Rhizobium leguminosarum* bv. trifolii with rice roots. Func Plant Biol 28:845–870

Yildirim E, Karlidag H, Turan M, Dursun A, Goktepe F (2011) Growth, nutrient uptake, and yield promotion of broccoli by plant growth promoting rhizobacteria with manure. Hort Sci 46:932–936

Yoon M-Y, Cha B, Kim J-C (2013) Recent trends in studies on botanical fungicides in agriculture. Plant Pathol J 29:1

Zarandi ME, Bonjar GHS, Dehkaei FP, Moosavi SAA, Farokhi PR, Aghighi S (2009) Biological control of rice blast (*Magnaporthe oryzae*) by use of *Streptomyces sindeneusis* isolate 263 in greenhouse. Am J Appl Sci 6:194–199

Zeng QG, Luo F, Zhang ZB, Yan RM, Zhu D (2012) Phosphate solubilizing rhizosphere bacterial T21 isolated from Dongxiang wild rice species promotes cultivated rice growth. Trans Tech Publ 167–175

Zhao J-L, Zhou L-G, Wu J-Y (2010) Promotion of *Salvia miltiorrhiza* hairy root growth and tanshinone production by polysaccharide–protein fractions of plant growth-promoting rhizobacterium *Bacillus cereus*. Pro Biochem 45:1517–1522

Zou C-S, Mo M-H, Gu Y-Q, Zhou J-P, Zhang K-Q (2007) Possible contributions of volatile-producing bacteria to soil fungistasis. Soil Biol Biochem 39:2371–2379

Zuo Y, Zhang F (2011) Soil and crop management strategies to prevent iron deficiency in crops. Plant Soil 339:83–95

Chapter 14
Plant Growth-Promoting Rhizobacteria-Induced Defense Against Insect Herbivores

Joseph Disi, Jocelyn Simmons and Simon Zebelo

Abstract Plant growth-promoting rhizobacteria (PGPR) improve plant health and productivity by providing protection to plants from diseases and pests and enhancing plant growth. PGPR induce systemic resistance (ISR) against microbial pathogens and herbivorous insects. There are limited studies that show the induction of systemic resistance in crop plants against insect pests. Commonly used PGPR genera in insect pest control include *Pseudomonas*, *Bacillus, Burkholderia, Xenorhabdus, Photorhabdus, Agrobacterium, Streptomyces,* etc. PGPR suppress the activity of insect pests by inducing systemic resistance that results in the production of secondary metabolites (terpenes, siderophores, hydrogen cyanide, etc.) and some display direct insect pathogenicity. This chapter focused on PGPR-induced defense against insect pest in field crops with emphasis on the mechanism of action involved against insect pests. PGPR-mediated biochemical and physical changes in the host plants that display insect pathogenicity, methods mixtures application, and challenges associated with their use of PGPR in sustainable agriculture.

Keywords PGPR · Induced systemic resistance · Entomopathogen · *Pseudomonas* · *Bacillus* · Insect pests

J. Disi
Department of Entomology, University of Georgia, Athens, GA 30602, USA
e-mail: joseph.disi@uga.edu

J. Simmons · S. Zebelo (✉)
Department of Agriculture Food and Resources Sciences,
University of Maryland Eastern Shore, Princess Anne, MD 21853, USA
e-mail: sazebelo@umes.edu

J. Simmons
e-mail: jasimmons@umes.edu

D. K. Maheshwari and S. Dheeman (eds.), *Field Crops: Sustainable Management by PGPR*, Sustainable Development and Biodiversity 23,
https://doi.org/10.1007/978-3-030-30926-8_14

14.1 Introduction

There are one million described insect species in our planet earth, but less than one percent of species are crop pests. Insect pests cause considerable losses in crop production. Crop producers typically rely on chemical insecticides to protect their crops against herbivorous insects. Approximately two million tons of insecticides applied to control insect pests' worldwide (Aktar et al. 2009). This intensive use of insecticides have negative impacts such as toxicity to humans and animals, environmental pollution, effects on nontarget species and development of resistance. The backlash of the negative impact of insecticides leads to the increasing demand from consumers to reduce the use of chemical insecticides in crop production and protection from deleterious pest and pathogens. Biological control using biopesticides are promising alternative to chemical insecticides. Particularly, the microbial pesticides have peculiar advantages, because they have unique mode of action. In many cases, they are species specific and have reduced toxicity. One such approach involves treating crops with Plant Growth-Promoting Rhizobacteria (PGPR) for induced defense against herbivores insects.

PGPR are specific strains of root-colonizing bacteria which can elicit increased rates of plant growth, suppress soil pathogens and induce systemic resistance (ISR) against diseases and insect pests. The effect of PGPR on insect pests could be indirect via ISR and/or direct as an entomopathogen. The indirect effect of PGPR against insects initiated through recognition of the microbes by the host plant, elicitation of specific hormonal signal pathways that might lead to the biosynthesis of defense-related chemical compounds, enzymes, protein, secondary metabolites, and volatile organic compounds (VOCs) against insect herbivores (Pineda et al. 2010, 2013; Van Oosten et al. 2008; Zamioudis and Pieterse 2012; Zebelo et al. 2016). Some PGPR strains exhibited direct insect pathogenicity. The most commonly used entomopathogenic bacteria, are *Bacillus thuringiensis* and *Photorhabdus/Xenorhabdus* species. These bacteria developed as an alternative to chemical pesticides. They are host specific and efficient to control insect pests. But these bacteria have two limitations. (1) They lack persistence in the environment and (2) their dependence in vectoring agents to infect hosts (e.g., nematodes). Interestingly, select PGPR strains display direct entomopathogenicity. Their high environmental persistence and ability to penetrate insect cuticle makes them the preferred candidate bio-insecticides for protection of plants against root-feeding insects (Kupferschmied et al. 2013) and nematodes (Wahla et al. 2012).

There are limited studies on PGPR-mediated plant–insect interaction. Zehnder et al. (1997a) elegantly demonstrated that PGPR could trigger ISR against insects using PGPR strain INR-7 (*Bacillus pumilus*) using Cucumber (*Cucumis sativus* L.) as a model system. *C. sativus* plants treated with INR-7 were less attractive to spotted cucumber beetle, *Diabrotica undecimpunctata* Howardi (Coleoptera: Chrysomelidae) and the striped cucumber beetle, *Acalyma vittatum* Fabricius (Coleoptera: Chrysomelidae), as a result, reduced the number of beetles recorded in INR-7-treated plants than the untreated control plants. In a similar study, Zehnder et al. (1997b) conclude that the reduced number of beetles on PGPR-treated plants was due to decreased production of curcubitacin (insect-feeding stimulant). Hanafi et al. (2007) reported that PGPR (*Bacillus subtilis*) treated tomato plants showed lower whitefly populations under greenhouse conditions. In an unrelated study jasmonic acid (JA) pathway involved in the reduction of whitefly population in PGPR-treated plants (Valenzuela-Soto et al. 2010). The reduction of insect pest population in PGPR-treated plants clearly showed that these bacteria might induce resistance in plants against important insect pests.

The genera of PGPR commonly used as biocontrol agents include *Pseudomonas, Bacillus, Burkholderia, Agrobacterium, Streptomyces*, etc. (Dey et al. 2014) either alone or in combination forming a consortium of strain, which triggers multiple beneficial effects to the growing plants. This chapter focused on the indirect and direct impact of PGPR on insect pests, induced defense against insect pest in plants.

14.2 PGPR-Mediated Plant–Insect Interactions

PGPR can enhance plant growth and health by fixing atmospheric nitrogen, increasing nutrient availability and uptake (Spaink 2000), involved in biosynthesis of essential plant growth-related hormones including Indole-3-acetic acid (IAA), cytokinins, auxins, gibberellins etc. (Contreras-Cornejo et al. 2009; van Loon 2007). Some PGPR strains promote plant growth via the biosynthesis of secondary metabolites and enzymes (Vacheron et al. 2013; Zhang et al. 2008). Rhizobacteria also increases plant health and trigger resistance to plant pathogens and insect herbivores by ISR (Hossain et al. 2016; Rashid and Chung 2017). Some of the PGPR promotes plant growth that could result in improved nutrient composition of the plants (Pandey et al. 2018) and this could affect insect performance (Schoonhoven and Dicke 2005). Insect pests might be attracted and take advantage of increased availability of nutrients in PGPR-treated plants (Schoonhoven and Dicke 2005). PGPR enhances plant tolerance through the regeneration of herbivore injured plant tissue as a result of increased nutrient and water uptake (Rashid and Chung 2017). Thus, PGPR augments plant resistance against insect pest, disease, and weeds by increasing nutrient availability and absorption, the production of secondary plant metabolites, and growth hormones, these phenomena either directly or indirectly induce defense in plants.

14.3 PGPR-Mediated ISR Against Insect Pests

At the heart of PGPR-mediated ISR is priming. Conrath (2011) defined priming as a physiological process by which plants prepare to initiate defense in plants against pathogens and insect herbivores. Pieterse et al. (2014) showed that PGPR prime plants are tolerant of various pathogens and insect herbivores. PGPR triggered ISR leads to jasmonic acid (JA) and ethylene (ET) in cross talk with salicylic acid (SA) independent and dependent pathways. Systemic Acquired Resistance (SAR) is the activation of pathogenesis-related (PR) proteins as a result of SA-dependent signaling pathway and necrotizing pathogens (Conrath et al. 2002; Hammerschmidt 2009).

PGPR triggered ISR involves jasmonic acid (JA) and ethylene (ET) independently or in cross talk with salicylic acid (SA)-dependent pathways. Through these signaling pathways, PGPR regulates plant hormones (JA, ET, and SA) that affect how insects interact with host plants (Shavit et al. 2013; Van Oosten et al. 2008). Chewing herbivorous trigger JA-pathways (Pineda et al. 2010). Zebelo et al. (2016) reported higher levels of JA and JA-related gene expression in PGPR-treated cotton plant that was injured by beet armyworm. The PGPR, *P. fluorescens* strain SS101 activates ISR against chewing insect through JA- and ET-dependent mechanisms (van de Mortel et al. 2012). Furthermore, higher expression of the JA/ET-dependent *ORA59* pathway than the JA-dependent *MYC2* pathway reported in Arabidopsis plants treated with *Pseudomonas simile* WCS417r was implicated in the induction of resistance against the cabbage moth larvae (Pangesti et al. 2016).

PGPR elicited ISR against phloem-feeding insects is dependent on both the JA/ET- and SA-signaling pathways (Niu et al. 2011). Valenzuela-Soto et al. (2010) demonstrated JA-dependent ISR and increased expression of JA-independent genes including terpenoid biosynthetic pathways genes, and photosynthetic genes by *Bacillus subtilis* against the phloem sucking insect whitefly on tomato plants (*Solanum lycopersicum*). *Arabidopsis* roots treated with *P. fluorescens* WCS417r were highly susceptible to the phloem-feeding aphid *Myzus persicae*, although they showed increased expression of *PDF1.2* and *LOX2* genes after insect attack (Pineda et al. 2012). These studies are examples of the role of different rhizobacteria genera, including *Pseudomonas* and *Bacillus,* have against phloem-feeding insects. This could be explained that microbe-associated molecular patterns (MAMPs) of different beneficial microbes recognized by plant root receptors leading to the production of specific hormonal signals. MAMPs of beneficial microbes including flagellin, secondary metabolites, and lipopolysaccharides activate MAMP-triggered immunity (MTI) that modulate hormonal signals in plants (Hermosa et al. 2012; Jacobs et al. 2011; Zamioudis and Pieterse 2012). For instance, lipopeptides of *B. amyloliquefaciens* S499 induced the expression of defense-related genes lipoxygenase D, and F (*LOXD*, *LOXF*)-induced ISR in tomato plants (Cawoy et al. 2014). The early event in rhizobacteria–plant interaction is the recognition of MAMPs of the rhizobacteria by plant root receptors and leads to the generation of a distinct ISR signal in the roots. The signal induced by MAMPs transport upward to shoots to trigger ISR in the leaves instantaneously activate the SA-, JA-, and ET-dependent signaling pathways. These signaling pathways expressed genes that regulate the production of defensive compounds such as enzymes, defensive proteins, and secondary metabolites (alkaloids, phenols, nonvolatile terpenes and VOCs). For example, the primed plants release VOCs upon herbivore damage (HD); these

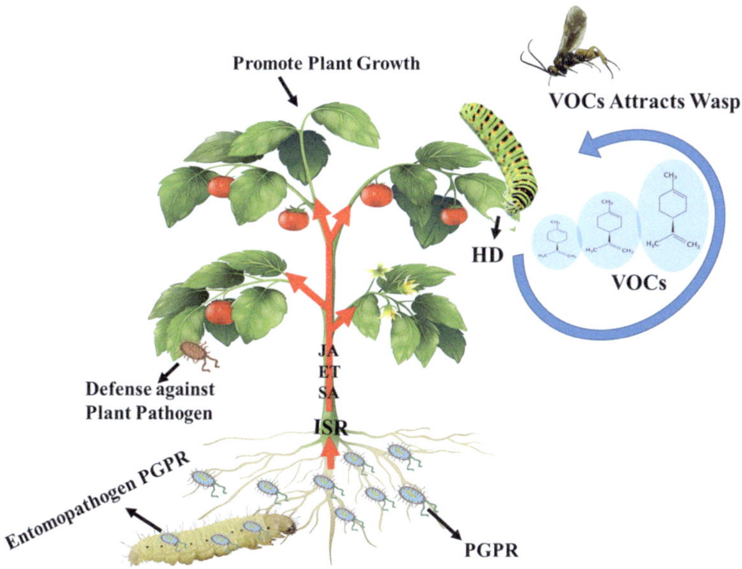

Fig. 14.1 PGPR prime the host plant to increase the defense against various pathogens and insect herbivores by the mechanism of induced systemic resistance (ISR) to produce the defensive compounds. The early event in rhizobacteria–plant interaction is the recognition of microbe-associated molecular patterns (MAMPs) of the rhizobacteria by plant root receptors and leads to the generation of a distinct ISR (red arrow) signal in the roots. The signal transport upward to shoots to trigger ISR in the leaves by instantaneously activating the SA, JA, and ET-dependent signaling pathways. These signaling pathways lead to the expression of genes that regulates the production of defensive compounds such as enzymes, defensive proteins, and secondary metabolites (alkaloids, phenols, nonvolatile terpenes and VOCs). For example, the primed plants release VOCs upon herbivore damage (HD), these VOCs might attract natural enemies of the herbivore (e.g., wasps) and pollinators or repel the herbivore directly. Some PGPR strains serves as an entomopathogen and causes disease root herbivores

VOCs might attract natural enemies of the herbivore (e.g., wasps) and pollinators or repel the herbivore directly (Fig. 14.1). Currently, there is no precise information on how the MAMPs of beneficial microbes modify phytohormone signaling pathways during plant–insect interactions.

14.3.1 PGPR-Mediated Antibiosis Compounds Biosynthesis

Antibiosis is one of the reported forms of PGPR-mediated host plant resistance mechanisms against herbivorous insects. Through elevated or compromised production of allelochemicals, insect-resistant crops reduce growth, inhibit reproduction, alter physiology, delay maturation, and induce various physical or behavioral abnormalities in insect herbivores that ultimately suppress insect pest abundance. Several studies show that PGPR triggers the biosynthesis of plant defense-related compounds against insects through various plant defense signaling pathways (Pangesti et al. 2016; van de Mortel et al. 2012). For example, plant hormones such as JA, ET, and SA discussed above regulate the key defense-related chemical compounds that include, but not limited to, flavonoids, lignin, and other secondary metabolites. Most of these defense-related compounds have been shown to have broad-spectrum properties, against plant pathogens and insect herbivores (Campos et al. 2014; Valenzuela-Soto et al. 2010; Zebelo et al. 2016). Biosynthesis of camalexin and glucosinolates is one way PGPR regulate plant defenses against plant pathogens and insect herbivores (Clay et al. 2009; Kim Jae et al. 2008; Mewis et al. 2005; Muller et al. 2010). Pangesti et al. (2016) reported that the *Arabidopsis* plants treated with PGPR activate JA/ET pathway that leads to the biosynthesis of camalexin and glucosinolates, which triggers ISR against chewing insects. Zebelo et al. (2016) showed that cotton plants inoculated with mixtures of PGPR *Bacillus* spp. increased transcript level of JA-related genes *GhLOX1, GhAOS,* and *GhOPR3*, and gossypol biosynthesis genes including the (+)-δ- cadinene synthase (CAD1) gene family (*Cdn1- A, CAD1-C1, Cdn1-C3,* and *Cdn1-C14*). Consequently, the level of gossypol was elevated in PGPR-treated plants and the herbivory by beet armyworm larvae was significantly reduced, suggesting that the high level of gossypol in PGPR-treated plants might have contributed to the induced resistance of cotton plants against beet armyworm larvae. Gossypol is a phenolic sesquiterpenoid aldehyde with the insecticidal property. High production of phenolic compounds (secondary metabolites) induces resistance against pathogens and insects (Sharma et al. 2009; Usha Rani and Jyothsna 2010) Remarkably high levels of gossypol in cotton reduces the growth and development of *Heliothis* and *Helicoverpa* (Noctuidae) larvae (Du et al. 2004; Stipanovic et al. 2006). JA treatment increases the level of gossypol in cotton plant, and this was correlated with reduced growth, and development of the mealybug, *Phenacoccus solenopsis* (Zhang et al. 2011). High level of phenolic compounds was reported in rice plants treated with PGPR, *P. fluorescens* WCS374r (De Vleesschauwer and Höfte 2009). Furthermore, flavonoids are commonly known as insect-feeding inhibitor (Schoonhoven and Dicke 2005). Arabidopsis plants treated with *Bacillus* and *Actinomycetes* increases the transcript levels of five transcription factors (*TT8, EGL3, MYB12, MYB114,* and *MYB113*) that regulates the genes for flavonoid biosynthesis pathways (Ali and McNear 2014). The transcription factor *MYC2* controls JA-signaling pathways upon herbivory, and *MYC2* also regulates flavonoids and anthocyanin biosynthesis pathways (De Vos et al. 2005; Dombrecht et al. 2007; Verhage et al. 2011).

14.3.2 Structural Modification of Plant Cells by PGPR

Proteins like lipoxygenase (LOX), lignin and lectins are associated with various defense-related processes, which include the formation of cell wall structure, stress adaptation, resistance to pathogens and insects in several crops (Pauwels et al. 2009; Xiaohong et al. 2012). Rice plants colonized by *P. fluorescens* triggers ISR against the leaf folder larvae by the activation of several enzymes including LOX, chitinases and trypsin inhibitors (Commare et al. 2002; Saravanakumar et al. 2007; Saravanakumar et al. 2008). Al Atalah et al. (2014) reported that the jacalin-related lectins display insecticidal activity against different types of insects. The lectin-induced in soybean plants during interaction with bacterial pathogens including *Xanthomonas axono podis* pv. Glycines, *P. syringae* pv. Tomato, and *B. amyloliquefaciens* KPS46 (Buensanteai et al. 2009). Lignin being a recalcitrant confers resistance to insect herbivores (Barakat et al. 2010). The lignin content in plant cell wall directly related with plant leaf hardiness when lignin content increases in the plant cell wall the leaf hardiness also increases for insect herbivores (Johnson et al. 2009). Peroxidase (POD) and polyphenol oxidase (PPO) catalyze the biosynthesis of lignin and different oxidative phenols that are involved in plant defense against insect pests (Bhonwong et al. 2009; Gulsen et al. 2010). For example, *P. fluorescens* strains Pf1, TDK1 and PY15 display ISR against the leaf folder larvae (*Cnaphalocrocis medinalis*) due to the activation of PPO in rice plants (Saravanakumar et al. 2008). The mechanism of induced resistance against insects by PGPR related to lignin biosynthesis remains unclear. However, PGPR-mediated plant defenses against insect herbivores can be induced through biochemical and physical changes in plants. For example, the induced production of a chemical, such as a flavonoid, is not only a chemical inhibitor of insects. Instead, it may be linked to physical modifications of the cell wall by lignification. PGPR modulated ISR might lead to the production of both chemical and physical barriers that might have antibiotic and antixenosis effect on insect herbivores.

14.3.3 Effect of PGPR in Sucking and Chewing Insects

Induced systemic resistance works by potentiating plants' innate immunity which is effective against insect herbivores. An insect with biting and chewing mouthparts typically induce responses similar to induced systemic resistance that is dependent upon jasmonic acid as a signaling molecule and may promote induced resistance against subsequent insect herbivores (Gatehouse 2002). Insects with piercing and sucking mouthparts tend to inflict limited tissue damage through feeding and are often able to evade the wound-induced defense response. These insects may activate defenses similar to systemically acquired resistance responses that are dependent upon salicylic acid as a signaling molecule (Walling 2000). PGPR mediated ISR might vary by insect species and their feeding guilds.

14.3.3.1 Sucking Insects

Induced systemic resistance by PGPR against sucking insects are plant or beneficial microbe specific. This effect also depends on whether the targeted insects are specialist or generalist feeders. For instance, commercial *Bacillus* species applied in mixtures in bell pepper was shown to induce tolerance to *Myzus persicae* Sulzer (Boutard-Hunt et al. 2009; Herman et al. 2008). Similarly, Fahimi et al. (2014) reported the adverse effect on the population of *Aphis gossypii* feeding on cucumber plants treated with *P. fluorescens* strain UTPF. Treatment of *Arabidopsis thaliana* with *P. fluorescens* strain WCS417r and then subsequently feeding the plant to the generalist aphid, *M. persicae* positively affected growth and development of the aphid, but no effect documented on the crucifer aphid *Brevicoryne brassicae* that fed on the same plant (Pineda et al. 2012). However, treatment of the aphid host plant (crucifer) with single and mixtures of *Bacillus* species suppressed the growth and development of *B. brassicae* (Gadhave et al. 2016a; Gadhave and Gange 2016). This indicates a plant-specific or microbial species effect. Valenzuela-Soto et al. (2010) showed that treatment with *Bacillus subtilis* reduced the development of *Bemisia tabaci* on tomato plants, contrary to a work by Shavit et al. (2013) who found that *B. tabaci* nymphs had higher survivorship after they fed on *Bacillus subtilis* WCS417r-treated tomato plants.

14.3.3.2 Chewing Insects

There are several examples of PGPR-mediated plant defenses on the behavior of chewing insects that cut across Lepidoptera, Coleoptera, and Diptera insect orders. Zehnder et al. (1997b) observed a reduction in the number of cucumber beetles (*Diabrotica undecimpunctata* Hawardi (Barber) on cucumber plants or seedlings treated with *Bacillus pumilus* strain INR-7 by reducing the production of phagostimulants. A similar effect was reported on other aboveground and belowground-feeding coleopterans in different field crops (Santos et al. 2014; Coy et al. 2017; Disi et al. 2018b). Studies suggested that PGPR modified behavior of root-feeding insects via enhanced emission of volatile organic compounds even though different PGPR genera used in these separate independent studies. In the study by Santos et al. (2014) root colonization by *Azospirillum braslense* negatively affected weight gain and preference by *Diabrotica* speciose on maize via increased emission of (*E*)-β-caryophyllene but root colonization did not cause increased growth of the plant. Similarly, treatment of maize seeds with *Bacillus pumilus* strain INR-7 had an adverse effect on weight gain and preference of *Diabrotica virgifera virgifera* but no growth promotion was recorded for this bacterial species (Disi et al. 2018b). Chiriboga et al. (2018) reported the enhanced expression of a gene involved in the production of (*E*)-β-caryophyllene which correlated with the significantly enhanced emission of (*E*)-β-caryophyllene in *Pseudomonas protegens* CHA0 and *Pseudomonas chlororaphis* PCL1391 treated maize exposed to infestation by *Diabrotica balteata* larvae.

Van Oosten et al. (2008) revealed adverse impact by *P. fluorescens* WCS417r on *Spodoptera exigua*, a generalist chewing herbivore. Pangesti et al. (2015a) reported that a generalist caterpillar, *Mamestra brassicae* weighed less upon feeding on roots of *A. thaliana* treated with *P. fluorescens* WCS417r, but there was no effect on the weight of the specialist *Pieris brassicae*. Application of *P. fluorescens* (Pf1 and TDK1) and *Beauveria bassiana* (B2) in combination with chitin affected the production of defense enzymes in groundnut plants and concomitantly reduced infestation by a leaf miner (*Aproaerema modicella*) and a pathogen (*Sclerotium rolfsii*) (Senthilraja et al. 2013). Zebelo et al. (2016) showed that mixture of *Bacillus* PGPR strains (Blend-8 and Blend-9) enhanced the level of gossypol in cotton leaves with consequence on growth and development of *S. exigua* larvae and pupae. Application of *Pseudomonas putida* and Rothia sp. on tomato seeds affected the production of proteins and defense-related enzymes in tomato plants after the plants infested with *Spodoptera litura* (Bano and Muqarab 2017). Disi et al. (2018a) reported reduction in emission of maize volatiles treated by single and mixture of *Bacillus* PGPR species that deterred *Ostrinia nubilalis* oviposition, suggesting that manipulation of secondary metabolites may be a common mechanism of PGPR-mediated ISR against insects (Zehnder et al. 1997a; Santos et al. 2014; Zebelo et al. 2016; Chiriboga et al. 2018).

14.4 The Entomopathogenicity of PGPR

Entomopathogen defined as the pathogens that can cause disease and kill insects. An entomopathogen could be fungus, bacteria, virus, nematode or protozoans. The entomopathogenic bacteria promote plant growth, enhance plant health via ISR and display insect pathogenicity. Kupferschmied et al. (2013) reviewed broadly about the potential insect pathogenicity of some PGPR strains. The non-PGPR entomopathogenic bacteria, mainly *Bacillus thuringiensis* (*Bt*), *Photorhabdus* sp., and *Xenorhabdus* sp. are commonly used as entomopathogens. Efficiently controls insect pests in laboratory, greenhouse and field conditions. For example, *Bt* is commercially available as topical sprays and has several advantages over chemical insecticides. *Bt* attacks specific insect species, and its application is considered to be environmentally sound and harmless to humans and other mammals (Kupferschmied et al. 2013). However, the use of *Bt* to control insect pests has limitations. *Bt* has limited persistence on the surface of plant leaves, this is due to sensitivity to solar irradiation as well as to the chemical environment on plant leaves, and unlike PGPR strains, *Bt* is not a competitive plant colonizer (Bizzarri and Bishop 2008; Raymond et al. 2010). Because the susceptible stages of the pest insects are during the early instar larvae, *Bt* provides only short-term crop protection in the field and requires precise application practices (Bravo et al. 2011). Similarly, the use of *Photorhabdus* sp. and *Xenorhabdus* sp. as biocontrol depends on nematode vector (*Steinernena* spp.) for insect infection. Intriguingly, certain strains of PGPR *Pseudomonas* spp. and *Streptomyces* spp. display insect pathogenicity and thus could be formulated to extend the present range of bio-insecticides for protection of plants against insect pests. These entomopathogenic PGPR strains have the remarkable ability to suppress soil-borne plant pathogens, promote plant growth, and induce systemic plant defenses against insect pests. Recently, at Auburn University, Department of Entomology and Plant pathology, Drs. Fadamiro and Kloepper screened more than three hundred strains of PGPR Bacillus spp. for their entomopathogenicity against beet armyworm, several strains showed toxicity against beet armyworm (unpublished data). *Pseudomonas* spp. and *Streptomyces* spp. are relatively well documented as an entomopathogen of several insect species.

14.4.1 **Pseusomonas** ***spp. as an Entomopathogen***

Pseudomonas fluorescens is an entomopathogeic PGPR used against termites (Devi and Kothamasi 2009), phytophagous ladybird beetles (Otsu et al. 2004), and aphids (Hashimoto 2002). *P. fluorescens* bacteria has been used successfully against herbivorous insects and as biological control of fungal plant pathogens (Commare et al. 2002; Karthiba et al. 2010; Péchy-Tarr et al. 2008). *P. fluorescens* infects the mid-gut of larvae, pupae, and adults within the Lepidopteran (Tang et al. 2012), Dipteran (Bansal et al. 2011; Corby-Harris et al. 2007), Coleopteran (Arias-Cordero et al. 2012; Saitou et al. 2009), Hemipteran species (Hashimoto 2002; Lacava et al. 2007), and Hymenopteran (Li et al. 2012; Mohr and Tebbe 2006).

Pseudomonas luminescens toxify insects through a single toxin gene called *makes caterpillars' floppy* (mcf). This mcf gene regulates a persuasive insect toxin Mcf1 (Péchy-Tarr et al. 2008). *Mcf1* is toxic to the midgut epithelial cells, and high expression leads to the floppy nature of the insect infected with *P. luminescens*. This potent insect toxin of *P. luminescens* might suppress immune response activity (Kupferschmied et al. 2013).

Similarly, the *mcf1*-related gene of *P. fluorescens* strain Pf-5 and CHA0, a part of the eight-gene cluster called *fit* for *P. fluorescens* insecticidal toxin (Péchy-Tarr et al. 2008). Kupferschmied et al. (2013) reported that the gene *fitD*, encodes for the actual insect toxin, fitABC and E genes encodes a type I secretion system and fitFGH genes encodes the regulatory proteins. Because the presence of potent Mcf1 toxin in the gene cluster in the Fit toxin, it is conceivable that *FitD* induces apoptosis in insect as in *P. luminescens*.

The Fit toxin gene detected in few PGPR strains, *P. protegens*, and *P. chlororaphis* showed high toxicity toward larvae of lepidopteran insects (Flury et al. 2016; Loper et al. 2012; Shen et al. 2013). Further, *P. protegens* strains CHA0 and Pf-5 were lethal to tobacco hornworm larvae, *Manduca sexta* and the greater wax moth *Galleria mellonella* (Péchy-Tarr et al. 2008). In a laboratory assay using artificial diet and leaves treated with *P. protegens* strain CHA0 and *P. chlororaphis* strain PCL1391 were shown insecticidal activity (Flury et al. 2016). Plants sprayed with a low concentration of CHA0 and Pf-5 strains that contain Fit toxin shown insecticidal activity against *Spodoptera littoralis* (cotton ballworm), *Heliothis virescens* (tobacco budworm), and *Plutella xylostella* (diamondback moth). (Kupferschmied et al. 2013). In contrast, fit-deficient *P. fluorescens* failed to toxicity against insects in the same assay (Ruffner et al. 2013). Ruffner et al. (2013) performed host specificity test and examined the potential side effects of the *Pseudomonads* toward beneficial insects; the Fit toxin had no oral toxicity toward the important pollinator, bumblebee *Bombus terrestris*. Further research might be needed to confirm host specify of diverse species of *Pseudomonads* to that of other beneficial arthropods.

The potential of entomopathogenic PGPR demonstrated impressively by feeding Chinese cabbage leaves containing suspension of GFP (green fluorescent protein)-tagged *P. protegens* CHA0 to larvae of the large cabbage white *Pieris brassicae* (Fig. 14.2a, b) (Kupferschmied et al. 2013). The bacteria colonize the insect gut and subsequently translocated into the hemocoel by so far unknown means, where they multiply and cause disease (Fig. 14.2b). Quick invasion of the bacteria into the insect blood system indicates the level of virulence of the microorganisms, which suggests that these bacteria should be considered for pest management (Kupferschmied et al. 2013).

Regardless of the insecticidal activity of Fit toxin, when Fit toxin gene knocks out from some isolates of *P. protegens* or *P. chlororaphis* strains, is not sufficient to reduce the toxic effects of Fit toxins to insects (Maria et al. 2013; Ruffner et al. 2013). Further studies might be required to discover the virulence factors in these insecticidal pseudomonads. Candidate virulence factors that might play a role in insect pathogenicity in some of these strains are the so-called toxin complexes (Tc), first identified in *P. luminescens*. They are large multimeric insecticidal protein complexes displayed on the surface of these bacteria (Ffrench-Constant et al. 2007). Remarkably, Tc-related genes found also in some of the strains of *P. chlororaphis* and *P. fluorescens* (Loper et al. 2012) but their role in insect pathogenicity yet to be investigated.

Despite the above successful findings of pseudomonads PGPR as entomopathogen to many insect pests, no insecticidal products exist on the market for biopesticides. Pseudomonads PGPR are successfully used as bio-fungicides in agriculture, specifically for crop protection (Ahmadzadeh et al. 2006; Siddiqui et al. 2008). Thus, these entomopathogenic *Pseudomonas* fit well into integrated pest management (IPM) programs (Dotta 2015).

14.4.2 **Streptomyces** *spp. as an Entomopathogen*

Another important PGPR with entomopathogenic characteristic are the Actinobacteria mostly in the genera Streptomyces. *Streptomyces* spp. are notable for their toxic activity against pathogens (e.g., Breza-Boruta et al. 2004) and herbivores especially the order Lepidoptera (Arasu et al. 2013; Kaur et al. 2014;

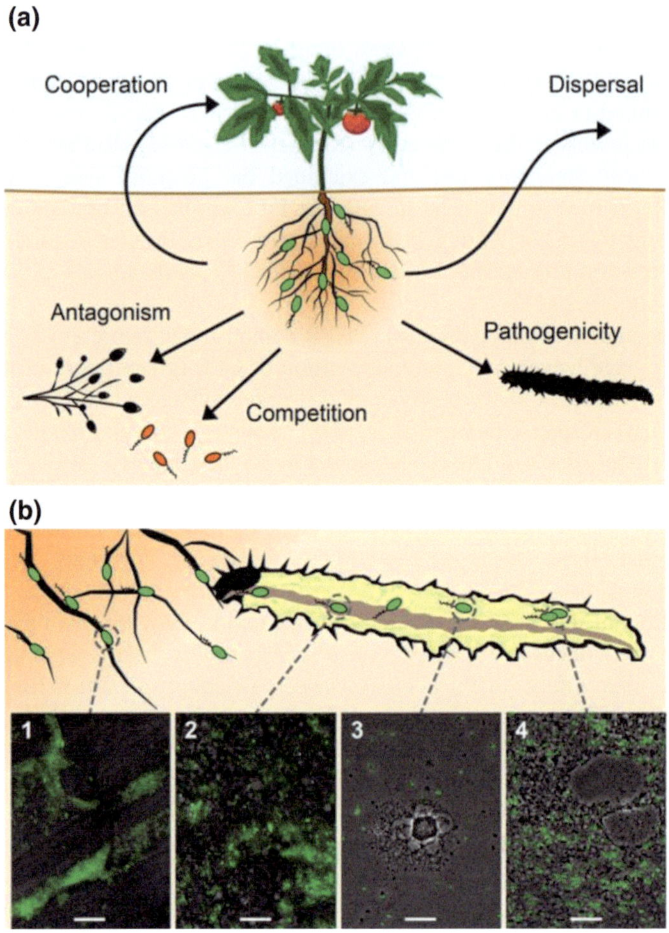

Fig. 14.2 Certain plant root-associated *Pseudomonas* bacteria exhibit insect pathogenicity as an additional trait to the well-studied biocontrol activity against phytopathogens. **a** The most important interactions of these plant-beneficial *pseudomonads* (in green) include cooperation with the plant host (growth promotion and induction of systemic resistance) and competition with and antagonism of soil-borne phytopathogens. In addition, they show insecticidal activity and can use insects as vectors for dispersal. **b** Certain strains of *Pseudomonas protegens* and *Pseudomonas chlororaphis* are capable of infecting and efficiently killing insect larvae after oral uptake. *P. protegens* strain CHA0 (here tagged with GFP for microscopical visualization) typically forms microcolonies on roots (1) of various plant species (here tomato). Following ingestion by herbivorous insects, the entomopathogenic *P. protegens* strain is able to colonize the midgut (2) of pest insect larvae (here the large cabbage white *Pieris brassicae*), possibly by competing with the intestinal microbiota. By a so far unknown mechanism CHA0 cells then cross the intestinal epithelial barrier and invade the hemocoel within less than 1 day after oral infection (3). Once in this body compartment, the bacteria proliferate, resist uptake and elimination by hemocytes and cause disease (4). Bars represent 10 μm (Adopted from Kupferschmied et al. (2013))

Vijayabharathi et al 2014; Goda et al. 2015; Sathya et al. 2016). Kaur et al. (2014) showed that high doses of secondary metabolites from *Streptomyces hydrogenas* DH16 increased mortality of *Spodoptera litura* (Fab.) larvae and lower doses prolonged development time in larvae and pupae. Polyketide metabolites from *Streptomyces* sp. AP-123 caused antifeedant and larvicidal activity on *Helicoverpa armigera* and *Spodoptera litura* larvae. Similar effects by *Streptomyces* sp. CAI-155 (Gopalakrishnan et al. 2012, 2016) and *Streptomyces griseoplanus* SAI-25 (Sathya et al. 2016) against *H. armigera* have also been documented on chickpea. These studies show that Actinobacteria have a strong potential as a biological control agent.

D-endotoxin proteins and Avermectins from *Bacillus thuringiensis* and *Streptomyces avermitilis*, respectively, are the two most toxic compounds against herbivores that have ever been identified from soil microbes. Of these two, avermectins is believed to have broad-spectrum activities against pathogens and herbivores and it is a gamma-amino butyric acid (GABA) receptor antagonist. It binds to the GABA receptor, leading to increased flow of chlorine into the muscle and subsequent muscle relaxation and eventual death of exposed organism (Ishaaya and Horowitz 1998). Since its discovery, insecticidal activities of Avermectins has been reported for insect pests from several orders and families (Lasota and Dybas 1991; Jacob and Sudini 2016; Hariprasad 2016). Negative effect of commercialized derivatives of Avermectins such as emamectin benzoate on *Cydia pomonella* and *Cydia molesta* (Ioriatti et al. 2009) and ivermectin on biology of the Indian Owlet-moth, *Spirama retorta* (Roychoudhury and Joshi 2011) is well documented. Ioriatti et al. (2009) reported that emamectin benzoate was effective at causing high mortalities to larva stages of codling moths both in laboratory and semi-field trials but minimal ovicidal effect was observed. Similarly, high mortality was recorded by invermectin on *S. retorta* a foliar pest of *Albizia* plants. However, a study showed that abamectin is very toxic to predators. Azod et al. (2016) reported 100% mortality of aphidophagous lady beetle, *Menochilus sexmaculatus* after feeding on the common pistachio psylla, *Agonoscena pistaciae*. Further studies are needed on a large scale to determine how best Avermectins and its derivatives can be integrated into current IPM systems as Coccinellids and other predators constitute a significant part of predators in agro landscapes.

14.5 PGPR Formulations for Insect Pest Control

PGPR formulations can come in ambient of different forms depending on their application. Although beneficial PGPR group of bacteria are used to ameliorate environmental degradation (Bashan et al. 2014), those for crop-specific agricultural use have been limited (Kloepper et al. 2004; Liu et al. 2016). The role of their formulation and application for agricultural needs broadly covered (Bashan et al. 2014). Depending on the availability of materials, formulations can be in liquid, solid dry powder, etc.(Arora et al. 2008; Atieno et al. 2012;

Bhattacharjya and Chandra 2013). The carrier materials can be organic (sawdust, vermicompost, peat, poultry manure, etc.) or inorganic (talc, perlite, clay soil, kaolin, etc.) or polymers (e.g., alginate), and these also depend on the need for the formulation (Bashan et al. 2002). Another determinant of the inoculum formulation depends on the characteristics of the bacteria.

There are only a few formulated PGPR inoculant for controlling insect pests in various field crops expect those commercially available for growth promotion and control of other foliar pathogens (Boutard-Hunt et al. 2009; Herman et al. 2008). However, the understanding of the "broad-spectrum" activity of many soil-inhabiting PGPR leads in this direction. The laboratory, greenhouse and field-based studies reported PGPR and some none-growth-promoting bacteria mediated plant–insect interactions with a focus on formulation media/carriers (Table 14.1). Further, as shown, most formulation was liquid and solid meant yet to address fundamental research. Only a few of the studies utilizing the various formulations [e.g., Gadhave et al. (2016a)] were conducted in the field, and very few showed carrier based formulations reported had good shelf-life (Sarma et al. 2011).

14.6 Application of PGPR Mixture

Application of PGPR depends on formulations (liquid, dry, slurry or pellets), the carrier material used for the formulations and the need-based requirement of farmers.. Given the complication of bacterial–plant–insect species specificity, PGPR is either applied singly or in the mixture (Domenech et al. 2006; Liu et al. 2016; Pangesti et al. 2013; Pineda et al. 2013; Pineda et al. 2010; Gadhave and Gange 2016; Raupach and Kloepper 1998). Certain PGPR strains promote growth of plants, but others induce resistance inplants. Targeting this discrepancy in the formulation of PGPR mixture should always inform the rationale or one of the most important reasons for formulating the product, but quite a few workers suggested that. PGPR mixture may provide broad-spectrum combination based qualities from individuals that make up the mixture with multifarious in nature. Several published works showed that application of PGPR mixture was more targeted on improvement of plant growth and yield (Cassán et al. 2009; Requena et al. 1997; Sandheep et al. 2013; Yadav and Verma 2014) except a few studies that demonstrated that a mixture of different microorganisms (*Azotobacter, Pseudomonas, and Bacillus*) enhanced their competitiveness compared to indigenous bacteria in the soil (e.g., Đalovic et al. 2013).

Table 14.1 Studies showing PGPR formulations used to test effects on insects: A few examples

PGPR strain(s)	Formulation medium/carrier	Plant species	Insect species tested	References
Pseudomonas putida 89B-61, *Serratia marsescens* 90–166, *Flavomonas oryzihatans* INR-5, *Bacillus pumilus* strain INR-7	Tryptic soybean broth (TSB)	Cucumber	*Diabrotica undecimpunctata howardi* and *Acalymma vittatum* (F.)	(Zehnder et al. 1997a)
Pseudomonas fluorescens PF1 and FP7, single and in mixture	Talc-based powder	Rice	*Cnaphalocrocis medinalis*	(Commare et al. 2002)
Pseudomonas fluorescens Pf1, TDK1 and PY15	Talc-based powder	Rice	*Cnaphalocrocis medinalis*	(Saravanakumar et al. 2007, 2008)
Bacillus subtilis strainGB03 Bacillus amyloliquefaciens IN937a	Commercial formulation	Pepper	*Myzus persicae*	(Herman et al. 2008)
Paenobacillus macerans GB122 and *Bacillus amyloliquefaciens* GB99	Commercial formulation	Pepper	*Myzus persicae*	(Boutard-Hunt et al. 2009)
Pseudomonas fluorescens (PGPR) *Beauveria bassiana* (Fungi)	Talc-based powder	Rice	*Cnaphalocrocis medinalis*	(Karthiba et al. 2010)
Bacillus subtilis BEB-DN (*Bs*DN)	Potato infusion medium	Tomato	*Bamisia tabaci*	(Valenzuela-Soto et al. 2010)
Pseudomonas fluorescens Pf1 and TDK1	Talc-based powder	Groundnut	*Aproaerema modicella*	(Senthilraja et al. 2013)
Pseudomonas fluorescens WCS417r	$MgSO_4$	*Arabdopsis*	Myzus persicae and *Diaeretiella rapae*	(Pineda et al. 2013)
Pseudomonas fluorescens WCS417r	$MgSO_4$	Tomato	*Bamisia tabaci*	(Shavit et al. 2013)
Pseudomonas fluorescens UTPF68, UTPF1, UTPF6 and CHA0	Methylcellulose solution	Cucumber	*Aphis gossypii*	(Fahimi et al. 2014)
Azospirillum brasilense	Nitro 1000	Maize	*Diabrotica speciosa*	(Santos et al. 2014)
Pseudomonas fluorescens WCS417r and *P. simiae* WCS417r	$MgSO_4$	*Arabdopsis*	*Mamestra brassicae, Pieris brassicae* and *Microplitis mediator*	(Pangesti et al. 2015a, b, 2016)

(continued)

Table 14.1 (continued)

PGPR strain(s)	Formulation medium/carrier	Plant species	Insect species tested	References
Pseudomonas putida and *Rothia* sp.	Water	Tomato	*Spodoptera litura*	(Bano and Muqarab 2017)
Bacillus cereus, *Bacillus subtilis* and *Bacillus amyloliquefaciens*	Saline water	Cucumber	*Brevicoryne brassicae*	(Gadhave and Gange 2016, 2016a, b)
Bacillus pumilis, Blend-8 and Blend-9 (both blends contained four strains)	Water	Cotton	*Spodoptera exigua*	(Zebelo et al. 2016)
Bacillus pumilis, MC1, MC2, MC3, MC4, Blend 8, Blend 18, Blend 19, Blend 20	Water	Bermudagrass	*Spodoptera frugiperda*	(Coy et al. 2017)
Bacillus velezensis YC7010	$MgSO_4$	*Arabdopsis*	*Myzus persicae*	(Rashid et al. 2017)
Bacillus pumilis, Blend-8 and Blend-9 (both blends contained four strains)	Water	Maize	*Ostrinia nubilalis* and *Diabrotica virgifera virgifera*	(Disi et al. 2018a, b)

14.7 Challenges Associated with the Use of PGPR in Sustainable Agriculture

The commercial success of PGPR strains needs economic and feasible market demand, consistent and broad-spectrum action, safety and stability, longer shelf-life, low capital costs and easy availability of career materials (Dey et al. 2014). The success and commercialization of PGPR strains depend on the collaborative work between the scientific institutions and chemical industries. These collaborative works mainly focused on the process of marketing include isolation of antagonist strains, screening, fermentation methods, mass production, formulation viability, toxicology, industrial linkages, quality control, and field efficacy. Though PGPR has potential use in sustainable agriculture and commercialization, the threat of certain PGPR (*P. aeruginosa, P. cepacia* and *B. cereus*) to impart toxicity to animals and pathogenic to human beings as opportunistic pathogens has to be clarified before large-scale acceptance (Nakkeeran et al. 2006). For example, volatile production of cyanide by some PGPR strains. Cyanide involves in growth promotion as well as a growth inhibition characteristic during PGPR–plant interactions. Moreover, cyanide acts as a biocontrol agent against specific plant pathogens (Ramette et al. 2006); likewise, cyanide can also cause adverse effects on plant growth and the environment (Solomos and Laties 1976). Potential PGPR strains must pass through several regulatory measures to be commercially viable. The PGPR strains should not pose any threat to human and animal health and should not be an environmental hazard. Unlike the commercialization of PGPR strains used to control plant pathogens, information concerning PGPR strains used to control insect pest is very limited. Despite that PGPR have been discovered and studied for last three decades, the widespread use of these products is yet to be seen. The use of PGPR as biocontrol of pests in sustainable agriculture remains underutilized.

14.8 Concluding Remarks and Future Perspectives

This chapter highlights PGPR mediated plant–insect interactions. The interaction between PGPR–plant–insect resulted to plant growth promotion and induced defenses against insect herbivores. The effect of PGPR-induced systemic resistance against microbial pathogens has been studied for many years, but relatively little is known about the effect of ISR on herbivorous insects. Inducing resistance against herbivores is not the only mechanism through which PGPR affect plant–insect interactions. Intriguingly, certain strains of plant root-colonizing *Pseudomonas* bacteria display insect pathogenicity. There is evidence that PGPR enhance the effectiveness of natural enemies by manipulating the production of VOCs, this can also decrease herbivore pressure (Disi et al. 2018b). It is important to note that even the direct and indirect effect of PGPR-mediated defenses decreased and the herbivore pressure increases, PGPR can enhance plant biomass and yield, increasing plant tolerance against insect herbivores. Moreover, PGPR-treated plants might attract other beneficial organisms, particularly pollinators and natural enemies of herbivores, the impact of PGPR on pollinators and natural enemies might need further studies.

Beneficial microbes have an enormous contribution that ranges from increasing nutrient and moisture uptake to induction of plant defense against plant pathogens and insect herbivores. Considering this vast importance beneficial microbes might also contribute to the bio-diversification of plants. Beneficial microbes modulate plant–insect interactions, and plant genotype is vital for the successes of these mutualistic interactions. It is recommended that breeders might include traits that enhance plant–microbe interactions in their selection process. Due to the extended persistence of beneficial microbes in the environment, they can be an integral part of sustainable integrated pest management programs.

PGPR has the potential to be used in integrated pest management (IPM), with the progress of agriculture toward sustainability, PGPR will find greater use as biocontrol agents. However, we should be genuine with thoughtfulness. Numerous studies have been done over the past several decades about the potential applications of a PGPR strains of biocontrol agents in managing a number of plant disease and pests, not much significant success achieved yet for application at field level and commercialization of PGPR products. Rigorous efforts will be required to demonstrate the benefits of the PGPR as biocontrol of plant disease and insect pests by conducting trials on farmers' field.

Conflict of Interest: The author(s) have no conflict of interest.

References

Ahmadzadeh MH, Afsharmanesh M, Javan-Nikkhah A, Sharifi T (2006) Identifi cation of some

Aktar MW, Sengupta D, Chowdhury A (2009) Impact of pesticides use in agriculture: their benefits and hazards. Interdiscip Toxicol 2:1–12. https://doi.org/10.2478/v10102-009-0001-7

Al Atalah B, Smagghe G, Van Damme EJM (2014) Orysata, a jacalin-related lectin from rice, could protect plants against biting-chewing and piercing-sucking insects. Plant Sci 221–222:21–28. https://doi.org/10.1016/j.plantsci.2014.01.010

Ali MB, McNear DH (2014) Induced transcriptional profiling of phenylpropanoid pathway genes increased flavonoid and lignin content in *Arabidopsis* leaves in response to microbial products. BMC Plant Biol 14:84. https://doi.org/10.1186/1471-2229-14-84

Arasu MV, Al-Dhabi NA, Saritha V, et al (2013) Antifeedant, larvicidal and growth inhibitory bioactivities of novel polyketide metabolite isolated from *Streptomyces* sp. AP-123 against *Helicoverpa armigera* and *Spodoptera litura*. BMC Microbiol 13. https://doi.org/10.1186/1471-2180-13-105

Arias-Cordero E, Ping L, Reichwald K, Delb H, Platzer M, Boland W (2012) Comparative evaluation of the gut microbiota associated with the below- and above-ground life stages (larvae and beetles) of the forest cockchafer, *Melolontha hippocastani*. PLOS ONE 7:e51557. https://doi.org/10.1371/journal.pone.0051557

Arora NK, Khare E, Naraian R, Maheshwari DK (2008) Sawdust as a superior carrier for the production of multipurpose bioinoculant using PGP *Rhizobial* and *Pseudomonad* strains and their impact on productivity of *Trifolium repense*. Curr Sci 95:90–94

Atieno M, Herrmann L, Okalebo R, Lesueur D (2012) Efficiency of different formulations of *Bradyrhizobium japonicum* and effect of co-inoculation of *Bacillus subtilis* with two different strains of *Bradyrhizobium japonicum*. World J Microbiol Biotechnol 28:2541–2550. https://doi.org/10.1007/s11274-012-1062-x

Azod F, Shahidi-Noghabi S, Mahdian K, Smagghe G (2016) Lethal and sublethal effects of spirotetramat and abamectin on predatory beetles (*Menochilus sexmaculatus*) via prey (*Agonoscena pistaciae*) exposure, important for integrated pest management in pistachio orchards. Belgian J Zool 146:113–122

Bano A, Muqarab R (2017) Plant defence induced by PGPR against *Spodoptera litura* in tomato (*Solanum lycopersicum* L.). Plant Biol (Stuttg) 19:406–412. https://doi.org/10.1111/plb.12535

Bansal R, Hulbert S, Schemerhorn B, Reese JC, Whitworth RJ, Stuart JJ, Chen M-S (2011) Hessian fly-associated bacteria: transmission, essentiality, and composition PLOS ONE 6: e23170. https://doi.org/10.1371/journal.pone.0023170

Barakat A, Bagniewska-Zadworna A, Frost CJ, Carlson JE (2010) Phylogeny and expression profiling of CAD and CAD-like genes in hybrid Populus (*P. deltoides* × *P. nigra*): evidence from herbivore damage for subfunctionalization and functional divergence. BMC Plant Biol 10:100. https://doi.org/10.1186/1471-2229-10-100

Bashan Y, de-Bashan LE, Prabhu SR, Hernandez J-P (2014) Advances in plant growth-promoting bacterial inoculant technology: formulations and practical perspectives (1998–2013). Plant and Soil 378:1–33. https://doi.org/10.1007/s11104-013-1956-x

Bashan Y, Hernandez J-P, Leyva LA, Bacilio M (2002) Alginate microbeads as inoculant carriers for plant growth-promoting bacteria. Biology and Fertility of Soils 35:359–368. https://doi.org/10.1007/s00374-002-0481-5

Bhattacharjya S, Chandra R (2013) Effect of inoculation methods of mesorhizobium ciceri and pgpr in chickpea (*Cicer areietinum* L.) on symbiotic traits, yields, nutrient uptake and soil properties. Legum Res 36:331–337

Bhonwong A, Stout MJ, Attajarusit J, Tantasawat P (2009) Defensive role of tomato polyphenol oxidases against cotton bollworm (*Helicoverpa armigera*) and beet armyworm (*Spodoptera exigua*). J Chem Eco 35:28–38. https://doi.org/10.1007/s10886-008-9571-7

Bizzarri MF, Bishop AH (2008) The ecology of *Bacillus thuringiensis* on the phylloplane: colonization from soil, plasmid transfer, and interaction with larvae of *Pieris brassicae*. Microb Ecol 56:133–139. https://doi.org/10.1007/s00248-007-9331-1

Boutard-Hunt C, Smart CD, Thaler J, Nault BA (2009) Impact of plant growth-promoting rhizobacteria and natural enemies on *Myzus persicae* (Hemiptera: Aphididae) infestations in pepper J Econ Entomol 102:2183–2191

Bravo A, Likitvivatanavong S, Gill SS, Soberón M (2011) *Bacillus thuringiensis*: a story of a successful bioinsecticide. Insect Biochem Mol Biol 41:423–431. https://doi.org/10.1016/j.ibmb.2011.02.006

Breza-Boruta B, Sadowski S, Paluszak Z (2004) Effect of Actinomycetes of *Streptomyces* genus isolated from potato cultivation on the pathogenic fungus *Rhizoctonia solani*. Phytopathol 31:33–43

Buensanteai N, Yuen GY, Prathuangwong S (2009) Priming, signaling, and protein production associated with induced resistance by *Bacillus amyloliquefaciens* KPS46. World J Microbiol Biotechnol 25:1275–1286. https://doi.org/10.1007/s11274-009-0014-6

Campos ML, Kang JH, Howe GA (2014) Jasmonate-triggered plant immunity. J Chem Ecol 40:657–675. https://doi.org/10.1007/s10886-014-0468-3

Cassán F, Perrig D, Sgroy V, Masciarelli O, Penna C, Luna V (2009) *Azospirillum brasilense* Az39 and *Bradyrhizobium japonicum* E109, inoculated singly or in combination, promote seed germination and early seedling growth in corn (*Zea mays* L.) and soybean (*Glycine max* L.). Eur J Soil Biol 45:28–35. https://doi.org/10.1016/j.ejsobi.2008.08.005

Cawoy H et al (2014) Plant defense stimulation by natural isolates of *bacillus* depends on efficient surfactin production. Mol Plant Microbe Interact 27:87–100. https://doi.org/10.1094/mpmi-09-13-0262-r

Chiriboga MX et al. (2018) Root-colonizing bacteria enhance the levels of (*E*)-beta-caryophyllene produced by maize roots in response to rootworm feeding. Oecologia 187:459–468. https://doi.org/10.1007/s00442-017-4055-5

Clay NK, Adio AM, Denoux C, Jander G, Ausubel FM (2009) Glucosinolate metabolites required for an *Arabidopsis* innate immune response. Science 323:95–101. https://doi.org/10.1126/science.1164627

Commare R, Nandakumar R, Kandan A, Suresh S, Bharathi M, Raguchander T, Samiyappan R (2002) *Pseudomonas fluorescens* based bio-formulation for the management of sheath blight disease and leaffolder insect in rice. Crop Prot 21:671–677. https://doi.org/10.1016/S0261-2194(02)00020-0

Conrath U (2011) Molecular aspects of defence priming. Trends Plant Sci 16:524–531. https://doi.org/10.1016/j.tplants.2011.06.004

Conrath U, Pieterse CM, Mauch-Mani B (2002) Priming in plant-pathogen interactions. Trends Plant Sci 7:210–216

Contreras-Cornejo HA, Macías-Rodríguez L, Cortés-Penagos C, López-Bucio J (2009) *Trichoderma virens*, a plant beneficial fungus, enhances biomass production and promotes lateral root growth through an auxin-dependent mechanism in arabidopsis. Plant Physiology 149:1579–1592. https://doi.org/10.1104/pp.108.130369

Corby-Harris V, Pontaroli AC, Shimkets LJ, Bennetzen JL, Habel KE, Promislow DE (2007) Geographical distribution and diversity of bacteria associated with natural populations of *Drosophila melanogaster*. Appl Environ Microbiol 73:3470–3479. https://doi.org/10.1128/AEM.02120-06

Coy M, Held DW, Kloepper JW (2017) Bacterial inoculant treatment of bermudagrass alters ovipositional behavior, larval and pupal weights of the fall armyworm (Lepidoptera: Noctuidae). Environ Entomol 46:831–838. https://doi.org/10.1093/ee/nvx102

Đalovic I, Josic D, Mrkovacki N, Bekavac RP, Purar B, Jocković Đ (2013) The competitiveness of *Azotobacter, Pseudomonas* and *Bacillus* applied as a mixture inoculum in rhizosphere of five maize genotypes assessed by genotyping and phenotyping methods. Agrosym 2013:382–386

De Vleesschauwer D, Höfte M (2009) Rhizobacteria-induced systemic resistance. In: Advances in Botanical Research, vol 51. Academic Press, pp 223–281. https://doi.org/10.1016/S0065-2296(09)51006-3

De Vos M et al (2005) Signal signature and transcriptome changes of arabidopsis during pathogen and insect attack. Mol Plant-Microbe Interact 18:923–937. https://doi.org/10.1094/mpmi-18-0923

Devi KK, Kothamasi D (2009) *Pseudomonas fluorescens* CHA0 can kill subterranean termite *Odontotermes obesus* by inhibiting cytochrome c oxidase of the termite respiratory chain. FEMS Microbiol Lett 300:195–200. https://doi.org/10.1111/j.1574-6968.2009.01782.x

Dey R, Pal KK, Tilak KVBR (2014) Plant growth promoting rhizobacteria in crop protection and challenges. In: Goyal A, Manoharachary C (eds) Future challenges in crop protection against fungal pathogens. Springer New York, New York, NY, pp 31–58. https://doi.org/10.1007/978-1-4939-1188-2_2

Disi JO, Zebelo S, Kloepper JW, Fadamiro H (2018a) Seed inoculation with beneficial rhizobacteria affects European corn borer (Lepidoptera: Pyralidae) oviposition on maize plants. Entomol Sci 21:48–58. https://doi.org/10.1111/ens.12280

Disi JO, Kloepper JW, Fadamiro HY (2018b) Seed treatment of maize with *Bacillus pumilus* strain INR-7 affects host location and feeding by Western corn rootworm, *Diabrotica virgifera virgifera*. J Pest Sci 91:515–522. https://doi.org/10.1007/s10340-017-0927-z

Dombrecht B, Xue GP, Sprague SJ, et al. (2007) MYC2 Differentially Modulates Diverse Jasmonate-Dependent Functions in *Arabidopsis*. Plant Cell 19:2225–2245. https://doi.org/10.1105/tpc.106.048017

Domenech J, Reddy MS, Kloepper JW, Ramos B, Gutierrez-Mañero J (2006) Combined application of the biological product ls213 with *bacillus, pseudomonas* or *chryseobacterium* for growth promotion and biological control of soil-borne diseases in pepper and tomato. Bio Control 51:245. https://doi.org/10.1007/s10526-005-2940-z

Dotta S (2015) Biopesticides: an ecofriendly approach for pest control. World J Pharm Pharm Sci 4(06):250–265

Du L, Ge F, Zhu S, Parajulee MN (2004) Effect of cotton cultivar on development and reproduction of *Aphis gossypii* (Homoptera: Aphididae) and its predator *propylaea japonica* (Coleoptera: Coccinellidae). J Econ Entomol 97:1278–1283. https://doi.org/10.1093/jee/97.4.1278

Fahimi A, Ashouri A, Ahmadzadeh M, Hoseini Naveh V, Asgharzadeh A, Maleki F, Felton GW (2014) Effect of PGPR on population growth parameters of cotton aphid. v. 47

Ffrench-Constant RH, Dowling A, Waterfield NR (2007) Insecticidal toxins from *Photorhabdus* bacteria and their potential use in agriculture. Toxicon 49:436–451. https://doi.org/10.1016/j.toxicon.2006.11.019

Flury P et al. (2016) Insect pathogenicity in plant-beneficial pseudomonads: Phylogenetic distribution and comparative genomics. vol 10. https://doi.org/10.1038/ismej.2016.5

Gadhave KR, Finch P, Gibson TM, Gange AC (2016a) Plant growth-promoting *Bacillus* suppress *Brevicoryne brassicae* field infestation and trigger density-dependent and density-independent natural enemy responses. J of Pest Sci 89:985–992. https://doi.org/10.1007/s10340-015-0721-8

Gadhave KR, Hourston JE, Gange AC (2016b) Developing soil microbial inoculants for pest management: can one have too much of a good thing? J Chem Ecol 42:348–356. https://doi.org/10.1007/s10886-016-0689-8

Gadhave KR, Gange AC (2016) Plant-associated *Bacillus* spp. alter life-history traits of the specialist insect *Brevicoryne brassicae* L. Agri For Entomol 18:35–42. https://doi.org/10.1111/afe.12131

Goda NF, El-Heneidy AH, Djelouah K, Hassan N (2015) Integrated pest management of the tomato leaf miner, *Tuta absoluta* (Meyrick) (Lepidoptera: Gelechiidae) in tomato fields in Egypt. 25:655–661

Gatehouse JA (2002) Plant resistance towards insect herbivores: a dynamic interaction. New Phytol 156:145–169

Gopalakrishnan S, Upadhyaya H, Vadlamudi S, Humayun P, Vidya MS, Alekhya G, Singh A, Vijayabharathi R, Bhimineni RK, Seema M, Rathore A, Rupela O (2012) Plant growth-promoting traits of biocontrol potential bacteria isolated from rice rhizosphere. SpringerPlus 1(1)

Gopalakrishnan S, Rajendran V, Arumugam S, Sharma HC, Vadlamudi S, Bhimineni RK, V Gonzalez S, Melø TM, Simic N (2016) Insecticidal activity of a novel fatty acid amide derivative from species against. Nat Prod Res 30(24):2760–2769

Gulsen O, Eickhoff T, Heng-Moss T, Shearman R, Baxendale F, Sarath G, Lee D (2010) Characterization of peroxidase changes in resistant and susceptible warm-season turfgrasses challenged by *Blissus occiduus*. Arthropod Plant Interact 4:45–55. https://doi.org/10.1007/s11829-010-9086-3

Hammerschmidt R (2009) Systemic acquired resistance. In: Advances in botanical research, vol 51. Academic Press, pp 173–222. https://doi.org/10.1016/S0065-2296(09)51005-1

Hanafi A, Traoré M, Schnitzler WH, Woitke M (2007) Induced resistance of tomato to whiteflies and pythium with the PGPR *Bacillus subtilis* in a soilless crop grown under greenhouse conditions. In: 2007 International Society for Horticultural Science (ISHS), Leuven, Belgium, pp 315–323. https://doi.org/10.17660/ActaHortic.2007.747.38

Hariprasad KV (2016) Recent advancement in the development of biopesticides by Actinomycetes for the control of insect pests. In: Arumugam S, Rajendran V (eds) Subramaniam G. Plant growth promoting Actinobacteria, A new avenue for enhancing the productivity and soil fertility of grain legumes, Springer, pp 47–52

Hashimoto Y (2002) Study of the bacteria pathogenic for aphids, isolation of bacteria and identification of insecticidal compound. AGRIS. Available via The Agriculture, Forestry and Fisheries Research Information Technology Center. http://www.affrc.go.jp/en/. Accessed 07 August 2018

Herman MAB, Nault BA, Smart CD (2008) Effects of plant growth-promoting rhizobacteria on bell pepper production and green peach aphid infestations in New York. Crop Prot 27:996–1002. https://doi.org/10.1016/j.cropro.2007.12.004

Hermosa R, Viterbo A, Chet I, Monte E (2012) Plant-beneficial effects of *Trichoderma* and of its genes. Microbiology 158:17–25. https://doi.org/10.1099/mic.0.052274-0

Hossain MT, Khan A, Chung EJ, Rashid MH, Chung YR (2016) Biological control of rice bakanae by an endophytic *Bacillus oryzicola* YC7007. Plant Pathol J 32:228–241. https://doi.org/10.5423/PPJ.OA.10.2015.0218

Ioriatti C, Anfora G, Angeli G, et al (2009) Toxicity of emamectin benzoate to *Cydia pomonella* (L.) and *Cydia molesta* (Busck) (Lepidoptera: Tortricidae): Laboratory and field tests. Pest Manag Sci 65:306–312. https://doi.org/10.1002/ps.1689

Ishaaya I, Horowitz AR (1998) Insecticides with novel modes of action: An overview. In: Ishaaya I, Degheele D (eds) Insecticides with novel modes of action: Mechanisms and application, eBook. Springer-Verlag, Berlin Heidelberg GmbH, pp 15–16

Jacob S, Sudini HK (2016) Indirect plant growth promotion in grain legumes: Role of Actinobacteria. In: Subramaniam G, Arumugam S, Rajendran V (Eds) Plant growth promoting Actinobacteria: A new avenue for enhancing the productivity and soil fertility of grain legumes, Springer pp 24

Jacobs S, Zechmann B, Molitor A et al. (2011) Broad-spectrum suppression of innate immunity is required for colonization of arabidopsis roots by the fungus *Piriformospora indica*. Plant Physio 156:726–740. https://doi.org/10.1104/pp.111.176446

Johnson MTJ, Smith SD, Rausher MD (2009) Plant sex and the evolution of plant defenses against herbivores. Proc Natl Acad Sci 106:18079–18084. https://doi.org/10.1073/pnas.0904695106

Karthiba L, Saveetha K, Suresh S, Raguchander T, Saravanakumar D, Samiyappan R (2010) PGPR and entomopathogenic fungus bioformulation for the synchronous management of leaffolder pest and sheath blight disease of rice Pest Manag Sci 66:555–564. https://doi.org/10.1002/ps.1907

Kaur T, Vasudev A, Sohal SK, Manhas RK (2014) Insecticidal and growth inhibitory potential of *Streptomyces hydrogenans* DH16 on major pest of India, *Spodoptera litura* (Fab.) (Lepidoptera: Noctuidae). BMC Microbiol 14(1)

Kim Jae H, Lee Byong W, Schroeder Frank C, Jander G (2008) Identification of indole glucosinolate breakdown products with antifeedant effects on *Myzus persicae* (green peach aphid) Plant J 54:1015–1026. https://doi.org/10.1111/j.1365-313X.2008.03476.x

Kloepper JW, Ryu CM, Zhang S (2004) Induced systemic resistance and promotion of plant growth by *Bacillus* spp. Phytopathology 94:1259–1266. https://doi.org/10.1094/PHYTO.2004.94.11.1259

Kumar Arora N (2014) Comparative study of different carriers inoculated with nodule forming and free living plant growth promoting bacteria suitable for sustainable agriculture. J Plant Pathol Microbiol 05:26–29. https://doi.org/10.4172/2157-7471.1000229

Kupferschmied P, Maurhofer M, Keel C (2013) Promise for plant pest control: root-associated pseudomonads with insecticidal activities. Front Plant Sci 4:287 https://doi.org/10.3389/fpls.2013.00287

Lacava T, Jennifer P, Dini AF, Francisco D-A, Luiz. RJ, A. MT (2007) Analysis of the bacterial community in glassy-winged sharpshooter heads. Entomol Res 37:261-266 https://doi.org/10.1111/j.1748-5967.2007.00123.x

Lasota JA, Dybas RA (1991) A Vermectins, a novel class of compounds: implications for use in arthropod pest control. Annu Rev Entomol 36:91–17. https://doi.org/10.1146/annurev.ento.36.1.91li H, Soares MA, Torres MS, Bergen M, White JF (2015) Endophytic bacterium, *Bacillus amyloliquefaciens*, enhances ornamental host resistance to diseases and insect pests. J Plant Interact 10:224–229. https://doi.org/10.1080/17429145.2015.1056261

Li X, Nan X, Wei C, He H (2012) The gut bacteria associated with camponotus japonicus mayr with culture-dependent and DGGE methods. Cur Microbiol 65:610–616. https://doi.org/10.1007/s00284-012-0197-1

Liu K, Garrett C, Fadamiro H, Kloepper JW (2016) Induction of systemic resistance in chinese cabbage against black rot by plant growth-promoting rhizobacteria. Biol Control 99:8–13. https://doi.org/10.1016/j.biocontrol.2016.04.007

Loper JE et al. (2012) Comparative genomics of plant-associated *Pseudomonas* spp.: insights into diversity and inheritance of traits involved in multitrophic interactions. PLOS Genetics 8: e1002784. https://doi.org/10.1371/journal.pgen.1002784

Maria P-T et al. (2013) Control and host-dependent activation of insect toxin expression in a root-associated biocontrol pseudomonad Environ Microbiol 15:736–750. https://doi.org/10.1111/1462-2920.12050

Mewis I, Appel HM, Hom A, Raina R, Schultz JC (2005) Major signaling pathways modulate arabidopsis glucosinolate accumulation and response to both phloem-feeding and chewing insects. Plant Physio 138:1149–1162. https://doi.org/10.1104/pp.104.053389

Mohr K, Tebbe C (2006) Diversity and phylotype consistency of bacteria in the guts of three bee species (Apoidea) at an oilseed rape field. Environ Microbiol 8:258–272. https://doi.org/10.1111/j.1462-2920.2005.00893.x

Molecular traits in fl uorescent Pseudomonads with antifungal activity. Iran J Biotech 4:245–253

Muller R, de Vos M, Sun JY, Sonderby IE, Halkier BA, Wittstock U, Jander G (2010) Differential effects of indole and aliphatic glucosinolates on lepidopteran herbivores. J Chem Ecol 36:905–913. https://doi.org/10.1007/s10886-010-9825-z

Nakkeeran S, Fernando WGD, Siddiqui ZA (2006) Plant growth promoting rhizobacteria formulations and its scope in commercialization for the management of pests and diseases. In: Siddiqui ZA (ed) PGPR: biocontrol and biofertilization. Springer Netherlands, Dordrecht, pp 257–296. https://doi.org/10.1007/1-4020-4152-7_10

Niu DD, Liu HX, Jiang CH, Wang YP, Wang QY, Jin H-L, Guo JH (2011) The Plant growth–promoting rhizobacterium *Bacillus cereus* AR156 induces systemic resistance in *Arabidopsis thaliana* by simultaneously activating salicylate- and jasmonate/ethylene-dependent signaling pathways. Mol Plant Microbe Interact 24:533–542. https://doi.org/10.1094/MPMI-09-10-0213

Otsu Y, Matsuda Y, Mori H et al. (2004) Stable phylloplane colonization by entomopathogenic bacterium *Pseudomonas fluorescens* KPM-018P and biological control of Phytophagous ladybird beetles *Epilachna vigintioctopunctata* (Coleoptera: Coccinellidae). Biocontrol Sci Technol 14:427–439. https://doi.org/10.1080/09583150410001683538

Pandey C, Bajpai VK, NegiYK, Maheshwari DK (2018) Effect of plant growth promoting *Bacillus* spp. on nutritional properties of *Amaranthus hypochondriacus* grains. Saudi J Biol Sci. https://doi.org/10.1016/j.sjbs.2018.03.003

Pangesti N, Pineda A, Dicke M, van Loon JJA (2015a) Variation in plant-mediated interactions between rhizobacteria and caterpillars: potential role of soil composition. Plant Biol 17:474–483. https://doi.org/10.1111/plb.12265

Pangesti N, Weldegergis BT, Langendorf B, van Loon JJ, Dicke M, Pineda A (2015b) Rhizobacterial colonization of roots modulates plant volatile emission and enhances the attraction of a parasitoid wasp to host-infested plants. Oecologia 178:1169–1180. https://doi.org/10.1007/s00442-015-3277-7

Pangesti N, Pineda A, Pieterse C, Dicke M, Van Loon J (2013) Two-way plant mediated interactions between root-associated microbes and insects: from ecology to mechanisms. Front Plant Sci 4. https://doi.org/10.3389/fpls.2013.00414

Pangesti N, Reichelt M, van de Mortel JE et al. (2016) Jasmonic acid and ethylene signaling pathways regulate glucosinolate levels in plants during rhizobacteria-induced systemic resistance against a leaf-chewing herbivore. J Chem Ecol 42:1212–1225. https://doi.org/10.1007/s10886-016-0787-7

Pauwels L, Inzé D, Goossens A (2009) Jasmonate-inducible gene: what does it mean? Trends Plant Sci 14:87–91. https://doi.org/10.1016/j.tplants.2008.11.005

Péchy-Tarr M, Bruck DJ, Maurhofer M et al. (2008) Molecular analysis of a novel gene cluster encoding an insect toxin in plant-associated strains of *Pseudomonas fluorescens*. Environ Microbiol10:2368–2386. https://doi.org/10.1111/j.1462-2920.2008.01662.x

Pieterse CM, Zamioudis C, Berendsen RL, Weller DM, Van Wees SC, Bakker PA (2014) Induced systemic resistance by beneficial microbes. Annu Rev Phytopathol 52:347–375. https:// https://doi.org/10.1146/annurev-phyto-082712-102340

Pineda A, Soler R, Weldegergis BT, Shimwela MM, JJ VANL, Dicke M (2013) Non-pathogenic rhizobacteria interfere with the attraction of parasitoids to aphid-induced plant volatiles via jasmonic acid signalling. Plant Cell Environ 36:393–404. https://doi.org/10.1111/j.1365-3040.2012.02581.x

Pineda A, Zheng SJ, van Loon JJ, Dicke M (2012) Rhizobacteria modify plant-aphid interactions: a case of induced systemic susceptibility. Plant Biol (Stuttg) 14 Suppl 1:83–90. https://doi.org/10.1111/j.1438-8677.2011.00549.x

Pineda A, Zheng SJ, van Loon JJ, Pieterse CM, Dicke M (2010) Helping plants to deal with insects: the role of beneficial soil-borne microbes. Trends Plant Sci 15:507–514. https://doi.org/10.1016/j.tplants.2010.05.007

Ramette A, Moënne-Loccoz Y, Défago G (2006) Genetic diversity and biocontrol potential of *fluorescent pseudomonads* producing phloroglucinols and hydrogen cyanide from Swiss soils naturally suppressive or conducive to *Thielaviopsis basicola*-mediated black root rot of tobacco. FEMS Microbiol Ecol 55:369–381. https://doi.org/10.1111/j.1574-6941.2005.00052.x

Rashid MH, Khan A, Hossain MT, Chung YR (2017) Induction of systemic resistance against aphids by endophytic *Bacillus velezensis* YC7010 via expressing PHYTOALEXIN DEFICIENT4 in *Arabidopsis*. Front Plant Sci 8:211. https://doi.org/10.3389/fpls.2017.00211

Rashid MH-O, Chung YR (2017) Induction of systemic resistance against insect herbivores in plants by beneficial soil microbes. Front Plant Sci 8:1816. https://doi.org/10.3389/fpls.2017.01816

Raupach GS, Kloepper JW (1998) Mixtures of plant growth-promoting rhizobacteria enhance biological control of multiple cucumber pathogens. Phytopathology 88:1158–1164. https://doi.org/10.1094/PHYTO.1998.88.11.1158

Raymond B, Johnston PR, Nielsen-LeRoux C, Lereclus D, Crickmore N (2010) *Bacillus thuringiensis*: an impotent pathogen? Trends Microbiol 18:189–194. https://doi.org/10.1016/j.tim.2010.02.006

Requena N, I J, M T, J B (1997) Interactions between plant-growth-promoting rhizobacteria (PGPR), arbuscular mycorrhizal fungi and *Rhizobium* spp. in the rhizosphere of *Anthyllis cytisoides*, a model legume for revegetation in mediterranean semi-arid ecosystems. New Phytol 136:667–677. https://doi.org/10.1046/j.1469-8137.1997.00786.x

Roychoudhury N, Joshi KC (2011) Effect of ivermectin on larval mortality and reproductive potential of Albizia foliage feeder, *Spirama retorta* Cramer (Lepidoptera: Noctuidae). J Pest Sci (2004) 84:117–122. https://doi.org/10.1007/s10340-010-0334-1

Ruffner B et al. (2013) Oral insecticidal activity of plant-associated pseudomonads. Environ Microbiol 15:751–763. https://doi.org/10.1111/j.1462-2920.2012.02884.x

Saitou K, Furuhata K, Kawakami Y, Fukuyama M (2009) Isolation of *Pseudomonas aeruginosa* from cockroaches captured in hospitals in japan, and their antibiotic susceptibility. Biocontrol Sci 14:155–159. https://doi.org/10.4265/bio.14.155

Sandheep AR, Asok AK, Shanavas J (2013) Combined inoculation of *Pseudomonas fluorescens* and *Trichoderma harzianum* for enhancing plant growth of vanilla (*Vanilla planifolia*) vol 16. https://doi.org/10.3923/pjbs.2013.580.584

Santos F et al. (2014) A novel interaction between plant-beneficial rhizobacteria and roots: colonization induces corn resistance against the root herbivore *Diabrotica speciosa*. PLOS ONE 9:e113280. https://doi.org/10.1371/journal.pone.0113280

Saravanakumar D, Kannappan M, Nallathambi L, Seetharaman S, Lingan R, Thiruvengadam R, Ramasamy S (2007) Pseudomonas-induced defence molecules in rice plants against leaffolder (*Cnaphalocrocis medinalis*) pest. Pest Manag Sci 63:714–721. https://doi.org/10.1002/ps.1381

Saravanakumar D, N L, B M, T R, S S, R S (2008) *Pseudomonas fluorescens* enhances resistance and natural enemy population in rice plants against leaffolder pest. J Appl Entomol 132:469–479. https://doi.org/10.1111/j.1439-0418.2008.01278.x

Sarma MVRK, Kumar V, Saharan K, et al. (2011) Application of inorganic carrier-based formulations of fluorescent pseudomonads and Piriformospora indica on tomato plants and evaluation of their efficacy. J Appl Microbiol 456–466

Sathya A, Vijayabharathi R, Kumari BR, et al (2016) Assessment of a diketopiperazine, cyclo (Trp-Phe) from *Streptomyces griseoplanus* SAI-25 against cotton bollworm, *Helicoverpa armigera* (Lepidoptera: Noctuidae). Appl Entomol Zool 51:11–20. https://doi.org/10.1007/s13355-015-0366-3

Schoonhoven. VLJJ, Dicke M (2005) Insect-plant biology Oxford, Oxford University Press

Senthilraja G, Anand T, Kennedy JS, Raguchander T, Samiyappan R (2013) Plant growth promoting rhizobacteria (PGPR) and entomopathogenic fungus bioformulation enhance the expression of defense enzymes and pathogenesis-related proteins in groundnut plants against leafminer insect and collar rot pathogen. Physiol Mol Plant Pathol 82:10–19. https://doi.org/10.1016/j.pmpp.2012.12.002

Sharma HC, Sujana G, Manohar Rao D (2009) Morphological and chemical components of resistance to pod borer, *Helicoverpa armigera* in wild relatives of pigeonpea. Arthropod-Plant Interact 3:151–161. https://doi.org/10.1007/s11829-009-9068-5

Shavit R, Ofek-Lalzar M, Burdman S, Morin S (2013) Inoculation of tomato plants with rhizobacteria enhances the performance of the phloem-feeding insect *Bemisia tabaci*. Front Plant Sci 4:306. https://doi.org/10.3389/fpls.2013.00306

Shen X, Hu H, Peng H, Wang W, Zhang X (2013) Comparative genomic analysis of four representative plant growth-promoting rhizobacteria in *Pseudomonas*. BMC Genomics 14:271. https://doi.org/10.1186/1471-2164-14-271

Siddiqui ZA, Shakeel U, Siddiqui S (2008) Biocontrol of wilt disease complex of pigeonpea by fluorescent pseudomonads and *Bacillus* spp. under pot and field conditions. Acta Phytopathol Entomol Hung 43:77–92

Solomos T, Laties GG (1976) Effects of cyanide and ethylene on the respiration of cyanide-sensitive and cyanide-resistant plant tissues. Plant Physio 58:47–50

Spaink HP (2000) Root nodulation and infection factors produced by rhizobial bacteria. Annu Rev Microbiol 54:257–288. https://doi.org/10.1146/annurev.micro.54.1.257

Stipanovic RD, Lopez JD, Dowd MK, Puckhaber LS, Duke SE (2006) Effect of racemic and (+)- and (−)-gossypol on the survival and development of *Helicoverpa zea* larvae. J Chem Ecol 32:959–968. https://doi.org/10.1007/s10886-006-9052-9

Tang X, Freitak D, Vogel H et al. (2012) Complexity and variability of gut commensal microbiota in polyphagous lepidopteran larvae. PLOS ONE 7:e36978. https://doi.org/10.1371/journal.pone.0036978

Usha Rani P, Jyothsna Y (2010) Biochemical and enzymatic changes in rice plants as a mechanism of defense. Acta Physiol Plant 32:695–701. https://doi.org/10.1007/s11738-009-0449-2

Vacheron J, Desbrosses G, Bouffaud M et al. (2013) Plant growth-promoting rhizobacteria and root system functioning. Front Plant Sci 4:356. https://doi.org/10.3389/fpls.2013.00356

Valenzuela-Soto JH, Estrada-Hernandez MG, Ibarra-Laclette E, Delano-Frier JP (2010) Inoculation of tomato plants (*Solanum lycopersicum*) with growth-promoting *Bacillus subtilis* retards whitefly *Bemisia tabaci* development. Planta 231:397–410. https://doi.org/10.1007/s00425-009-1061-9

Van de Mortel JE, de Vos RCH, Dekkers E et al. (2012) Metabolic and transcriptomic changes induced in Arabidopsis by the rhizobacterium *Pseudomonas fluorescens* SS101. Plant Physiol 160:2173–2188. https://doi.org/10.1104/pp.112.207324

Van der Ent S, Van Hulten M, Pozo MJ et al (2009) Priming of plant innate immunity by rhizobacteria and beta-aminobutyric acid: differences and similarities in regulation. New Phytol 183:419–431

Van Loon LC (2007) Plant responses to plant growth-promoting rhizobacteria. Eur J Plant Pathol 119:243–254. https://doi.org/10.1007/s10658-007-9165-1

Van Oosten VR, Bodenhausen N, Reymond P, Van Pelt JA, Van Loon LC, Dicke M, Pieterse CM (2008) Differential effectiveness of microbially induced resistance against herbivorous insects in *Arabidopsis*. Mol Plant Microbe Interact 21:919–930. https://doi.org/10.1094/MPMI-21-7-0919

Verhage A, Vlaardingerbroek I, Raaijmakers C, Van Dam N, Dicke M, Van Wees S, Pieterse C (2011) Rewiring of the jasmonate signaling pathway in *Arabidopsis* during insect herbivory. Front Plant Sci 2. https://doi.org/10.3389/fpls.2011.00047

Vijayabharathi R, Kumari BR, Sathya A, et al (2014) Biological activity of entomopathogenic Actinomycetes against lepidopteran insects (Noctuidae: Lepidoptera). Can J Plant Sci 94:759–769. https://doi.org/10.4141/cjps2013-298

Wahla V, Maheshwari DK, Bajpai VK (2012) Nematicidal fluorescent pseudomonads for the in vitro and in vivo suppression of root knot (Meloidogyne incognita) of Capsicum annuum L. Pest Manag Sci 68(8):1148–1155

Walling LL (2000) The myriad plant responses to herbivores. J of Plant Growth Regul 19:195–216

Xiaohong T et al. (2012) The rice hydroperoxide lyase OsHPL3 functions in defense responses by modulating the oxylipin pathway. Plant J 71:763–775. https://doi.org/10.1111/j.1365-313X.2012.05027.x

Yadav J, Verma JP (2014) Effect of seed inoculation with indigenous Rhizobium and plant growth promoting rhizobacteria on nutrients uptake and yields of chickpea (*Cicer arietinum* L.) Eur J Soil Biol 63:70–77. https://doi.org/10.1016/j.ejsobi.2014.05.001

Zamioudis C, Pieterse CM (2012) Modulation of host immunity by beneficial microbes. Mol Plant Microbe Interact 25:139–150. https://doi.org/10.1094/MPMI-06-11-0179

Zebelo S, Song Y, Kloepper JW, Fadamiro H (2016) Rhizobacteria activates (+)-delta-cadinene synthase genes and induces systemic resistance in cotton against beet armyworm (*Spodoptera exigua*). Plant Cell Environ 39:935–943. https://doi.org/10.1111/pce.12704

Zehnder G, Kloepper J, Tuzun S et al (1997a) Insect feeding on cucumber mediated by rhizobacteria-induced plant resistance. Entomol Exp Appl 83:81–85

Zehnder G, Kloepper J, Yao C, Wei G (1997b) Induction of systemic resistance in cucumber against cucumber beetles (Coleoptera: Chrysomelidae) by plant growth-promoting rhizobacteria. J Econ Entomol 90:391–396. https://doi.org/10.1093/jee/90.2.391

Zhang H, Xie X, Kim MS, Kornyeyev DA, Holaday S, Pare PW (2008) Soil bacteria augment *Arabidopsis* photosynthesis by decreasing glucose sensing and abscisic acid levels in planta. Plant J 56:264–273. https:// https://doi.org/10.1111/j.1365-313x.2008.03593.x

Zhang P, Zhu X, Huang F, Liu Y, Zhang J, Lu Y, Ruan Y (2011) Suppression of jasmonic acid-dependent defense in cotton plant by the mealybug *Phenacoccus solenopsis*. PLoS One 6: e22378. https://doi.org/10.1371/journal.pone.0022378

Chapter 15
Potential Effect of Plant Growth-Promoting Rhizobacteria (PGPR) on Enhancing Protection Against Viral Diseases

Ahmed R. Sofy, Mahmoud R. Sofy, Ahmed A. Hmed and Noha K. El-Dougdoug

Abstract Plant viruses spread around the globe and have been considered one of the most critical plant pathogens, leading to severe economic losses in crop productivity and yield quality. Unlike pests, fungi, and bacteria, no direct control methods can use against viruses. Managing to plant viral diseases depends primarily on the genetic resistance of host plants and their environment, as well as on the performance of synthetic pesticides to control vectors, an essential strategy for managing viral diseases. Effective plant viral disease pesticides are available, but because residual poisoning persists, they are not considered useful in a long-term solution because of environmental hazards and public health problems. So, new ways were appealed to complement existing strategies to manage the viral disease for better and more sustainable viral disease control. The use of bioinoculants is one of the ways to protect crops that can reduce viral infection to enhance plant growth, resulting in a significant economic return for growers. In recent years, PGPR-systemic resistance to plant viruses has trended toward viral disease management, although many PGPR-ISR studies have centered on several pathogens of fungi and bacteria. This chapter will address the spectrum of PGPR-mediated ISR

A. R. Sofy (✉) · M. R. Sofy · A. A. Hmed
Botany and Microbiology Department, Faculty of Science, Al-Azhar University, Nasr City, Cairo 11884, Egypt
e-mail: ahmed_sofy@azhar.edu.eg

M. R. Sofy
e-mail: mahmoud_sofy@azhar.edu.eg

N. K. El-Dougdoug
Botany and Microbiology Department, Faculty of Science, Benha University, Benha, Egypt

D. K. Maheshwari and S. Dheeman (eds.), *Field Crops: Sustainable Management by PGPR*, Sustainable Development and Biodiversity 23,
https://doi.org/10.1007/978-3-030-30926-8_15

against some plant viruses including banana bunchy top virus, bean common mosaic virus strain blackeye cowpea mosaic, bean yellow mosaic virus, bitter gourd yellow mosaic virus, cucumber green mottle mosaic virus, cucumber mosaic virus, papaya ringspot virus, pepper mild mottle virus, potato virus X, potato virus Y, sunflower necrosis virus, tobacco mosaic virus, tobacco necrosis virus, tomato chlorotic spot virus, tomato mosaic tobamovirus, tomato mottle virus, tomato spotted wilt virus, tomato yellow leaf curl virus, urdbean leaf crinkle virus, and watermelon mosaic virus.

Keywords Plant viruses · Biocontrol · Crops management · PGPR · SAR · ISR

15.1 Introduction

Biotic and abiotic stresses adversely affect plant growth parameters, quality, and quantity based on the plants and stage(s), where average productivity of plant can be reduced between 65 and 87% (Gursoy et al. 2012). Plant viruses have occurred worldwide and are the most critical plant pathogens responsible for severe economic losses in the productivity and quality of many crops (Balconi et al. 2012). In contrast to pests, fungi, or bacteria, no direct control methods are evolved against viruses so far. Management of plant viral diseases depends mainly on host plants genetic resistance and their environment as well as synthetic pesticides to control vectors where it is an essential strategy in viral management (Srinivasan and Mathivanan 2009). Effective plant viral disease pesticides are available, but because residual poisoning persists, they are not considered useful in a long-term solution because of environmental health hazards (Srinivasan and Mathivanan 2009). The ever-increasing costs of pesticides, on the one hand, and consumer demand for food without pesticides, on the other, have resulted in seeking replacements for these chemical products (Gerhardson 2002). On the other hand, there are also specific diseases such as caused by certain viruses and viroids that have few, ineffective or inexistent chemical solutions (El-Dougdoug et al. 2012a; Gerhardson 2002; Sofy et al. 2013a, 2012, 2014b). It is, therefore, investigate for plant viral diseases management by inducing natural defenses of plants, e.g., systemically acquired resistance (SAR) (Ryals et al. 1994). In recent years, plant growth-promoting bacteria (PGPR)-systemic resistance to plant viruses has trended in the management of viral diseases, despite the fact that many PGPR-ISR studies have centered on several pathogens of fungi and bacteria (Kloepper et al. 2004a, b; van Loon et al. 1998), but available literature revealed the limited information on viral disease management by using PGPR and other beneficial microorganisms. Furthermore, some PGPRs stimulate plant growth, resulting in significant economic returns for growers (Babalola 2010).

15.2 PGPR-Induced Systemic Resistance Against Plant Viruses

There are two types of pathogen-induced resistance [Induced Systemic Resistance (ISR) and Systemic Acquired Resistance (SAR)] defined as "the activation of the host plant's chemical or physical defense mechanism by an inducer" that leads to multiple pathogens being controlled (Kloepper 1993; Kloepper et al. 1992). The ISR's expression in response to the pathogen inoculation challenge is similar to that of SAR due to reduced disease severity or reduced number of diseased plants (Van Loon and Bakker 2005). This reduction is often related to a reduction in pathogen growth and reduction of tissue invasion in induced tissues, showing that the plant can better withstand the pathogen (Van Loon 2000; Van Loon and Bakker 2005).

PGPRs are saprophytic bacterial microorganisms that are free-living in the rhizosphere and colonize the root system aggressively (Ramjegathesh et al. 2013). They can provide plants with beneficial effects through additional secretion such as vitamin, hormone, growth factors that help enhances plant growth and productivity (Babalola 2010). Several bacteria including the species of *Aeromonas*, *Agrobacterium*, *Arthrobacter*, *Alcaligenes*, *Azospirillum*, *Azoarcus*, *Azotobacter*, *Bacillus*, *Burkholderia*, *Bradyrhizobium*, *Comamonas*, *Cyanobacteria* (predominantly *Anabaena* and *Nostoc*), *Enterobacter*, *Gluconacetobacter*, *Herbaspirillum*, *Klebsiella*, *Paenibacillus*, *Pseudomonas*, *Rhizobium*, *Serratia*, *Variovorax*, *Streptomyces*, and *Xanthomonas* have been reported as PGPR (Vessey 2003). Some of these genera such as *Azoarcus*, *Burkholderia*, *Gluconacetobacter*, and *Herbaspirillum* established in cells and tissues of higher plants thus are called include endophytic species (Vessey 2003). However, the majority of PGPR strains registered are *Pseudomonas* and *Bacillus* species (Ramamoorthy et al. 2001). PGPRs are regarded as inducing systemic resistance, which can reduce the severity of some diseases in crop plants, termed as PGPR-induced systemic resistance (Kloepper et al. 1992).

Salicylic acid (SA) is a key signaling molecule that works locally in intracellular signal transduction (Shirasu et al. 1997). It may enhance the release of H_2O_2 and H_2O_2-derived active oxygen and induce activities of defense-related genes (Shirasu et al. 1997). Two downstream signaling pathways of salicylic acid reported as the defense mechanism, where the first pathway triggers fungal and bacterial resistance through *PR* and *NPR1* genes expression, and the other stimulates resistance to viral infection through alternative oxidase (Murphy et al. 1999). Since hormones primarily regulate plant defense against viruses, first relying on salicylic acid and second on jasmonic acid (Alazem and Lin 2015). On the other hand, due to the link between SA-mediated defense and siRNA antiviral mechanism, salicylic acid is considered to be extremely important for the resistance (local & systemic), where it participates in basal immune responses and *R*-gene resistance (Alamillo et al. 2006; Alazem and Lin 2015; Baebler et al. 2014; Beris et al. 2018). The defense response based on salicylic acid involves mitogen-activated kinase activation leading to increased *NPR1* regulation, thereby induces *PR* genes transcription (Beckers et al.

2009; Gkizi et al. 2016; Hirt et al. 2013; Kohler et al. 2002; Yi et al. 2015). Further, the treatment with salicylic acid or biologically active salicylic acid analogs led to the expression of the gene *NtRDRP1* (Beris et al. 2018). The gene *NtRDRP1* triggered following infection with many viruses is an essential element in the defense of plants against viral infection (Beris et al. 2018; Xie et al. 2001). PGPR activates ISR in salicylate-dependent and -independent manners and intersects somewhat with the jasmonate/ethylene pathway (Niu et al. 2011; Ryu et al. 2004; van Wees et al. 2000). Multiple strains of rhizobacteria were observed to produce salicylic acid (Van Loon and Bakker 2005). There are two criteria for clarifying whether these strains elicit ISR via SAR pathway activation depending on SA; first, *PRs* induction must be related to ISR; second, induced systemic resistance and *PRs* induction must be nullified in NahG plants (Van Loon and Bakker 2005). Experiments by De Meyer et al. (1999) with transformed tobacco of NahG (SA hydroxylase gene transgenic) revealed that salicylic acid accumulation in plants resulting from induced resistance of *Pseudomonas aeruginosa* strain 7NSK2 is due to its expression, but not due to its induction, and is therefore identical to SAR induced by tobacco mosaic virus (TMV), which is also not expressed in NahG tobacco (Gaffney et al. 1993). The characteristic, salicylic acid-stimulating *PRs* in tobacco plants were expressed when ISR was elicited against TNV via *Pseudomonas fluorescens* strain CHA0 (Maurhofer et al. 1994). Similarly, the expression of biosynthetic genes of salicylic acid (*pchA* & *pchB*) in the salicylic acid negative, root-colonizing *P. fluorescens* strain P3 considerably enhanced its capacity to trigger resistance against TNV in tobacco, confirming that salicylic acid induces SAR against viruses (Maurhofer et al. 1998). Unlike salicylic acid-producing *Pseudomonas fluorescens* strain CHA0, when the genes (*pchA* & *pchB*) were introduced, salicylic acid production increased but didn't enhance ISR in tobacco against TNV (Maurhofer et al. 1998). On the other hand, root application of *Bacillus subtilis* strain G1 triggered ISR against TMV by activating the signaling defense genes (*PR-1a* and *PR-1b*), and regulatory genes (*NPR1* and *Coi1*) indicating salicylic acid signaling pathway activation (Wang et al. 2009). However, unlike SAR, resistance induced by *P. aeruginosa* strain 7NSK2 during the TMV challenge time was not linked to the expression of *PR1a* (De Meyer et al. 1999). So, De Meyer et al. (1999) suggested that treatment with strain 7NSK2 would only enhance expression of the systemic tissue defense gene, which also explains why its resistance level is lower than in case of SAR. On the other hand, SA elicitation and ethylene signaling indicate expression patterns of encoding the osmotin-like protein (*CaPR5*) and the encoding of putative antifungal protein (*CaPR4*) (Choi and Hwang 2014). Similarly, the characteristic, SA-inducible PRs (*CaPR4*, *CaPR5*, and *CaPR10*) expressed in pepper plants when ISR was triggered by *Bacillus amyloliquefaciens* strain 5B6 against CMV (Lee and Ryu 2016). Beris et al. (2018) experiments also showed that after treatment with *B. amyloliquefaciens* strain MBI600, the salicylic acid signaling pathway induction in tomato plants and various patterns of gene expression of tomato defense-related genes have been detected against potato virus Y (PVY) and tomato spotted wilt virus (TSWV) infection. Beris et al. (2018) revealed this through transcriptional analysis of a group of genes connected with SA-related defense [*SlRdR1* (Xie et al. 2001), *SlNPR1* (Kohler et al. 2002), *SlPR1b.1* (Block et al. 2005)] or JA-related defense [*SlLoxD* (Hu et al. 2015) and *SlCOI1* (Xie et al. 1998)], priming [*SlNPR1* (Yi et al. 2015), *SlMPK3* (Beckers et al. 2009)], or RNAi basal defense against viruses [*SlRdR1* (Xie et al. 2001)] (Beris et al. 2018). In comparison, *B. amyloliquefaciens* strain EXTN-1 treated *Arabidopsis* wild-type Col-0 plants resulted in the simultaneous activation of two representative molecular markers (*PR-1* & *PDF1.2*), indicating that strain EXTN-1 induces systemic resistance through SA-dependent and JA-dependent pathways (Ahn et al. 2002). Similarly, experiments of Park et al. (2006) with *B. vallismortis* strain EXTN-1 in transgenic tobacco have proved that there was activation of *PR-la* and *PDF1.2* defense genes upon treatment with EXTN-1 indicating involvement of a salicylic acid (SA) and jasmonic acid (JA) dependent pathway, where *PR-la* and *PDF1.2* genes frequently used as SA and JA signaling indicators, respectively (Reymond and Farmer 1998). Indeed, tobacco plants treatment with *B. amyloliquefaciens* strain EXTN-1 resulted in earlier and increased expression of defense-associated genes (HMGR, PAL, and *PR-1a*) in the presence of pepper mild mottle virus infection at non-inoculated, upper leaves; hence, resistance phase established by EXTN-1 treatment developed locally and systemically (Ahn et al. 2002). On the other hand, treatment with *B. amyloliquefaciens* strain EXTN-1 alone did not induce such strong gene activation (Ahn et al. 2002). 3-hydroxy-3-methylglutaryl CoA reductase (HMGR), phenylalanine ammonia-lyase (PAL), and *PR-1a* genes perceived as the reliable molecular markers, wherein tobacco defense system(s) via salicylic acid-dependent pathway was activated or not (Ahn et al. 2002). Since the expression of these genes was induced within a short time after TMV infection under the tobacco *N* gene-activated state (Kang et al. 1998). Further, mRNA accumulations of these genes also induced by the salicylic acid treatment (Klessig et al. 1994).

Conversely, Ryu et al. (2004) reported that *Arabidopsis thaliana* is protected by *Serratia marcescens* strain 90-166 from cucumber mosaic virus (CMV) through a virus protection signaling pathway in which *NPR1*- and SA are independent, but JA-dependent. Since the severity of CMV symptoms in NahG and Col-0 plants has been reduced with strain 90-166 and its SA-deficient mutant 90-166-1441, in addition to the absence of induction of the *PR-1* gene through strain 90-166, which is generally used as a salicylic acid signaling indicator, indicating that resistance has occurred via a salicylic acid-independent pathway against CMV (Ryu et al. 2004).

Hydrogen peroxide (H_2O_2) is a systemic source of the early molecular signal that strengthens apoptotic tissue and causes cell infection apoptosis (Ahn et al. 2011). H_2O_2 is a standard response to plant pathogens, a key signal that stimulates defensive responses (Zhang et al. 2009). PGPR-triggered ISR is also related to activation for increased cell defensive responses to pathogen attacks, like accelerated accumulation of H_2O_2 (Conrath et al. 2002; Niu et al. 2011; Van Wees et al. 2008). Li et al. (2016a) reported that *Enterobacter asburiae* BQ9 initiated the tomato plants to accelerate and improve the ability to activate cellular defense responses systemically (production of H_2O_2) triggered only via the attack of tomato yellow leaf curl virus (TYLCV). A large number of ISR-related defense enzymes include PAL, polyphenol oxidase (PPO), ascorbate peroxidase (ASC), peroxidase

(POD), superoxide dismutase (SOD), glutathione reductase (GR), lipoxygenase (LOX), catalase (CAT), chitinase, β-1,3-glucanase, and proteinase inhibitors (Van Loon and Bakker 2005). Furthermore, these enzymes give rise to molecules liberation that generates the primary steps in resistance induction, because of phenolic compounds and phytoalexins (Chen et al. 2000; Sofy et al. 2014a, b, 2018a, b; Vanitha and Umesha 2011). The phenolic compounds are oxidized by peroxidase into their quinone derivatives, which inhibits viruses through viral RNA inactivation (Lamb and Dixon 1997).

In *Phaseolus vulgaris, Bacillus globisporus* and *Pseudomonas fluorescens*–treated leaves showed increased activity of peroxidase and β-1,3-glucanase, respectively in pathogen-inoculated leaf sheaths, tobacco necrosis virus (TNV) (Shoman et al. 2003). On the other hand, polyacrylamide-gel electrophoretic analysis detected two similar induced proteins in *Phaseolus vulgaris* leaves treated with culture filtrates of *Streptomyces gibsonii* and *P. fluorescens* compared to the control (water-treated), which may play a role in defense mechanism in P. vulgaris plants against TNV (Shoman et al. 2003). PAL has a vital function in multiple secondary metabolites synthesis, such as phenylpropanoids and phenols, as well as salicylic acid derivatives and lignin, which imparts immunity to plants and stimulates PGPR-triggered resistance (Gerasimova et al. 2005; Harish et al. 2008; Li et al. 2016a). Since secondary metabolites accumulation is intended to limit the invasion of viruses (Lian et al. 2011). So, induction of defense enzymes (POD, PPO, and PAL), and *PR* proteins by application of *P. fluorescens* strains can improve plants to be resistant to banana bunchy top virus (BBTV) (Harish et al. 2008) and tomato spotted wilt virus (TSWV) (Kandan et al. 2005). Likewise, in *Bacillus*-treated tobacco plants challenged with tomato spotted wilt virus, the defense enzyme (POD, PPO, and PAL) amount and *PR* proteins have been increased significantly compared to control (Lian et al. 2011). Damayanti et al. (2007) reported that *Bacillus* sp. strain I-6, *Bacillus cereus* strain I-35, and *Brevibacterium sanguinis* strain I-16 increased the POD activity in hot pepper plants after TMV inoculation, suggested that these rhizobacteria might able to enhance plant's defense response through high POD activity. Similarly, increased total phenols and activity of defense enzymes (POD & PPO) observed in *Streptomyces griseorebens* and *S. cavourensis*-treated cucumber plants challenged with *Cucumber mosaic virus* (Shafie et al. 2016).

Also, *Rhizobium leguminosarum* bv. *viceae* (mixtures of ICARDA-441 and ARC-202) induced systemic resistance against bean yellow mosaic virus (BYMV) via seed treatment, where enhanced levels of defense enzymes activities (peroxidase and polyphenol oxidase), total phenols, and free proline observed in faba bean (Sofy et al. 2014a). Further stated five unique (polypeptide markers) induced in *R. leguminosarum* bv. *viceae* (ICARDA-441 + ARC-202)-treated faba bean plants inoculated with the BYMV (Sofy et al. 2014a). One of the first plant responses to bacterial, fungal and viral infections is active oxygen species production, whereas superoxide dismutase, a significant antioxidant enzyme, provides cell protection in plants from active oxygen species by converting $O2^-$ to H_2O_2 and O_2 (Fridovich 1986; Mehdy 1994; Vanacker et al. 2000). Recently Li et al. (2016a) noted that

Enterobacter asburiae BQ9 generated systemic acquired resistance in tomato plants against TYLCV, where the activities of PAL, POD, CAT, and SOD were significantly increased in TYLCV-challenged *E. asburiae* BQ9-pretreated plants. Furthermore, *PR1* (*PR1a* and *PR1b*) gene transcriptions of BQ9-treated tomato plants, which were challenged by TYLCV, were faster and broader than untreated plants. Li et al. (2016a, b further noted that the replication of cucumber green mottle mosaic virus (CGMMV) in pretreated cucumber plants with *Stenotrophomonas maltophilia* strain HW2 was delayed for more than three days, as well as increased expression of both defense-associated genes (*PR1* and *PR5*) and defense-associated enzymes (CAT and SOD), resulting in increased cucumber resistance.

15.3 Viral Protection Spectrum via PGPR

Lately, research on the ISR-mediated PGPR scope against viruses in various plants has gained significance. Several species of (PGPR) are used as microbiological inoculants for protecting plants from virus infection and enhancing crop yield as shown in the Table (15.1).

15.3.1 Banana Bunchy Top Virus

Banana bunchy top virus (BBTV), a member of the *Nanoviridae* family (*Babuvirus* genus), causes the bunchy top disease, a serious banana virus disease (*Musa* spp., *Musaceae*) (Dale 1987; Harding et al. 1991). It is transmitted in a persistent circulative and non-replicative manner by the aphid vector *Pentalonia nigronervosa* Coq. (Hu et al. 1996). The disease is difficult to eradicate and manage, where there are no possible strategies currently available to protect these plants against BBTV completely (Harish et al. 2008). Nevertheless, another way of integral management of this disease is for ISR to use in vitro virus-free micropropagated banana plantlets with rhizobacteria (PGPR) and endophyte bacteria (PGPE) to strengthen the plant against the virus (Harish et al. 2008). For instance, the application of *Streptomyces chibaensis* culture filtrate ten days prior to BBTV inoculation highly inhibited BBTV infection (Hewedy et al. 2008). In field conditions, the treatment of plants with a mixture formulated as (EPB5 + EPB22 + Pf1 + CHA0) was also very successful in limiting BBTV, where the mixture consisted of one strain of endophytic *Pseudomonas* (EPB5), one strain of endophytic *Bacillus* (EPB22), and two strains of rhizobacterial *P. fluorescens* (Pf1 & CHA0) (Harish et al. 2008). Further, Kavino et al. (2007a, b) indicated the effective use of a mixture beneficial microbes (EPB22 + Pf1 + CHA0) in reducing the disease incidence of BBTV in tissue culture banana plantlets. Also, according to Kavino et al. (2009), the two chitin-formulated *P. fluorescens* strains (Pf1 and CHA0) have demonstrated their effectiveness in BBTV control. In another study, under greenhouse and field

Table 15.1 Management of plant viruses using PGPR

Virus	Plant	Potential PGPR	References
Banana bunchy top virus (BBTV)	Banana (*Musa* spp.)	*Streptomyces chibaensis*	(Hewedy et al. 2008)
		Pseudomnas (EPB5) + *Bacillus* (EPB22) + *P. fluorescens* (Pf1 + CHA0)	(Harish et al. 2008)
		Bacillus (EPB22) + *P. fluorescens* (Pf1 + CHA0)	(Kavino et al. 2007a, b)
		P. fluorescens (Pf1 + CHA0)	(Kavino et al. 2009)
		Bacillus (EPB22) + *P. fluorescens* (Pf1)	(Harish et al. 2009a)
Bean common mosaic virus strain blackeye cowpea mosaic (BCMV-BlCM)	Cowpea (*Vigna unguiculata*)	*B. subtilis* (GB03) + *B. pumilus* (T4)	(Shankar et al. 2009)
Bean yellow mosaic virus (BYMV)	Faba bean (*Vicia faba*)	*P. fluorescens* (FB11)	(Elbadry et al. 2006)
		*R. leguminosarum*bv. *viceae* (FBG05)	
		*R. leguminosarum*bv. *viceae* (ICARDA-441 + ARC-202)	(Sofy et al. 2014a)
		Microbien (*Azotobacter* sp. + *Asosprillum* sp. + *B. megaterium* + *P. fluorescens* + *R. leguminosorum*)	(Hilal et al. 2016)
Bitter gourd yellow mosaic virus (BGYMV)	Bitter gourd (*Momordica charantia*)	*P. chlororaphis*	(Rajinimala et al. 2009)
		P. fluorescens	
Cucumber green mottle mosaic virus (CGMMV)	Cucumber (*Cucumis sativus*)	*Stenotrophomonas maltophilia* (HW2)	(Li et al. 2016b)
	Watermelon (*Citrullus lanatus*)	*P. oleovorans* (KBPF-004)	(Kim et al. 2017)
Cucumber mosaic virus (CMV)	Cucumber (*Cucumis sativus*)	*P. syringae* pv. *lachrymans*	(Bergstrom et al. 1982)
		P. fluorescens	(Raupach et al. 1996)
		Serratia marcescens	
		B. pumilus (SE49) + *B. amyloliqefaciens* (IN937a)	(Jetiyanon and Kloepper 2002)
		B. pumilus (SE49 + IN937b)	(Jetiyanon et al. 2003; Jetiyanon and Kloepper 2002)
		B. pumilus (SE34 + IN937b)	(Jetiyanon et al. 2003)
		Streptomyces violaceusniger	(Galal 2006)
		Azotobacter chroococcum	(El-Borollosy and Oraby 2012)
		P. fluorescens	
		Streptomyces calvus	(El-Dougdoug et al. 2012b)
		Streptomyces canarius,	
		Streptomyces vinaceusdrappus	
		Streptomyces nogalater	
		Streptomyces viridosporus	
		Streptomyces griseorebens	(Shafie et al. 2016)
		Streptomyces cavourensis	

(continued)

Table 15.1 (continued)

Virus	Plant	Potential PGPR	References
	Tomato (*Solanum lycopersicon* L.)	*Pseudomonas fluorescens*	(Raupach et al. 1996)
		Serratia marcescens	
		B. amyloliquefaciens (IN937a)	(Zehnder et al. 2000, 2001)
		B. subtilus (IN937b)	
		B. pumilus (SE34)	
		B. amyloliquefaciens (IN937a) + *B. subtilis* (GB03)	(Murphy et al. 2003)
		Azospirillum lipoferum (MRB16) + *A. brasilienses* (SP7) + *A. brasilienses* (N040) + *Anabena oryzae* Fritsch	(Dashti et al. 2007)
		P. aeruginosa + *Stenotrophomonas rhizophilia*	(Dashti et al. 2012)
	Pepper (*Capsicum annuum*)	*B. amyloliquefaciens* strain 5B6	(Lee and Ryu 2016)
	Tobacco (*Nicotiana tabacum*)	*B. pumilus* (SE34)	(Kloepper et al. 2004b)
		Pseudomonas chlororaphis (O6)	(Ryu et al. 2007a)
		Paenibacillus lentimorbus (B-30488)	(Kumar et al. 2016)
	Arabidopsis thaliana	*Serattia marcescens* (90-166)	(Ryu et al. 2004)
		B. pumilus (SE34)	
		B. subtilis (GB03) + *B. amyloliquefaciens* (IN937a)	(Ryu et al. 2007b)
Papaya ringspot virus (PRSV)	Squash	*B. pumilus* (SE34) + *B. amyloliquefaciens* (IN937a)	(Abdalla et al. 2017)
		B. pumilus (SE34) + *B. sphaericus* (SE56) + *B. amyloliquefaciens* (IN937a)	
Pepper mild mottle virus (PMMoV)	Tobacco (*Nicotiana tabacum*)	*B. amyloliquefaciens* (EXTN-1)	(Ahn et al. 2002)
	Pepper (*Capsicum annuum*)	*P. oleovorans* (KBPF-004)	(Kim et al. 2017)
Potato virus X (PVX)	Tobacco (*Nicotiana tabacum* L. cv. White Burley)	*Streptomyces afghanensis*	(Hussein 1992)
	Potato (*Solanum tuberosum*L.) variety Daeseo	*B. vallismortis* (EXTN-1)	(Park et al. 2006)
Potato virus Y (PVY)	*Chenopodium quinoa*	*Streptomyces erythraeus* (QS01)	(Mohamed and Galal 2005)
		Streptomyces erythraeus (QS02)	
		Streptomyces naganishii (QS03)	
		Streptomyces michigansis (QS04)	
	Potato (*Solanum tuberosum* L. variety Daeseo)	*B. vallismortis* (EXTN-1)	(Park et al. 2006)
	Tomato (*Solanum lycopersicon* L.)	*B. amyloliquefaciens* (MBI600)	(Beris et al. 2018)

(continued)

Table 15.1 (continued)

Virus	Plant	Potential PGPR	References
Sunflower necrosis virus (SNV)	Sunflower (*Helianthus annuus* L.)	PGPMC-1 [*Streptomyces fradiae* (MML1042) + *B. licheniformis* (MML2501) + *Bacillus* (MML2551) + *P. aeruginosa* (MML2212)]	(Srinivasan and Mathivanan 2009)
Tobacco mosaic virus (TMV)	Tobacco (*Nicotiana tabacum*)	*P. syringae*	(Loebenstein and Lovrekovich 1966)
		B. uniflagellatus	(Mann 1969)
	Datura metel	*Streptomyces rochei*	(Mansour et al. 1988)
	Tobacco (*Nicotiana tabacum*)		
	Tobacco (*Nicotiana tabacum* L. cv. White Burley)	*Streptomyces afghanensis*	(Hussein 1992)
	Tobacco (*Nicotiana tabacum* L. cv. White Burley)	*P. aeruginosa* (7NSK2)	(De Meyer et al. 1999)
	Tobacco (*Nicotiana glutinosa*)	*Streptomyces erythraeus* (QS01)	(Mohamed and Galal 2005)
		Streptomyces erythraeus (QS02)	
		Streptomyces naganishii (QS03)	
		Streptomyces michigansis (QS04)	
	Hot Pepper	*B. cereus* (I-35)	(Damayanti et al. 2007)
	Tomato (*Lycopersicon esculentum*)	*Pseudomonas* spp. (B-25)	(Kirankumar et al. 2008)
	Tobacco (*N. tabacum* cv. NC89)	*B. subtilis* (G1)	(Wang et al. 2009)
		B. amyloliquefaciens (FZB24)	
	Tobacco (*Nicotiana tabacum*)	*B. subtilis* (SW1)	(Lian et al. 2011)
		B. pumilus (EN16)	
	Tobacco (*N. tabacum* cv. Xanthi-nc)	*P. chlororaphis* (O6)	(Park et al. 2012)
	Tobacco (*Nicotiana tabacum*)	*B. amyloliquefaciens* (Ba33)	(Shen et al. 2013)
	Tobacco (*N. tabacum*cv. Xanthi-nc)	*P. oleovorans* (KBPF-004)	(Kim et al. 2017)
Tobacco necrosis virus (TNV)	Tobacco (*Nicotiana tabacum*)	*P. fluorescens* (CHA0)	(Maurhofer et al. 1994, 1998)
		P. fluorescens (P3)	(Maurhofer et al. 1998)
	Bean (*Phaseolus vulgaris*)	*P. fluorescens*	(Shoman et al. 2003)
		Streptomyces gibsonii	
Tomato chlorotic spot virus	Tomato (*Solanum lycopersicon* L.)	*B. amyloliquefaciens* (IN937a)	

(continued)

Table 15.1 (continued)

Virus	Plant	Potential PGPR	References
(TCSV)		*B. pumilus* (SE34) + *B. amyloliquefaciens* (IN937a)	(Abdalla et al. 2017)
		B. pumilus (SE34) + *B. sphaericus* (SE56) + *B. amyloliquefaciens* (IN937a)	
Tomato mosaic tobamovirus (ToMV)	Tobacco (*Nicotiana tabacum* L. cv. White Burley)	*Streptomyces afghanensis*	(Hussein 1992)
	Datura metel	*P. fluorescens* 2	(Megahed et al. 2013)
		B. circulans	
Tomato mottle virus (ToMoV)	Tomato (*Lycopersicon esculentum*)	*B. amyloliquefaciens* (IN 937a)	(Murphy et al. 2000; Zehnder et al. 2001)
		B. subtilis (IN 937b)	
		B. pumilus (SE34)	
Tomato spotted wilt virus (TSWV)	Tomato (*Lycopersicon esculentum* cv. Co-3)	*P. fluorescens* (CHA0)	(Kandan et al. 2002; 2005)
		P. fluorescens (CHA0 + CoT-1)	
		P. fluorescens (CHA0 + CoT-1 + CoP-1)	
	Tomato (*Solanum lycopersicon* L.)	*B. amyloliquefaciens* (MBI600)	(Beris et al. 2018)
Tomato yellow leaf curl virus (TYLCV)	Tomato (*Lycopersicon esculentum* cv. Hezuo 903)	*Enterobacter asburiae* (BQ9)	(Li et al. 2016a)
Urdbean leaf crinkle virus (ULCV)	Blackgram (*Vigna mungo*)	*P. fluorescens* (pf1)	(Karthikeyan et al. 2009)
		P. fluorescens (CHA0)	
Watermelon mosaic virus (WMV)	Pumpkin (*Cucurbita maxima*)	*B. pumilus* 293 (B2)	(Elbeshehy et al. 2015)

conditions, mixture application (EPB22 + Pf1) increased yield by reducing the BBTV incidence (Harish et al. 2009a). Since bacterized plants significantly activate PR proteins, defense enzymes (PPO, POD, PAL, chitinase, and ß-1,3-glucanase) and phenolic compounds, thereby inducing resistance to BBTV (Harish et al. 2009a, b; Kavino et al. 2007b).

15.3.2 *Bean Common Mosaic Virus Strain Blackeye Cowpea Mosaic*

Bean common mosaic virus strain blackeye cowpea mosaic (BCMV-BlCM), a member of the family *Potyviridae* (genus *Potyvirus*) in cowpea (*Vigna unguiculate*, *Fabaceae*) is seed-transmitted, where the frequency of transmission is as high as

31%, and economic yield losses encountered, whenever virus-infected seed is used (Puttaraju et al. 2002; Zettler and Evans 1972). Also, BCMV-BlCM is aphid-transmitted in the field through *Aphis crassivora* and *Myzus persicae* in a nonpersistent manner (Murphy et al. 1987). Both aphid- and seeds-transmitted can lead to high yield losses, requiring the use of chemical control in aphids or the production of virus-free seeds or disease management through induced systemic resistance using PGPR (Hao et al. 2003). Shankar et al. (2009) used seven PGPR strains [*Bacillus pumilus* (T4), *B. amyloliquefaciens* (IN937a), *B. pumilus* (INR7), *B. subtilis* (SE34), *B. subtilis* (IN937b), *B. subtilis* (GB03), and *Brevibacillus brevis* (IPC11)] as promising inducers against BCMV-BlCM. All PGPR strains showed a significant increase in seed germination and seedling vigor and a decrease in the incidence of BCMV compared to untreated control under both greenhouse and field conditions (Shankar et al. 2009). Since under greenhouse conditions, the strains (GB03 and T4) provided 41 and 42% protection, respectively, in the treated cowpea seeds against BCMV-BlCM, whereas the strain (GB03) provided 34% under field conditions (Shankar et al. 2009). On the other hand, application of a combination of the two strains (GB03 + T4) registered the highest protection of 69 and 62% against BCMV-BlCM under greenhouse and field conditions, respectively (Shankar et al. 2009).

15.3.3 Bean Yellow Mosaic Virus

Bean yellow mosaic virus (BYMV), a member of the *Potyviridae* family (*Potyvirus* genus), is transmitted by many aphid species in a nonpersistent manner, one of the most frequently found and probably most damaging viruses affecting the production of field-grown legumes and some non-legumes (Bos 1970; Boswell and Gibbs 1983; Edwardson and Christie 1991; Fauquet et al. 2005). Faba bean (*Vicia faba* L., *Fabaceae*) is considered the most important nutritious popular food crop in Egypt and other countries around the world (Elbadry et al. 2006; Sofy et al. 2014a). Makkouk et al. (2003) recorded BYMV in 89% of the samples collected from surveyed Egyptian faba bean fields with high-level BYMV symptoms (80–100% infection). Controlling of BYMV is difficult due to its quite broad host range and its ability to be transmitted in a nonpersistent manner by many aphid species and also through seeds in certain legume species (Berlandier et al. 1997; Frison et al. 1990). Elbadry et al. (2006) investigated ISR in faba bean against BYMV through seed bacterization with *Rhizobium leguminosarum* bv. *viceae* strain FBG05, and *Pseudomonas fluorescens* strain FB11 individually or in combination. The results showed a decrease in the concentration of BYMV (ELISA) and the percentage of disease incidence (PDI) in the BYMV-challenged plants treated individually with strains (FBG05 and FB11) compared to non-bacterized and control-challenged plants (Elbadry et al. 2006). Co-inoculant preparation of the two strains (FBG05 + FB11) revealed an insignificant additional reduction of the ELISA value or PDI compared to the strain (FB11) alone which caused a marked reduction

(Elbadry et al. 2006). Also, Sofy et al. (2014a)_ENREF_44 investigated the mixture of two *R. leguminosarum* bv. *viceae* strains (ICARDA-441 + ARC-202) that triggered ISR against BYMV-induced bean yellow mosaic disease with significant increases in the levels of abscisic acid, salicylic acid, jasmonic acid in addition to total phenol, free proline and the activities of peroxidase and polyphenol oxidase enzymes compared to the non-bacterized and controlled challenged plants. Hilal et al. (2016) also observed induced systemic protection using five formulations of PGPR [Biogen (*Azotobacter* sp.), Cerealin (*Azotobacter* sp. + *Asosprillum* sp. + *Bacillus polymixa*), Microbien (*Azotobacter* sp. + *Asosprillum* sp. + *B. megaterium* +*Pseudomonas fluorescens*+ *Rhizobium leguminosorum*), Nitrobien (*Azotobacter* sp. + *Asosprillium* sp.) and Rhizobacterin (*Azotobacter* sp. + *Asosprillum* sp. + *R. leguminosarum*)] as foliar spraying or seed soaking of faba bean against *Botrytis fabae* and BYMV. Interestingly, Microbien's foliar treatment significantly reduced the incidence and severity of the disease for both *Botrytis fabae* and BYMV (Hilal et al. 2016). Therefore, PGPR strain could be used in the same crop to cause resistance to several pathogens (Ramamoorthy et al. 2001).

15.3.4 Bittergourd Yellow Mosaic Virus

Bitter gourd (*Momordica charantia* L., *Cucurbitaceae*) is considered an ancient species native to tropical Asia and Africa (Behera et al. 2010). It is widespread in China, Malaysia, India, and tropical Africa (Behera et al. 2010). Bitter gourd yellow mosaic virus (BGYMV), a member of the *Geminiviridae* family (*Begomovirus* genus) is one of the viruses that affect the bitter gourd causing severe yield loss due to its vector whitefly *Bemisia tabaci* (Rajinimala et al. 2005, 2009). Induction of systemic disease resistance in bitter gourd plants against BGYMV is one of the methods used for controlling the disease (Rajinimala et al. 2003, 2009). Seed treatment with *P. chlororaphis* and *P. fluorescens* has consistently reduced the disease incidence at 45 days after sowing (DAS), and further, it is significantly reduced at 75 DAS compared to inoculated untreated control (Rajinimala et al. 2009). The mean plant height of *P. chlororaphis*- and *P. fluorescens*-treated plants at 75 DAS was higher than in the inoculated control, as well as significantly increased phenol content, peroxidase activity, and polyphenol oxidase activity (Rajinimala et al. 2009).

15.3.5 Cucumber Green Mottle Mosaic Virus

Ainsworth (1935) first described the cucumber green mottle mosaic virus (CGMMV), a member of the *Virgoviridae* family (*Tobamovirus* genus), when it was found to cause cucumber disease in England. It also affects other vegetable crops such as squash, pumpkin, zucchini, and gherkin, and the fruit cucurbits of watermelon and melon (Dombrovsky et al. 2017). As a result of increased trade between different regions, CGMMV spread rapidly through mechanical means, seeds, soil, pollen, and other propagating materials (Liu et al. 2014, 2016c; Reingold et al. 2015). Li et al. (2016b) used *Stenotrophomonas maltophilia* strain HW2 to control CGMMV, where it demonstrated its ability to promote the growth of cucumber plants, colonize effectively in cucumber rhizosphere, and its effectiveness against CGMMV was good. Li et al. (2016b) found that the replication of CGMMV in pretreated cucumber plants with the strain (HW2) delayed for more than three days and that the expression of viral protein genes in the leaf was inhibited compared to the control. Additionally, the expression of both defense-associated genes (*PR1* and *PR5*) and defense-associated enzymes (CAT & SOD) increased by the strain (HW2) resulting in increased cucumber resistance (Li et al. 2016b). On the other hand, Kim et al. (2017) evaluated the antiviral activity of *Pseudomonas oleovorans* strain KBPF-004 in CGMMV seed transmission compared to strain ATCC 8062 (control strain). Since virus-infected seeds harvested from infected watermelon plants were treated with cell-free culture supernatant of each strain before planting compared to untreated CGMMV-infected seeds (Kim et al. 2017). Kim et al. (2017) found that strain KBPF-004 reduced the rate of GGMMV seed transmission to 15.8% compared to 59.7% for strain ATCC 8062, indicating that the viability of GGMMV was adversely affected by strain KBPF-004.

15.3.6 Cucumber Mosaic Virus

Cucumber mosaic virus (CMV), a member of the *Bromoviridae* family (*Cucumovirus* genus), has a host range of more than 1200 plants in over 100 families worldwide, including crops, vegetables, ornamentals, and woody plants, where it is transmitted in a nonpersistent manner by at least 75 species of aphid (El-Dougdoug et al. 2014a; Megahed et al. 2012; Mochizuki and Ohki 2012; Palukaitis et al. 1992; Sofy and Soliman 2011; Tóbiás et al. 1982).

Several authors studied the induction of systemic disease resistance against CMV in plants such as cucumber, tomato, pepper, tobacco, and *A. thaliana* as follows. Bergstrom et al. (1982) were the first to show that the cucumber leaves treated with *P. syringae* pv. *lachrymans* induced systemic resistance to CMV. The number of chlorotic and primary lesions in CMV-inoculated decreased and systemic mosaic symptoms delayed (Bergstrom et al. 1982). Also, seed treatment of cucumber or tomato with *S. marcescens* 90-166, and *P. fluorescens* 89B-27 reduced the mean number of CMV symptomatic plants resulting in systemic resistance (Raupach et al. 1996). Protection against CMV PGPR seed treatment has either eliminated the development of viral symptoms in cucumber cotyledons or reduced disease severity in tomato (Raupach et al. 1996). Similarly, *S. marcescens* strain 90-166 and *B. pumilus* strain SE34 systemically protect *A. thaliana*, by reducing the severity of the CMV symptoms and significantly reducing its accumulation in systemically infected leaves (Ryu et al. 2004). Further, Ryu et al. (2007a) noted that the *GacS* global regulator's role in the rhizosphere-competent *P. chlororaphis* strain O6 is relevant in stimulating growth promotion and inducing resistance in tobacco cv. Samsun. El-Borollosy and Oraby (2012) recommended the use of *Azotobacter chroococcum* followed by *P. fluorescens* to promote cucumber growth and induce systemic resistance against CMV. Whereas, two types of treatment were performed: (i) healthy cucumber plants were sprayed and subsequently inoculated with CMV after spraying at intervals (5 and 10 days) and (ii) healthy cucumber seeds were irrigated every 3 days–15 days with two hundred ml of each bacterial culture or supernatants and afterward inoculated with CMV (El-Borollosy and Oraby 2012).

Several authors characterized the role of *Streptomyces* isolates as PGPR in ISR against CMV. Galal (2006) found that the foliage treatment with culture filtrates of the five *Streptomyces* strains (*S. violarus*, *S. nasri* strain H35, *Streptomyces* sp., *S. aureofaciens*, and *S. violaceuisniger*) resulted in 50–85% reduction of the mosaic symptoms. Application of bacterial filtrate before CMV inoculation showed a higher level of antiviral activity than after CMV inoculation when the most favorable incubation period was after 6 h of incubation (Galal 2006). On the other hand, seeds soaking in the bacterial filtrate for 2 h achieved the highest viral inhibition (Galal 2006). Of all the strains tested, *S. violaceusniger* significantly inhibited the virus with the highest percentage (Galal 2006). El-Dougdoug et al. (2012b) also identified five Egyptian *Streptomyces* isolates (*S. calvus*, *S. canarius*, *S. vinaceusdrappus*, *S. nogalater*, and *S. viridosporus*) capable of producing an antiviral component in culture filtrate that was not phytotoxic and effective in local and systemic control of CMV infection. Both *S. canarius* and *S. viridosporus* showed a maximum reduction in the percentage of local lesions produced on the *Chenopodium amaranticolor* by CMV (El-Dougdoug et al. 2012b). Recently, Shafie et al. (2016) reported that foliage treatment with *S. griseorebens* and *S. cavourensis* culture filtrates resulted in a significant reduction in the severity of CMV infection disease and inhibitory effect when applied 48 h before virus inoculation.

On the other hand, due to its long shelf life, stability, and efficiency, several *Bacillus* species have been widely assessed as practical biological management agents (Kloepper et al. 2004a). Lee and Ryu (2016) demonstrated that foliar application of *B. amyloliquefaciens* strain 5B6 mediated ISR against CMV in pepper. Since the relative RNA content of CMV coat protein was considerably reduced in a 3-year field trial through the treatment with the strain (5B6) compared to water control, as demonstrated by qRT-PCR (Lee and Ryu 2016). On the other hand, Jetiyanon and Kloepper (2002) demonstrated that eleven mixtures of PGPR and one single strain treatment significantly reduced CMV of cucumber. Since seed treatment with [*B. pumilus* strain (SE49) + *Bacillus amyloliqefaciens* strain (IN937a)], [*B. pumilus* strains (SE49 +IN937b)], (*B. pumilus* strains (SE34 + IN937b)], [*B. pumilus* strain (SE49) + *B. sphaericus* strain (SE56)], [*B. amyloliqefaciens* strain (IN937a) + *B. pumilus* strain (IN937b)], [*B. pumilus* strain (INR7) + *B. amyloliqefaciens* strain (IN937a)], [*B. pumilus* strains (INR7 + T4)],

[*B. pumilus* strains (IN937b + T4)], [*B. pumilus* strain (IN937b) + *B. sphaericus* strain (SE56)], [*B. amyloliqefaciens* strain (IN937a) alone], [*B. pumilus* strains (IN937b + INR7)] and [*B. pumilus* strains (SE34 + T4)], reduced mean numbers of CMV symptomatic plants as a result to induced systemic resistance with 0.8, 0.8, 1.0, 1.0, 1.2, 1.4, 1.4, 1.6, 1.6, 1.6, 1.8, and 2, respectively, compared to the sodium phosphate buffer control treatment (3.0) (Jetiyanon and Kloepper 2002). In another study, Jetiyanon et al. (2003) reported that consortia (strain SE49 + strain IN937b) and (strain SE34 + strain IN937b) offered nearly 80% suppression of disease, while consortia (strain IN937a + strain IN937b, strain INR7 + strain T4, and strain IN937a alone) offered nearly 60% suppression of disease compared to the control.

Zehnder et al. (2000, 2001) identified PGPR strains that protect tomato plants under greenhouse and field conditions from CMV systemic infection. Initially, 26 PGPR strains were evaluated in greenhouse experiments to induce systemic resistance to CMV in tomatoes (Zehnder et al. 2000, 2001). Since four PGPR strains [*B. pumilus* (SE34), *B. subtilus* (IN937b), *B. amyloliquefaciens* (IN937a), and *Kluyvera cryocrescens* (IN114)] selected for field trials and their effectiveness assessed on the basis of symptomatic plant percentages ranged from 32 to 58% of the most effective PGPR treatments compared to 88 to 98% of the challenged control (Zehnder et al. 2000, 2001). In field experiments, three PGPR strains (*B. amyloliquefaciens* strain IN937a, *B. subtilus* strain IN937b, and *B. pumilus* strain SE34) reduced the development of CMV symptoms and the incidence of disease (Zehnder et al. 2000, 2001). Likewise, Murphy et al. (2003) evaluated six combined formulations with the carrier chitosan that included *Bacillus subtilis* strain GB03 together with one of *B. amyloliquefaciens* (IN937a), *B. pumilus* (SE34), *B. pumilus* (INR7), *B. pumilus* (T4), or *B. subtilis* (IN937b) as promising inducers against CMV in tomato. A mixture of PGPR [*B. amyloliquefaciens* (IN937a) + *B. subtilis* (GB03)] showed the ability to protect tomato from CMV (Murphy et al. 2003). This mixture also triggered ISR against CMV and encouraged plant growth in *Arabidopsis thaliana* (Ryu et al. 2007b). Likewise, *B. pumilus* strain SE34 elicited ISR against CMV in tobacco (Kloepper et al. 2004b). Kumar et al. (2016) suggested that *P. lentimorbus* strain B-30488 induce resistance against CMV in tobacco (*N. tabacum* cv. White Burley). The strain (B-30488) was isolated from cow's milk, which increased plant strength, while the accumulation of RNA virus and virulence in CMV infected tobacco plants systemic leaves decreased significantly by about 12-fold (91%) compared to control plants (Kumar et al. 2016). The strain (B-30488) was isolated from cow's milk, where the strain increased plant strength, while the accumulation of RNA virus and virulence in CMV infected tobacco plant systemic leaves decreased significantly by 91% compared to control plants (Kumar et al. 2016).

On the other hand, the mixing culture of *Azospirillum lipoferum* strain MRB16, *A. brasilienses* strain SP7, *A. brasilienses* strain N040, and *Anabena oryzae* Fritsch showed the ability to protect tomato against CMV containing viral satellite RNA (CARNA 5) leading to increased fruit yield both in the greenhouse and in the field (Dashti et al. 2007). Also, Dashti et al. (2012) observed that a PGPR mixture containing *P. aeruginosa* and *S. rhizophilia*, with a mild strain of CMV associated with viral satellite RNA (CMV-KU1) enhanced the growth of tomato plants protected from the severe CMV-16 with 91.3% disease prevention.

15.3.7 Papaya Ringspot Virus

Papaya ringspot virus (PRSV) a member of the *Potyviridae* family (*Potyvirus* genus), causes a destructive disease and is the main restrictive agent in the cultivation of cucurbit or papaya throughout the world, where it is transmitted in a nonpersistent manner by multiple species of aphids (Gonsalves et al. 2010; Tripathi et al. 2008). The PRSV has two types in which the first type called PRSV-P affects both cucurbit and papaya and the second type only affects cucurbit but does not affect papaya and is called PRSV-W, previously known as watermelon mosaic virus-1 (Bateson et al. 2002; Gonsalves et al. 2010; Tripathi et al. 2008). Quite recently, Abdalla et al. (2017) reported that mixing either two PGPR strains [*B. pumilus* (SE34) + *B. amyloliquefaciens* (IN937a)] or three strains [*B. pumilus* (SE34) + *B. sphaericus* (SE56) + *B. amyloliquefaciens* (IN937a)] significantly reduced the severity of PRSV-W disease in squash plants 3 and 6 weeks after virus inoculation compared to infected untreated control. Since soil drenching is the more effective method observed by Abdalla et al. (2017), then root dipping and seed coating treatment.

15.3.8 Pepper Mild Mottle Virus

Pepper mild mottle virus (PMMoV), a member of the *Virgaviridae* family (*Tobamovirus* genus), is one of the most important pathogens of pepper crops, where infection can reach 100% in the field (Green 2003; Wetter and Conti 1988). PMMoV is easily transmitted by grafting, mechanical contact and PMMoV-infected seed coats (Genda et al. 2005; Svoboda et al. 2006). *B. amyloliquefaciens* strain EXTN-1 trigger ISR in tobacco against PMMoV through salicylate-dependent and jasmonate-dependent pathways (Ahn et al. 2002). On the other hand, Kim et al. (2017) evaluated the antiviral activity of *Pseudomonas oleovorans* strain KBPF-004 in PMMoV seed transmission compared to strain ATCC 8062 (control strain). Since virus-infected seeds harvested from infected pepper plants were treated with cell-free culture supernatant of each strain before planting compared to untreated PMMoV -infected seeds (Kim et al. 2017). Kim et al. (2017) found that strain KBPF-004 reduced the rate of PMMoV seed transmission to 15.5% compared to 61.9% for strain ATCC 8062, indicating that the viability of PMMoV was adversely affected by strain KBPF-004.

15.3.9 Potato Virus X

Potato virus X (PVX), a member of the *Alphaflexiviridae* family (*Potexvirus* genus), is a major pathogen of potato crops worldwide, infecting a wide range of hosts, notably within the *Solanaceae* family (Aboul-Ata et al. 2011; King et al. 2011). It has been included among pathogens with significant economic impact and can cause significant economic losses in synergistic co-infection with potato virus Y (Khurana and Singh 1988; Vance 1991). Hussein (1992) reported that both concentrated metabolites and their acetone extract of *Streptomyces afghanensis* (Egyptian isolate) inhibited the development of local lesions caused by PVX-infected *N. tabacum* L. cv. White Burley, where the maximum antiviral effect was observed 2 h after infection. On the other hand, the potato plants treated (seed treatment) with *Bacillus vallismortis* strain EXTN-1 increased the yield to untreated control by up to 45% (Park et al. 2006).

15.3.10 Potato Virus Y

Potato virus Y (PVY), a member of the *Potyviridae* family (*Potyvirus* genus), is a serious pathogen infecting several important crop species in the night shad family, including potato, tomato, tobacco, and pepper, resulting in significant yield losses and quality degradation (El-Dougdoug et al. 2014b; Glais et al. 2002; Sofy et al. 2013b). Mohamed and Galal (2005) observed that both the culture filtrates with pellets (cells of streptomycetes) of four Egyptian halotolerant *Streptomyces* isolates (*S. erythraeus* strain QS01, *S. erythraeus* strain QS02, *S. naganishii* strain QS03, and *S. michigansis* strain QS04) isolated from Qaroon Lake showed a decrease in the number of local necrotic lesions produced by PVY on *Chenopodium quinoa*. On the other hand, potato plants treated with *Bacillus vallismortis* strain EXTN-1 increased yields of up to 45% and chlorophyll content compared to the untreated control by protecting plants against potato virus Y (Park et al. 2006). Also, the application of *B. amyloliquefaciens* strain MBI600 decreased the accumulation of PVY during early infection and overdue detection of PVY in apical leaves (Beris et al. 2018). Where, in addition to the length of the tomato plants treated with the strain (MBI600), the fresh and dry weight after 30 days of inoculation was significantly higher than that of water and BTH treatments (Beris et al. 2018).

15.3.11 Sunflower Necrosis Virus

Sunflower necrosis virus (SNV) is a strain of tobacco streak virus (TSV), a member of the *Bromoviridae* family (*Ilarvirus* genus) that has recently caused significant crop loss in sunflower in India (Chavhan et al. 2017; Ravi et al. 2001). Although

Ilarvirus transmitted through seeds (van Regenmortel et al. 2000), SNV transmission not confirmed by sunflower seeds (Srinivasan and Mathivanan 2009). Insecticides are mainly used to control the SNV carrier (thrips) for disease management as observed by Srinivasan and Mathivanan (2009). On the other hand, the application of two PGPR microbial consortia [PGPMC-1 (*Streptomyces fradiae* strain MML1042 + *Bacillus licheniformis* strain MML2501 + *Bacillus* strain MML2551 + *P. aeruginosa* strain MML2212), and PGPMC-2 (strains MML2501 + MML2551 + MML2212)] to the sunflower plants in two different formulations in powder and liquid forms resulted in increased efficiency of ISR against SNV (Srinivasan and Mathivanan 2009). In addition to a significant reduction in disease, plants treated with PGPMC-1 liquid formulation registered an increase in seed germination, plant height and yield parameters with a higher income and benefit–cost ratio (Srinivasan and Mathivanan 2009).

15.3.12 Tobacco Mosaic Virus

Tobacco mosaic virus (TMV), a member of the *Virgoviridae* family (*Tobamovirus* genus), is a global plant virus infecting many horticultural crops and one of the most destructive plant diseases, resulting in severe losses (Sutic et al. 1999). Several studies have shown that individual PGPR strains could result in ISR against TMV as follows. Loebenstein and Lovrekovich (1966) noted that, injecting heat-killed cells of *P. syringae* into the intracellular spaces of *N. tabacum* var. Samsun NN leaves 0–7 days before inoculation with TMV reduced TMV lesions to 3–12% compared with water-injected control leaves. Whereas heat-killed bacteria were injected 5–72 h after inoculation with TMV, the number of lesions decreases by 86% and the area per lesion by 95–33% (Loebenstein and Lovrekovich 1966). Knowledge of this kind of resistance came mainly from the work of several authors including Loebenstein and Lovrekovich (1966). Also, De Meyer et al. (1999) showed that root colonization with *P. aeruginosa* strain 7NSK2 triggered ISR in tobacco against TMV, which is phenotypically similar to SAR TMV. *Pseudomonas* spp. strain B-25 treatment significantly improved growth components and fruit yield of TMV-challenged tomato plants, whereas at 60 DAT (days after transplanting); plant height, the number of branches, number of leaves, total biomass, chlorophyll content, and fruit yield were increased by 44.8, 69.9, 59.5, 52.2, 158.2 and 102%, respectively, over challenged control (TMV only) (Kirankumar et al. 2008). Also, treatment with the strain (B-25) increased NPK uptake in TMV-challenged tomato plants, where N, P, and K uptake were increased by 78.6, 84.5 and 80.3%, respectively, and higher than challenged control (Kirankumar et al. 2008). *P. chlororaphis*strain strain O6 has been shown to have ISR activity against cucumber mosaic virus, where the global *GacS* regulator is considered necessary for strain (O6) to stimulate ISR determinants (Ryu et al. 2007a). The nature of antiviral components in O6 against TMV identified by Park et al. (2012) as a cyclic peptide with molecular formula $C_{39}H_{67}N_9O_{12}S$ composed of seven different amino acids and called "Peptamine." Peptamine (1000 μg/ml) exhibited more than 95% of disease suppression on the leaves treated with the mixture of TMV (Park et al. 2012). Recently, Kim et al. (2017) reported that *P. oleovorans* strain KBPF-004 cell-free culture diluted supernatant (1/20) treated *N. tabacum* cv. Xanthi-nc leaves showed the highest, i.e., 96% antiviral activity against mechanical transmission of TMV, whereas strain ATCC 8062 showed 19.7% compared to untreated leaf regions (control).

Long back, Mann (1969) showed that the applied culture of *Bacillus uniflagellatus* used as a soil drench method induces systemic resistance to TMV and reduces the number of lesions. Similarly, Fahmy and Mohamed (1984) found that the culture filtrate of certain microorganisms containing *B. subtilis* reduced the local lesions produced by TMV on the tobacco cv. Samsun. Damayanti et al. (2007) reported that *Bacillus* sp. strain I-6, *B. cereus* strain I-35, and *Brevibacterium sanguinis* strain I-16 protected hot pepper plants from TMV and attained maximally by strain 1-15. Wang et al. (2009) assessed that individual strains (*B. amyloliquefaciens* strains FZB24 and FZB42 and *B. subtilis* strains G1 and B3) could induce resistance against TMV and decrease the severity of the disease 28 days post-inoculation. However, the effect of *B. amyloliquefaciens* (FZB24) and *B. subtilis* (G1) treatments was most significant (Wang et al. 2009). Furthermore, western blot analysis confirmed that tobacco plants treated with *B. amyloliquefaciens* (FZB24) and *B. subtilis* (G1) contained a lower amount of the TMV coat protein compared to tobacco plants treated with *B. amyloliquefaciens* (FZB42) and *B. subtilis* (B3) (Wang et al. 2009). Tobacco plants treated with *B. pumilus* strain EN16 or *B. subtilis* strain SW1 also offered protection of 52% and 71%, respectively, at 14 d of inoculation (Lian et al. 2011). The amount of virus detected by ELISA decreased in tobacco plants treated with EN16 or SW1 (Lian et al. 2011). Further, after 5 d and 7 d intervals between PGPR treatment and TMV inoculations, respectively, strain EN16 and SW1 induce optimal resistance (Lian et al. 2011). Shen et al. (2013) recommended the use of *B. amyloliquefaciens* strain Ba33 as a soil disinfectant and an antiviral agent against TMV, where Ba33 treatments decreased the number of local necrotic lesions and disease index. The best treatment was spraying *N. tabacum* with strain Ba33 simultaneously with TMV inoculation in field trials comparable to ningnanmycin registered and established as an antiviral agent in tobacco (Shen et al. 2013).

Mansour et al. (1988) Observed inhibition of TMV (from *Datura metel* leaves) using *Streptomyces rochei*, while *S. rimosus*or and *S. gougerotti* caused weak inhibition. Both the concentrated metabolites and their acetone extract of *Streptomyces afghanensis* (Egyptian isolate) inhibited the development of local lesions caused by TMV-infected *N. tabacum* L. cv. White Burley, where the maximum antiviral effect was observed 2 h after infection (Hussein 1992). Similarly, the culture filtrates with cell pellets of four Egyptian halotolerant *Streptomyces* isolates (*S. erythraeus* strain QS01, *S. erythraeus* strain QS02, *S. naganishii* strain QS03, and *S. michigansis* strain QS04) showed a decrease in the number of local necrotic lesions produced by TMV on *N. glutinosa* (Mohamed and Galal 2005).

15.3.13 Tobacco Necrosis Virus

Tobacco necrosis virus (TNV), a member of the *Tombusviridae* family (*Necrovirus* genus), is a plant virus has an extensive host range including many cultivated species, where it is infectious to 298 species of 167 genera belonging to 54 families (Edwardson and Christie 1997). The virus occurs in plant roots and is transmitted in a nonpersistent manner by the chytrid fungus vector *Olpidium*, especially *O. brassicae* and *O. bornavirus* (Fry and Campbell 1966; Hull 2002; Sutic et al. 1999). The soil application of *P. fluorescens* strain CHA0 which produces salicylic acid naturally under iron-limited conditions exhibits induced systemic protection against TNV in tobacco, where the incidence of leaf necrosis in tobacco plants is reduced after inoculation with TNV (Maurhofer et al. 1994, 1998). While the introduction of the salicylic acid biosynthetic genes (*pchA* and *pchB*) into *P. fluorescens* strain CHA0 that produces salicylic acid increased salicylic acid production in tobacco rhizosphere and in vitro, it did not improve ISR in tobacco against TNV (Maurhofer et al. 1998). On the other hand, the introduction of salicylic acid biosynthetic genes into *P. fluorescens* strain P3 which does not produce salicylic acid gives this strain the ability to produce salicylic acid in vitro and significantly enhances its capability to ISR in tobacco against TNV (Maurhofer et al. 1998). Foliage treatment with *B. globisporus*, *P. fluorescens*, or *S. gibsonii* culture filtrate profoundly reduced the local lesions number in *Phaseolus vulgaris*, where *P. fluorescens* and *S. gibsonii* showed the highest inhibitory effect of TNV infection by 91.5% and 97.2%, respectively (Shoman et al. 2003).

15.3.14 Tomato Chlorotic Spot Virus

Tomato chlorotic spot virus (TCSV), a member of the *Bunyaviridae* family (*Tospovirus* genus), is transmitted by thrips (*Thysanoptera*: *Thripidae*) in a persistent and propagative manner, where the most efficient TCSV vectors are the dark form of *Frankliniella schultzei* followed by *Frankliniella occidentalis* (King et al. 2011; Martínez et al. 2018; Peters et al. 1996; Polston et al. 2013). TCSV causes diseases associated with significant losses in overall yield and quality in both agricultural and ornamental crops (Pappu et al. 2009; Polston et al. 2013). Application of *Bacillus amyloliquefaciens* strain IN937a reduced the severity of TCSV disease in tomato plants by almost 50% compared to infected untreated control (Abdalla et al. 2017). Whereas soil drenching application of mixture of either two PGPR strains [*Bacillus pumilus* (SE34) + *B. amyloliquefaciens* (IN937a)] or three strains [*B. pumilus* (SE34) + *B. sphaericus* (SE56) + *B. amyloliquefaciens* (IN937a)] showed the highest reduction of TCSV disease severity at 3 and 6 weeks after virus inoculation compared to infected untreated control in both first and second trials (Abdalla et al. 2017).

15.3.15 Tomato Mosaic Tobamovirus

Tomato mosaic tobamovirus (ToMV), a member of the *Virgaviridae* family (*Tobamovirus* genus), is found in tomatoes around the world where it can be seed-borne and spread very quickly through mechanical transmission by grafting or contaminated tools and workers, resulting in significant losses (Adams et al. 2012; Broadbent 1965). Egyptian isolate of *Streptomyces afghanensis* has an antiviral effect against ToMV, that inhibited local lesion produced on *N. tabacum* L. cv. White Burley (Hussein 1992). Recently, Megahed et al. (2013) mentioned that the liquid culture, cells and culture filtrate of individual *P. fluorescens* 2, and *B. circulans* imparted ISR on tomato plants and reduced the ToMV symptoms and ToMV local lesions produced on *Datura metel* as ToMV indicator host. The seed treatment with an individual liquid culture, microbial culture filtrate and microbial cells of *P. fluorescens* 2 profoundly reduced the local lesions number by 49.2, 57.7, and 58.5%, respectively, compared to infected untreated control, while *B. circulans* reduced the local lesions number by 42.3, 46.8, and 47.3%, respectively, compared to infected untreated control (Megahed et al. 2013). Interestingly, *P. fluorescens* proved better than *B. circulans* during field trials (Megahed et al. 2013).

15.3.16 Tomato Mottle Virus

Tomato mottle virus (ToMoV), a member of the *Geminiviridae* family (*Begomovirus* genus), transmitted through whitefly B biotype, *Bemisia tabaci* (Gennadius), was a primary limiting agent of tomato yields (Kring et al. 1991; Polston et al. 1993; Simone et al. 1990). ToMoV management was limited due to the ability of its vector to acquire insecticide resistance and lack of genetically resistant tomatoes (Denholm et al. 1996; Stansly et al. 1991). In 1997, Murphy et al. (2000) and Zehnder et al. (2001) noted that the severity ratings of ToMoV disease were significantly less in the plants treated with powder-based treatments of *B. amyloliquefaciens* strain IN937a, and *B. subtilus* strain IN937b compared to the control. Consequently, in all powder-based treatments, the Southern blot analysis showed a lower percentage of ToMoV-infected tomato plants compared to plants undergoing seed treatment singly or control treatment (Murphy et al. 2000; Zehnder et al. 2001). Further, Murphy et al. (2000) and Zehnder et al. (2001) used *B. pumilus* strain SE34 instead of *Bacillus amyloliquefaciens* strain IN937a, where the severity symptom ratings of ToMoV were notably reduced in plants treated as seed treatment alone with *B. subtilis* (IN937b), in addition to plants treated as powder treatment alone and seed + powder treatment with *B. pumilus* (SE34) compared to control (Murphy et al. 2000; Zehnder et al. 2001).

15.3.17 Tomato Spotted Wilt Virus

Tomato spotted wilt virus (TSWV), a member of the *Bunyaviridae* family (*Tospovirus* genus), ranks among the top 10 most economically important plant viruses worldwide with a broad host range of up to 1000 plant species (Mumford et al. 1996; Naidu et al. 2008; Parrela et al. 2003). TSWV transmitted by western flower thrips, *Frankliniella occidentalis* (Pergande) as the primary vector in a persistent and propagative manner (Ogada et al. 2013; Whitfield et al. 2005), as well as transmitted by multiple species of thrips (Ullman et al. 1997). It causes severe damages to crops grown in greenhouses, and open fields in all climate zones (German et al. 1992; Peters et al. 1996). Treatment of seeds, seedlings, soil or foliar applications with *P. fluorescens* strains (CHA0, CoT-1, and CoP-1) either alone or in mixtures has triggered ISR against TSWV infection in tomato under both greenhouse and field conditions (Kandan et al. 2002, 2005). They found that the mixture of (CHA0 + CoT-1 + CoP-1) decreased disease maximum by 84%, followed by the strain (CHA0) individually or in combined with the strain (CoT-1) by 80% compared to the untreated control plants (Kandan et al. 2002, 2005). This is why *P. fluorescens* strains treated tomato plants demonstrated increased growth promotion compared to untreated control plants under both greenhouse and field conditions (Kandan et al. 2002, 2005). On the other hand, Beris et al. (2018) demonstrated that under two different environmental conditions, the application of *B. amyloliquefaciens* strain MBI600 as foliar, drench, or soil amendment reduced the TSWV incidence by up to 80%.

15.3.18 Tomato Yellow Leaf Curl Virus

Tomato yellow leaf curl virus (TYLCV), a member of the *Geminiviridae* family (*Begomovirus* genus), is a plant virus spread throughout the world and can be found in most places where tomatoes are grown and is considered one of the most destructive plant diseases in the world leading to severe losses (Navas-Castillo et al. 2011; Rybicki 2015; Sofy et al. 2017). TYLCV transmitted by the whitefly *Bemisia tabaci* (Hemiptera; *Aleyrodidae*) in a circulative and persistent manner (Gotz et al. 2012). *Enterobacter asburiae* strain BQ9-elicited ISR against tomato yellow leaf curl disease of TYLCV-induced tomato was investigated by Li et al. (2016a) who observed a significant 42% decrease in the disease even after 45 days of inoculation leading to long-term plant protection against TYLCV, in addition to increased growth parameters compared to mock treatment plants (Li et al. 2016a). Moreover, viral load in BQ9-pretreated tomato plants was reduced after 9 days by nearly 1/3 of viral control (Li et al. 2016a). In addition to the production of H_2O_2, the expression of both defense-associated genes and defense enzymes was associated with BQ9-induced ISR against TYLCV (Li et al. 2016a).

15.3.19 Urdbean Leaf Crinkle Virus

Urdbean leaf crinkle virus (ULCV) has not been assigned to any genus and family until now and is a plant virus that causes urdbean leaf crinkle disease (ULCD) which is also considered to be very severe among viral diseases (Gautam et al. 2016; Ravinder Reddy et al. 2005). The disease is more severe in blackgram than mungbean under field conditions, where the virus is transmitted by seed, grafting, and sap inoculation, as well as by insects, whiteflies, and aphids (Gautam et al. 2016). Karthikeyan et al. (2009) reported that foliar- and soil-applied *P. fluorescens* (strain pf1 or strain CHA0) 24 h prior to virus inoculation triggered ISR against ULCV in blackgram, significantly reducing ULCV infection. The strain (Pf1) was highly effective in reducing the percentage of ULCV infection by 9.33% for foliar application or soil application compared to 74.50% for the control (Karthikeyan et al. 2009). Whereas, foliar- and soil-applied strain (CHA0) reduced the percentage of ULCV infection by 17.31%, and 20%, respectively, compared to the control by 77.16% (Karthikeyan et al. 2009). Furthermore, the strain (Pf1) increased the activity of POD, PAL, and PPO and induced the phenolics accumulation in blackgram plants against ULCV, thus helping the host plants to resist the disease (Karthikeyan et al. 2009).

15.3.20 Watermelon Mosaic Virus

Watermelon mosaic virus (WMV), a member of the *Potyviridae* family (*Potyvirus* genus), is a plant virus mostly distributed in temperate and Mediterranean regions with broader host range than most *Potyviruses*, resulting in severe losses to all cucurbit plants (Desbiez et al. 2009; Desbiez and Lecoq 2008; Moradi 2011). A strategic approach to plant management of watermelon mosaic virus disease is based on the use of insecticides to control its vectors (whites and aphids) and the cross-protection of genetically engineered plants (Elbeshehy et al. 2015; Lomonossoff 1995). Elbeshehy et al. (2015) investigated the induction of systemic disease resistance in pumpkin against WMV by soil application with two PGPR strains (*B. subtilis* 281 strain B1 and *B. pumilus* 293 strain B2) either individually or in combination. They found that the strain (B2) suppressed disease by approximately 77.7% significantly higher than plants treated with the strain (B1) and mixture (B1 + B2) that suppressed disease by 33.3% and 66.6%, respectively (Elbeshehy et al. 2015). Furthermore, in plants treated with the strain (B2) or a mixture of strains (B1 + B2), the concentration of WMV measured by ELISA was 0.16 and 0.34, respectively, which was significantly lower compared to plants treated with strain (B1) and untreated WMV control than 0.75 and 1.366, respectively (Elbeshehy et al. 2015).

15.4 Conclusion

Viral diseases are difficult to eradicate and manage, where there are currently no strategies to completely protect plants from viruses. PGPRs are considered to induce systemic resistance, which can reduce the severity of certain diseases in crop plants, known as PGPR-induced systemic resistance. Although PGPR-ISR studies have centered on several pathogens of fungi and bacteria, several species of PGPR are used as microbial inoculants to protect plants from virus infection and improve crop yield. Since once resistance induced, it will provide nonspecific protection against viruses. This chapter deals with several studies highlighting the importance of several PGPR species against the viral infection of different plants. This approach is, therefore, one of the methods of environmentally friendly prevention and sustainable development.

Conflict of Interest The author(s) have no conflict of interest.

References

Abdalla OA, Bibi S, Zhang S (2017) Application of plant growth-promoting rhizobacteria to control Papaya ringspot virus and Tomato chlorotic spot virus. Arch Phytopathol Plant Prot 50:584–597. https://doi.org/10.1080/03235408.2017.1352248

Aboul-Ata A-AE et al (2011) Diagnosis and control of cereal viruses in the Middle East. Adv Virus Res 81:33–61. https://doi.org/10.1016/b978-0-12-385885-6.00007-9

Adams MJ, Heinze C, Jackson AO, Kreuze JF, Macfarlane SA, Torrance L (2012) Tobamovirus. In: King AMQ, Adams MJ, Carstens EB, Lefkowitz EJ (eds) Virus taxonomy: ninth report of the international committee on taxonomy of viruses. Elsevier/Academic Press, London, United Kingdom, pp 1153–1156

Ahn IP, Lee SW, Kim MG, Park SR, Hwang DJ, Bae SC (2011) Priming by rhizobacterium protects tomato plants from biotrophic and necrotrophic pathogen infections through multiple defense mechanisms. Mol Cells 32:7–14. https://doi.org/10.1007/s10059-011-2209-6

Ahn IP, Park K, Kim CH (2002) Rhizobacteria-induced resistance perturbs viral disease progress and triggers defense-related gene expression. Mol Cells 13:302–308

Ainsworth GC (1935) Mosaic diseases of the cucumber. Ann Appl Biol 22:55–67. https://doi.org/10.1111/j.1744-7348.1935.tb07708.x

Alamillo JM, Saénz P, García JA (2006) Salicylic acid-mediated and RNA-silencing defense mechanisms cooperate in the restriction of systemic spread of *Plum pox virus* in tobacco. Plant J 48:217–227. https://doi.org/10.1111/j.1365-313X.2006.02861.x

Alazem M, Lin N-S (2015) Roles of plant hormones in the regulation of host-virus interactions. Mol Plant Pathol 16:529–540. https://doi.org/10.1111/mpp.12204

Babalola OO (2010) Beneficial bacteria of agricultural importance. Biotechnol Lett 32:1559–1570. https://doi.org/10.1007/s10529-010-0347-0

Baebler Š et al (2014) Salicylic acid is an indispensable component of the Ny-1 resistance-gene-mediated response against *Potato virus Y* infection in potato. J Exper Bot 65:1095–1109. https://doi.org/10.1093/jxb/ert447

Balconi C, Stevanato P, Motto M, Biancardi E (2012) Breeding for biotic stress resistance/tolerance in plants. In: Ashraf M, Öztürk M, Ahmad M, Aksoy A (eds) Crop production for agricultural improvement. Springer, Dordrecht, pp 57–114. https://doi.org/10.1007/978-94-007-4116-4_4

Bateson MF, Lines RE, Dale JL, Ha CV, Revill P, Gibbs AJ, Chaleeprom W (2002) On the evolution and molecular epidemiology of the potyvirus *Papaya ringspot virus*. J General Virol 83:2575–2585. https://doi.org/10.1099/0022-1317-83-10-2575

Beckers GJM, Jaskiewicz M, Liu Y, Underwood WR, He SY, Zhang S, Conrath U (2009) Mitogen-Activated Protein Kinases 3 and 6 are required for full priming of stress responses in *Arabidopsis thaliana*. Plant Cell Online 21:944–953. https://doi.org/10.1105/tpc.108.062158

Behera TK, Behera S, Bharathi LK, John KJ, Simon PW, Staub JE (2010) Bitter gourd: botany, horticulture, breeding. In: Janick J (ed) Horticultural reviews, vol 37. https://doi.org/10.1002/9780470543672.ch2

Bergstrom GC, Johnson MC, Kuc J (1982) Effect of local infection of cucumber by *Colletotrichum lagenarium*, *Pseudomonas lachrymans*, or *Tobacco necrosis virus* on systemic resistance to *Cucumber mosaic virus*. Phytopathology 72:922–926. https://doi.org/10.1094/Phyto-72-922

Beris D, Theologidis I, Skandalis N, Vassilakos N (2018) *Bacillus amyloliquefaciens* strain MBI600 induces salicylic acid dependent resistance in tomato plants against *Tomato spotted wilt virus* and *Potato virus Y*. Sci Rep 8:10320. https://doi.org/10.1038/s41598-018-28677-3

Berlandier FA, Thackray DJ, Jones RAC, Latham LJ, Cartwright L (1997) Determining the relative roles of different aphid species as vectors of cucumber mosaic and bean yellow mosaic viruses in lupins. Ann Appl Biol 131:297–314. https://doi.org/10.1111/j.1744-7348.1997.tb05158.x

Block A, Schmelz E, O'Donnell PJ, Jones JB, Klee HJ (2005) Systemic acquired tolerance to virulent bacterial pathogens in tomato. Plant Physiol 138:1481–1490. https://doi.org/10.1104/pp.105.059246

Bos L (1970) Bean yellow mosaic virus. In: CMI/AAB descriptions of plant viruses, No 40, p 4

Boswell KF, Gibbs AJ (1983.) Viruses of legumes 1983. Description and Keys, from VIDE. Australian National University, Canberra, Australia

Broadbent L (1965) The epidemiology of tomato mosaic. Ann Appl Biol 56:177–205. https://doi.org/10.1111/j.1744-7348.1965.tb01227.x

Chavhan RL, Verma MK, Hinge VR, Kadam US, Kokane SB, Chakrabarty PK (2017) Sequence analysis of coat protein and molecular profiling of *Sunflower necrosis virus* (SNV) strains from Indian subcontinent. J Plant Biochem Biotechnol 27:28–35. https://doi.org/10.1007/s13562-017-0412-z

Chen C, BÉLanger RR, Benhamou N, Paulitz TC (2000) Defense enzymes induced in cucumber roots by treatment with plant growth-promoting rhizobacteria (PGPR) and *Pythium aphanidermatum*. Physiol Mol Plant Pathol 56:13–23. https://doi.org/10.1006/pmpp.1999.0243

Choi HW, Hwang BK (2014) Molecular and cellular control of cell death and defense signaling in pepper. Planta 241:1–27. https://doi.org/10.1007/s00425-014-2171-6

Conrath U, Pieterse CMJ, Mauch-Mani B (2002) Priming in plant–pathogen interactions. Trends Plant Sci 7:210–216. https://doi.org/10.1016/s1360-1385(02)02244-6

Dale JL (1987) Banana bunchy top: an economically important tropical plant virus disease. Adv Virus Res 33:301–325. https://doi.org/10.1016/s0065-3527(08)60321-8

Damayanti TA, Pardede H, Mubarik NR (2007) Utilization of root-colonizing bacteria to protect hot-pepper against *Tobacco mosaic tobamovirus*. HAYATI J Biosci 14:105–109. https://doi.org/10.4308/hjb.14.3.105

Dashti NH, Ali NY, Cherian VM, Montasser MS (2012) Application of plant growth-promoting rhizobacteria (PGPR) in combination with a mild strain of *Cucumber mosaic virus* (CMV) associated with viral satellite RNAs to enhance growth and protection against a virulent strain of CMV in tomato. Can J Plant Pathol 34:177–186. https://doi.org/10.1080/07060661.2012.685495

Dashti NH, Montasser MS, Ali NY, Bhardwaj RG, Smith DL (2007) Nitrogen biofixing bacteria compensate for the yield loss caused by viral satellite RNA associated with *Cucumber mosaic virus* in tomato. Plant Pathol J 23:90–96. https://doi.org/10.5423/ppj.2007.23.2.090

De Meyer G, Audenaert K, Höfte M (1999) *Pseudomonas aeruginosa* 7NSK2-induced systemic resistance in tobacco depends on in *planta* salicylic acid accumulation but is not associated with PR1a expression. Eur J Plant Pathol 105:513–517. https://doi.org/10.1023/a:1008741015912

Denholm I, Cahil M, Byrne FJ, Devonshire AL (1996) Progress with documenting and combating insecticide resistance in *Bemisia*. In: Gerling D, Mayer RT (eds) *Bemisia* 1995: taxonomy, biology, damage, control and management. Intercept Ltd, Andover, Hants, United Kingdom, pp 577–603

Desbiez C, Joannon B, Wipf-Scheibel C, Chandeysson C, Lecoq H (2009) Emergence of new strains of *Watermelon mosaic virus* in South-eastern France: evidence for limited spread but rapid local population shift. Virus Res 141:201–208. https://doi.org/10.1016/j.virusres.2008.08.018

Desbiez C, Lecoq H (2008) Evidence for multiple intraspecific recombinants in natural populations of *Watermelon mosaic virus* (WMV, *Potyvirus*). Arch Virol 153:1749–1754. https://doi.org/10.1007/s00705-008-0170-2

Dombrovsky A, Tran-Nguyen LTT, Jones RAC (2017) *Cucumber green mottle mosaic virus*: rapidly increasing global distribution, etiology, epidemiology, and management. Ann Rev Phytopathol 55:231–256. https://doi.org/10.1146/annurev-phyto-080516-035349

Edwardson JR, Christie RG (1991) CRC handbook of viruses affecting legumes. CRC Press Inc, Boca Raton, FL, USA

Edwardson JR, Christie RG (1997) Viruses infecting peppers and other solanaceous crops. University of Florida, Ginesville, USA

El-Borollosy AM, Oraby MM (2012) Induced systemic resistance against *Cucumber mosaic cucumovirus* and promotion of cucumber growth by some plant growth-promoting rhizobacteria. Ann Agric Sci 57:91–97. https://doi.org/10.1016/j.aoas.2012.08.001

El-Dougdoug KA, Dawoud RA, Rezk AA, Sofy AR (2012a) Incidence of fruit trees viroid diseases by tissue pint hbridization in Egypt. Int J Virol 8:114–120. https://doi.org/10.3923/ijv.2012.114.120

El-Dougdoug KA, Ghaly MF, Taha MA (2012b) Biological control of *Cucumber mosaic virus* by certain local *Streptomyces* isolates: inhibitory effects of selected five Egyptian isolates. Int J Virol 8:151–164. https://doi.org/10.3923/ijv.2012.151.164

El-Dougdoug KA, Sofy AR, Hameed GA, Dawood RA (2014a) Intraspecific diversity of *Cucumber mosaic cucumoviridae* in Egypt. Int J Virol 10:94–102. https://doi.org/10.3923/ijv.2014.94.102

El-Dougdoug KA, Sofy AR, Mousa AA, Refaey EE (2014b) Monitoring variability responses of cultivated potato varieties infected with *Potato virus Y* pepper isolate. Egypt J Virol 11:82–101

Elbadry M, Taha RM, Eldougdoug KA, Gamal-Eldin H (2006) Induction of systemic resistance in faba bean (Vicia faba L.) to B*ean yellow mosaic potyvirus* (BYMV) via seed bacterization with plant growth promoting rhizobacteria. J Plant Dis Prot 113:247–251. https://doi.org/10.1007/bf03356189

Elbeshehy EKF, Youssef SA, Elazzazy AM (2015) Resistance induction in pumpkin *Cucurbita maxima* L. against *Watermelon mosaic potyvirus* by plant growth-promoting rhizobacteria. Biocontrol Sci Technol 25:525–542. https://doi.org/10.1080/09583157.2014.994198

Fahmy FG, Mohamed MS (1984) Effect of onion root exudates, sulphide compounds and culture filtrates of certain microorganisms on infectivity of tobacco mosaic and potato viruses. Egypt J Phytopathol 16:65–69

Fauquet CM, Mayo MA, Maniloff J, Desselberger U, Ball LA (2005) Virus taxonomy: classification and nomenclature of viruses. In: Eighth report of the international committee on taxonomy of viruses. Elsevier/Academic Press, London, p 1259

Fridovich I (1986) Biological effects of the superoxide radical. Arch Biochem Biophys 247:1–11. https://doi.org/10.1016/0003-9861(86)90526-6

Frison EA, Bos L, Hamilton RI, Mathur SB, Taylor JD (1990) FAO/IBPGR technical guidelines for the safe movement of legume germplasm. Food and agriculture organisation of the United Nations and International Board for plant genetic resources, Rome, Italy

Fry PR, Campbell RN (1966) Transmission of a *Tobacco necrosis virus* by Olpidium brassicae. Virology 30:517–527. https://doi.org/10.1016/0042-6822(66)90127-9

Gaffney T et al (1993) Requirement of salicylic acid for the induction of systemic acquired resistance. Science 261:754–756. https://doi.org/10.1126/science.261.5122.754

Galal AM (2006) Induction of systemic acquired resistance in cucumber plant against *Cucumber mosaic cucumovirus* by local *Streptomyces* strains. Plant Pathol J 5:343–349. https://doi.org/10.3923/ppj.2006.343.349

Gautam NK, Kumar K, Prasad M (2016) Leaf crinkle disease in urdbean (*Vigna mungo* L. Hepper): An overview on causal agent, vector and host. Protoplasma 253:729–746. https://doi.org/10.1007/s00709-015-0933-z

Genda Y, Sato K, Nunomura O, Hirabayashi T, Ohnishi J, Tsuda S (2005) Immunolocalization of *Pepper mild mottle virus* in *Capsicum annuum* seeds. J General Plant Pathol 71:238–242. https://doi.org/10.1007/s10327-005-0189-0

Gerasimova NG, Pridvorova SM, Ozeretskovskaya OL (2005) Role of L-phenylalanine ammonia Lyase in the induced resistance and susceptibility of sotato plants. Appl Biochem Microbiol 41:103–105. https://doi.org/10.1007/s10438-005-0019-3

Gerhardson B (2002) Biological substitutes for pesticides. Trends Biotechnol 20:338–343. https://doi.org/10.1016/s0167-7799(02)02021-8

German TL, Ullman DE, Moyer JW (1992) Tospoviruses: diagnosis, molecular biology, phylogeny, and vector relationships. Ann Rev Phytopathol 30:315–348. https://doi.org/10.1146/annurev.py.30.090192.001531

Gkizi D, Lehmann S, L'Haridon F, Serrano M, Paplomatas EJ, Métraux J-P, Tjamos SE (2016) The innate immune signaling system as a regulator of disease resistance and induced systemic resistance activity against *Verticillium dahliae*. Mol Plant-Microbe Interact 29:313–323. https://doi.org/10.1094/mpmi-11-15-0261-r

Glais L, Tribodet M, Kerlan C (2002) Genomic variability in *Potato potyvirus Y* (PVY): evidence that $PVY^{N}W$ and PVY^{NTN} variants are single to multiple recombinants between PVY^{O} and PVY^{N} isolates. Arch Virol 147:363–378. https://doi.org/10.1007/s705-002-8325-0

Gonsalves D, Tripathi S, Carr JB, Suzuki JY (2010) *Papaya ringspot virus*. Plant Health Instr. https://doi.org/10.1094/phi-i-2010-1004-01

Gotz M et al (2012) Implication of *Bemisia tabaci* heat shock protein 70 in begomovirus-whitefly interactions. J Virol 86:13241–13252. https://doi.org/10.1128/jvi.00880-12

Green SK (2003) *Pepper mild mottle virus*. In: Pernezny K, Roberts PD, Murphy JF, Goldberg NP (eds) Compendium of pepper diseases. American Phytopathological Society, St. Paul, MN, pp 32–33

Gursoy M, Balkan A, Ulukan H (2012) Ecophysiological responses to stresses in plants: a general approach. Pak J Biol Sci PJBS 15:506–516. https://doi.org/10.3923/pjbs.2012.506.516

Hao NB, Albrechtsen SE, Nicolaisen M (2003) Detection and identification of the blackeye cowpea mosaic strain of Bean common mosaic virus in seeds of *Vigna unguiculata* spp. from North Vietnam. Australas Plant Pathol 32:505–509. https://doi.org/10.1071/ap03052

Harding RM, Burns TM, Dale JL (1991) Virus-like particles associated with banana bunchy top disease contain small single-stranded DNA. J General Virol 72:225–230. https://doi.org/10.1099/0022-1317-72-2-225

Harish S, Kavino M, Kumar N, Balasubramanian P, Samiyappan R (2009a) Induction of defense-related proteins by mixtures of plant growth promoting endophytic bacteria against *Banana bunchy top virus*. Biol Control 51:16–25. https://doi.org/10.1016/j.biocontrol.2009.06.002

Harish S, Kavino M, Kumar N, Samiyappan R (2009b) Differential expression of pathogenesis-related proteins and defense enzymes in banana: interaction between endophytic bacteria, Banana bunchy top virus and Pentalonia nigronervosa. Biocontrol Sci Technol 19:843–857. https://doi.org/10.1080/09583150903145000

Harish S, Kavino M, Kumar N, Saravanakumar D, Soorianathasundaram K, Samiyappan R (2008) Biohardening with plant growth promoting rhizosphere and endophytic bacteria induces systemic resistance against *Banana bunchy top virus*. Appl Soil Ecol 39:187–200. https://doi.org/10.1016/j.apsoil.2007.12.006

Hewedy MA, Husseiny SM, Salama MI, Nour-Eldein HA, Abdelsalam SM (2008) Reduction in *Banana bunchy top virus* infution rate by using *Streptomyces chibaensi* filtrates. Pak J Biotechnol 5:51–61

Hilal AA, Shafie RM, El-Sharkawy HHA (2016) Interaction between *Bean yellow mosaic virus* and *Botrytis fabae* on faba bean and the possibility of their control by plant growth-promoting rhizobacteria. Egypt J Phytopathol 44:81–97

Hirt H et al (2013) Dual regulation of gene expression mediated by extended MAPK activation and salicylic acid contributes to robust innate immunity in *Arabidopsis thaliana*. PLoS Genet 9: e1004015. https://doi.org/10.1371/journal.pgen.1004015

Hu JS, Wang M, Sether D, Xie W, Leonhardt KW (1996) Use of polymerase chain reaction (PCR) to study transmission of *Banana bunchy top virus* by the banana aphid (*Pentalonia nigronervosa*). Ann Appl Biol 128:55–64. https://doi.org/10.1111/j.1744-7348.1996.tb07089.x

Hu T, Hu Z, Zeng H, Qv X, Chen G (2015) Tomato lipoxygenase D involved in the biosynthesis of jasmonic acid and tolerance to abiotic and biotic stress in tomato. Plant Biotechnol Rep 9:37–45. https://doi.org/10.1007/s11816-015-0341-z

Hull R (2002) Matthew's plant virology. 4th edn. Academic Press, San Diego

Hussein ME (1992) The effect of an Egyptian isolate of *Streptomyces afghanensis* on some plant viruses. Acta Virol 36:479–482

Jetiyanon K, Fowler WD, Kloepper JW (2003) Broad-spectrum protection against several pathogens by PGPR mixtures under field conditions in Thailand. Plant Dis 87:1390–1394. https://doi.org/10.1094/PDIS.2003.87.11.1390

Jetiyanon K, Kloepper JW (2002) Mixtures of plant growth-promoting rhizobacteria for induction of systemic resistance against multiple plant diseases. Biol Control 24:285–291. https://doi.org/10.1016/S1049-9644(02)00022-1

Kandan A, Commare RR, Nandakumar R, Ramiah M, Raguchander T, Samiyappan R (2002) Induction of phenylpropanoid metabolism by *Pseudomonas fluorescens* against tomato spotted wilt virus in tomato. Folia Microbiol 47:121–129. https://doi.org/10.1007/bf02817669

Kandan A, Ramiah M, Vasanthi VJ, Radjacommare R, Nandakumar R, Ramanathan A, Samiyappan R (2005) Use of *Pseudomonas fluorescens*-based formulations for management of *Tomato spotted wilt virus* (TSWV) and enhanced yield in tomato. Biocontrol Sci Technol 15:553–569. https://doi.org/10.1080/09583150500088546

Kang MK, Park KS, Choi D (1998) Coordinated expression of defense-related genes by TMV infection or salicylic acid treatment in tobacco. Mol Cells 8:388–392

Karthikeyan G, Doraisamy S, Rabindran R (2009) Pseudomonas fluorescens mediated systemic resistance against *Urdbean leaf crinkle virus* in blackgram (*Vigna mungo*). Arch Phytopathol Plant Prot 42:201–212. https://doi.org/10.1080/03235400600982519

Kavino M, Harish S, Kumar N, Samiyappan R (2009) Rhizobacteria-mediated growth promotion of banana Leads to protection against *Banana bunchy top virus* under field conditions. Acta Hortic 69–76. https://doi.org/10.17660/actahortic.2009.828.5

Kavino M, Harish S, Kumar N, Saravanakumar D, Damodaran T, Samiyappan R (2007a) Potential implications of biopriming in banana (*Musa* spp) plantlets against *Banana bunchy top virus* (BBTV). J Plant Interact 2:149–158. https://doi.org/10.1080/17429140701586365

Kavino M, Harish S, Kumar N, Saravanakumar D, Damodaran T, Soorianathasundaram K, Samiyappan R (2007b) Rhizosphere and endophytic bacteria for induction of systemic resistance of banana plantlets against bunchy top virus. Soil Biol Biochem 39:1087–1098. https://doi.org/10.1016/j.soilbio.2006.11.020

Khurana SMP, Singh MN (1988) Yield loss potential of potato viruses X and Y inIndian potatoes. J Indian Potato Assoc 15:27–29

Kim N-G et al (2017) *Pseudomonas oleovorans* strain KBPF-004 culture supernatants reduced seed transmission of *Cucumber green mottle mosaic virus* and *Pepper mild mottle virus*, and remodeled aggregation of 126 kDa and subcellular localization of movement protein of *Pepper mild mottle virus*. Plant Pathol J 33:393–401. https://doi.org/10.5423/ppj.oa.03.2017.0047

King AMQ, Adams MJ, Carstens EB, Lefkowitz EJ (2011) Virus taxonomy: classification and nomenclature of viruses. In: Ninth report of the international committee on taxonomy of viruses. Elsevier/Academic Press, London

Kirankumar R, Jagadeesh KS, Krishnaraj PU, Patil MS (2008) Enhanced growth promotion of tomato and nutrient uptake by plant growth-promoting rhizobacterial isolates in presence of *Tobacco mosaic virus* pathogen. Karnataka J Agric Sci 21:309–311

Klessig DF, Malamy J, Hennig J, Sanchez-Casas P, Indulski J, Grynkiewicz G, Chen Z (1994) Induction, modification, and perception of the salicylic acid signal in plant defence. Biochem Soc Symp 60:219–229

Kloepper JW (1993) Plant growth promoting rhizobacteria as biological control agents. In: Metting FB (ed) Soil microbial ecology— applications in agricultural and environmental management. Dekker, New York, pp 255–274

Kloepper JW, Reddy MS, Rodríguez-Kabana R, Kenney DS, Kokalis-Burelle N, Martinez-Ochoa N (2004a) Application for rhizobacteria in transplant production and yield enhancement. Acta Hortic 219–229. https://doi.org/10.17660/actahortic.2004.631.28

Kloepper JW, Ryu C-M, Zhang S (2004b) Induced systemic resistance and promotion of plant growth by *Bacillus* spp. Phytopathology 94:1259–1266. https://doi.org/10.1094/phyto.2004.94.11.1259

Kloepper JW, Tuzun S, Kuć JA (1992) Proposed definitions related to induced disease resistance. Biocontrol Sci Technol 2:349–351. https://doi.org/10.1080/09583159209355251

Kohler A, Schwindling S, Conrath U (2002) Benzothiadiazole-induced priming for potentiated responses to pathogen infection, wounding, and infiltration of water into leaves requires the *NPR1/NIM1* gene in *Arabidopsis*. Plant Physiol 128:1046–1056. https://doi.org/10.1104/pp.010744

Kring JB, Schuster DJ, Price JF (1991) Sweetpotato whitefly vectored geminivirus on tomato in Florida. Plant Dis Note 75:1186

Kumar S et al (2016) *Paenibacillus lentimorbus* inoculation enhances tobacco growth and extenuates the virulence of *Cucumber mosaic virus*. PLoS ONE 11:e0149980. https://doi.org/10.1371/journal.pone.0149980

Lamb C, Dixon RA (1997) The oxidative burst in plant disease resistance. Ann Rev Plant Physiol Plant Mol Biol 48:251–275. https://doi.org/10.1146/annurev.arplant.48.1.251

Lee GH, Ryu C-M (2016) Spraying of leaf-colonizing *Bacillus amyloliquefaciens* protects pepper from *Cucumber mosaic virus*. Plant Dis 100:2099–2105. https://doi.org/10.1094/pdis-03-16-0314-re

Li H et al (2016a) Control of *Tomato yellow leaf curl virus* disease by *Enterobacter asburiae* BQ9 as a result of priming plant resistance in tomatoes. Turk J Biol 40:150–159. https://doi.org/10.3906/biy-1502-12

Li H, Huang W, Xu L, Zhou X, Liu H, Cheng Z (2016b) *Stenotrophomonas maltophilia* HW2 enhanced cucumber resistance against Cucumber green mottle mosaic virus. J Plant Biol 59:488–495. https://doi.org/10.1007/s12374-016-0246-6

Li J-X, Liu S-S, Gu Q-S (2016c) Transmission efficiency of *Cucumber green mottle mosaic virus* via seeds, soil, pruning and irrigation water. J Phytopathol 164:300–309. https://doi.org/10.1111/jph.12457

Lian L, Xie L, Zheng L, Lin Q (2011) Induction of systemic resistance in tobacco against *Tobacco mosaic virus* by *Bacillus* spp. Biocontrol Sci Technol 21:281–292. https://doi.org/10.1080/09583157.2010.543667

Liu HW, Luo LX, Li JQ, Liu PF, Chen XY, Hao JJ (2014) Pollen and seed transmission of *Cucumber green mottle mosaic virus* in cucumber. Plant Pathol 63:72–77. https://doi.org/10.1111/ppa.12065

Loebenstein G, Lovrekovich L (1966) Interference with *Tobacco mosaic virus* local lesion formation in tobacco by injection heat-killed cells of *Pseudomonas syringae*. Virology 30:587–591. https://doi.org/10.1016/0042-6822(66)90162-0

Lomonossoff GP (1995) Pathogen-derived resistance to plant viruses. Ann Rev Phytopathol 33:323–343. https://doi.org/10.1146/annurev.py.33.090195.001543

Makkouk KM, Rizkallah L, Kumari SG, Zaki M, Enein RA (2003) First record of *Chickpea chlorotic dwarf virus* (CpCDV) affecting faba bean (*Vicia faba*) crops in Egypt. Plant Pathol 52:413–413. https://doi.org/10.1046/j.1365-3059.2003.00834.x

Mann EW (1969) Inhibition of *Tobacco mosaic virus* by a bacterial extract. Phytopathology 59:658–662

Mansour FA, Soweha HE, Shaaban SA, Mohamadin AH (1988) Studies on the antiviral activity of some bacterial isolates belonging to streptomycetes. Egypt J Bot 31:167–183

Martínez RT, de Almeida MMS, Rodriguez R, de Oliveira AS, Melo FL, Resende RO (2018) Identification and genome analysis of *Tomato chlorotic spot virus* and dsRNA viruses from coinfected vegetables in the Dominican Republic by high-throughput sequencing. Virol J 15:24. https://doi.org/10.1186/s12985-018-0931-9

Maurhofer M, Hase C, Meuwly P, Métraux JP, Défago G (1994) Induction of systemic resistance of tobacco to *Tobacco necrosis virus* by the root-colonizing *Pseudomonas fluorescens* strain CHA0: influence of the gacA gene and of pyoverdine production. Phytopathology 84:139–146. https://doi.org/10.1094/Phyto-84-139

Maurhofer M, Reimmann C, Schmidli-Sacherer P, Heeb S, Haas D, Défago G (1998) Salicylic acid biosynthetic genes expressed in *Pseudomonas fluorescens* strain P3 improve the induction of systemic resistance in tobacco against *Tobacco necrosis virus*. Phytopathology 88:678–684. https://doi.org/10.1094/phyto.1998.88.7.678

Megahed AA, El-Dougdoug KA, Othman BA, Lashin SM, Ibrahim MA, Sofy AR (2012) A new Egyptian satellite strain of *Cucumber mosaic cucumovirus*. Int J Virol 8:240–257. https://doi.org/10.3923/ijv.2012.240.257

Megahed AA, El-Dougdoug KA, Othman BA, Lashin SM, Ibrahim MA, Sofy AR (2013) Induction of resistance in tomato plants against *Tomato mosaic tobamovirus* using beneficial microbial isolates. Pak J Biol Sci 16:385–390. https://doi.org/10.3923/pjbs.2013.385.390

Mehdy MC (1994) Active oxygen species in plant defense against pathogens. Plant Physiol 105:467–472

Mochizuki T, Ohki ST (2012) *Cucumber mosaic virus*: viral genes as virulence determinants. Mol Plant Pathol 13:217–225. https://doi.org/10.1111/j.1364-3703.2011.00749.x

Mohamed SH, Galal AM (2005) Identification and antiviral activities of some halotolerant *Streptomycetes* isolated from Qaroon lake. Int J Agric Biol 7:747–753

Moradi Z (2011) Diagnosis and molecular variability of *Watermelon mosaic virus* isolates from North, East, North-east and North-west regions of Iran. Asian J Plant Pathol 5:115–125. https://doi.org/10.3923/ajppaj.2011.115.125

Mumford RA, Barker I, Wood KR (1996) The Biology of the Tospoviruses. Ann Appl Biol 128:159–183. https://doi.org/10.1111/j.1744-7348.1996.tb07097.x

Murphy AM, Chivasa S, Singh DP, Carr JP (1999) Salicylic acid-induced resistance to viruses and other pathogens: a parting of the ways? Trends Plant Sci 4:155–160. https://doi.org/10.1016/s1360-1385(99)01390-4

Murphy JF, Reddy MS, Ryu C-M, Kloepper JW, Li R (2003) Rhizobacteria-mediated growth promotion of tomato leads to protection against *Cucumber mosaic virus*. Phytopathology 93:1301–1307. https://doi.org/10.1094/phyto.2003.93.10.1301

Murphy JF, W. BO, W. W (1987) Characterization of a *Blackeye cowpea mosaic virus* strain from South Carolina. Plant Dis 71:243–248. https://doi.org/10.1094/pd-71-0243

Murphy JF, Zehnder GW, Schuster DJ, Sikora EJ, Polston JE, Kloepper JW (2000) Plant growth-promoting rhizobacterial mediated protection in tomato against *Tomato mottle virus*. Plant Dis 84:779–784. https://doi.org/10.1094/pdis.2000.84.7.779

Naidu RA, Sherwood JL, Deom CM (2008) Characterization of a vector-non-transmissible isolate of *Tomato spotted wilt virus*. Plant Pathol 57:190–200. https://doi.org/10.1111/j.1365-3059.2007.01707.x

Navas-Castillo J, Fiallo-Olivé E, Sánchez-Campos S (2011) Emerging virus diseases transmitted by whiteflies. Ann Rev Phytopathol 49:219–248. https://doi.org/10.1146/annurev-phyto-072910-095235

Niu D-D, Liu H-X, Jiang C-H, Wang Y-P, Wang Q-Y, Jin H-L, Guo J-H (2011) The plant growth–promoting rhizobacterium *Bacillus cereus* AR156 induces systemic resistance in *Arabidopsis thaliana* by simultaneously activating salicylate- and jasmonate/rthylene-dependent signaling pathways. Mol Plant-Microbe Interact 24:533–542. https://doi.org/10.1094/mpmi-09-10-0213

Ogada PA, Maiss E, Poehling HM (2013) Influence of *Tomato spotted wilt virus* on performance and behaviour of western flower thrips (*Frankliniella occidentalis*). J Appl Entomol 137:488–498. https://doi.org/10.1111/jen.12023

Palukaitis P, Roossinck MJ, Dietzgen RG, Francki RI (1992) *Cucumber mosaic virus*. Adv Virus Res 41:281–348

Pappu HR, Jones RAC, Jain RK (2009) Global status of tospovirus epidemics in diverse cropping systems: Successes achieved and challenges ahead. Virus Res 141:219–236. https://doi.org/10.1016/j.virusres.2009.01.009

Park JY, Yang SY, Kim YC, Kim J-C, Dang QL, Kim JJ, Kim IS (2012) Antiviral peptide from *Pseudomonas chlororaphis* O6 against *Tobacco mosaic virus* (TMV). J Korean Soc Appl Biol Chem 55:89–94. https://doi.org/10.1007/s13765-012-0015-2

Park KS, Paul D, Ryu KR, Kim EY, Kim YK (2006) *Bacillus vallismortis* strain EXTN-1 mediated systemic resistance against *Potato virus Y* and *X* in the field. Plant Pathol J 22:360–363. https://doi.org/10.5423/ppj.2006.22.4.360

Parrela G, Gognalons P, Gebre-Selassiè K, Vovlas C, Marchoux G (2003) An update of the host range of *Tomato spotted wilt virus*. J Plant Pathol 85:227–264

Peters D, Wijkamp I, van de Wetering F, Goldbach R (1996) Vector relations in the transmission and epidemiology of tospoviruses. Acta Hortic 29–43. https://doi.org/10.17660/actahortic.1996.431.2

Polston JE, Hiebert E, McGovern RJ, Stansly PA, Schuster DJ (1993) Host range of *Tomato mottle virus*, a new geminivirus infecting tomato in Florida. Plant Dis 77:1181–1184. https://doi.org/10.1094/pd-77-1181

Polston JE, Wood E, Palmateer AJ, Zhang S (2013) *Tomato chlorotic spot virus*. University of Florida Institute of Food and Agricultural Sciences. http://edis.ifas.ufl.edu/pp306

Puttaraju HR, Prakash HS, Shetty HS (2002) Contribution of seedborne *Blackeye cowpea mosaic potyvirus* to disease dynamics and loss of yield. Trop Sci 42:147–152

Rajinimala N, Rabimdran R, Ramiah M, Kamalakannan A, Mareeswari P (2005) Virus-vector relationship of *Bittergourd yellow mosaic virus* and the whitefly *Bemisia tabaci* Genn. Acta Phytopathol Entomol Hung 40:23–30. https://doi.org/10.1556/APhyt.40.2005.1-2.3

Rajinimala N, Rabindran R, Ramaiah M (2009) Management of *Bittergourd yellow mosaic virus* (BGYMV) by using virus inhibiting chemical, biocontrol agents, antiviral principles (AVP) and insecticide. Arch Phytopathol Plant Prot 42:738–750. https://doi.org/10.1080/03235400701390729

Rajinimala P, Rabindran R, Ramaiah M, Nagarajan P, Varanavasiappan S PGPR mediated disease resistance in bitter gourd against *Bitter gourd yellow mosaic virus*. In: Proceedings of the 6th international PGPR workshop, Calicut, India 5–10 Oct 2003. pp 548–552. (Abstracts and Short Papers)

Ramamoorthy V, Viswanathan R, Raguchander T, Prakasam V, Samiyappan R (2001) Induction of systemic resistance by plant growth promoting rhizobacteria in crop plants against pests and diseases. Crop Prot 20:1–11. https://doi.org/10.1016/s0261-2194(00)00056-9

Ramjegathesh R, Samiyappan R, Raguchander T, Prabakar K, Saravanakumar D (2013) Plant–PGPR interactions for pest and disease resistance in sustainable agriculture. In: Maheshwari D (ed) Bacteria in agrobiology: disease management. Springer, Berlin, pp 293–320. https://doi.org/10.1007/978-3-642-33639-3_11

Raupach GS, Liu L, Murphy JF, Tuzun S, Kloepper JW (1996) Induced systemic resistance in cucumber and tomato against *Cucumber mosaic cucumovirus* using plant growth-promoting rhizobacteria (PGPR). Plant Dis 80:891–894. https://doi.org/10.1094/pd-80-0891

Ravi KS, Buttgereitt A, Kitkaru AS, Deshmukh S, Lesemann DE, Winter S (2001) Sunflower necrosis disease from India is caused by an *Ilarvirus* related to *Tobacco streak virus*. Plant Pathology 50:800-800. https://doi.org/10.1046/j.1365-3059.2001.00623.x

Ravinder Reddy C, Tonapi VA, Varanasiappan S, Navi SS, Jayarajan R (2005) Influence of plant age on infection and symptomatological studies on *Urdbean leaf crinkle virus* in urdbean (*Vigna mungo*). Int J Agric Sci 1:1–6

Reingold V, Lachman O, Blaosov E, Dombrovsky A (2015) Seed disinfection treatments do not sufficiently eliminate the infectivity of *Cucumber green mottle mosaic virus* (CGMMV) on cucurbit seeds. Plant Pathol 64:245–255. https://doi.org/10.1111/ppa.12260

Reymond P, Farmer EE (1998) Jasmonate and salicylate as global signals for defense gene expression. Curr Opin Plant Biol 1:404–411. https://doi.org/10.1016/s1369-5266(98)80264-1

Ryals J, Uknes S, Ward E (1994) Systemic acquired resistance. Plant Physiol 104:1109–1112

Rybicki EP (2015) A top ten list for economically important plant viruses. Arch Virol 160:17–20. https://doi.org/10.1007/s00705-014-2295-9

Ryu C-M, Murphy JF, Mysore KS, Kloepper JW (2004) Plant growth-promoting rhizobacteria systemically protect *Arabidopsis thaliana* against *Cucumber mosaic virus* by a salicylic acid and NPR1-independent and jasmonic acid-dependent signaling pathway. Plant J 39:381–392. https://doi.org/10.1111/j.1365-313X.2004.02142.x

Ryu C-M, Kang BR, Han SH, Cho SM, Kloepper JW, Anderson AJ, Kim YC (2007a) Tobacco cultivars vary in induction of systemic resistance against *Cucumber mosaic virus* and growth promotion by *Pseudomonas chlororaphis* O6 and its *gacS* mutant. Eur J Plant Pathol 119:383–390. https://doi.org/10.1007/s10658-007-9168-y

Ryu C-M, Murphy JF, Reddy MS, Kloepper JW (2007b) A two-strain mixture of rhizobacteria elicits induction of systemic resistance against *Pseudomonas syringae* and *Cucumber mosaic virus* coupled to promotion of plant growth on *Arabidopsis thaliana*. J Microbiol Biotechnol 17:280–286

Shafie RM, Hamed AH, El-Sharkawy HHA (2016) Inducing systemic resistance against *Cucumber mosaic cucumovirus* using *Streptomyces* spp. Egypt J Phytopathol 44:127–142

Shankar ACU, Nayaka SC, Niranjan-Raj S, Kumar HB, Reddy MS, Niranjana SR, Prakash HS (2009) Rhizobacteria-mediated resistance against the blackeye cowpea mosaic strain of *Bean common mosaic virus* in cowpea (*Vigna unguiculata*). Pest Manag Sci 65:1059–1064. https://doi.org/10.1002/ps.1791

Shen L, Wang F, Liu Y, Qian Y, Yang J, Sun H (2013) Suppression of *Tobacco mosaic virus* by *Bacillus amyloliquefaciens* strain Ba33. J Phytopathol 161:293–294. https://doi.org/10.1111/jph.12058

Shirasu K, Nakajima H, Rajasekhar VK, Dixon RA, Lamb C (1997) Salicylic acid potentiates an agonist-dependent gain control that amplifies pathogen signals in the activation of defense mechanisms. Plant Cell 9:261–270. https://doi.org/10.1105/tpc.9.2.261

Shoman SA, Abd-Allah NA, El-Baz AF (2003) Induction of resistance to *Tobacco necrosis virus* in bean plants by certain microbial isolates. Egypt J Biol 5:10–18

Simone GW, Brown JK, Hiebert E, Cullen RE (1990) New geminivirus epidemic in Florida tomatoes and peppers. Phytopathology 80:1063

Sofy AR, Soliman AM (2011) Molecular identification of a *Cucumber mosaic virus* subgroup I Egyptian isolate from geranium based on bioinformatics analysis of *CP* gene sequence Egypt. Egypt J Virol 8:178–194

Sofy AR, Mousa AA, Soliman AM, El-Dougdoug KA (2012) The limiting of climatic factors and predicting of suitable habitat for citrus gummy bark disease occurrence using GIS. Int J Virol 8:165–177. https://doi.org/10.3923/ijv.2012.165.177

Sofy AR, Mahfouze SA, El-Enany MA (2013a) Isozyme markers for response of wild potato species to potato spindle tuber viroid Egyptian isolate. World Appl Sci J 27:1010–1022. https://doi.org/10.5829/idosi.wasj.2013.27.08.13711

Sofy AR, Mousa AA, Refaey EE, El-Dougdoug KA (2013b) Weeds as a reservoir systemic for *Potato potyvirus Y* (PVY) in Egypt. Egypt J Virol 10:183–203

Sofy AR, Attia MS, Sharaf AMA, El-Dougdoug KA (2014a) Potential impacts of seed bacterization or *Salix* extract in faba bean for enhancing protection against *Bean yellow mosaic disease*. Nat Sci 12:67–82

Sofy MR, Sharaf AMA, Noufl M, Sofy AR (2014b) Physiological and biochemical responses in *Cucurbita pepo* Leaves associated with some elicitors-induced systemic resistance against *Zucchhini yellow mosaic virus*. Int J Modern Botany 4:61–74

Sofy AR, El-Dougdoug KA, Mousa AA, Refaey EE (2017) Impact of two TYLCV Egyptian isolates on metabolic and antioxidant activities in some tomato cultivars. Int J Adv Res Biol Sci 4:110–133

Sofy MR, Sharaf AMA, El-Nosary ME, Sofy AR (2018a) Potential impacts of yeast extract in *Cucumis Ssativus* for enhancing protection against cucumber mosaic disease. New York Sci J 11:77–83

Sofy MR, Sharaf AMA, El-Nosary ME, Sofy AR (2018b) *Salix alba* extract induces systemic resistance in *Cucumis sativus* infected by *Cucumber mosaic virus*. Nat Sci 16:107–113

Srinivasan K, Mathivanan N (2009) Biological control of *Sunflower necrosis virus* disease with powder and liquid formulations of plant growth promoting microbial consortia under field conditions. Biol Control 51:395–402. https://doi.org/10.1016/j.biocontrol.2009.07.013

Stansly PA, Schuster DJ, Leibee GL (1991) Management strategies for the sweetpotato whitefly. In: Vavrina CS (ed) Proceedings Florida Tomato Institute 1991. University of Florida, IFAS, Vegetable Crops Special Series SS-VEC-01, pp 20–42

Sutic DD, Ford RE, Tosic MT (1999) Handbook of plant virus diseases. CRC Press, New York

Svoboda J, Červená G, Rodová J, Jokeš M (2006) First report of *Pepper mild mottle virus* in pepper seeds produced in the Czech Republic. Plant Prot Sci 42:34–37. https://doi.org/10.17221/2694-pps

Tóbiás I, Maat DZ, Huttinga H (1982) Two Hungarian isolates of *Cucumber mosaic virus* from sweet pepper (*Capsicum annuum*) and melon (*Cucumis melo*): identification and antiserum preparation. Neth J Plant Pathol 88:171–183. https://doi.org/10.1007/bf02140880

Tripathi S, Suzuki JY, Ferreira SA, Gonsalves D (2008) *Papaya ringspot virus*-P: characteristics, pathogenicity, sequence variability and control. Mol Plant Pathol 9:269–280. https://doi.org/10.1111/j.1364-3703.2008.00467.x

Ullman DE, Sherwood JL, German TL (1997) Thrips as vectors of plant pathogens. Thrips as crop pests. Lewis T edn. CAB International, New York, NY, pp 539–565

Van Loon LC (2000) Systemic induced resistance. In: Slusarenko AJ, Fraser RSS, van Loon LC (eds) Mechanisms of resistance to plant diseases. Springer, Dordrecht, pp 521–574. https://doi.org/10.1007/978-94-011-3937-3_13

Van Loon LC, Bakker PAHM (2005) Induced systemic resistance as a mechanism of disease suppression by rhizobacteria. In: Siddiqui ZA (ed) PGPR: Biocontrol and biofertilization. Springer, Dordrecht, pp 39–66. https://doi.org/10.1007/1-4020-4152-7_2

van Loon LC, Bakker PAHM, Pieterse CMJ (1998) Systemic rsistance induced by rhizosphere bacteria. Ann Rev Phytopathol 36:453–483. https://doi.org/10.1146/annurev.phyto.36.1.453

van Regenmortel MHV et al (2000) Virus taxonomy: classification and nomenclature of viruses. In: Seventh report of the international committee on taxonomy of viruses. Elsevier/Academic Press, London, p 1162 van Wees SCM, de Swart EAM, van Pelt JA, van Loon LC, Pieterse CMJ (2000) Enhancement of induced disease resistance by simultaneous activation of salicylate- and jasmonate-dependent defense pathways in *Arabidopsis thaliana*. Proc Natl Acad Sci 97:8711–8716. https://doi.org/10.1073/pnas.130425197

Van Wees SCM, Van der Ent S, Pieterse CMJ (2008) Plant immune responses triggered by beneficial microbes. Curr Opin Plant Biol 11:443–448. https://doi.org/10.1016/j.pbi.2008.05.005

Vanacker H, Carver TLW, Foyer CH (2000) Early H_2O_2 accumulation in mesophyll cells leads to induction of glutathione during the hyper-sensitive response in the barley-powdery mildew interaction. Plant Physiol 123:1289–1300. https://doi.org/10.1104/pp.123.4.1289

Vance VB (1991) Replication of *Potato virus X* RNA is altered in coinfections with *Potato virus Y*. Virology 182:486–494. https://doi.org/10.1016/0042-6822(91)90589-4

Vanitha SC, Umesha S (2011) *Pseudomonas fluorescens* mediated systemic resistance in tomato is driven through an elevated synthesis of defense enzymes. Biologia Plant 55:317–322. https://doi.org/10.1007/s10535-011-0045-3

Vessey JK (2003) Plant growth promoting rhizobacteria as biofertilizers. Plant Soil 255:571–586. https://doi.org/10.1023/a:1026037216893

Wang S, Wu H, Qiao J, Ma L, Liu J, Xia Y, Gao X (2009) Molecular mechanism of plant growth promotion and induced systemic resistance to *Tobacco mosaic virus* by *Bacillus* spp. J Microbiol Biotechnol 19:1250–1258. https://doi.org/10.4014/jmb.0901.008

Wetter C, Conti M (1988) Pepper mild mottle virus. In: CMI/AAB descriptions of plant viruses, no. 330

Whitfield AE, Ullman DE, German TL (2005) Tospovirus-thrips interactions. Ann Rev Phytopathol 43:459–489. https://doi.org/10.1146/annurev.phyto.43.040204.140017

Xie DX, Feys BF, James S, Nieto-Rostro M, Turner JG (1998) *COI1*: an Arabidopsis gene required for jasmonate-regulated defense and fertility. Science 280:1091–1094. https://doi.org/10.1126/science.280.5366.1091

Xie Z, Fan B, Chen C, Chen Z (2001) An important role of an inducible RNA-dependent RNA polymerase in plant antiviral defense. Proc Natl Acad Sci 98:6516–6521. https://doi.org/10.1073/pnas.111440998

Yi SY, Min SR, Kwon S-Y (2015) NPR1 is instrumental in piming for the enhanced flg22-induced MPK3 and MPK6 activation. Plant Pathol J 31:192–194. https://doi.org/10.5423/ppj.nt.10.2014.0112

Zehnder GW, Murphy JF, Sikora EJ, Kloepper JW (2001) Application of rhizobacteria for induced resistance. Eur J Plant Pathol 107:39–50. https://doi.org/10.1023/a:1008732400383

Zehnder GW, Yao C, Murphy JF, Sikora ER, Kloepper JW (2000) Induction of resistance in tomato against *Cucumber mosaic cucumovirus* by plant growth-promoting rhizobacteria. Biocontrol 45:127–137. https://doi.org/10.1023/a:1009923702103

Zettler FW, Evans IR (1972) *Blackeye cowpea mosaic virus* in Florida: host range and incidence in certified cowpea seed. In: Proceedings of the Florida State horticultural society, pp 99–101

Zhang S, Yang X, Sun M, Sun F, Deng S, Dong H (2009) Riboflavin-induced priming for pathogen defense in *Arabidopsis thaliana*. J Integr Plant Biol 51:167–174. https://doi.org/10.1111/j.1744-7909.2008.00763.x

Chapter 16
Conclusion

Piyush Pandey

Abstract The beneficial microorganisms, majorly PGPR are helpful in the improvement of plant growth and enhanced crop productivity. The future of organic agriculture is bound to be dominated by the application of PGPR. This book has been brilliantly carved out with the information on the mechanisms of interactions, and cross talk of PGPR with plants, that will be useful in designing the strategies of their application. Application of PGPR for biocontrol, as biopesticides, is another domain that has translated in profitable industry, yet the information on their use against viral infections is fascinating. Further, phytoremediation of heavy metal from contaminated soil has been discussed in this book. In fact, the sustainable management of field crops by PGPR triumph its merit, for ensuring the food security in eco-friendly manner, and this book is a valuable collection of information for its application and success.

Keywords PGPR · Biocontrol · Plant immunity · Cereals crops · Horticulture crops

"*A reasonable agriculture would do its best to emulate nature. Rather than change the earth to suit a crop… it would diversify its crops to suit the earth*"—this quote by Klinkenborg (2013) reflects the necessity of agricultural technologies to proceed toward green and sustainable agri practices. This book, has been complemented with elaborated discussions, which has come up as an excellent cumulative effort of scientists from different expertise, and single goal, i.e. promotion of PGPR for better crop productivity and management.

A couple of chapters is focused on using different combinations of organisms like various rhizobacteria, or PGPR with mycorrhizae, for the improvement of plant growth. In fact, the term consortium was given in the last part of nineteenth century for symbiotically living two or more bacteria, but the applications of such relations in soil is recently becoming popular among users, which is supposed to provide

P. Pandey (✉)
Department of Microbiology, Assam University, Silchar 788011, Assam, India
e-mail: piyushddn@gmail.com

D. K. Maheshwari and S. Dheeman (eds.), *Field Crops: Sustainable Management by PGPR*, Sustainable Development and Biodiversity 23,
https://doi.org/10.1007/978-3-030-30926-8_16

better results. Not only it provides relatively more functional genes in system, but also ensures the stability and survival of inoculated PGPR in new environment (Pandey and Maheshwari 2007).

There are another two contributions which discuss the use of PGPR in vegetables. While one of these is specifically focused on cucurbitaceous vegetables, while another one describes the significance of PGPR in productivity and disease management in tomato, pepper, melon, radish, and lettuce. In fact, the importance of horticultural crops in food safety is less realized, but these crops are important not only for economic reasons but also as a source of nutrition to large population. PGPR shall play a significant role in improving the quality of produce, along with economic benefits.

The cross talk between PGPR and host plant have been a very interesting aspect for research. It has been realized that the chemical communications between soil bacteria and host plants may induce the plants to tolerate abiotic stress such as salt and drought stress. In fact, differentially expressed stress proteins had been identified in the treated host plants and expression of stress-specific genes had been found to increase by 1.5-fold in *PGPR*-treated plants (Lim and Kim 2013). Such facts make use of PGPR even more interesting where the direct mechanisms of plant growth are further strengthened by the indirect benefits to plants. Similarly, phytoremediation of heavy metal contaminated soil is coming up as another applied aspect of PGPR (Kotoky and Pandey 2018), and this required a systematic discussion, which has been incorporated in this book.

The PGPR are versatile organisms. Their competitive physiological skills of PGPR in rhizosphere also provide them the ability to remove some phytopathogenic parasites from the soil. This indirect mechanism of growth promotion has been utilized for the formulation of PGPR-based biopesticides. Considering the benefits of biological control, it is appropriate to understand the molecular and physiological mechanisms of interactions, to obtain maximum benefits (Backer et al. 2018). The biological control of fungal phytopathogens has been relatively studied more, yet it's stimulating and thought-provoking to read a chapter on the protection of plants from viral diseases, primarily by the mechanism of induced systemic resistance. There might be several checkpoints to understand the underlying mechanism, such as de novo production of pathogenesis-related proteins, modifications in cell wall composition and/or synthesis of phytoalexins, yet there is a possibility to find many other compounds, which are likely to exist but have not been identified (Heil and Bostock 2002).

This book will be useful not only to the researchers, but also to each and every stakeholder that contributes toward food security and green agriculture. The enriched efforts of contributors and editorial team have resulted in a volume, which systematically describes to different issues, and applications of PGPR in sustainable agriculture. Needless to mention, that maybe, such volumes will be needed to place the PGPR technology in field to its potential, but this particular book stands on its merit for the information and contents, which will be useful to all.

References

Backer R, Rokem JS, Ilangumaran G, Lamont J, Praslickova D, Ricci E, Subramanian S et al (2018) Plant growth-promoting rhizobacteria: context, mechanisms of action, and roadmap to commercialization of biostimulants for sustainable agriculture. Front Plant Sci 9:1473. https://doi.org/10.3389/fpls.2018.01473

Heil M, Bostock RM (2002) Induced systemic resistance (ISR) against pathogens in the context of induced plant defences. Annal Bot 89(5):503–512

Klinkenborg V (2013) More Scenes from the Rural Life. Princeton Architectural Press, p 109

Kotoky R, Pandey P (2018) Plant-microbe symbiosis as an instrument for the mobilization and removal of heavy metals from contaminated soils—a realistic approach. Curr Biotechnol 7:71

Lim JH, Kim SD (2013) Induction of drought stress resistance by multi-functional PGPR *Bacillus licheniformis* K11 in pepper. Plant Pathol J 29(2):201–208

Pandey P, Maheshwari DK (2007) Bioformulation of *Burkholderia* sp. MSSP with a multispecies consortium for growth promotion of *Cajanus cajan*. Can J Microbiol 53:213–222

Index

D. K. Maheshwari and S. Dheeman (eds.), *Field Crops: Sustainable Management by PGPR*, Sustainable Development and Biodiversity 23,
https://doi.org/10.1007/978-3-030-30926-8

L
Lactobacillus, 79, 81, 85
Lactuca sativa, 234
Leaching, 50, 52, 58
Lead, 254
Legumes, 146
Leguminosae, 146
Leifsonia, 148, 150–153
Lettuce, 234
Lipid peroxidation, 254
LOX2, 388
Luffa, 70
Lupinus, 261, 263
Lycopersicon esculentum, 223, 226

M

N

L

M
Maize, 222, 230, 240, 340, 343
MAMP-triggered immunity, 388
Mandua, 108, 113, 114
Mcf1-related gene, 394
Medicago, 257, 260, 261
Medicago sativa, 202
Medicago truncatula, 199
Melon, 70, 77, 84, 85, 88, 89, 92, 93, 230
Mentha sps, 239
MerA, 259, 268–270
MerB, 259, 269, 270
MerC, 259
Mercury, 253–261, 263–270
Mercury pollution, 255
MerD, 259
MerF, 259
*Mer*operon, 258, 269–271
MerP

MIX
Papier aus verantwortungsvollen Quellen
Paper from responsible sources
FSC® C105338

If you have any concerns about our products,
you can contact us on
ProductSafety@springernature.com

In case Publisher is established outside the EU,
the EU authorized representative is:
Springer Nature Customer Service Center GmbH
Europaplatz 3, 69115 Heidelberg, Germany

Printed by Libri Plureos GmbH
in Hamburg, Germany